Key Concepts in DNA Repair

Edited by **Nas Wilson**

R CALLISTO REFERENCE

New York

Published by Callisto Reference,
106 Park Avenue, Suite 200,
New York, NY 10016, USA
www.callistoreference.com

Key Concepts in DNA Repair
Edited by Nas Wilson

International Standard Book Number: 978-1-63239-439-2 (Hardback)

Contents

Preface

The main aim of this book is to educate learners and enhance their research focus by presenting diverse topics covering this vast field. This is an advanced book which compiles significant studies by distinguished experts in the area of analysis. This book addresses successive solutions to the challenges arising in the area of application, along with it; the book provides scope for future developments.

The book provides an overview of current knowledge and discusses some of the unanswered questions at the forefront of research on the key concepts in DNA Repair. This book is a collection of several chapters divided into various distinct sections. Each chapter has been researched by leading experts in their respective fields. The range of the book varies from the DNA damage response and DNA repair methods to transformational characteristics of DNA repair, providing a picture of present perceptions of the DNA repair procedure to roles of DNA repair gene mutations in carcinogenesis and neurodegeneration. This book is a collection of contemporary works on DNA injury and the associated cellular response.

It was a great honour to edit this book, though there were challenges, as it involved a lot of communication and networking between me and the editorial team. However, the end result was this all-inclusive book covering diverse themes in the field.

Finally, it is important to acknowledge the efforts of the contributors for their excellent chapters, through which a wide variety of issues have been addressed. I would also like to thank my colleagues for their valuable feedback during the making of this book.

Editor

Part 1

DNA Damage Response

A Recombination Puzzle Solved: Role for New DNA Repair Systems in *Helicobacter pylori* Diversity/Persistence

Ge Wang and Robert J. Maier

Department of Microbiology, University of Georgia, Athens
Georgia

1. Introduction

1.1 Helicobacter pylori pathogenesis

Helicobacter pylori is a gram-negative, slow-growing, microaerophilic, spiral bacterium. It is one of the most common human gastrointestinal pathogens, infecting almost 50% of the world's population [1]. Peptic ulcer disease is now approached as an infectious disease, and *H. pylori* is responsible for the majority of duodenal and gastric ulcers [2]. There is strong evidence that *H. pylori* infection increases the risk of gastric cancer [3], the second most frequent cause of cancer-related death. *H. pylori* infections are acquired by oral ingestion and is mainly transmitted within families in early childhood [2]. Once colonized, the host can be chronically infected for life, unless *H. pylori* is eradicated by treatment with antibiotics.

H. pylori is highly adapted to its ecologic niche, the human gastric mucosa. The pathogenesis of *H. pylori* relies on its persistence in surviving a harsh environment, including acidity, peristalsis, and attack by phagocyte cells and their released reactive oxygen species [4]. *H. pylori* has a unique array of features that permit entry into the mucus, attachment to epithelial cells, evasion of the immune response, and as a result, persistent colonization and transmission. Numerous virulence factors in *H. pylori* have been extensively studied, including urease, flagella, BabA adhesin, the vacuolating cytotoxin (VacA), and the cag pathogenicity island (cag-PAI) [5]. In addition to its clinical importance, *H. pylori* has become a model system for persistent host-associated microorganisms [6]. How *H. pylori* can adapt to, and persist in, the human stomach has become a problem of general interest in both microbial physiology and in pathogenesis areas.

1.2 Genetic diversity of *H. pylori*

H. pylori displays exceptional genetic variability and intra-species diversity [7]. Allelic diversity is obvious as almost every unrelated isolate of *H. pylori* has a unique sequence when a sequenced fragment of only several hundred base pairs is compared among strains for either housekeeping or virulence genes [8-10]. Approximately 5% nucleotide divergence is commonly observed at the majority of gene loci between pairs of unrelated *H. pylori* strains [11]. *H. pylori* strains also differ considerably in their gene contents, the genetic macro-diversity. The two sequenced strains 26695 and J99 share only 94% of their genes, whereas approximately 7% of the genes are unique for each strain [12, 13]. Supporting studies using whole-genome microarray detected numerous genomic changes in the paired sequential isolates of *H. pylori* from the same patient [14, 15].

Mechanisms proposed to account for the observed genetic variability include mainly the high inherent mutation rate and high frequency of recombination [16]. The spontaneous mutation rate of the majority of *H. pylori* strains lies between 10^{-5} and 10^{-7} [17]. This is several orders of magnitude higher than the average mutation rate of *Escherichia coli*, and similar to that of *E. coli* strains defective in mismatch repair functions (mutator strains) [18]. While mutation is essential for introducing sequence diversity into the species, a key role in generating diversity is played by recombination.

H. pylori is naturally competent for DNA transformation, and has a highly efficient system for recombination of short-fragment involving multiple recombination events within a single locus [19, 20]. A special apparatus homologous to type IV secretion system (T4SS, encoded by *comB* locus) is dedicated to a DNA uptake role [21, 22] and a composite system involving proteins at the *comB* locus and ComEC mediates two-step DNA uptake in *H. pylori* [23]. T4SS systems are known to transport DNA and proteins in other bacteria, but *H. pylori* is the only species known to use a T4SS for natural competence [24]. Unlike several other bacterial species, *H. pylori* does not require specific DNA sequences for uptake of related DNA [25]. Instead, numerous and efficient restriction modification systems take over the function as a barrier to horizontal gene transfer from foreign sources [26, 27].

Population genetic analyses of unrelated isolates of *H. pylori* indicated that recombination was extremely frequent in *H. pylori* [9, 28]. There is evidence that humans are occasionally infected with multiple genetically distinct isolates and that recombination between *H. pylori* strains can occur in humans [29, 30]. Using mathematical modeling approaches on sequence data from 24 pairs of sequential *H. pylori* isolates, Falush et al. [31] estimated that the mean size of imported fragments was only 417 bp, much shorter than that observed for other bacteria. The recombination rate per nucleotide was estimated as 6.9×10^{-5}, indicating that every pair of strains differed on average by 114 recombination events. Compared to other bacteria studied in this way [32-34], the recombination frequency within *H. pylori* is extraordinarily high. The *H. pylori* genome also has extensive repetitive DNA sequences that are targets for intragenomic recombination [35].

2. Overview of DNA repair in *H. pylori*

Oxidative DNA damage represents a major form of DNA damage. Among the many oxidized bases in DNA, 8-oxo-guanine is a ubiquitous biomarker of DNA oxidation [36]. In addition, acid (low pH) conditions may result in DNA damage via depurination [37]. *H. pylori* survives on the surface of the stomach lining for the lifetime of its host and causes a chronic inflammatory response. Several lines of evidence suggest that *H. pylori* is exposed to oxidative damage soon after infection [38, 39]. Under physiological conditions, *H. pylori* is thought to frequently suffer oxidative and acid stress [40, 41]. In addition to diverse oxidant detoxification enzymes (e.g. superoxide dismutase, catalse, and peroxiredoxins) [42] and potent acid avoidance mechanisms (mainly urease) [43], efficient DNA repair systems are required for *H. pylori* to survive in the host.

2.1 DNA repair systems in *H. pylori*

The whole genome sequences of *H. pylori* revealed it contains several DNA repair pathways that are common to many bacterial species, while it lacks other repair pathways or contains only portions of them. *H. pylori* encodes the homologues of all four members of the nucleotide excision repair (NER) pathway; these are UvrA, UvrB, UvrC, and UvrD, all of which are well conserved in bacteria. NER deals with DNA-distorting lesions, in which an excinuclease removes a 12- to 13- nucleotide segment from a single strand centered around the lesion; the resulting gap is then filled in by repair synthesis [44]. Loss of *uvrB* in *H. pylori* was shown to confer sensitivity to UV light, alkylating agents and low pH, suggesting that the *H. pylori* NER pathway is functional in repairing a diverse array of DNA lesions [45]. *H. pylori* UvrD was shown to play a role in repairing DNA damage and limiting DNA recombination, indicating it functions to ultimately maintain genome integrity [46].

The methyl-directed mismatch repair system (MMR), consisting of MutS1, MutH, and MutL, is conserved in many bacteria and eukaryotes, and it plays a major role in maintaining genetic stability. MMR can liberate up to 1000 nucleotides from one strand during its function to correct a single mismatch arising during DNA replication [47]. Notably, MMR does not exist in *H. pylori*, contributing to the high mutation rates observed in *H. pylori* [17]. *H. pylori* has a MutS homologue that belongs to the MutS2 family. *H. pylori* MutS2 was shown to bind to DNA structures mimicking recombination intermediates and to inhibit DNA strand exchange, thus it may play a role in maintaining genome integrity by suppressing homologous and homeologous DNA recombination [48]. In addition, *H. pylori* MutS2 appears to play a role in repairing oxidative DNA damage, specifically 8-oxo-guanine [49].

Damaged bases can be repaired by a variety of glycosylases that belong to the base excision repair (BER) pathway. All glycosylases can excise a damaged base resulting in an apurinic/apyrimidinic (AP) site, while some of them additionally nick the DNA deoxyribose-phosphate backbone (via an AP lyase activity). *H. pylori* harbors the glycosylase genes *ung, mutY, nth*, and *magIII*, whereas several other genes appear to be absent from the *H. pylori* genome, e.g. *tag, alkA*, and *mutM*. The *H. pylori* endonuclease III (*nth* gene product), which removes oxidized pyrimidine bases, was shown to be important in establishing long-term colonization in the host [50]. The *H. pylori* MutY glycosylase is functional in removing adenine from 8-oxoG:A mispair, and the loss of MutY leads to attenuation of the colonization ability [51-53].

To repair DNA double strand breaks and blocked replication forks, *H. pylori* is equipped with an efficient system of DNA recombinational repair, which is the main focus of this review (See section 4).

2.2 *H. pylori* response to DNA damage

Many bacteria encode a genetic program for a coordinated response to DNA damage called the SOS response. The best known *E. coli* SOS response is triggered when RecA binds ssDNA, activating its co-protease activity towards LexA, a transcriptional repressor [54]. Cleavage of LexA results in transcriptional induction of genes involved in DNA repair, low-fidelity polymerases, and cell cycle control. However, the *H. pylori* genome contains neither a gene for LexA homolog nor the genes for low-fidelity polymerases, and an SOS response pathway seems to be absent in *H. pylori* [12, 13].

To define pathways for an *H. pylori* DNA damage response, Dorer et al. [55] used cDNA based microarrays to measure transcriptional changes in cells undergoing DNA damage. In both ciprofloxacin treated cells and the Δ*addA* (a major DNA recombination gene, see section 4.4 below) mutant cells, the same set of genes were induced which include genes required for energy metabolism, membrane proteins, fatty acid biosynthesis, cell division, and some translation factors, although the contribution of these genes to survival in the face of DNA damage is not understood. No DNA repair genes, a hallmark of the SOS response, were induced in either the antibiotic-treated cells or the recombination gene deleted strain. Surprisingly, several genes involved in natural competence for DNA transformation (*com* T4SS components *comB3*, *comB4* and *comB9*) were induced significantly. Indeed, natural transformation frequency was shown to be increased under DNA damage conditions. Another DNA damage-induced gene was a lysozyme-encoding gene. Experimental evidence was provided that a DNA damage-induced lysozyme may target susceptible cells in culture and provide a source of DNA for uptake [55]. Taken together, DNA damage (mainly DSBs in their experiments) induces the capacity for taking up DNA segments from the neighboring cells of the same strain (homologous) or co-colonizing strain (homeologous) that may be used for recombinational DNA repair.

3. Mechanisms of DNA recombinational repair known in model bacteria

Although the bulk of DNA damage affects one strand of a duplex DNA segment, occasionally both DNA strands opposite each other are damaged; the latter situation necessitates recombinational repair using an intact homologous DNA sequence [56, 57]. DNA double-strand breaks (DSB) occur as a result of a variety of physical or chemical insults that modify the DNA (e.g. DNA strands cross-links). In addition, if a replication fork meets damaged bases that cannot be replicated, the fork can collapse leading to a DSB. In *E. coli*, 20-50% of replication forks require recombinational repair to overcome damage [58].

Homologous recombinational repair requires a large number of proteins that act at various stages of the process [56]. The first stage, **pre-synapsis**, is the generation of 3' single-stranded (ss) DNA ends that can then be used for annealing with the homologous sequence on the sister chromosome. In *E. coli*, the two types of two-strand lesions (double strand end and daughter strand gap) are repaired by two separate pathways, RecBCD and RecFOR, respectively [57]. The second and most crucial step in DNA recombination is the introduction of the 3' DNA overhang into the homologous duplex of the sister chromosome, termed **synapsis**. This is performed by RecA in bacteria. RecA binds to ssDNA in an ATP-dependent manner, and RecA-bound ssDNA (in a right-handed helix structure) can invade homologous duplex DNA and mediate strand annealing, accompanied by extrusion of the other strand that can pair with the remaining 5' overhang of the DSB (called D-loop formation).

During DNA recombination, the single stranded DNA (ssDNA) is always coated (protected) by ssDNA-binding protein (SSB), which has a higher affinity to ssDNA than RecA. RecA needs to be loaded (during pre-synapsis stage), either by RecBCD or RecFOR, onto the generated ssDNA that is coated with SSB. During the third step in recombination, **post-synapsis**, RecA-promoted strand transfer produces a four-stranded exchange, or Holliday junctions (HJ) [59]. The RecG and RuvAB helicases are two pathways that process the branch migration of HJ. Finally, RuvC resolves HJ in an orientation determined by RuvB, and the remaining nicks are sealed by DNA ligase.

Several other genes (recJ, recQ, recN) are also required for recombination, although their functions are unclear [60, 61]. Single stranded exonuclease RecJ and RecQ helicase are sometimes needed to enlarge the gap for RecFOR to act [62]. RecN, RecO, and RecF were found to be localized to distinct foci on the DNA in *Bacillus subtilis* cells after induction of DSBs [63]. These proteins form active repair centers at DSBs and recruit RecA, initiating homologous recombination. RecN was shown to play an important role in repairing DSBs, probably coordinating alignment of the broken segments with intact duplexes to facilitate recombination [64].

4. DNA recombinational repair factors in *H. pylori*

While some genes that are predicted to be involved in DNA recombinational repair, including *recA, recG, recJ, recR, recN, and ruvABC*, were annotated from the published *H. pylori* genome sequences, many genes coding for the components that are involved in the pre-synapsis stage, such as RecBCD, RecF, RecO, and RecQ, were missing. Considering that *H. pylori* is highly genetic diverse with a high recombination frequency, this has been a big puzzle over the past decade. Recent studies revealed the existence of both pathways, AddAB (RecBCD-like) and RecRO, for initiation of DNA recombinational repair in *H. pylori*. In the following sections we will summarize the current understanding of DNA recombinational repair in *H. pylori* by reviewing the literature accumulated in recent years.

4.1 The central recombination protein RecA

The RecA protein is a central component of the homologous recombination machinery and of the SOS system in most bacteria. The relatively small RecA protein contains many functional domains including different DNA-binding sites and an ATP-binding site. *E. coli* RecA has also coprotease activities for the LexA repressor and other factors involved in SOS response. However, *H. pylori* genome does not contain a LexA homolog and an SOS response pathway is likewise absent in *H. pylori*. Thus, a coprotease activity may be dispensable for the *H. pylori* RecA protein. Nevertheless, RecA is required for DNA damage response observed in *H. pylori*, although the underlying mechanism is unclear [55].

Before the genome era, the roles of *H. pylori* RecA in DNA recombination and repair have been studied genetically [65, 66]. *H. pylori* RecA (37.6 kDa protein) is highly similar to known bacterial RecA proteins. The *H. pylori* recA mutants were severely impaired in their ability to survive treatment with DNA damaging agents such as UV light, methyl methanesulfonate, ciprofloxacin, and metronidazole. *H. pylori* RecA also played a role in survival at low pH in a mechanism distinct from that mediated by urease [66]. Disruption of *recA* in *H. pylori* abolished general homologous recombination [65]. Interestingly, *H. pylori* RecA protein is subject to posttranslational modifications that result in a slight shift in its electrophoretic mobility [67]. One putative mechanism for RecA modification is protein glycosylation. *H. pylori* RecA protein was shown to be membrane associated, but this association is not dependent on the posttranslational modification. The RecA modification is required for full activity of DNA repair [67].

In recent years, the phenotypes of *H. pylori* recA mutants have been further characterized in comparison with other mutants. Among the mutants of DNA recombination and repair genes, *recA* mutants displayed the most severe phenotypes. For example, *recA* mutants were much more sensitive to UV or Gamma radiation than the *recB* or *recO* single mutants, and were similar to the *recBO* double mutant [68-70]. The *recA* mutants completely lost the ability to undergo natural transformation [68-70]. The intra-genomic recombination frequency of the *recA* mutant was also much lower than that of the *recR* or *recB* single mutants [68, 71]. Finally, the *recA* mutants completely lost the ability to colonize mouse stomachs [69]. In competition experiments (mixed infection with wild type and mutant strains), *recA* mutant bacteria were never recovered, while some *addA* or *addB* mutant bacteria were recovered from mouse stomachs.

4.2 Post-synapsis proteins RuvABC and RecG

In addition to the synapsis protein RecA, the genes for post-synapsis proteins (RuvABC and RecG) are also well conserved among bacteria [72]. Genes for RuvABC proteins are present in *H. pylori*, thus *H. pylori* seems to be able to restore Holliday Junctions in a similar way to *E. coli*. RuvC is a Holliday junction endonuclease that resolves recombinant joints into nicked duplex products. A *ruvC* mutant of *H. pylori* was more sensitive (compared to the wild type) to oxidative stress and other DNA damaging agents including UV light, mitomycin C, levofloxacin and metronidazole [73]. As Macrophage cells are known to produce an oxidative burst to kill bacterial pathogens, the survival of *H. pylori ruvC* mutant within macrophages was shown to be 100-fold lower than that of the wild type strain [73]. Furthermore, mouse model experiments revealed that the 50% infective dose of the *ruvC* mutant was approximately 100-fold higher than that of the wild-type strain. Although the *ruvC* mutant was able to establish colonization at early time points, infection was spontaneously cleared from the murine gastric mucosa over long periods (36 to 67 days) [73]. This was the first experimental evidence that DNA recombination processes are important for establishing and maintaining long-term *H. pylori* infection. Further studies suggested that RuvC function and, by inference, recombination facilitate bacterial immune evasion by altering the adaptive immune response [74], although the underlying mechanisms remain obscure.

RuvAB proteins are involved in the branch migration of Holliday junctions. The annotated *H. pylori* RuvB (HP1059) showed extensive homology (52% sequence identity) to *E. coli* RuvB, particularly within the helicase domains. However, unlike in *E. coli*, *ruvA*, *ruvB*, and *ruvC* are located in separate regions of the *H. pylori* chromosome, which may predict possible functional differences. In contrast to *E. coli ruvB* mutants, which have moderate susceptibility to DNA damage, the *H. pylori ruvB* mutant has intense susceptibility to UV, similar to that of a *recA* mutant [75]. Similarly, the *H. pylori ruvB* mutant has a significantly diminished MIC (minimal inhibitory concentration) for ciprofloxacin, an agent that blocks DNA replication fork progression, to the same extent as the *recA* mutant. In agreement with these repair phenotypes, the *ruvB* mutant has almost completely lost the ability of natural transformation of exogenous DNA (frequency of <10⁻⁸), similar to the *recA* mutant. In an assay measuring the intra-genomic recombination (deletion frequency between direct repeats), the *ruvB* mutants displayed significantly (four- to sevenfold) lower deletion frequencies than the background level. All four phenotypes of the *ruvB* mutant suggested that *H. pylori* RuvAB is the predominant pathway for branch migration in DNA recombinational repair [75].

In *E. coli*, an alternative pathway processing branch migration of Holliday junctions is the RecG helicase. In marked contrast to *E. coli*, *H. pylori recG* mutants do not have defective DNA repair, as measured by UV-light sensitivity and ciprofloxacin susceptibility [76]. Furthermore, *H. pylori recG* mutants have increased frequencies of intergenomic recombination and deletion, suggesting that branch migration and Holliday junction resolution are more efficient in the absence of RecG function [75, 76]. Thus, the effect of *H. pylori* RecG seems to be opposite to that of the RuvAB helicase. In the RuvABC pathway, the RuvC endonuclease nicks DNA, catalyzing Holliday junction resolution into double-stranded DNA. Although the resolvase in the RecG pathway has not been completely elucidated, it has been hypothesized that RusA may serve this function in *E. coli* [77]. By introducing *E. coli* *rusA* into *H. pylori* *ruvB* mutants, the wild-type phenotypes for DNA repair and recombination were restored [75]. A hypothesis was proposed that RecG competes with RuvABC for DNA substrates but initiates an incomplete repair pathway (due to the absence of the RecG resolvase RusA) in *H. pylori*, interfering with the RuvABC repair pathway [75].

4.3 *H. pylori* RecN

Bacterial RecN is related to the SMC (structure maintenance of chromosome) family of proteins in eukaryotes, which are key players in a variety of chromosome dynamics, from chromosome condensation and cohesion to transcriptional repression and DNA repair [78]. SMC family proteins have a structural characteristic of an extensive coiled-coil domain located between globular domains at the N- and C-termini that bring together Walker A and B motifs associated with ATP-binding [79]. *E. coli* RecN is strongly induced during the SOS response and was shown to be involved in RecA-mediated recombinational repair of DSBs [64]. In *Bacillus subtilis*, RecN was shown to be recruited to DSBs at an early time point during repair [63, 80, 81]. In vitro, RecN was shown to bind and protect 3' ssDNA ends in the presence of ATP [82].

In the published *H. pylori* genome sequence [12], HP1393 was annotated as a *recN* gene homolog. The *H. pylori* *recN* mutant is much more sensitive to mitomycin C, an agent that predominantly causes DNA DSBs, indicating RecN plays an important role in DSB repair in *H. pylori* [83]. In normal laboratory growth conditions, an *H. pylori* *recN* mutant does not show a growth defect, but its survival is greatly reduced under oxidative stress which resembles the *in vivo* stress condition. While very little fragmented DNA was observed in either wild type or *recN* mutant strain when cells were cultured under normal microaerobic conditions; after oxidative stress treatment the *recN* mutant cells had a significantly higher proportion of the DNA as fragmented DNA than did the wild type [83]. Similar roles of RecN in protection against oxidative damage have been demonstrated in *Neisseria gonorrhoeae* [84, 85]. In addition, the *H. pylori* *recN* mutant is much more sensitive to low pH than the wild type strain, suggesting that RecN is also involved in repair of acid-induced DNA damage [83]. This could be relevant to its physiological condition, as *H. pylori* appears to colonize an acidic niche on the gastric surface [41].

As mentioned in the sections above, loss of *H. pylori* RecA, RuvB or RuvC functions results in a great decrease of DNA recombination frequency. Similarly, the *H. pylori* *recN* mutant has a significant decrease of DNA recombination frequency, suggesting that RecN is a critical factor in DNA recombinational repair [83]. In contrast, loss of UvrD or MutS2 in *H. pylori* resulted in an increase of DNA recombination frequency [46, 48]. Suppression of DNA recombination by UvrD or MutS2, and facilitation of DNA recombination by RecN, may play a role in coordinating DNA repair pathways. Recombinational repair could be mutagenic due to homeologous recombination or cause rearrangement due to recombination with direct repeat sequences. In addition, recombinational repair systems are much more complex and require more energy to operate, compared to nucleotide excision repair (NER) and base excision repair (BER) systems. Thus UvrD, as a component of NER, and MutS2 as a likely component of a BER (8-oxoG glycosylase) system [49], both suppress DNA recombination. Both NER and BER systems would be expected to continuously function in low stress conditions. Under a severe stress condition when large amounts of

DSBs are formed, RecN perhaps recognizes DSBs and recruits proteins required for initiation of DNA recombination.

The role of *H. pylori* RecN *in vivo* has been demonstrated, as the *recN*-disrupted *H. pylori* cells are less able to colonize hosts than wild type cells [83]. However, the mouse colonization phenotype of the *recN* strain seems to be less severe than those observed for the *recA* or *ruvC* mutants. In contrast to RecA or RuvC which are major components of DNA recombination machinery, RecN is a protein specific for repairing DSBs by linking DSB recognition and DNA recombination initiation. It was proposed that the attenuated ability to colonize mouse stomachs by *recN* cells was mainly due to the strain's failure to repair DSBs through a DNA recombinational repair pathway.

4.4 AddAB helicase-nuclease

DNA helicases play key roles in many cellular processes by promoting unwinding of the DNA double helix [86]. Bacterial genomes encode a set of helicases of the DExx family that fulfill several, sometimes overlapping functions. Based on the sequence homology, bacterial RecB, UvrD, Rep, and PcrA were classified as superfamily I (SF1) helicases [86-88]. In the well-studied *E. coli*, RecBCD form a multi-functional enzyme complex that processes DNA ends resulting from a double-strand break. RecBCD is a bipolar helicase that splits the duplex into its component strands and digests them until encountering a recombinational hotspot (Chi site). The nuclease activity is then attenuated and RecBCD loads RecA onto the 3' tail of the DNA [89]. Another bacterial enzyme complex AddAB, extensively studied in *Bacillus subtilis*, has both nuclease and helicase activities similar to those of RecBCD enzyme [90, 91].

The genes for RecBCD or AddAB were missing in the published *H. pylori* genome [12, 13]. However, HP1553 from strain 26695 was annotated as a gene encoding a putative helicase [12], and the corresponding gene from strain J99 was annotated as *pcrA* [13]. Amino acid sequence alignment of HP1553 to *E. coli* RecB (or to *B. subtilis* AddA) revealed 24% identity (to both heterologous systems) at the N-terminal half (helicase domain), and no significant homology at the C-terminal half (including nuclease domain). Thus, HP1553 could be a RecB (or AddA)-like helicase [69, 92]. Furthermore, by using the highly conserved AddB nuclease motif "GRIDRID" in BLAST search, HP1089 was identified as the putative AddB homolog [69]. Now it is accepted that HP1553 and HP1089 are termed *addA* and *addB* respectively in *H. pylori* with a reminder that previous *recB* [20, 68, 70, 92] was the equivalent of *addA* [69, 71, 93]. Both genes *addA* and *addB* are present in 56 *H. pylori* clinical isolates from around the world [94]; thus they are considered core genes that are not strain variable.

The biochemical activities of *H. pylori* AddAB helicase-nuclease have been demonstrated [69]. Cytosolic extracts from wild-type *H. pylori* showed detectable ATP-dependent nuclease activity with ds DNA substrate, while the *addA* and *addB* mutants lack this activity. Cloned *H. pylori* *addA* and *addB* genes express ATP-dependent exonuclease in *E. coli* cells. These genes also conferred ATP-dependent DNA unwinding (helicase) activity to an *E. coli* *recBCD* deletion mutant, indicating that they are the structural genes for this enzyme [69]. The roles of individual (helicase, exonuclease) activity of the AddA and AddB in DNA repair, recombination, and mouse infection have been further studied by site-directed mutagenesis approach [93].

H. pylori *addA* and *addB* mutant strains showed heightened sensitivity to mitomycin C and the DNA gyrase inhibitor ciprofloxacin, both of which lead to DNA ds breaks [69, 92]. The level of sensitivity was similar to that seen for a *recA* mutant, but more severe than for the *recN* mutant. It is thus concluded that AddAB plays a major role in the repair of DNA ds breaks [69, 92]. On the other hand, the *addA* and *addB* mutants were markedly less sensitive to UV irradiation than a *recA* mutant, suggesting that AddAB does not play a major role in repair of UV damage in *H. pylori* [69]. AddA was shown to be important for *H. pylori* protection against oxidative stress-induced damage, as the *addA* mutant cells were significantly more sensitive to oxidative stress and contained a large amount of fragmented DNA [92]. Furthermore, loss of AddA resulted in reduced frequencies of apparent gene conversion between homologous genes encoding outer membrane proteins (*babA* to *babB*) [69]. Finally, it was shown that the *addA* and *addB* mutant strains display a significantly attenuated ability to colonize mouse stomachs, in both competition experiments and during single-strain infections [69, 92].

While *addA* and *addB* are adjacent in the chromosome in most bacteria, including other epsilon Proteobacteria, this is not the case in *H. pylori*. However, the phenotypes of *H. pylori addA* and *addB* mutants are indistinguishable. Thus, it was proposed [69] that the AddA and AddB act together in a complex, as do the RecBCD polypeptides and AddAB polypeptides of other bacteria. If so, the control of the unlinked *H. pylori addA* and *addB* genes to maintain the proper stoichiometry of the two polypeptides remains an interesting question.

Regarding the role of *H. pylori* AddA in DNA recombination during natural transformation, conflicting results were reported from different studies. The *addA* (note: it was named *recB* in certain references) mutant showed enhanced [68, 70], decreased [20, 71, 92], or no change [27, 69] in transformation frequency. Indeed, a high degree of variability (>100-fold) in transformation frequency in *H. pylori* was observed between different strains and different experiments. The use of different assay systems may partly explain the discrepancy in transformation results. For example, the total genomic DNA from antibiotic-resistant strain was used for the transformation assay in certain studies, while in others the defined linear DNA fragments of small size [92]. Use of the transformation frequency as an indicator of DNA recombination frequency is based on the assumption that the wild type *H. pylori* and its isogenic *rec* strains are equally competent for DNA uptake. However, it is now known that this assumption is not valid because DNA damage triggers genetic exchange in *H. pylori* [55]. *H. pylori addA* mutant cells suffered more DNA damage [92], and have an enhanced competence for DNA uptake [55]. Thus, the accumulation of unrepaired DNA damage and subsequent poor growth, as well as unknown strain differences, could be the main cause of the high degree of variability in *H. pylori* transformation frequency [27].

4.5 *H. pylori* RecRO pathway

RecFOR is a highly conserved DNA recombination pathway in bacteria, and is mainly used for ssDNA gap repair [72]. In the published *H. pylori* genome sequences, only the *recR* gene was annotated [12, 13]. Although RecF historically served as a reference for RecFOR pathway, it is absent from genomes of many bacteria including *H. pylori* [72]. By bioinformatics analysis, Marsin et al [68] identified HP0951 as a novel RecO orthologue, although its sequence identity with the *E. coli* protein is lower than 15%. Recent studies in *E. coli* indicated that RecOR in the absence of RecF can perform recombination by loading RecA [95, 96]. Whereas the RecO protein can displace ssDNA-binding protein (SSB) and bind to ssDNA, RecR is the key component for loading RecA onto ssDNA [95, 97]. Likely, the RecRO pathway (with no RecF) is present in *H. pylori*.

The *recR* and *recO* mutants showed marked sensitivity to DNA damaging agents metronidazole and UV light, indicating roles of RecR and RecO in DNA repair. Unlike the *addA* (*recB*) mutant, the *recR* and *recO* mutants did not show significant sensitivity to ionizing radiation (IR) and to mitomycin C [68, 71], suggesting that RecRO pathway is not responsible for repairing DNA damage induced by these agents, most likely double strand breaks. This is in contrast to *E. coli* where the RecFOR pathway sometimes substitutes for the RecBCD pathway and in *Deinococcus radiodurance* where the RecFOR pathway plays a major role in double strand break repair [98, 99]. On the other hand, *H. pylori recR* and *recO* mutants were shown to be much more sensitive to oxidative stress and to acid stress than the wild type strain [71], indicating that *H. pylori* RecRO pathway is involved in repairing DNA damage induced by these stress conditions. The *addA recO* double mutant (deficient in both AddAB and RecRO pathways) was significantly more sensitive to atmospheric oxygen than the *recO* single mutant, indicating that both RecRO and AddAB pathways are important for survival of oxidative damage. Similar roles of the RecBCD and the RecFOR pathways for survival of oxidative damage were also observed in *E. coli* [57, 100] and in *Neisseria gonorrhoeae* [84]. In those bacteria, however, the RecBCD appeared to be the predominant (over the RecFOR) repair pathway for oxidative damage. Our results suggest that the two pathways in *H. pylori* play similarly important roles in repairing oxidative stress-derived DNA damage [71]. In accordance with the sensitivity to oxidative and acid stress in vitro, *H. pylori recR* and *recO* mutants were shown to be less able to colonize mouse stomachs [71]. Furthermore, the mouse colonization ability of the *addA recO* double mutant was significantly lower than that of the *addA* or *recO* single mutant. Therefore, both AddAB- and RecRO-mediated DNA recombinational repair in *H. pylori* play an important role in bacterial survival and persistent colonization in the host.

Although differing results regarding the effect of *addA* gene on transformation frequency were reported by different research groups, it was agreed that the RecRO-pathway is not involved in recombination of exogenous DNA into the *H. pylori* genome in the process of transformation [68, 71]. The RecRO pathway is known to have a major role in intragenomic recombination at repeat sequences [101]. Using an assay to assess the deletion frequency resulting from recombination on direct repeat sequences (358 bp long), Marsin et al [68] showed that the *recR* and *recO* mutants exhibited a statistically significantly lower deletion frequency than the wild type strain, suggesting a role of RecRO in intragenomic recombination. Recently we adopted a similar assay using DNA constructs (deletion cassettes) that contain identical repeat sequences of different length (IDS100 and IDS350) [71]. The results indicated that the intra-genomic recombination of 100 bp-long direct repeat sequences in *H. pylori* is partially dependent on RecR and RecA, yet a large portion of the recombination event is RecA-independent. This is basically in agreement (with small variance) with the results of Aras et al [35] who reported that the repeat sequences of 100 bp or shorter recombined through a RecA-independent pathway. For the deletion cassette containing repeat sequences of 350 bp in length, inactivation of *recR* or *recA* resulted in a significant 4-fold or 35-fold decrease respectively in deletion frequency, indicating that RecR plays a significant role in recombination of IDS350, while this recombination was highly dependent on RecA.

5. Concluding remarks and perspectives

Severe *Helicobacter pylori*-mediated gastric diseases are associated with the bacterium's persistence in the host and its adaptability to host differences, which in turn is associated with its remarkable genetic variability. DNA recombination is an extraordinarily frequent event in *H. pylori*, and this manifests itself into a bacterium with unusual flexibility in stress-combating enzymes, repair mechanisms, and other adaptability characteristics. Nearly every *H. pylori* recombination-related gene studied thus far by a gene directed mutant analysis approach has documented they are individually important in stomach colonization ability; this underscores the importance of these recombination repair processes in bacterial survival in the host. It is well recognized that homologous DNA recombination is a special system in bacteria for repairing stalled replication forks and double strand breaks, while generating genetic diversity as an advantageous byproduct [102]. *H. pylori* may be an especially fruitful organism in which to learn the ultimate boundaries in roles of recombination repair enzymes, as *H. pylori* is subject to intense and prolonged host mediated stress and it displays an enormous genetic diversity.

Substantial progress has been made recently in unraveling the complex systems of DNA recombinational repair in *H. pylori*. As expected, whole genome sequencing has been a powerful tool to aid in identifying recombination-related proteins in *H. pylori*. For example, *recA*, *recR*, *recN*, and *ruvABC* were identified and confirmed to play important roles in *H. pylori* as could be expected from results for other bacteria. Some recombination-related proteins (e.g. MutS2, RecG), however, play unique roles in *H. pylori*. Most of the genes for the major components of the two pre-synapsis pathways (RecBCD and RecFOR) were not annotated from *H. pylori* genome sequences, which drove researchers' interest to search for additional novel systems required for *H. pylori* DNA recombinational repair. Recent studies revealed the existence of both pathways, AddAB and RecRO, in *H. pylori*. Although they display a limited level of sequence homology to the known recombination enzymes, both AddAB and RecRO were shown to play important roles in *H. pylori* DNA recombinational repair, conferring resistance to oxidative and acid stress.

The major components of DNA recombinational repair machinery in *H. pylori* are listed in Table 1. *H. pylori* RecN protein may recognize DNA double strand breaks and recruits AddAB helicase-nuclease complex for further processing. While not being involved in repair of DNA double strand breaks, *H. pylori* RecRO proteins play a major role in intra-genomic recombination at repeat sequences. Both pre-synapsis pathways (AddAB and RecRO) require RecA for catalyzing DNA strand exchange (synapsis) and *H. pylori* RuvABC is the predominant pathway for DNA branch migration and Holliday Junction resolution (post-synapsis). Although the major functions of these components are similar to those observed in model bacteria, some novel attributes of these components have been discovered, which may be related to the highly-specific lifestyle of *H. pylori*. Additional new components that work synergistically with these pathways could be found in this unique bacterium via future biochemical and genetic approaches.

6. Acknowledgements

The work on *H. pylori* DNA repair in our laboratory was supported by NIH grant R21AI076569 and by the University of Georgia Foundation.

Gene	HP # (a)	Activity / function	Main phenotypes of mutant (b)	reference
recN	1393	Initiates DSB-induced recombination.	Sensitive to DSB damage; Sensitive to oxidative stress; Attenuated mouse colonization.	[83]
recJ	0348	5'-3' ssDNA exonuclease.	Not studied experimently.	
addA	1553	AddAB Helicase-nuclease;	Sensitive to DSB damage;	[69, 92]
addB	1089	Initiates DSB-induced recombination.	Sensitive to oxidative stress; Attenuated mouse colonization.	
recR	0925	RecRO recombination pathway; Initiates ssDNA gap repair.	Not sensitive to DSB damage; Sensitive to oxidative stress; Attenuated mouse colonization.	[68, 71]
recO	0951			
recA	0153	DNA recombinase; Catalyzes DNA pairing and strand exchange.	Sensitive to DNA damaging agents; Decreased recombination frequency; Defective mouse colonization.	[65, 66, 69]
recG	1523	Holiday junction helicase.	Not sensitive to DNA damaging agents; Increased recombination frequency.	[76]
ruvA	0883	Holliday junction recognition.	Not studied experimently.	
ruvB	1059	Holiday junction helicase.	Sensitive to DNA damaging agents; Decreased recombination frequency.	[75]
ruvC	0877	Holliday junction resolvase.	Sensitive to DNA damaging agents; Decreased recombination frequency; Attenuated mouse colonization.	[73]

(a) HP# refers to the gene number in the genome sequence of strain 26695 [12].
(b) DSB (double strand breaks) damage refers to those damages caused e.g. by ionizing radiation, mitomycin C, or ciprofloxacin.

Table 1. *H. pylori* genes involved in DNA recombinational repair

7. References

[1] Dunn BE, Cohen H, Blaser MJ. Helicobacter pylori. Clin Microbiol Rev 1997;10(4):720-41.
[2] Suerbaum S, Michetti P. Helicobacter pylori infection. N Engl J Med 2002;347(15):1175-86.
[3] Uemura N, Okamoto S, Yamamoto S, *et al.* Helicobacter pylori infection and the development of gastric cancer. N Engl J Med 2001;345(11):784-9.
[4] McGee DJ, Mobley HL. Mechanisms of Helicobacter pylori infection: bacterial factors. Curr Top Microbiol Immunol 1999;241:155-80.
[5] Kusters JG, van Vliet AH, Kuipers EJ. Pathogenesis of Helicobacter pylori infection. Clin Microbiol Rev 2006;19(3):449-90.
[6] Kang J, Blaser MJ. Bacterial populations as perfect gases: genomic integrity and diversification tensions in Helicobacter pylori. Nat Rev Microbiol 2006;4(11):826-36.
[7] Suerbaum S. Genetic variability within Helicobacter pylori. Int J Med Microbiol 2000;290(2):175-81.
[8] Falush D, Wirth T, Linz B, *et al.* Traces of human migrations in Helicobacter pylori populations. Science 2003;299(5612):1582-5.
[9] Suerbaum S, Smith JM, Bapumia K, Morelli G, Smith NH, Kunstmann E, Dyrek I, Achtman M. Free recombination within Helicobacter pylori. Proc Natl Acad Sci U S A 1998;95(21):12619-24.

[10] Kansau I, Raymond J, Bingen E, *et al.* Genotyping of Helicobacter pylori isolates by sequencing of PCR products and comparison with the RAPD technique. Res Microbiol 1996;147(8):661-9.

[11] Wang G, Humayun MZ, Taylor DE. Mutation as an origin of genetic variability in Helicobacter pylori. Trends Microbiol 1999;7(12):488-93.

[12] Tomb JF, White O, Kerlavage AR, *et al.* The complete genome sequence of the gastric pathogen Helicobacter pylori. Nature 1997;388(6642):539-47.

[13] Alm RA, Ling LS, Moir DT, *et al.* Genomic-sequence comparison of two unrelated isolates of the human gastric pathogen Helicobacter pylori. Nature 1999;397(6715):176-80.

[14] Salama N, Guillemin K, McDaniel TK, Sherlock G, Tompkins L, Falkow S. A whole-genome microarray reveals genetic diversity among Helicobacter pylori strains. Proc Natl Acad Sci U S A 2000;97(26):14668-73.

[15] Israel DA, Salama N, Krishna U, Rieger UM, Atherton JC, Falkow S, Peek RM, Jr. Helicobacter pylori genetic diversity within the gastric niche of a single human host. Proc Natl Acad Sci U S A 2001;98(25):14625-30.

[16] Kraft C, Suerbaum S. Mutation and recombination in Helicobacter pylori: mechanisms and role in generating strain diversity. Int J Med Microbiol 2005;295(5):299-305.

[17] Bjorkholm B, Sjolund M, Falk PG, Berg OG, Engstrand L, Andersson DI. Mutation frequency and biological cost of antibiotic resistance in Helicobacter pylori. Proc Natl Acad Sci U S A 2001;98(25):14607-12.

[18] Horst JP, Wu TH, Marinus MG. Escherichia coli mutator genes. Trends Microbiol 1999;7(1):29-36.

[19] Lin EA, Zhang XS, Levine SM, Gill SR, Falush D, Blaser MJ. Natural transformation of helicobacter pylori involves the integration of short DNA fragments interrupted by gaps of variable size. PLoS Pathog 2009;5(3):e1000337.

[20] Kulick S, Moccia C, Didelot X, Falush D, Kraft C, Suerbaum S. Mosaic DNA imports with interspersions of recipient sequence after natural transformation of Helicobacter pylori. PLoS One 2008;3(11):e3797.

[21] Hofreuter D, Odenbreit S, Haas R. Natural transformation competence in Helicobacter pylori is mediated by the basic components of a type IV secretion system. Mol Microbiol 2001;41(2):379-91.

[22] Karnholz A, Hoefler C, Odenbreit S, Fischer W, Hofreuter D, Haas R. Functional and topological characterization of novel components of the comB DNA transformation competence system in Helicobacter pylori. J Bacteriol 2006;188(3):882-93.

[23] Stingl K, Muller S, Scheidgen-Kleyboldt G, Clausen M, Maier B. Composite system mediates two-step DNA uptake into Helicobacter pylori. Proc Natl Acad Sci U S A 2010;107(3):1184-9.

[24] Alvarez-Martinez CE, Christie PJ. Biological diversity of prokaryotic type IV secretion systems. Microbiol Mol Biol Rev 2009;73(4):775-808.

[25] Saunders NJ, Peden JF, Moxon ER. Absence in Helicobacter pylori of an uptake sequence for enhancing uptake of homospecific DNA during transformation. Microbiology 1999;145 (Pt 12):3523-8.

[26] Aras RA, Small AJ, Ando T, Blaser MJ. Helicobacter pylori interstrain restriction-modification diversity prevents genome subversion by chromosomal DNA from competing strains. Nucleic Acids Res 2002;30(24):5391-7.

[27] Humbert O, Dorer MS, Salama NR. Characterization of Helicobacter pylori factors that control transformation frequency and integration length during inter-strain DNA recombination. Mol Microbiol 2011;79(2):387-401.

[28] Go MF, Kapur V, Graham DY, Musser JM. Population genetic analysis of Helicobacter pylori by multilocus enzyme electrophoresis: extensive allelic diversity and recombinational population structure. J Bacteriol 1996;178(13):3934-8.

[29] Kersulyte D, Chalkauskas H, Berg DE. Emergence of recombinant strains of Helicobacter pylori during human infection. Mol Microbiol 1999;31(1):31-43.

[30] Salama NR, Gonzalez-Valencia G, Deatherage B, Aviles-Jimenez F, Atherton JC, Graham DY, Torres J. Genetic analysis of Helicobacter pylori strain populations colonizing the stomach at different times postinfection. J Bacteriol 2007;189(10):3834-45.

[31] Falush D, Kraft C, Taylor NS, Correa P, Fox JG, Achtman M, Suerbaum S. Recombination and mutation during long-term gastric colonization by Helicobacter pylori: estimates of clock rates, recombination size, and minimal age. Proc Natl Acad Sci U S A 2001;98(26):15056-61.

[32] Achtman M, Zurth K, Morelli G, Torrea G, Guiyoule A, Carniel E. Yersinia pestis, the cause of plague, is a recently emerged clone of Yersinia pseudotuberculosis. Proc Natl Acad Sci U S A 1999;96(24):14043-8.

[33] Guttman DS, Dykhuizen DE. Clonal divergence in Escherichia coli as a result of recombination, not mutation. Science 1994;266(5189):1380-3.

[34] Linz B, Schenker M, Zhu P, Achtman M. Frequent interspecific genetic exchange between commensal Neisseriae and Neisseria meningitidis. Mol Microbiol 2000;36(5):1049-58.

[35] Aras RA, Kang J, Tschumi AI, Harasaki Y, Blaser MJ. Extensive repetitive DNA facilitates prokaryotic genome plasticity. Proc Natl Acad Sci U S A 2003;100(23):13579-84.

[36] Martinez GR, Loureiro AP, Marques SA, et al. Oxidative and alkylating damage in DNA. Mutat Res 2003;544(2-3):115-27.

[37] Foster JW, Bearson B. Acid-sensitive mutants of Salmonella typhimurium identified through a dinitrophenol lethal screening strategy. J Bacteriol 1994;176(9):2596-602.

[38] Algood HM, Cover TL. Helicobacter pylori persistence: an overview of interactions between H. pylori and host immune defenses. Clin Microbiol Rev 2006;19(4):597-613.

[39] Chaturvedi R, Cheng Y, Asim M, et al. Induction of polyamine oxidase 1 by Helicobacter pylori causes macrophage apoptosis by hydrogen peroxide release and mitochondrial membrane depolarization. J Biol Chem 2004;279(38):40161-73.

[40] Ding SZ, Minohara Y, Fan XJ, et al. Helicobacter pylori infection induces oxidative stress and programmed cell death in human gastric epithelial cells. Infect Immun 2007;75(8):4030-9.

[41] Scott DR, Marcus EA, Wen Y, Oh J, Sachs G. Gene expression in vivo shows that Helicobacter pylori colonizes an acidic niche on the gastric surface. Proc Natl Acad Sci U S A 2007;104(17):7235-40.

[42] Wang G, Alamuri P, Maier RJ. The diverse antioxidant systems of Helicobacter pylori. Mol Microbiol 2006;61(4):847-60.

[43] Pflock M, Kennard S, Finsterer N, Beier D. Acid-responsive gene regulation in the human pathogen Helicobacter pylori. J Biotechnol 2006;126(1):52-60.

[44] Sancar A. Mechanisms of DNA excision repair. Science 1994;266(5193):1954-6.

[45] Thompson SA, Latch RL, Blaser JM. Molecular characterization of the Helicobacter pylori uvr B gene. Gene 1998;209(1-2):113-22.

[46] Kang J, Blaser MJ. UvrD helicase suppresses recombination and DNA damage-induced deletions. J Bacteriol 2006;188(15):5450-9.

[47] Modrich P. Mismatch repair, genetic stability, and cancer. Science 1994;266(5193):1959-60.

[48] Pinto AV, Mathieu A, Marsin S, Veaute X, Ielpi L, Labigne A, Radicella JP. Suppression of homologous and homeologous recombination by the bacterial MutS2 protein. Mol Cell 2005;17(1):113-20.

[49] Wang G, Alamuri P, Humayun MZ, Taylor DE, Maier RJ. The Helicobacter pylori MutS protein confers protection from oxidative DNA damage. Mol Microbiol 2005;58(1):166-76.

[50] O'Rourke EJ, Chevalier C, Pinto AV, Thiberge JM, Ielpi L, Labigne A, Radicella JP. Pathogen DNA as target for host-generated oxidative stress: role for repair of bacterial DNA damage in Helicobacter pylori colonization. Proc Natl Acad Sci U S A 2003;100(5):2789-94.

[51] Eutsey R, Wang G, Maier RJ. Role of a MutY DNA glycosylase in combating oxidative DNA damage in Helicobacter pylori. DNA Repair (Amst) 2007;6(1):19-26.

[52] Mathieu A, O'Rourke EJ, Radicella JP. Helicobacter pylori genes involved in avoidance of mutations induced by 8-oxoguanine. J Bacteriol 2006;188(21):7464-9.

[53] Huang S, Kang J, Blaser MJ. Antimutator role of the DNA glycosylase mutY gene in Helicobacter pylori. J Bacteriol 2006;188(17):6224-34.

[54] Butala M, Zgur-Bertok D, Busby SJ. The bacterial LexA transcriptional repressor. Cell Mol Life Sci 2009;66(1):82-93.

[55] Dorer MS, Fero J, Salama NR. DNA damage triggers genetic exchange in Helicobacter pylori. PLoS Pathog 2010;6(7):e1001026.

[56] Cromie GA, Connelly JC, Leach DR. Recombination at double-strand breaks and DNA ends: conserved mechanisms from phage to humans. Mol Cell 2001;8(6):1163-74.

[57] Kuzminov A. Recombinational repair of DNA damage in Escherichia coli and bacteriophage lambda. Microbiol Mol Biol Rev 1999;63(4):751-813, table of contents.

[58] Cox MM, Goodman MF, Kreuzer KN, Sherratt DJ, Sandler SJ, Marians KJ. The importance of repairing stalled replication forks. Nature 2000;404(6773):37-41.

[59] Liu Y, West SC. Happy Hollidays: 40th anniversary of the Holliday junction. Nat Rev Mol Cell Biol 2004;5(11):937-44.

[60] Fernandez S, Ayora S, Alonso JC. Bacillus subtilis homologous recombination: genes and products. Res Microbiol 2000;151(6):481-6.

[61] Skaar EP, Lazio MP, Seifert HS. Roles of the recJ and recN genes in homologous recombination and DNA repair pathways of Neisseria gonorrhoeae. J Bacteriol 2002;184(4):919-27.

[62] Courcelle J, Donaldson JR, Chow KH, Courcelle CT. DNA damage-induced replication fork regression and processing in Escherichia coli. Science 2003;299(5609):1064-7.

[63] Kidane D, Sanchez H, Alonso JC, Graumann PL. Visualization of DNA double-strand break repair in live bacteria reveals dynamic recruitment of Bacillus subtilis RecF,

RecO and RecN proteins to distinct sites on the nucleoids. Mol Microbiol 2004;52(6):1627-39.

[64] Meddows TR, Savory AP, Grove JI, Moore T, Lloyd RG. RecN protein and transcription factor DksA combine to promote faithful recombinational repair of DNA double-strand breaks. Mol Microbiol 2005;57(1):97-110.

[65] Schmitt W, Odenbreit S, Heuermann D, Haas R. Cloning of the Helicobacter pylori recA gene and functional characterization of its product. Mol Gen Genet 1995;248(5):563-72.

[66] Thompson SA, Blaser MJ. Isolation of the Helicobacter pylori recA gene and involvement of the recA region in resistance to low pH. Infect Immun 1995;63(6):2185-93.

[67] Fischer W, Haas R. The RecA protein of Helicobacter pylori requires a posttranslational modification for full activity. J Bacteriol 2004;186(3):777-84.

[68] Marsin S, Mathieu A, Kortulewski T, Guerois R, Radicella JP. Unveiling novel RecO distant orthologues involved in homologous recombination. PLoS Genet 2008;4(8):e1000146.

[69] Amundsen SK, Fero J, Hansen LM, Cromie GA, Solnick JV, Smith GR, Salama NR. Helicobacter pylori AddAB helicase-nuclease and RecA promote recombination-related DNA repair and survival during stomach colonization. Mol Microbiol 2008;69(4):994-1007.

[70] Marsin S, Lopes A, Mathieu A, Dizet E, Orillard E, Guerois R, Radicella JP. Genetic dissection of Helicobacter pylori AddAB role in homologous recombination. FEMS Microbiol Lett 2010;311(1):44-50.

[71] Wang G, Lo LF, Maier RJ. The RecRO pathway of DNA recombinational repair in Helicobacter pylori and its role in bacterial survival in the host. DNA Repair (Amst) 2011; 10: 373-379.

[72] Rocha EP, Cornet E, Michel B. Comparative and evolutionary analysis of the bacterial homologous recombination systems. PLoS Genet 2005;1(2):e15.

[73] Loughlin MF, Barnard FM, Jenkins D, Sharples GJ, Jenks PJ. Helicobacter pylori mutants defective in RuvC Holliday junction resolvase display reduced macrophage survival and spontaneous clearance from the murine gastric mucosa. Infect Immun 2003;71(4):2022-31.

[74] Robinson K, Loughlin MF, Potter R, Jenks PJ. Host adaptation and immune modulation are mediated by homologous recombination in Helicobacter pylori. J Infect Dis 2005;191(4):579-87.

[75] Kang J, Blaser MJ. Repair and antirepair DNA helicases in Helicobacter pylori. J Bacteriol 2008;190(12):4218-24.

[76] Kang J, Tavakoli D, Tschumi A, Aras RA, Blaser MJ. Effect of host species on recG phenotypes in Helicobacter pylori and Escherichia coli. J Bacteriol 2004;186(22):7704-13.

[77] Sharples GJ, Ingleston SM, Lloyd RG. Holliday junction processing in bacteria: insights from the evolutionary conservation of RuvABC, RecG, and RusA. J Bacteriol 1999;181(18):5543-50.

[78] Hirano T. At the heart of the chromosome: SMC proteins in action. Nat Rev Mol Cell Biol 2006;7(5):311-22.

[79] Haering CH, Lowe J, Hochwagen A, Nasmyth K. Molecular architecture of SMC proteins and the yeast cohesin complex. Mol Cell 2002;9(4):773-88.

[80] Kidane D, Graumann PL. Dynamic formation of RecA filaments at DNA double strand break repair centers in live cells. J Cell Biol 2005;170(3):357-66.

[81] Sanchez H, Kidane D, Castillo Cozar M, Graumann PL, Alonso JC. Recruitment of Bacillus subtilis RecN to DNA double-strand breaks in the absence of DNA end processing. J Bacteriol 2006;188(2):353-60.

[82] Sanchez H, Alonso JC. Bacillus subtilis RecN binds and protects 3'-single-stranded DNA extensions in the presence of ATP. Nucleic Acids Res 2005;33(7):2343-50.

[83] Wang G, Maier RJ. Critical role of RecN in recombinational DNA repair and survival of Helicobacter pylori. Infect Immun 2008;76(1):153-60.

[84] Stohl EA, Seifert HS. Neisseria gonorrhoeae DNA recombination and repair enzymes protect against oxidative damage caused by hydrogen peroxide. J Bacteriol 2006;188(21):7645-51.

[85] Stohl EA, Criss AK, Seifert HS. The transcriptome response of Neisseria gonorrhoeae to hydrogen peroxide reveals genes with previously uncharacterized roles in oxidative damage protection. Mol Microbiol 2005;58(2):520-32.

[86] Singleton MR, Wigley DB. Modularity and specialization in superfamily 1 and 2 helicases. J Bacteriol 2002;184(7):1819-26.

[87] Petit MA, Ehrlich D. Essential bacterial helicases that counteract the toxicity of recombination proteins. Embo J 2002;21(12):3137-47.

[88] Lestini R, Michel B. UvrD controls the access of recombination proteins to blocked replication forks. Embo J 2007;26(16):3804-14.

[89] Singleton MR, Dillingham MS, Gaudier M, Kowalczykowski SC, Wigley DB. Crystal structure of RecBCD enzyme reveals a machine for processing DNA breaks. Nature 2004;432(7014):187-93.

[90] Kooistra J, Haijema BJ, Hesseling-Meinders A, Venema G. A conserved helicase motif of the AddA subunit of the Bacillus subtilis ATP-dependent nuclease (AddAB) is essential for DNA repair and recombination. Mol Microbiol 1997;23(1):137-49.

[91] Yeeles JT, Dillingham MS. A dual-nuclease mechanism for DNA break processing by AddAB-type helicase-nucleases. J Mol Biol 2007;371(1):66-78.

[92] Wang G, Maier RJ. A RecB-like helicase in Helicobacter pylori is important for DNA repair and host colonization. Infect Immun 2009;77(1):286-91.

[93] Amundsen SK, Fero J, Salama NR, Smith GR. Dual nuclease and helicase activities of Helicobacter pylori AddAB are required for DNA repair, recombination, and mouse infectivity. J Biol Chem 2009;284(25):16759-66.

[94] Gressmann H, Linz B, Ghai R, et al. Gain and loss of multiple genes during the evolution of Helicobacter pylori. PLoS Genet 2005;1(4):e43.

[95] Sakai A, Cox MM. RecFOR and RecOR as distinct RecA loading pathways. J Biol Chem 2009;284(5):3264-72.

[96] Handa N, Morimatsu K, Lovett ST, Kowalczykowski SC. Reconstitution of initial steps of dsDNA break repair by the RecF pathway of E. coli. Genes Dev 2009;23(10):1234-45.

[97] Inoue J, Honda M, Ikawa S, Shibata T, Mikawa T. The process of displacing the single-stranded DNA-binding protein from single-stranded DNA by RecO and RecR proteins. Nucleic Acids Res 2008;36(1):94-109.

[98]   Ivancic-Bace I, Peharec P, Moslavac S, Skrobot N, Salaj-Smic E, Brcic-Kostic K. RecFOR
        function is required for DNA repair and recombination in a RecA loading-deficient
        recB mutant of Escherichia coli. Genetics 2003;163(2):485-94.

[99]   Bentchikou E, Servant P, Coste G, Sommer S. A major role of the RecFOR pathway in
        DNA double-strand-break repair through ESDSA in Deinococcus radiodurans.
        PLoS Genet;6(1):e1000774.

[100]   Imlay JA, Linn S. Mutagenesis and stress responses induced in Escherichia coli by
        hydrogen peroxide. J Bacteriol 1987;169(7):2967-76.

[101]   Galitski T, Roth JR. Pathways for homologous recombination between chromosomal
        direct repeats in Salmonella typhimurium. Genetics 1997;146(3):751-67.

[102]   Cox MM. Historical overview: searching for replication help in all of the rec places.
        Proc Natl Acad Sci U S A 2001;98(15):8173-80.

2

The Role of DDB2 in Regulating Cell Survival and Apoptosis Following DNA Damage - A Mini-Review

Chuck C.-K. Chao
Department of Biochemistry and Molecular Biology
Graduate Institute of Biomedical Sciences, Chang Gung University, Taiwan
Republic of China

1. Introduction

Nucleotide excision repair (NER) represents a central cellular process for the removal of structurally and chemically diverse DNA lesions [Friedberg et al., 2006]. Mutations in genes involved in NER are associated with rare autosomal recessive syndromes such as xeroderma pigmentosum (XP), a condition characterized by sensitivity to UV light, neurological abnormalities, and a propensity to develop skin cancer (Cleaver, 2005). The observation that cells from XP subgroup E (XP-E cells XP2RO and XP3RO) are defective in recognizing damaged DNA and performing NER highlighted the physiological importance of the protein termed DNA damage-binding protein, or DDB [Chu & Chang, 1988]. The DDB protein, sometimes also referred to as UV-DDB due to its high affinity and specificity for UV-damaged DNA, contains two principal subunits, DDB1 and DDB2 [Grossman, 1976; Keeney et al., 1993; Takao et al., 1993]. The DDB protein complex also binds to non-UV-damaged DNA, like cisplatin-modified DNA, although with much lower affinity. Although the history of DDB spans more than two decades, the complete understanding of its physiological functions remains to be clarified. The activity of DDB has been repeatedly described in crude mammalian cell extracts by electrophoretic mobility shift assays or filter-binding assays performed by different laboratories since the first report of its discovery [Feldberg & Grossman, 1976]. Notably, micro-injections of DDB complexes into the nucleus of XP-E cells restored NER activity [Keeney et al., 1994], supporting the notion that DDB participates in chromatin NER. The *DDB1* gene from simian cells was the first *DDB* gene to be identified [Takao et al., 1993]. The human *DDB1* and *DDB2* genes were subsequently sequenced [Dualan et al., 1995; Lee et al., 1995]. Soon after, DNA sequencing from Linn's laboratory revealed that *DDB2* is mutated in XP-E cells which lack DDB activity [Nichols et al., 1996; Tang & Chu, 2002]. The predicted DDB2 protein sequence was shown to contain several functional domains, including WD40 repeats, post-translation modification sites (e.g. acetylation, phosphorylation, and ubiquitination), DDB1- and DNA-binding sites, as well as a DWD box. Notably, in a majority of XP-E cell lines, DDB2 was found to be altered at domains other than the one required for binding DNA. Thus, DDB appears to be regulated at several levels in UV-irradiated cells, including by transcriptional activation of DDB2 mRNA, post-translational modification, translocation to the nucleus, complex formation, and proteolytic degradation of DDB2 protein through ubiquitination [for a recent review, see Sugasawa, 2010]. Notably, 60% of chromatin-bound DDB2 is degraded within 4 hrs of UV irradiation. After 48 hrs, DDB2 mRNA levels increase several fold above the level seen in non-irradiated cells [Nichols et al., 2000; Rapic-Otrin et al., 2002]. Interestingly, the majority of UV-induced DNA photoproducts in human cells are repaired by this time [Mitchell et al., 1985].

2. DDB2 recognizes DNA damage during global genome NER

NER removes diverse DNA lesions, ranging from UV-induced cyclobutane pyrimidine dimers (CPD) and 6-4 pyrimidine-pyrimidone photoproducts (6-4PP) to a variety of bulky adducts formed by environmental carcinogens. Mammalian NER comprises global genome NER (GG-NER) and transcription-coupled NER (TC-NER). These two processes involve similar but distinct repair proteins that process DNA damage and chromatin proteins like histones may significantly regulate the activity of repair proteins (reviewed by Friedberg et al., 2006). One such multiprotein complex involved in GG-NER and containing both DDB1 and DDB2 is closely related to a complex containing DDB1 and the Cockayne syndrome group A (CSA) protein in TC-NER. In GG-NER, DNA is initially surveyed for lesions by XP group C (XPC) protein-RAD23B (Sugasawa et al., 1998) and the UV-DDB complex (Fitch et al., 2003; Moser et al., 2005; Sugasawa et al., 2005). DDB2 binds to DDB1 to form the DDB complex which may recognize UV-induced DNA damage and recruit proteins of the NER pathway to initiate GG-NER (Hwang et al., 1999; Tang et al., 2000). The DDB complex preferentially binds to UV-induced CPD, 6-4PP, apurinic sites, and short mismatches (Fujiwara et al., 1999; Kulaksiz et al., 2005; Sugasawa et al., 2005; Wittschieben et al., 2005). While XPC functions as a versatile factor that senses abnormal DNA structures, DDB appears to recognize more specific types of lesions, particularly UV-induced 6-4PP, whereas binding to CPD is much weaker but nonetheless detectable [Payne & Chu, 1994]. Strikingly, structural analysis of DDB bound to DNA duplex containing 6-4PP has revealed that the DDB2 subunit is responsible for the interaction, and this subunit induces the movement of the two affected bases into a binding pocket, therefore indicating that DDB has evolved to specifically recognize dinucleotide lesions, like UV photolesions [Figure 1; Scrima et al., 2008]. Furthermore, accumulating evidence has confirmed the existence of multiple forms of DDB2 mRNA splicing variants, including isoforms D1 and D2, which do not interact with DDB1, but inhibit UV-damaged DNA repair (Inoki et al., 2004). DDB2 is ubiquitously expressed in human tissues, with the highest level being found in corneal endothelium and the lowest level in the brain. Isoform D1 is highly expressed in brain and heart tissues, whereas isoforms D2, D3, and D4 are weakly expressed in these tissues (Inoki et al., 2004). Interestingly, repair of DNA damage induced by UV light appears to be less active in brain and heart tissues which are naturally protected against UV irradiation and express high levels of isoform D1.

3. DDB2 links DNA repair to protein ubiquitination

Another breakthrough that links protein ubiquitination with GG-NER is the finding that DDB is part of an ubiquitin ligase (E3) complex. Epitope-tagged DDB2 purified from cells was found in complex with CUL4A, ROC1, DDB1, and the COP9 signalosome [Groisman et al., 2003]. Besides its function as part of the DDB-protein complex, DDB2 may function as a substrate-recognition module within the CUL4A ubiquitination complex. CUL4 is one of

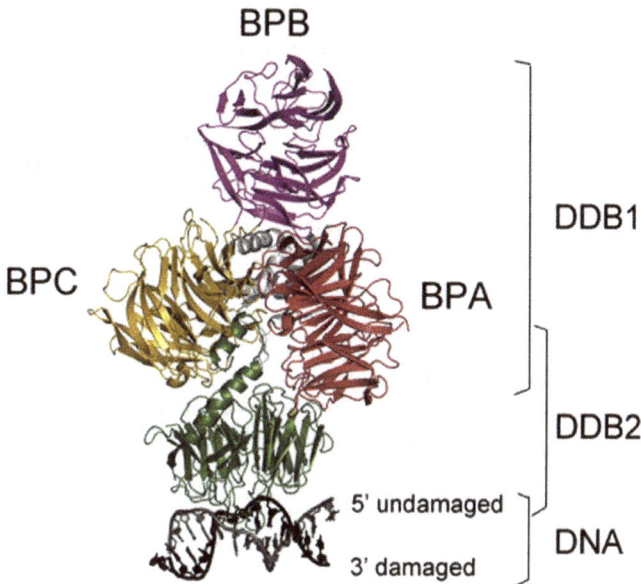

Fig. 1. Overall structure of the DDB1-DDB2-DNA complex. Ribbon representation of the DDB-DNA^{6-4PP} complex: DDB2; DDB1-BPA; DDB1-BPB; DDB1-BPC; DDB1-CTD. The DNA^{6-4PP} damaged and undamaged DNA strands are depicted in black and gray, respectively. DNA binding is carried out exclusively by the DDB2 subunit via its WD40 domain. The DDB1 structure consists of three WD40 β-propeller domains (BPA, BPB, and BPC) and a C-terminal helical domain (CTD, shown at the center). DDB2 binds to an interface between the DDB1 propellers BPA and BPC, where its helix-loop-helix motif inserts into a cavity formed by the two propellers. The structures reveal the molecular mechanism underlying high-affinity recognition of UV lesions (damaged DNA strand) that are refractory to detection by XPC. The structures also suggest a mechanism for the assembly of the DDB-CUL4 ubiquitin ligase in chromatin and provide a framework for understanding the ubiquitination of proteins proximal to damage sites. [For detail, see Scrima et al., 2008].

three founding cullins that are conserved from yeast to humans. A large number of E3 ubiquitin-protein ligase complexes are part of the DCX proteins (short for DDB1-CUL4-X-box). Components of the CUL4-DDB-ROC1 (also known as CUL4-DDB-RBX1) include CUL4A or CUL4B, DDB1, DDB2, and RBX1 (Chen et al., 2001; Groisman et al., 2003). Other CUL4-DDB-ROC1 complexes may also exist in which DDB2 is replaced by a subunit that targets an alternative substrate. These targeting subunits are generally known as DCAF proteins (short for DDB1- and CUL4-associated factor) or CDW (short for CUL4-DDB1-associated WD40-repeat; for reviews, see Lee & Zhou, 2007; Jackson & Xiong, 2009; Sugasawa, 2009). Many CUL4 complexes are involved in chromatin regulation and are frequently hijacked by viruses (reviewed by Jackson & Xiong, 2009). The DDB1-CUL4-ROC1 complex may ubiquitinate histones H2A, H3, and H4 at sites of UV-induced DNA damage (Wang et al., 2006; Kapetanaki et al., 2006; Guerrero-Santoro et al., 2008). The ubiquitination of histones may facilitate their removal from the nucleosome and promote assembly of NER components for subsequent DNA repair. Furthermore, the DDB1-CUL4-ROC1 complex ubiquitinates XPC and DDB2, which may enhance DNA binding by XPC and promote NER (El-Mahdy et al., 2006; Sugasawa et al., 2005). Structural analysis support the notion that CUL4 uses DDB1 as a large β-propeller protein and as a linker to interact with a subset of WD40 proteins like DDB2, which serves as substrate receptors, forming as many as 90 E3 complexes in mammals [Jackson & Xiong, 2009]. Taken together, these results indicate that DDB complex is a component of the CUL4A-based ubiquitin ligase DDB1-CUL4A[DDB2], and that DDB2 may coordinate the ubiquitination of various proteins at DNA damage sites during GG-NER.

In addition, CUL4B also binds to UV-damaged chromatin as a part of the DDB1-CUL4B[DDB2] E3 ligase in the presence of functional DDB2. Nevertheless, CUL4B is localized in the nucleus and facilitates the transfer of DDB1 into the nucleus independently of DDB2 [Guerrero-Santoro et al., 2008]. Notably, DDB1-CUL4B[DDB2] is more efficient than DDB1-CUL4A[DDB2] in mono-ubiquitinating histone H2A in vitro, suggesting that the DDB1-CUL4B[DDB2] E3 ligase may have a distinctive function in modifying the chromatin structure at sites of UV lesions and promoting efficient GG-NER. Intriguingly, the CSA protein, a WD40 motif protein defective in a complementation group of Cockayne's syndrome, forms a similar E3 complex in place of DDB2 at damage sites during TC-NER. Although not detected in the DDB2 and CSA complex, CUL4B is highly expressed in mammalian cells, and the two CUL4 isoforms CUL4A and CUL4B appear to be redundant, at least for some cellular functions [Higa et al., 2003; Hu et al., 2004].

4. DDB2 inhibits apoptosis in cultured cell lines and *Drosophila*

Although the regulation of the DDB2 gene is complex, evidence on the biological function of DDB2 in response to apoptotic stimuli has accumulated. Evidence from biochemical experiments has shown how DDB2 interacts with proteins, DNAs, and RNAs. Most strikingly, structural studies using X-ray crystallography support the evidence of biochemical studies, as seen for example with GG-NER. Nevertheless, a complete understanding of the biological roles of DDB2 remains to be fully elucidated. To assess this question, we explored the role of DDB2 in regulating UV sensitivity in both human cells and *Drosophila* [Sun et al., 2010]. As such, a full-length DDB2 open reading frame sequence was overexpressed in cells that express low or no DDB2. Conversely, DDB2 expression was suppressed in cells that endogenously express high levels of DDB2 by stable expression of full-length anti-sense cDNA. Using this strategy, we found that DDB2 displays a protective role against UV irradiation and cell surface death receptor signaling in both cisplatin-selected human HeLa cells and hamster V79 cells [Sun et al., 2002a; Sun et al., 2002b; Sun & Chao, 2005a]. Furthermore, cFLIP expression was upregulated by DDB2 in a dose- and time-dependent manner in HeLa cells, a process associated with inhibition of apoptosis [Sun & Chao, 2005a]. Inhibition of cFLIP by anti-sense oligonucleotides substantially inhibited apoptosis induced by UV irradiation and death receptor signaling in HeLa and other cell lines. Importantly, the protective effect of DDB2 was only detected in cells in which cFLIP is elicited during apoptotic stimuli. In contrast, DDB2 did not show a protective effect against apoptotic stimuli in human cell lines in which cFLIP expression was not induced [Sun et al., 2010]. A transcription reporter assay also showed that DDB2 induces the transcription of cFLIP in a p38/MAPK-dependent manner [Sun & Chao, 2005b], suggesting that the DDB2/cFLIP pathway may be active in specific cell conditions [Figure 2]. Surprisingly, overexpression of a DDB2 mutant (82TO) that does not significantly enhance DDB activity (Nichols et al., 1996), also protected HeLa cells from both UV- and Fas-induced cell death (Sun et al., 2002a; Sun & Chao, 2005a), suggesting that the protection effect of DDB2 may be independent of its DNA repair activity. Furthermore, ectopic expression of human DDB2 in *Drosophila* dramatically reduced UV-induced animal death compared to control GFP expression. On the other hand, expression of DDB2 in *Drosophila* failed to rescue a different type of apoptosis induced by the genes *reaper* or *eiger* [Sun et al., 2010]. Depletion of DDB2 in HeLa cells did not affect apoptosis induced by cisplatin or mitomycin C (Sun et al., 2002a). In addition, overexpression or inhibition of DDB2 in HeLa cells only slightly affected cisplatin-induced caspase-8 signaling and apoptosis (Sun & Chao, 2005a), probably due to the observation that cisplatin primarily induces mitochondrial apoptotic signaling (Gonzalez et al., 2001). These observations suggest that the modulation of apoptosis by DDB2 may be unique.

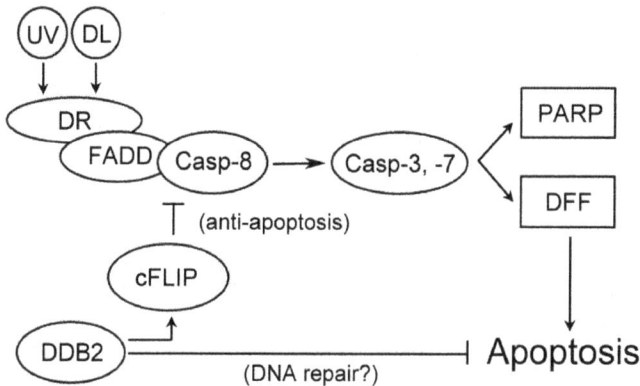

Fig. 2. Model illustrating the role of DDB2 in regulating non-DNA damage-induced apoptosis. An anti-apoptotic effect is proposed for DDB2 against death ligand- or UV-induced stress through cFLIP up-regulation. DDB2 transactivation of cFLIP is required to enhance their apoptosis-inhibitory function. UV- or death receptor-induced apoptosis is attenuated by the up-regulated cFLIP; consequently, activation of initiator caspases (3 and 7), cleavage of protein substrates (PARP and DFF), and apoptosis are inhibited. DDB2 may also attenuate UV-induced apoptosis through repair of DNA damage. However, evidence from protective DDB2 mutants suggests possible alternative pathways. DL, death ligands; DR, death receptors. [Modified from Sun and Chao, 2005a]

Cross-resistance to UV was found in cisplatin-selected cells, which overexpress DDB2 [Chu & Chang, 1990; Chao et al., 1991]. DDB2 is a transcriptional partner of E2F1; however, the target of DDBs/E2F1 has not been identified (Hayes et al., 1998; Shiyanov et al., 1999). We found that the overexpression of DDB2 increases the expression of cFLIP at both the mRNA and protein levels in resistant cells in which DDB2 has been genetically suppressed [Sun and Chao, 2005a]. E2F1 was also shown to regulate the expression of cFLIP (Stanelle et al., 2002). Therefore, cFLIP may represent the first potential target of DDB2/E2F1. E2F1 promotes TNF-induced apoptosis by stabilizing the TRAF2 protein (Phillips et al., 1999). However, the possibility that DDB2/E2F1 may co-activate cFLIP expression suggests a possible dual role for E2F1 in regulating cell survival and death. Additional overexpression of E2F1 does not increase endogenous cFLIP expression more than overexpression of DDB2 alone (Peng, 2008). Thus, the increased level of E2F1 observed in resistant cells is not enough to support the apoptotic resistance mediated by DDB2-cFLIP. Although induction of cFLIP by DDB2 is required to protect cells against UV-induced apoptosis, at least in HeLa cells, we could not exclude the possibility that other genes are also involved in mediating the anti-apoptotic effect of DDB2.

5. Ectopic expression of DDB2 induces apoptosis in DDB2-deficient cells

An extensive review of XP-E and DDB has been presented by Itoh who focused on XP-E and DDB2 as well as the classification of photosensitive diseases [Itoh, 2006]. Surprisingly, XP-E cell strains proved to be abnormally resistant to UV irradiation and possessed reduced caspase-3 activity. Since the apoptotic defect in XP-E strains could be rescued by exogenous p53 expression, DDB2 was also proposed to regulate p53-mediated apoptotic pathway after UV irradiation in human primary cell strains [Itoh et al, 2000; 2003]. Cells from DDB2-knockout mice also showed abnormal resistance and impaired p53 response to UV irradiation similar to human XP-E cell strains [Itoh et al., 2004]. Furthermore, a recent study has demonstrated that mouse embryonic fibroblasts and human HeLa that express DDB2 shRNA are resistant to apoptosis induced by a variety of DNA-damaging agents despite the activation of p53 and other pro-apoptotic genes [Stoyanova et al., 2009]. Also, these DDB2-deficient cells are resistant to E2F1-induced apoptosis, probably due to the observation that these cells undergo p21Waf1/Cip1-associated cell cycle arrest following DNA damage. Notably, DDB2 targets p21Waf1/Cip1 for proteolysis and this process involves Mdm2 in a manner that is distinct from the p53-regulatory activity of Mdm2 [Stoyanova et al., 2009]. These results suggest a new regulatory loop involving DDB2, Mdm2, and p21Waf1/Cip1 that is critical in determining the cellular fate between apoptosis and cell cycle arrest (for DNA repair) in response to DNA damage. The existence of this regulatory loop may be strengthened by showing that forced expression of DDB2 renders XP-E or DDB2-deficient cells sensitive to apoptotic stimuli.

6. Cancer-prone DDB2-deficient mice

DDB2-knockout mice have been shown to be prone to cancer formation [Itoh et al., 2004]. Importantly, mice with single DDB2 allele knockout showed enhanced skin cancer following UV-B exposure, suggesting that DDB2 heterozygotes may be predisposed to skin cancer [Itoh et al., 2004]. In addition, XP mouse models were reported to be prone to the formation of papillomas induced by 7,12-dimethylbenz[a]anthracene (DMBA) [de Bohr et al., 1999; Nakane et al., 1995; de Vries et al., 1995], a carcinogen that produces bulky DNA adducts usually repaired by the NER system. On the other hand, p53-knockout mice are prone to spontaneous tumors [Donehower et al., 1992; Jacks et al., 1994], but not to tumors induced by DMBA or 12-O-tetradecanoyl-phorbol-13-acetate (TPA) [Kemp et al.,1993]. Taken together, these observations suggest that DDB2 may be involved in cancer formation through p53-mediated pathways. However, it is unclear whether re-introducing DDB2 in DDB2-knockout mice may prevent cancer formation.

7. Concluding remarks and future perspectives

The various results cited above suggest that the genetic integrity or gene expression status of the cells may be critical in determining the regulatory effects of DDB2 in response to apoptotic stimuli. The level of DDB2, p53, E2F1, and other proteins such as anti-apoptotic cFLIP and cell-cycle arrest p21, for instance, should be considered. The pro-apoptotic activity of p53 could vary between primary and cultured cell lines. For example, p53 activity in HeLa cells is hijacked by the human papillomavirus (HPV) E6 protein, a process that weakens apoptotic signaling in these cells. High levels of DDB2 may up-regulate and potentiate p53 activity by up-regulating apoptotic proteins in p53-normal cells. As such, HeLa cells, which harbor nearly null-p53 activity and additional anti-apoptotic cFLIP activity elicited by DDB2, may become resistant to apoptosis in response to cytotoxic DNA damage. These cellular responses are not surprising if the cultured cell lines were transformed by viruses or chemical means. Unfortunately, the cell lines used for the studies mentioned above are often treated this way. Furthermore, the expression of DDB2 isoforms, including the inhibitory D1 isoform, is often overlooked and the differential expression of such isoforms may dictate the cellular responses observed. Accordingly, alternative splicing of DDB2 transcripts and alteration of these genetic factors by other means in cell lines must be considered while evaluating the role of DDB2 in regulating apoptosis. In fact, there is no evidence so far that the apoptotic resistance of DDB2-defective XP-E, DDB2-knockout mouse cells, or DDB2-deficient human cells could be rescued by re-introducing DDB2 expression. In this sense, DDB2 is required to suppress apoptosis, but it does not suffice to be apoptotic. Furthermore, DDB2 as a proteasome component can target various proteins, such as p21 which is involved in cell cycle arrest, subsequently dysregulating cell cycle arrest during stress repair and leading to apoptosis. The cisplatin-selected HeLa cells used in our study do not display G1 arrest following mild, repairable DNA damage [Lin-Chao & Chao, 1994], which may explain the negligible, pro-apoptotic influence of DDB2 found by others [Stoyanova et al., 2009]. Therefore, an updated model is proposed in Figure 3, in

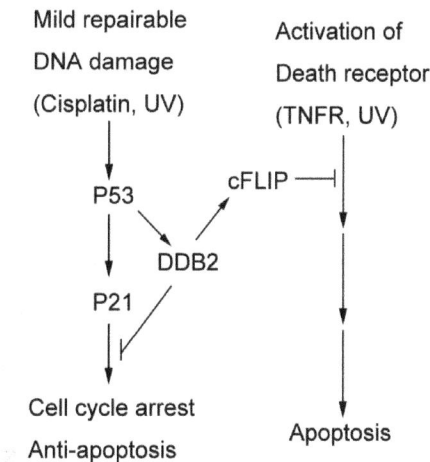

Fig. 3. Updated model for the regulation of DNA damage-induced apoptosis by DDB2. In this model, DNA damage applied to cells was mild and reached repairable level, leading to inhibition of apoptosis and cell cycle arrest for stress repair. The regulatory effect of DDB2 can be pro-apoptotic in cells experiencing mild DNA damage through p21 degradation which is targeted by DDB2. On the other hand, DDB2 can also be anti-apoptotic in cells harboring non-DNA damage apoptotic stimuli (e.g., death receptor) with up-regulation of anti-apoptotic cFLIP. Accordingly, the final outcome may be influenced by intrinsic mutations or extrinsic viral hijacking that can impair checkpoint for G1 arrest via p53 and p21.

which the regulatory effect of DDB2 can be either pro-apoptotic in cells that respond to mild DNA damage or anti-apoptotic in cells that respond to non-DNA damage apoptotic stimuli and that show up-regulation of the anti-apototic cFLIP. Notably, we found that human DDB2 may play a protective role against UV irradiation in the fruit fly *Drosophila* which does not express DDB2 as seen in the DDB2-defective cultured cell models. Therefore, the seemingly contrasting results mentioned above may be explained by our models, and primary cell cultures which are more representative of in vivo situations may represent a better choice for future studies of the biological functions of DDB2.

8. Acknowledgements

The author would like to thank the members of his research group for helpful discussions. The author also appreciates the help of Jan Martel during preparation of the manuscript. This review is partly supported by the National Science Council (Taiwan), Chang Gung University, and the Foundation for the Advancement of Outstanding Scholarship.

9. References

de Bohr, J., van Steeg, H., Berg, R.J.W., Garssen, J., de Wit, J., van Oostrum, C.T.M., Beems, R.B., van der Horst, G.T., van Kreijl, C.F., de Gruijl, F.R., Bootsma, D., Hoeijmakers, J.H., Weeda, G. (1999). Mouse model for the DNA repair/basal transcription disorder trichothiodystrophy reveals cancer predisposition. *Cancer Res.*, 59, 3489-3494

Chao, C.C., Huang, S.L., Huang, H.M., and Lin-Chao, S. (1991). Cross-resistance to UV radiation of a cisplatin-resistant human cell line: overexpression of cellular factors that recognize UV-modified DNA. *Mol. Cell. Biol.*, 11, 2075-2080

Chen, X., Zhang, Y., Douglas, L., Zhou, P. (2001). UV-damaged DNA-binding proteins are targets of CUL-4A-mediated ubiquitination and degradation. *J. Biol. Chem.*, 276, 48175-48182

Chu, G., Chang, E. (1988). Xeroderma pigmentosum group E cells lack a nuclear factor that binds to damaged DNA. *Science*, 242, 564-567

Chu, G., Chang, E. (1990). Cisplatin-resistant cells express increased levels of a factor that recognizes damaged DNA. *Proc. Natl. Acad. Sci. USA*, 87, 3324-3327

Cleaver, J.E. (2005) Cancer in xeroderma pigmentosum and related disorders of DNA repair. *Nat Rev Cancer* 5, 564-573

Donehower, L.A., Harvey, M., Slagle, B.L., McArthur, M.J., Montogomery, Jr. C.A., Butel, J.S., Bradley A. (1992). Mice deficient for p53 are developmentally normal but susceptible to spontaneous tumours. *Nature*, 356, 215-221

Dualan, R., Brody, T., Keeney, S., Nichols, A.F., Admon, A., Linn, S. (1995). Chromosomal localization and cDNA cloning of the genes (DDB1 and DDB2) for the p127 and p48 subunits of a human damage specific DNA-binding protein. *Genomics*, 29, 62-69

El-Mahdy, M.A., Zhu, Q., Wang, Q.E., Wani, G., Praetorius-Ibba, M., Wani, A.A. (2006). Cullin 4A-mediated proteolysis of DDB2 protein at DNA damage sites regulates in vivo lesion recognition by XPC. *J. Biol. Chem.*, 281: 13404–13411

Feldberg, R.S., Grossman, L. (1976). A DNA binding protein from human placenta specific for ultraviolet damaged DNA. *Biochemistry*, 15, 2402-2408

Fitch, M.E., Nakajima, S., Yasui, A., Ford, J.M. (2003). In vivo recruitment of XPC to UV-induced cyclobutane pyrimidine dimers by the DDB2 gene product. *J. Biol. Chem.*, 278, 46906-46910

Friedberg, E.C., Walker, G.C., Siede, W., Wood, R.D., Schultz, R.A., Ellenberger, T. (2006). *DNA repair and mutagenesis.* Second Edition, ASM Press, Washington, DC

Fujiwara, Y., Masutani, C., Mizukoshi, T., Kondo, J., Hanaoka, F., Iwai, S. (1999). Characterization of DNA recognition by the human UV-damaged DNA-binding protein. *J. Biol. Chem.*, 274, 20027-20033

Gonzalez, V.M., Fuertes, M.A., Alonso, C., Perez, J.M. (2001). Is cisplatin-induced cell death always produced by apoptosis? *Mol. Pharmacol.*, 59, 657–663

Groisman, R., Polanowska, J., Kuraoka, I., Sawada, J., Saijo, M., Drapkin, R., Kisselev, A.F., Tanaka, K., Nakatani, Y. (2003). CSA complexes is differentially regulated by the COP9 signalosome in response to DNA damage. *Cell*, 113, 357-367

Guerrero-Santoro, J., Kapetanaki, M.G., Hsieh, C.L., Gorbachinsky, I., Levine, A.S., Rapic-Otrin, V. (2008). The cullin 4B-based UV-damaged DNA-binding protein ligase binds to UV-damaged chromatin and ubiquitinates histone H2A. *Cancer Res.*, 68, 5014-5022

Hayes, S., Shiyanov, P., Chen, X., Raychaudhuri, P. (1998). DDB, a putative DNA repair protein, can function as a transcriptional partner of E2F1. Mol. Cell Biol.,18, 240–249

Higa, L.A., Mihaylov, I.S., Banks, D.P., Zheng, J., Zhang, H. (2003). Radiation-mediated proteolysis of CDT1 by CUL4-ROC1 and CSN complexes constitutes a new checkpoint. *Nat. Cell Biol.*, 5, 1008-1015

Hu, J., McCall, C.M., Ohta, T., Xiong, Y. (2004). Targeted ubiquitination of CDT1 by the DDB1-CUL4A-ROC1 ligase in response to DNA damage. *Nat. Cell Biol.*, 6, 1003-1009

Hwang, B.J., Ford, J.M., Hanawalt, P.C., Chu, G. (1999). Expression of the p48 xeroderma pigmentosum gene is p53-dependent and is involved in global genomic repair. *Proc. Natl. Acad. Sci. U.S.A.*, 96, 424-428

Inoki, T., Yamagami, S., Inoki, Y., Tsuru, T., Hamamoto, T., Kagawa, Y., Mori, T., Endo, H. (2004). Human DDB2 splicing variants are dominant negative inhibitors of UV-damaged DNA repair. *Biochem. Biophys. Res. Commun.*, 314, 1036-1043

Itoh, T., Linn, S., Ono, T., Yamaizumi, M. (2000). Reinvestigation of the classification of five cell strains of xeroderma pigmentosum group E with reclassification of three of them. *J. Invest. Dermatol.*, 114,1022-1029

Itoh, T., O'Shea, C., Linn, S. (2003). Impaired regulation of tumor suppressor p53 caused by mutations in the xeroderma pigmentosum DDB2 gene: mutual regulatory interactions between p48 DDB2 and p53. *Mol. Cell. Biol.*, 23, 7540-7553

Itoh, T., Cado, D., Kamide, Y., Linn, S. (2004). DDB2 disruption leads to skin tumors and resistance to apoptosis after exposure to ultraviolet light but not a chemical carcinogen. *Proc. Natl. Acad. Sci. USA*, 101, 2052-2057

Itoh, T. (2006). Xeroderma pigmentosum group E and DDB2, a smaller subunit of damage-specific DNA binding protein: Proposed classification of xeroderma pigmentosum, Cockayne syndrome, and ultraviolet-sensitive syndrome. *.J Dermatol. Sc.i*, 41, 87-96

Jacks, T., Remington, L., Williams, B.O., Schmitt, E.M., Halachmi, S., Bronson, R.T., Weinberg, R.A. (1994).Tumor spectrum analysis in p53-mutant mice. *Curr Biol*, 4,1-7

Jackson, S., Xiong, Y. (2009). CRL4s: the CUL4-RING E3 ubiquitin ligases. *Trends Bochem. Sci.*, 34, 562-570

Kamarajan, P., Sun, N.-K., Chao, C.C.-K. (2003). Upregulation of FLIP in cisplatin-selected HeLa cells causes cross-resistance to Fas death signaling. *Biochem. J.*, 376, 253-260

Kapetanaki, M.G., Guerrero-Santoro, J., Bisi, D.C., Hsieh, C.L., Rapic-Otrin, V., Levine, A.S. (2006). The DDB1-CUL4ADDB2 ubiquitin ligase is deficient in xeroderma pigmentosum group E and targets histone H2A at UV-damaged DNA sites. *Proc. Natl. Acad. Sci. U.S.A.*, 103, 2588-2593

Keeney, S., Chang, G.J., Linn, S. (1993). Characterization of a human DNA damage binding protein implicated in xeroderma pigmentosum E. *J. Biol. Chem.*, 268, 21293-21300.

Keeney, S., Eker, A.P.M., Brody, T., Vermeulen, W., Bootsma, D., Hoeijmakers, J.H.J., Linn, S. (1994). Correction of the DNA repair defect in xeroderma pigmentosum group E by injection of a DNA damage binding protein. *Proc. Natl. Acad. Sci. U.S.A.*, 91, 4053-4056

Kemp, C.J., Donehower, L.A., Bradley, A., Balmain, A. (1993). Reduction of p53 gene dosage does not increase initiation or promotion but enhances malignant progression of chemically induced skin tumors. *Cell*, 74, 813-822

Kulaksiz, G., Reardon, J.T., Sancar, A. (2005). Xeroderma pigmentosum complementation group E protein (XPE/DDB2): purification of various complexes of XPE and analyses of their damaged DNA binding and putative DNA repair properties. *Mol. Cell. Biol.*, 25, 9784-9792

Lee, J., Zhou, P. (2007). DCAFs, the missing link of the CUL4-DDB1 ubiquitin ligase. *Molecular Cell*, 26, 775-780

Lee, T.H., Elledge, S.J., Butel, J.S. (1995). Hepatitis B virus X protein interacts with a probable cellular DNA repair protein., *J. Virol.*, 69, 1107–1114

Lin-Chao, S., Chao, C.C.-K. (1994). Reduced inhibition of DNA synthesis and G2 arrest in the cell cycle progression of resistant HeLa cells in response to cis-diamminedichloroplatinum(II). *J. Biomed. Sci.*, 1, 131-138

Mitchell, D.L., Haipek, C.A., Clarkson, J.M. (1985). (6-4) Photoproducts are removed from the DNA of UV-irradiated mammalian cells more efficiently than cyclobutane pyrimidine dimers. *Mutat. Res.*, 143, 109–112

Moser, J., Volker, M., Kool, H., Alekseev, S., Vrieling, H., Yasui, A., van Zeeland, A.A., Mullenders, L.H. (2005). The UV-damaged DNA binding protein mediates efficient targeting of the nucleotide excision repair complex to UV-induced photo lesions. *DNA Repair*, 4, 571–582

Nakane, H., Takeuchi, S., Yuba, S., Siajo, M., Nakatsu, Y., Murai, H., Nakatsuru, Y., Ishikawa, T., Hirota, S., Kitamura, Y., Kato, Y., Tsunoda, Y., Miyauchi, H., Horio, T., Tokunaga, T., Matsunaga, T., Nikaido, O., Nishimune, Y., Okada a, Y., Tanaka, K. (1995). High incidence of ultraviolet-B- or chemical-carcinogen induced skin tumours in mice lacking the xeroderma pigmentosum group A gene. *Nature*, 377,165-168

Nichols, A.F., Ong, P., Linn, S., (1996). Mutations specific to the xeroderma pigmentosum group-E DDB(−) phenotype. *J. Biol. Chem.*, 271, 24317-24320

Nichols, A.F., Itoh, T., Graham, J.A., Liu, W., Yamaizumi, M., Linn, S. (2000). Human damage-specific DNA-binding protein p48: characterization of XPE mutations and regulation following UV irradiation. *J. Biol. Chem.*, 275, 21422–21428

Payne, A., Chu, G., (1994). Xeroderma pigmentosum group E binding factor recognizes a broad spectrum of DNA damage. *Mutat. Res.,* 310, 89–102

Peng, K.-Y. (2008). *Effects of DDB2 overexpression on UV toxicity and cell growth.* Master thesis, Chang Gung University, Taiwan

Phillips, A.C., Ernst, M.K., Bates, S., Rice, N.R., Vousden, K.H. (1999). E2F-1 potentiates cell death by blocking antiapoptotic signaling pathways. *Mol. Cell* 4, 771–781

Rapic-Otrin, V., McLenigan, M.P., Bisi, D.C., Gonzalez, M., Levine, A.S. (2002). Sequential binding of UV DNA damage binding factor and degradation of the p48 subunit as early events after UV irradiation. *Nucleic Acids Res.,* 30, 2588–2598

Scrima, A., Konickova, R., Czyzewski, B.K., Kawasaki, Y., Jeffrey, P.D., Nakatani, R.Y., Iwai, S., Pavletich, N.P., Thoma, N.H. (2008). Structural basis of UV DNA damage recognition by the DDB1-DDB2 complex. *Cell,* 135, 1213–1223

Shiyanov, P., Hayes, S.A., Donepudi, M., Nichols, A.F., Linn, S., Slagle, B.L., Raychaudhuri, P. (1999).The naturally occurring mutants of DDB are impaired in stimulating nuclear import of the p125 subunit and E2F1-activated transcription. *Mol. Cell Biol.,* 19, 4935–4943

Stanelle, J., Stiewe, T., Theseling, C.C., Peter, M., Putzer, B.M. (2002). Gene expression changes in response to E2F1 activation. *Nucleic Acids Res.,* 30,1859–1867

Stoyanova, T., Roy, N., Kopanja, D., Bagchi, S., Raychaudhuri, P. (2009). DDB2 decides cell fate following DNA damage. *Proc Natl Acad Sci USA,* 106, 10690–10695

Sugasawa, K., Ng, J.M., Masutani, C., Iwai, S., van der Spek, P.J., Eker, A.P., Hanaoka, F., Bootsma, D., and Hoeijmakers, J.H. (1998). Xeroderma pigmentosum group C protein complex is the initiator of global genome nucleotide excision repair. *Mol. Cell,* 2, 223–232

Sugasawa, K., Okuda, Y., Saijo, M., Nishi, R., Matsuda, N., Chu, G., Mori, T., Iwai, S., Tanaka, K., Tanaka, K., Hanaoka, F. (2005). UV-induced ubiquitylation of XPC protein mediated by UV-DDB-ubiquitin ligase complex. *Cell,* 121, 387–400

Sugasawa, K. (2009). UV-DDB: A molecular machine linking DNA repair with ubiquitination. *DNA Repair,* 8, 969-972

Sugasawa, K. (2010). Regulation of damage recognition in mammalian global genomic nucleotide excision repair. *Mutat. Res.,* 685, 29–37

Sun, C.-L., Chao, C.C.-K. (2005a). Cross-resistance to death ligand-induced apoptosis, in cisplatin-selected HeLa cells, associated with overexpression of DDB2 and subsequent induction of cFLIP. *Mol. Pharmacol.,* 67, 1307-1314

Sun, C.-L., Chao, C.C.-K. (2005b). Potential attenuation of p38 signaling by DDB2 as a factor in acquired TNF resistance. *Int. J. Cancer,* 115, 383-389

Sun, N.-K., Kamarajan, P., Huang, H., Chao, C.C.-K. (2002a). Restoration of UV sensitivity in UV-resistant HeLa cells by antisense-mediated depletion of damaged DNA-binding protein 2 (DDB2). *FEBS Lett.,* 512, 168-172

Sun, N.-K., Lu, H.-P., Chao, C.C.-K. (2002b). Overexpression of damaged DNA-binding protein 2 (DDB2) potentiates UV resistance in hamster V79 cells. *Chang Gung Med. J.,* 25, 686-695

Sun, N.-K., Sun, C.-L., Lin, C.-H., Pai, L.-M., Chao, C.C.-K. (2010). Damaged DNA-binding protein 2 (DDB2) protects against UV irradiation in human cells and *Drosophila. J. Biomed. Sci.,* 17, 27

Takao, M., Abramic, M., Moos, M., Otrin, V.R., Wootton, J.C., McLenigan, M., Levine, A.S., Protic', M. (1993). A 127 kDa component of a UV-damaged DNA-binding complex, which is defective in some xeroderma pigmentosum group E patients, is homologous to a slime-mold protein, *Nucleic Acids Res.* 21 (1993) 4111–4118.

Tang, J.Y., Hwang, B.J., Ford, J.M., Hanawalt, P.C., Chu, G. (2000). Xeroderma pigmentosum p48 gene enhances global genomic repair and suppresses UV-induced mutagenesis. *Mol. Cell*, 5, 737-744

Tang, J., Chu, G. (2002). Xeroderma pigmentosum complementation group E and UV-damaged DNA-binding protein. *DNA Repair*, 1, 601–616

de Vries, A., van Oostrom, C.T.M., Hofhuis, F.M.A., Dortant, P.M., Berg, R.J.W., de Gruiji, F.R., Wester, P.W., van Kreijl, C.F., Capel, P.J., van Steeg, H., Verbeek, S.J. (1995). Increased susceptibility to ultraviolet-B and carcinogens of mice lacking the DNA excision repair gene XPA. *Nature* 377, 169-173

Wang, H., Zhai, L., Xu, J., Joo, H.-Y., Jackson, S., Erdjument-Bromage, H., Tempst, P., Xiong, Y., Zhang, Y. (2006). Histone H3 and H4 ubiquitylation by the CUL4-DDB-ROC1 ubiquitin ligase facilitates cellular response to DNA damage. *Mol. Cell*, 22, 383-394

Wittschieben, B.O., Iwai, S., Wood, R.D. (2005). DDB1-DDB2 (xeroderma pigmentosum group E) protein complex recognizes a cyclobutane pyrimidine dimer, mismatches, apurinic/apyrimidinic sites, and compound lesions in DNA. *J. Biol. Chem.*, 280, 39982-39989

3

RloC: A Translation-Disabling tRNase Implicated in Phage Exclusion During Recovery from DNA Damage

Gabriel Kaufmann et al.*
Tel Aviv University
Israel

1. Introduction

Bacteria respond to DNA damage by inducing the expression of numerous proteins involved in DNA repair and the reversible arrests of DNA replication and the cell division cycle (Fernandez De Henestrosa *et al*, 2000). This general rule may be violated by a conserved bacterial protein termed RloC (Davidov & Kaufmann, 2008). RloC combines structural-functional properties of two unrelated proteins (i) the universal DNA-damage-responsive/DNA-repair protein Rad50/SbcC (Williams *et al*, 2007) and (ii) the translation-disabling, phage-excluding anticodon nuclease (ACNase) PrrC (Blanga-Kanfi *et al*, 2006). These seemingly conflicting features may be reconciled in a model where RloC is mobilized as an antiviral back-up function during recovery from DNA damage (Davidov & Kaufmann, 2008), when DNA restriction, the cell's primary immune system is temporarily shut-off (Thoms & Wackernagel, 1984). Another intriguing feature of RloC is its ability to excise its substrate's wobble nucleotide (Davidov & Kaufmann, 2008). This harsh lesion is expected to encumber reversal by phage enzymes that repair the tRNA nicked by PrrC (Amitsur *et al*, 1987). Evaluating RloC's salient features and purported role requires prior description of its more familiar distant homolog PrrC and a DNA-damage-sensing device RloC shares with Rad50/SbcC. We conclude with an account of cellular RNA and DNA repair tools related to the phage tRNA repair mechanism that counteracts PrrC and may be frustrated by RloC.

2. PrrC – A potential phage-excluding tool counteracted by tRNA repair enzymes

2.1 A host-phage survival cascade yields an RNA repair pathway

RNA repair may seem unnecessary because damaged RNA molecules can be readily replenished by re-synthesis. Yet, there exist situations where RNA repair could be the preferred or only possible option. A case in point is presented by an RNA repair pathway triggered by the ACNase PrrC. This conserved bacterial protein was detected in quest of roles of two phage T4-encoded enzymes: 3'-phosphatase/5'-polynucleotide kinase (PseT/Pnk,

---

* Elena Davidov, Emmanuelle Steinfels-Kohn, Ekaterina Krutkina, Daniel Klaiman, Tamar Margalit, Michal Chai-Danino and Alexander Kotlyar
Tel Aviv University, Israel

henceforth Pnk) (Richardson, 1965;Becker & Hurwitz, 1967;Cameron & Uhlenbeck, 1977) and RNA ligase 1 (Rnl1, Silber *et al*, 1972;Ho & Shuman, 2002). The combined activities of Pnk and Rnl1 seemed tailored to fix RNA nicks, converting 3'-phosphoryl or 2',3'-cyclic phosphate and 5'-OH cleavage ends into 3'→5' phosphodiester linkages (Kaufmann & Kallenbach, 1975;Amitsur *et al*, 1987). Suggested alternative roles in DNA metabolism (Novogrodsky *et al*, 1966;Depew & Cozzarelli, 1974) were assigned in later years to a related eukaryal DNA kinase-phosphatase essential for genome stability and a possible therapeutic target in cancer cells rendered resistant to genotoxic drugs (Weinfeld *et al*, 2011).

Pnk and Rnl1 are dispensable for T4 growth on common *E. coli* laboratory strains but required on a rare host encoding the optional locus *prr* (*pnk* and *rnl1* restriction) (Depew & Cozzarelli, 1974; Sirotkin *et al*, 1978; Runnels *et al*, 1982; Jabbar & Snyder, 1984). Mutating a minuscule T4 orf termed *stp* (*s*uppressor of *t*hree-*p*rime *p*hosphatase) abrogates *prr* restriction (Depew & Cozzarelli, 1974;Depew *et al*, 1975;Chapman *et al*, 1988;Penner *et al*, 1995). These facts reinforced the notion that Pnk and Rnl1 cooperate in RNA nick repair. They also led to the detection of the *prr*-encoded latent ACNase comprising the core ACNase PrrC and PrrC's silencing partner, the associated type Ic DNA restriction-modification (R-M) system EcoprrI (Levitz *et al*, 1990;Linder *et al*, 1990;Amitsur *et al*, 1992;Tyndall *et al*, 1994). EcoprrI and PrrC are also genetically linked, the ACNase core gene *prrC* is flanked by the genes encoding the three R-M subunit types *hsdMSR/prrABD* (Fig. 1A).

Type I R-M systems to which EcoprrI belongs recognize with their HsdS subunit a bipartite target containing a variable 6-8nt long spacer such as EcoprrI's CCAN$_7$RTGC (Tyndall *et al*, 1994). HsdS associates with two HsdM protomers to form a site-specific DNA methylase (HsdM$_2$S). Further attachment of two HsdR protomers yields a full-fledged R-M protein (HsdR$_2$M$_2$S). The R-M protein ignores a fully methylated target and readily methylates a hemi-methylated one. A fully unmodified target, usually of foreign DNA, induces the helicase domains of the HsdR protomers to pump-in DNA flanking the target sequence at the expense of ATP hydrolysis. This translocation and consequent DNA looping go on until an obstacle is encountered and cleavage occurs, usually far away from the specific recognition site. The type I R-M proteins are divided into families by antigenic cross-reactivity, subunit interchangeability and sequence similarity. PrrC is invariably linked to type Ic family members while RloC may interact with type Ia or the distantly related type III R-M proteins. For detailed coverage of DNA restriction and anti-restriction the readers are encouraged to consult relevant reviews (Murray, 2000;Dryden *et al*, 2001;Youell & Firman, 2008;Janscak *et al*, 2001).

EcoprrI normally silences PrrC's ACNase activity in the uninfected cell (Fig. 1B). The significance of this masking interaction is indicated by the "double-edged" nature of the T4 encoded peptide Stp, mutations in which suppress *prr* restriction. Thus, Stp inhibits EcoprrI's DNA restriction, probably its intended function; and activates the latent ACNase, its host co-opted task (Penner *et al*, 1995). Once activated PrrC nicks cellular tRNA^{Lys} 5' to the wobble base, yielding 2', 3'-cyclic phosphate and 5'-OH termini. Since T4 shuts-off host transcription (Mathews, 1994) and does not encode tRNA^{Lys} (Schmidt & Apirion, 1983) the lesion inflicted by PrrC could disable T4 late translation and contain the infection (Sirotkin *et al*, 1978). However, T4 overcomes also this hurdle by using Pnk and Rnl1 to resuscitate the damaged tRNA^{Lys}. Pnk heals the cleavage termini, converting them into a 3'-OH and 5'-P pair that Rnl1 seals (Amitsur *et al*, 1987)(Fig. 1B). In other words, this host-phage survival cascade gave rise to an RNA repair pathway. The ability of the *prr*-encoded latent ACNase to restrict only tRNA repair-deficient phage invokes the possible existence of a "smarter" ACNase able to encumber phage reversal. Later we ask if RloC could be one.

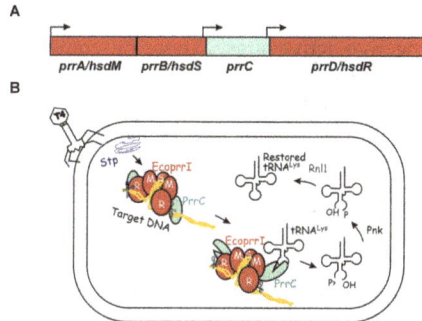

Fig. 1. A host-phage survival cascade gives rise to an RNA repair pathway. A. The optional host locus *prr* comprises the core ACNase gene *prrC* and flanking genes encoding the type Ic DNA R-M protein EcoprrI that silences PrrC's ACNase activity. Arrows mark transcription start sites. B. Cleavage-ligation of tRNA^{Lys} in phage T4 infected *E. coli prr$^+$*. T4's anti-DNA restriction factor Stp inhibits EcoprrI and activates the latent ACNase. The resultant disruption of tRNA^{Lys} is reversed by the T4's tRNA repair enzymes Pnk and Rnl1.

Nested *prr* loci where *prrC* intervenes a type Ic *hsd* locus (Fig 1A) appear sporadically in distantly related bacteria. They are present in some strains of a given species but not in others, as would a niche-function (Blanga-Kanfi *et al*, 2006). They abound among *Proteobacteria*, are less frequent in *Bacteroidetes* and *Firmicutes*, rare in *Actinobacteria* and apparently absent from *Cyanobacteria*. PrrC's phylogenic tree does not match the bacterial, unlike the associated type Ic R-M protein, which only rarely teams with PrrC. In contrast, a stand-alone *prrC* gene has not been detected so far. These facts hint that PrrC can be readily transmitted by horizontal gene transfer (HGT), possibly from a *prr* donor to an *hsd* acceptor. The dependence of PrrC's function on its detoxifying partner, the linked R-M system is indicated also by their coincident inactivation in a *Neisseria meningitidis* strain (Meineke and Shuman, pers. comm.). This addiction and the similar ACNase activities of various PrrC orthologs examined (Davidov & Kaufmann, 2008;Meineke *et al*, 2010) further suggest that PrrC acts in general as a translation-disabling, antiviral contingency mobilized when the linked R-M system is compromised.

The host-phage survival cascade depicted in Fig. 1B entails some caveats. Namely, the DNA of T4 and related phages incorporates 5-hydromethylcytosine (5-HmC) instead of cytosine and 5-HmC is further glucosylated at the DNA level (Morera *et al*, 1999). Due to this hyper-modification the phage DNA is refractory to many DNA restriction nucleases (Miller *et al*, 2003b) including EcoprrI and, hence, need not be protected from them by Stp. Moreover, a T4 mutant with unmodified cytosine in its DNA succumbs to EcoprrI's restriction, notwithstanding Stp's presence. The failure of Stp to protect this EcoprrI-sensitive mutant can be accounted for by the delayed-early schedule of its expression, a few minutes after the onset of the infection (Jabbar & Snyder, 1984;David *et al*, 1982). Due to these reasons EcoprrI's DNA restriction and Stp's anti-restriction activities were investigated using surrogate lambdoid phages (Jabbar & Snyder, 1984;Penner *et al*, 1995). Yet, the conservation of Stp's sequence among T4-like phages (Penner *et al*, 1995) http://phage.ggc.edu/, indicates that this anti-DNA restriction factor provides selective advantage, e.g., preventing nucleases related to EcoprrI from cleaving nascent, not yet glucosylated progeny DNA.

The importance of Pnk and Rnl1 as PrrC's countermeasures is suggested by the following observations. First, docking tRNA on the crystal structure of T4 Pnk or Rnl1 places the anticodon loop at their respective active sites. These outcomes have been taken to indicate that both Pnk and Rnl1 evolved to repair a disrupted anticodon loop (Galburt et al, 2002;El Omari K. et al, 2006). Second, T4-related phages expected to infect prr-encoding bacteria feature both Pnk and Rnl1 (Miller et al, 2003a;Blondal et al, 2005;Blondal et al, 2003) whereas T4-related cyanophages, which are less likely to encounter prr, lack these tRNA repair proteins (http://phage.ggc.edu/).

2.2 PrrC's functional organization

PrrC comprises a regulatory motor domain occupying the N-proximal two thirds of its 396aa polypeptide (EcoPrrC). The remaining part constitutes the ACNase domain (Fig. 2A). The N-domain resembles ATP Binding Cassette (ABC) ATPases. These are universal motor components found in membrane-spanning transporters and in soluble proteins engaged in DNA repair, translation and related functions (Hopfner & Tainer, 2003). PrrC's N-domain differs from typical ABC ATPases in certain sequence attributes and in its unusual nucleotide specificity. The ABC ATPase motifs found in it partake in binding and hydrolysis of the nucleotide triphosphate moiety (Chen et al, 2003). However, the nucleobase recognizing motif of many transporter ABC ATPases termed A- or Y-loop (Ambudkar et al, 2006) is missing from PrrC. On the other hand, PrrC contains between its Walker A and Q-loop motifs a unique 16-residue motif rich in aromatic, acidic and other hydrophilic residues (Fig. 2A). This PrrC Box motif is highly degenerate (or rudimental) in RloC and is missing from other ABC ATPases and any other protein in the public database (Amitsur et al, 2003;Blanga-Kanfi et al, 2006). The PrrC Box candidates as a Y-loop substitute, responsible perhaps for PrrC's unusual specificity, the ability to simultaneously interact with its two different effector nucleotides GTP and dTTP (Blanga-Kanfi et al, 2006; unpublished data).

PrrC's ACNase domain harbors a catalytic ACNase triad (Arg^{320}-Glu^{324}-His^{356} in EcoPrrC) shared also by most RloC's orthologs except for a few cases where Glu is replaced by Asp. By analogy with the catalytic triad of RNase T1 (Gerlt, 1993;Steyaert, 1997), in the PrrC/RloC triad Glu and His could function as respective general base and acid catalysts while Arg could stabilize the pentameric transition state phosphate. The ACNase domain contains also residues implicated in recognition of the substrate's anticodon. Mutating one of them, EcoPrrC's Asp^{287} impairs the reactivity of the natural substrate and enhances that of analogs with a hypomodified or heterologous wobble base. These compensations hint that Asp^{287} interacts with the wobble base modifying side chain (Meidler et al, 1999;Jiang et al, 2001;Jiang et al, 2002).

When PrrC is expressed by itself it exhibits overt (core) ACNase activity. This core activity purifies with an oligomeric PrrC form, possibly a dimer of dimers. The N-domains of each dimer are expected to create two nucleotide binding sites (NBS) at their anti-parallel dimerization interfaces, as do typical ABC ATPases (Hopfner et al, 2000;Chen et al, 2003). In contrast, the ACNase C-domains are thought to dimerize in parallel, judged from the (i) behavior of a peptide mimic of a PrrC region implicated in the recognition of the tRNA substrate and (ii) ability of single to-Cys replacements in an overlapping PrrC region to induce disulphide-bond-dependent subunit dimerization (Klaiman et al, 2007).

Accordingly, the PrrC dimer of dimers assumes a phosphofructokinase-like topology (Schirmer & Evans, 1990) (Fig. 2B).

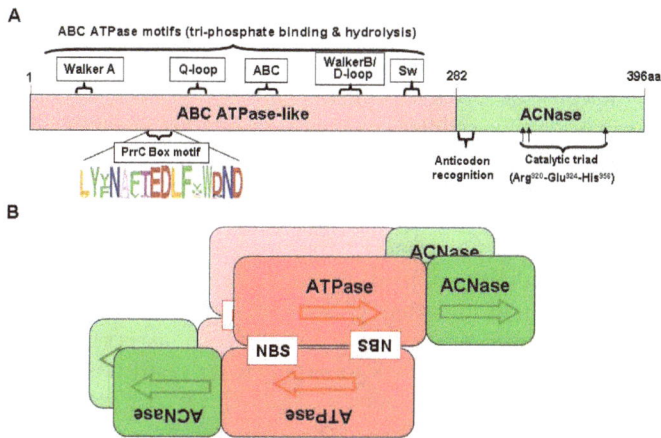

Fig. 2. Functional structure and possible quaternary organization of PrrC. **A.** PrrC's N-proximal ABC-ATPase domain features motifs involved in binding and hydrolysis of the nucleotide's triphosphate moiety (Walker A, Q-loop, ABC signature (ABC), Walker B, D-loop and linchpin Switch region (SW) but not the nucleobase recognizing Y-loop motif. The unique PrrC Box motif shown in WebLogo format, a putative functional substitute of the Y-loop, could confer the unusual GTP/dTTP specificity of PrrC. **B.** Antiparallel dimerization of the N-domains (Moody & Thomas, 2005) and anticipated parallel dimerization of the C-domains (Klaiman *et al*, 2007) suggest that PrrC assumes a phosphofructokinase-like quaternary topology (Schirmer & Evans, 1990). NBS – nucleotide binding site.

2.3 Players in PrrC's silencing and activation

As mentioned, PrrC's toxic activity is normally silenced, being unleashed only during phage infection. The requisite switches are provided in the case of *Eco*PrrC by its silencing partner Ecoprrl, the phage T4-encoded anti-DNA restriction factor Stp and the motor domains of the ACNase protein itself. Insights into the underlying mechanisms were provided by the discrepant behaviors of the latent ACNase holoenzyme and the core ACNase activity of the unassociated PrrC. Thus, *in vitro* activation of the latent ACNase requires besides the Stp peptide, the DNA tethered to Ecoprrl, GTP hydrolysis and the presence of dTTP. In contrast, the overt activity of the core ACNase is refractory to Stp, DNA and GTP but rapidly decays without dTTP (Amitsur *et al*, 2003;Blanga-Kanfi *et al*, 2006). These differences have been taken to indicate that Stp triggers the activation of the latent ACNase, GTP hydrolysis drives conformational changes needed to turn it on while the binding of dTTP stabilizes the ACNase once activated. The possible role of Ecoprrl's DNA ligand is discussed later in this section.

GTP and dTTP probably exert their respective ACNase activating and stabilizing functions by interacting with PrrC's N-domains. This is suggested by their binding to full-sized PrrC protein or PrrC's isolated N-domains with vastly differing affinities (mM- and μM-range, respectively) and without displacing each other (Amitsur *et al*, 2003;Blanga-Kanfi *et al*, 2006; and unpublished data). This unusual specificity distinguishes PrrC from its distant homolog RloC and other ABC ATPase-containing proteins, which bind and hydrolyze ATP or GTP (Guo *et al*, 2006) and are not expected to avidly bind dTTP (our unpublished data).

The biological significance of PrrC's idiosyncratic interaction with dTTP has been hinted at by the dramatic increase in the cellular level of dTTP early in phage T4 infection, when the ACNase is induced (Amitsur *et al*, 2003;Blanga-Kanfi *et al*, 2006). The increased level of dTTP benefits the phage by safeguarding effective and faithful replication of its AT-rich DNA. In fact, delaying dTTP's accretion by mutating T4's dCMP deaminase (Cd) elicits a mutator phenotype indicated by increased frequency of AT$\rightarrow$GC transitions (Sargent & Mathews, 1987). The Cd deficiency, and, by implication, the consequent delay in dTTP's accretion, also reduce 2-3 fold the extent of the PrrC-mediated cleavage of tRNA^{Lys}. This partial inhibition does not suffice to suppress *prr* restriction but is synthetically suppressive with a leaky *stp* mutation that also fails to suppress *prr* restriction by itself (Klaiman & Kaufmann, 2011). Thus, dTTP's accretion is another T4 contraption expatiated by the bacterial host, in that case to stabilize the activated ACNase.

PrrC's ability to "gauge" changes in dTTP's level could benefit its host also by precluding the toxicity of any free PrrC molecules that could arise in the uninfected cell due to their translation in excess over EcoprrI or dissociation from the latent holoenzyme. Their excessive translation may be stochastic or programmed to saturate the silencing partner. PrrC's dissociation from the latent holoenzyme may be accidental or due to EcoprrI's disruption in response to DNA damage (Restriction Alleviation, RA) (Makovets *et al*, 2004) (see also section 3.6). Free PrrC's cytotoxicity has been indicated by the coincident inactivation of *prrC* and linked *hsd* genes, by the self-limiting expression of free PrrC (Meidler *et al*, 1999;Blanga-Kanfi *et al*, 2006) and the rapid *in vivo* inactivation of the core ACNase (Amitsur *et al*, 2003). The ACNase enhancing effects of dTTP's accretion during phage T4 infection (Klaiman & Kaufmann, 2011) and *in vitro* stabilization of the core ACNase by dTTP (Amitsur *et al*, 2003) suggest that the *in vivo* instability of the core ACNase owes to the relatively low dTTP level in the uninfected cell. Although this level far exceeds that needed to stabilize the core ACNase *in vitro*, the actual level availed to PrrC in the cell could be prohibitively low due to localization of the nucleotide pools (Wheeler *et al*, 1996). In sum, we propose that PrrC's ability to gauge dTTP's level not only stabilizes its activated form but also confines the toxicity of this ACNase to the viral target.

Yet another player in PrrC's regulation is the DNA tethered to EcoprrI (Amitsur *et al*, 2003). Its possible role is suggested by three observations. First, short nonspecific ssDNA oligonucleotides avidly bind PrrC and competitively inhibit its ACNase activity (Fig. 3A and unpublished results), hinting that ssDNA encountered by PrrC in the uninfected cell helps silence the ACNase. Second, the type Ic DNA R-M protein EcoR124I unwinds short DNA stretches flanking its target sequence (van Noort *et al*, 2004;Stanley *et al*, 2006), suggesting a possible source for the putative ACNase-inhibiting ssDNA. Third, within a latent ACNase complex tethered to an EcoprrI DNA ligand PrrC was UV-crosslinked to DNA regions flanking EcoprrI's recognition site (Fig. 3B). These facts underlie a model where DNA unwound by EcoprrI helps silence PrrC and its rewinding due to Stp's interaction with EcoprrI unleashes the ACNase (Fig. 3C).

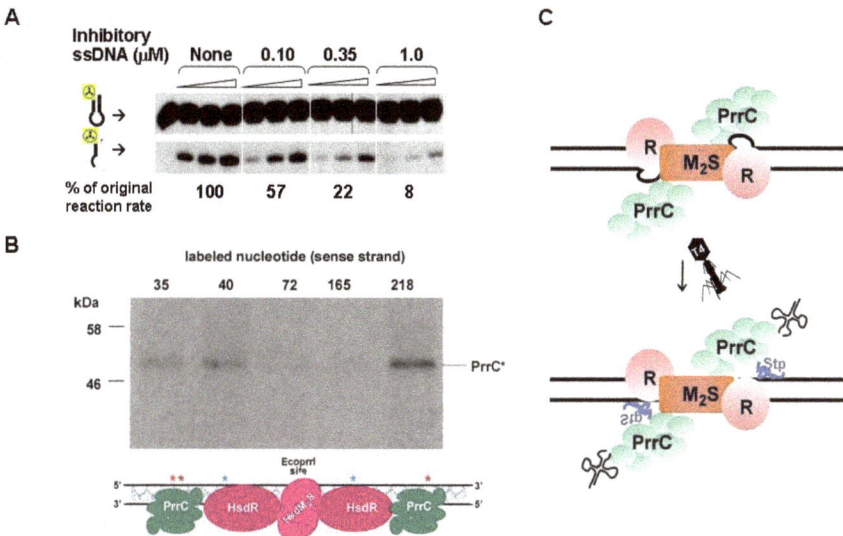

Fig. 3. DNA tethered to EcoprrI could figure in PrrC's regulation. A. ssDNA inhibits PrrC ACNase. PrrC ACNase was assayed using a 5'-^{32}P labeled anticodon stem loop substrate and increasing levels of a nonspecific 17nt PCR primer. B. PrrC contacts DNA regions flanking EcoprrI's target. A 249bp DNA fragment with a near-central EcoprrI site was singly ^{32}P-labeled at specific sites and tethered to the EcoprrI-PrrC complex. Following UV-irradiation, DNase I digestion, the photo-labeled PrrC was immunoprecipitated, separated by SDS-PAGE and monitored by autoradiography. Brown and blue asterisks indicate sites PrrC did or did not crosslink to, respectively. C. In this model DNA unwound by EcoprrI silences PrrC and its rewinding due to Stp's interaction with EcoprrI unleashes the ACNase.

3. RloC - A translation-disabling and potential DNA-damage-sensing protein

3.1 Functional organization

RloC is a conserved bacterial protein that shares PrrC's overall organization into a motor N-domain and ACNase C-domain (Fig. 4) (Davidov & Kaufmann, 2008). However, RloC is about twice as large, its orthologs ranging in size between 650 to 900 residues compared to 350-420 with PrrC. This increase is mainly due to a long coiled-coil forming sequence inserted between RloC's Walker A and ABC signature motifs. This coiled-coil sequence contains near its center a loop featuring the conserved zinc-hook motif CXXC. A similar coiled-coil insert in an ABC ATPase head-domain characterizes the universal DNA-damage-checkpoint/DNA-repair protein Rad50/SbcC (Hopfner *et al*, 2002;Connelly *et al*, 1998). Rad50's insert protrudes from the ATPase head-domain as an antiparallel coiled-coil presenting the zinc-hook motif at its apex. The apical ends of two such protrusions dimerize by coordinating Zn^{++} to their four cysteines. This zinc-hook linkage can arise intra-molecularly, connecting the two coiled-coil protrusions of the same Rad50 dimer. Alternatively, when Rad50's ATPase head-domains are bound to DNA the two protrusions straighten. In this form they can dimerize only inter-molecularly, bridging in this manner distant DNA molecules (Moreno-Herrero *et al*, 2005). Other proteins belonging to the SMC (Structure Maintenance of Chromosomes) super-family exhibit similar DNA bridging activity but link their coiled-coil protrusions via apical hydrophobic domains (Hirano, 2005). RloC is the only known protein other than Rad50/SbcC with a coiled-coil/zinc-hook containing ABC-ATPase domain. Therefore, cellular functions imparted by Rad50/SbcC may provide clues to RloC's.

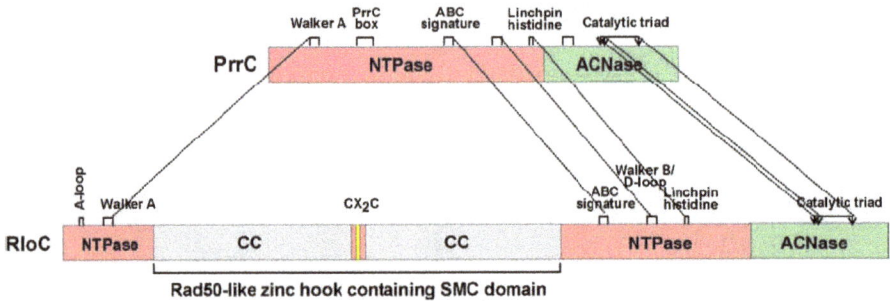

Fig. 4. RloC and PrrC share the same functional organization. The alignment of *Gka*RloC and *Eco*PrrC sequences reveals shared ABC ATPase and ACNase motifs and presence in RloC's N-domain of a large coiled-coil (CC) stretch interrupted by a loop containing the zinc hook motif CXXC (adapted from ref. 4).

3.2 RloC's occurrence and genomic attributes

RloC genes appear in major bacterial phyla except for *Cyanobacteria*. They are often encased within a cryptic mobile element as a single cargo gene. This pattern and a phylogenic tree not matching the bacterial suggest that RloC is readily transmitted by HGT, like PrrC. RloC's genes are also sporadically distributed but they occur ~3-fold more frequently than PrrC's. These facts suggest that the niche function RloC provides is more beneficial to its bacterial host.

RloC was originally identified as one of various open reading frames that intervene type Ia *hsd* loci in different *Campylobacter jejuni* strains (Restriction Linked Orf, Miller *et al*, 2005). This fact and the overall resemblance to PrrC could be taken to indicate that RloC is a related ACNase also silenced by an associated Hsd protein (Davidov & Kaufmann, 2008). Yet, only ~10% of the identified RloC orthologs turned out to be linked to type Ia or the distantly related type III DNA R-M system. Nonetheless, other genomic attributes suggested that the majority of the non-linked RloC orthologs team with an R-M system *in trans*. First, most bacteria encoding them encode also a suitable R-M system while in those lacking it RloC often features poor ATPase or ACNase motifs, as if inactivated. Second, some *rloC* genes are flanked by a cryptic *hsd* locus, a full-fledged homologue of which exists elsewhere in the genome, hinting that a past Hsd-RloC interaction *in cis* was superseded by one *in trans*. Third, RloC is occasionally linked to an ArdC-like anti-DNA restriction factor (Belogurov *et al*, 2000) with or without an adjacent R-M system, suggesting its possible regulation by an R-M system in either case. Fourth, non-linked *rloC* and *hsd* genes of one species, but not their respective flanking genes can be missing both from related, syntenic species [e.g., *Acinetobacter* sp. ADP1 *rloC* and *hsd* (ACIAD0152, ACIAD3430-2) but not flanking genes are missing from various *A. baumannii* strains] (http://www.cns.fr/agc/ microscope/mage/viewer.php?S_id=36&wwwpkgdb =aa12fda27bb61b62ac34913acfd35916.)

The role ascribed to the R-M proteins in RloC's ACNase regulation need not contradict the existence of additional or alternative switches provided by the coiled-coil/zinc-hook insert. For example, silencing of the ACNase function by the latter device could be advantageous when RloC is introduced by HGT into a new host. Namely, silencing by a pre-existing R-M system could require a highly promiscuous interaction between the two partners. The possibility that RloC is endowed with an internal ACNase silencing mechanism agrees with properties of the ortholog encoded by the thermophile *Geobacillus kaustophilus* (*Gka*RloC) to be described in the following sections.

3.3 RloC wobble-nucleotide-excising activity

Due to its potential toxicity, RloC's ACNase activity was expected to be as unstable as PrrC's (Blanga-Kanfi *et al*, 2006). Indeed, among several RloC orthologs investigated, only *Gka*RloC proved sufficiently stable to warrant its *in vitro* characterization (Davidov & Kaufmann, 2008). Yet, even *Gka*RloC's ACNase is intrinsically unstable. Its *in vitro* activity is highest at 25°C and undetectable at 45°C (our unpublished results) although *G. kaustophilus* grows optimally at 65°C (Takami *et al*, 2004). When expressed in *E. coli Gka*RloC preferentially cleaved tRNA[Glu]. However, identifying RloC's natural substrate must await physiological studies. This reservation is based on the experience gained with PrrC, the over-expression of which results in cleavages of secondary substrates that overwhelm the natural (Meidler *et al*, 1999).

A more striking difference between RloC and PrrC is the ability of the former to cleave its tRNA substrates successively, first 3' and then 5' to the wobble position (Davidov & Kaufmann, 2008). Such an excision reaction using as a substrate yeast tRNA[Glu] radiolabeled 3' to the wobble base is shown in Fig. 5. The incision of this substrate 3' to the wobble base yields a labeled 5' fragment containing residues 1-34. This intermediate is further cleaved immediately upstream, yielding the labeled wobble-nucleotide. Under these *in* vitro conditions *Gka*RloC inadvertently incises the substrate also 5' to the wobble base but this reaction yields a dead-end product that is not further cleaved. This is indicated by the accumulation of this product when the overall reaction declines; and of RloC to cleave it when generated by PrrC, which normally cleaves its substrates 5' to the wobble position. Such a 5' incision product of GkaRloC is not detected *in vivo* and, therefore, is considered an *in vitro* artifact. The excision of the wobble nucleotide has been observed with different tRNA and anticodon-stem-loop substrates and was catalyzed also by a mesophilic RloC species of *E. coli* APECO1 (Davidov & Kaufmann, 2008; unpublished data).

3.4 RloC may frustrate phage reversal

The harsh lesion inflicted by *Gka*RloC could render this ACNase a more potent antiviral device than PrrC. Namely, RloC could perform the successive cleavages of its substrate in a processive manner, i.e., without releasing the incision intermediate. The phage tRNA repair enzymes would in that case process and ligate back the fragments lacking the wobble nucleotide and yield a defective product. Conversely, if *Gka*RloC's incision intermediate were accessible, the repair enzymes would faithfully restore the original tRNA substrate. Simulated *in vitro* encounters between *Gka*RloC and T4 Pnk or both tRNA repair enzymes indicated that a sizable fraction of its incision intermediate was occluded from the repair enzymes (Davidov & Kaufmann, 2008; and unpublished data). It is possible that under physiological conditions RloC's would more effectively occlude its incision intermediate.

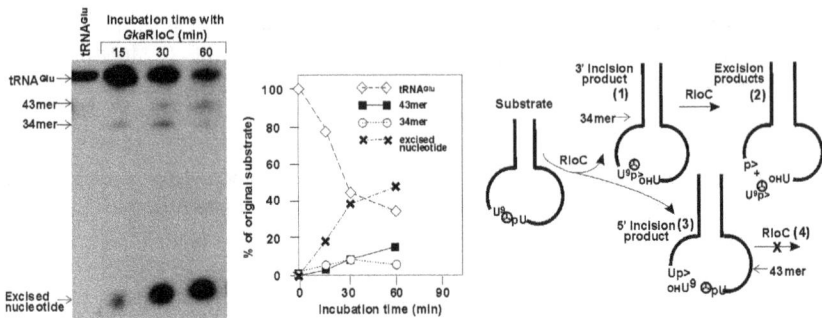

Fig. 5. RloC excises the wobble nucleotide. Yeast tRNA^{Glu} ^{32}P-labeled 3' to the wobble base was incubated with GkaRloC. The 34mer resulting from incision 3' to the wobble base is further cleaved, yielding the wobble nucleotide. The 43mer resulting from incision 5' to the wobble base is a dead-end product that is not further cleaved. It is considered an *in vitro* artifact, as explained in the text. In the cartoon depicting these reactions the substrate is schematically represented by the anticodon stem loop outline. ⊕ marks the labeled phosphate. U^9 is the modified wobble base 5-methoxycarbonylmethyl-2-thiouridine. (mcm^{5}s^{2}U).

Moreover, repeated cleavage-ligation cycles would diminish the proportion of any incision intermediate ligated back by phage enzymes. On the other hand, the existence of tRNA repair enzymes that more efficiently extract RloC's incision intermediate and generate perhaps repair products immune to re-cleavage (Chan *et al*, 2009b) cannot be excluded. Clearly, whether RloC does frustrate phage reversal remains to be examined in situations closer to the natural.

3.5 RloC's DNA bridging domain regulates its ACNase

RloC's second striking feature is the coiled-coil/zinc-hook insert in its ABC ATPase head-domain. The presence of this structure raised the possibility that RloC is endowed with Rad50-like DNA bridging activity and uses such a faculty to respond to DNA damage cues by turning on its ACNase. That RloC is in fact endowed with DNA bridging activity is indicated by an electrophoresis mobility shift experiment and by scanning force microscopy (AFM) imaging. In the first experiment we compared GkaRloC constructs with an intact or mutated zinc-hook. The first protein aggregated a dsDNA probe that the second only bound (Fig. 6). Their discrepant behavior suggests that the aggregation was due to the formation of zinc-hook-dependent DNA bridges. Preliminary AFM imaging data reinforce this assumption (Fig. 7).

That RloC's ACNase is regulated by the protein's coiled-coil/zinc-hook and ATPase head-domain is indicated by several observations. First, mutating RloC's zinc-hook dramatically enhances its ACNase activity *in vivo* and *in vitro* (Davidov & Kaufmann, 2008). Second, GkaRloC's ACNase activity is modestly enhanced by ATP and further stimulated when the protein is also tethered to DNA (Fig. 8). In contrast, DNA alone has no effect on the ACNase and the residual ACNase activity seen without added ATP is abolished by the non-hydrolyzable analog AMP-PNP. Presumably, RloC's interaction with DNA turns on its ATPase to drive conformational changes that activate the ACNase. Interestingly, mutating the zinc-hook renders the ACNase refractory to these various agents, uncoupling the ACNase from the protein's internal controls (not shown). Together, these facts suggest that RloC's mode of interaction with DNA, which is sensed by its coiled-coil/zinc-hook monitoring device and relayed by the ATPase (Fig. 9), determines if the protein's ACNase will be silenced or turned on.

The ability to activate *Gka*RloC's ACNase by ATP hydrolysis in the presence of tethered DNA is in stark contrast with the behavior of PrrC's ACNase. As mentioned, PrrC's ACNase is activated by nucleotide hydrolysis only when associated with its silencing partner EcoprrI. However, its unassociated form exhibits overt ACNase activity refractory to nucleotide hydrolysis. This discrepancy raises the possibility that RloC's ACNase can be regulated by the internal device of the protein, the coiled-coil/zinc-hook and the ATPase domain that harbors this structure.

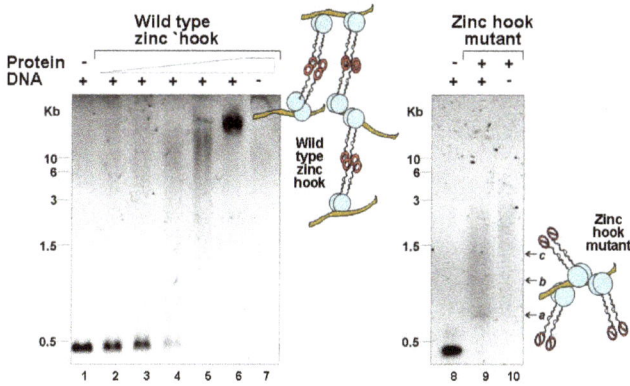

Fig. 6. *Gka*RloC aggregates DNA in a zinc-hook-dependent manner. A 485bp DNA fragment was incubated with increasing levels of GkaRloC's ACNase-null mutant E696A (lanes 2-6) or with its ZH mutant derivative E696A-C291G (lane 9). Lanes 1 and 8 contain only DNA, 7,10 only the indicated protein. The cartoons depict the assumed bridged DNA aggregate formed by E696A (right) and the simpler complex formed by E696A-C291G (left). The ACNase-null mutation allows high level expression and facilitates the isolation of the RloC proteins.

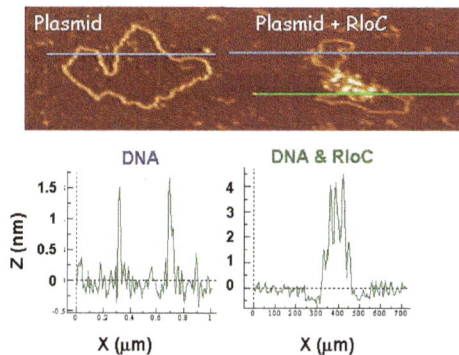

Fig. 7. AFM images of plasmid pUC19 (DNA) and its complex with RloC-E696A (DNA and RloC). Blue lines stretch over pure DNA regions, green lines also over regions containing the bound protein. Regions transected by the green line feature virtual heights both of the DNA alone (~1.5nm) and of the presumptive RloC-DNA complexes (~4.5nm).

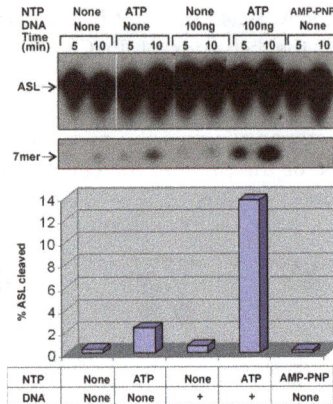

Fig. 8. *Gka*RloC's ATPase and tethered DNA cooperatively regulate its ACNase function. *Gka*RloC's ACNase activity was assayed using as a substrate a 5'-^{32}P labeled anticodon-stem-loop analog corresponding to mammalian tRNA^{Lys3} (ASL). The reaction was performed in the absence or presence of 2mM of ATP and/or 10ng/μl of BstE II digested λ DNA, or in the presence of the non-hydrolyzable ATP analog AMP-PNP. The 7mer is a radiolabeled fragment resulting from the final excision reaction

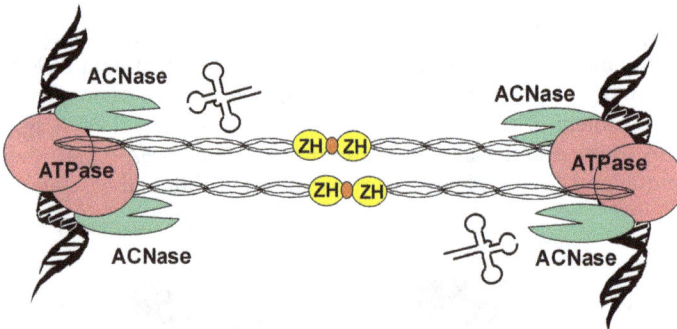

Fig. 9. RloC's anticipated DNA bridging activity. By analogy with Rad50, RloC bridges DNA through Zn^{++} (orange circles) coordinated at zinc-hook (ZH) dimerization interfaces (yellow circles) at the apical tips of the coiled-coils protruding from the DNA-borne ATPase head domains (pink circles). The status of the bound DNA sensed by RloC determines if its ATPase will be activated and drive structural changes needed to switch on the ACNase domains (split green ovals) toward tRNA cleavage.

3.6 Is RloC a suicidal DNA-damage-responsive device?

If RloC can be regulated by its internal devices, what role plays the anticipated interaction of RloC with a DNA R-M protein? Do these external and internal devices cooperate or act separately, responding to the same or different environmental cues? The present state of RloC's research does not permit us to distinguish between these possibilities, let alone assign to this protein specific biological functions. However, cues provided by Rad50/SbsC, the only other known coiled-coil/zinc-hook containing entity, may facilitate the formulation of useful guiding hypotheses. Here it will suffice to briefly summarize pertinent features of this universal DNA-damage-responsive, DNA-repair protein. For comprehensive coverage several recent reviews are suggested (Hirano, 2006;Stracker & Petrini, 2011;Williams *et al*, 2010;Paull, 2010) as well as relevant chapters in this book.

Archaeal Rad50 and the bacterial SbcC counterparts associate with the respective dimeric DNases Mre11 or SbcD. The eukaryal Rad50-Mre11 complex (MR) further associates with an adapter protein termed Nbs1 (Xrs2 in yeast), which links the ternary complex to key DNA damage checkpoints. The ternary MRN complex controls key sensing, signaling, regulating, and effecter responses triggered by DNA double-strand breaks (DSB). These responses include the activation of master regulators such as ATM as well as roles in homologous recombinational repair (HRR), microhomology-mediated end joining (MMEJ) and, occasionally, non-homologous end-joining (NHEJ). Rad50 figures in these transactions as a DNA-bridging SMC protein, using its coiled-coil/zinc-hook and ATPase to properly orient the DNA molecules it bridges and its associated protein partners (Hirano, 2005;Stracker & Petrini, 2011;Williams *et al*, 2010;Paull, 2010;Stracker & Petrini, 2011). As mentioned, Rad50's coiled-coils bend when the ATPase domains are free and stretches when tethered to DNA (van *et al*, 2003;Moreno-Herrero *et al*, 2005). This flexibility also allows the linked ATPase domain to communicate nucleotide binding and DNA ligand signals across distances and between components of the complex. These transmissions depend, among others, on the binding of Mre11 to the coiled-coil portion closest to the ATPase domain, which positions the DNase to resect DSB ends (Williams *et al*, 2011).

Rad50's bacterial homologue SbcC may likewise exert its function as a DNA bridging protein, directing SbcD to cleave hairpin structures that impede DNA replication and initiate DSB that drive HRR (Darmon *et al*, 2010;Storvik & Foster, 2011;White *et al*, 2008). Interestingly, over-expressed in *E. coli*, *SbcC* co-localizes with the replication factory whereas SbcD is dispersed throughout the cytoplasm. Their discrepant behaviors underlie the proposal that at its low, natural level SbcC constantly checks the replication fork for misfolded DNA, recruiting SbcD only when repair is required. A different distribution in *B. subtilis* suggests that in this organism SbcCD partakes also in NHEJ (Mascarenhas *et al*, 2006; Darmon *et al*, 2007).

*Gka*RloC could use its DNA bridging activity (Figs. 5, 6) to monitor the status of cellular DNA molecules like Rad50 and SbcC. However, there is no evidence that RloC associates with a DNase corresponding to Mre11 or SbcD. On the other hand, RloC's regulatory domain, Rad50/SbcC's counterpart is uniquely appended to the translation-disabling ACNase domain. It is tempting to speculate therefore that the ACNase C-domain interacts with the regulatory N-domain in a manner analogous to Mre11's, i.e., tethers to the proximal portion of the coiled-coil fiber emerging from the ATPase head-domain. Such a contact could help transduce DNA damage signals sensed by RloC's DNA monitoring device and relayed by the ATPase to the ACNase effecter domain. The existence of such a signal transduction pathway agrees with the effects of RloC's zinc-hook mutations, ATPase and tethered DNA on its ACNase function (Davidov & Kaufmann, 2008) (Fig 8).

The suggestions that RloC's ACNase is activated in response to DNA damage and, consequently, arrests translation may seem self-contradictory. After all, bacteria normally respond to DNA insults by enhancing the synthesis of DNA repair and other stress responsive proteins (Fernandez De Henestrosa *et al*, 2000). This apparent contradiction may be reconciled by considering the phenomenon of DNA restriction alleviation (RA) (Thoms & Wackernagel, 1984). RA is enacted in response to genotoxic stress as a protective measure intended to prevent degradation of self DNA. In the best documented RA case, the restriction subunit HsdR of the type Ia R-M protein EcoKI is degraded by the protease ClpXP (Makovets *et al*, 2004). In the case of the type Ic protein EcoR124I, RA may entail dissociation or functional occlusion of the HsdR subunit (Youell & Firman, 2008). RA prevents the degradation of fully unmodified portions of the cellular DNA synthesized during the recovery from DNA damage, mainly by HRR. In fact, exposure of an RA-deficient mutant to DNA damage causes DSB and eventual cell death (Cromie *et al*, 2001;Makovets *et al*, 2004;Blakely & Murray, 2006).

RA exacts also a price. Namely, inactivation of the cell's primary immune system renders it highly vulnerable to phage infection (Yamagami & Endo, 1969;Blakely & Murray, 2006). In theory, RloC could benefit its host in this situation by acting as an antiviral back-up device, mobilized when the cell is infected by a phage during recovery from DNA damage. The activation of RloC under these circumstances would prevent the spread of the phage to other members of the vulnerable bacterial population. In this regard RloC could resemble PrrC, which fails to rescue the cell in which it is turned on but can contain the infection. However, the proposed mode of RloC's activation calls for combined inputs of DNA damage and phage infection. Namely, phage infection alone would be offset by the functional DNA restriction nuclease while DNA damage alone would be effectively dealt with by the SOS response (Friedberg *et al*, 2006). It is noteworthy that exposure of an RloC encoding species to mytomycin C did not induce detectable ACNase activity (unpublished results).

Clearly, the above model raises more questions than it attempts to answer. For example, how does the anticipated RloC-Hsd interaction fit in this scheme? Do the genotoxic and viral stress signals cooperate or act separately? Can RloC frustrate phage encoded tRNA repair? To address these issues it will be necessary to employ experimental systems based on natural RloC-encoding hosts and cognate T4-like phages that activate RloC and encode a tRNA repair system.

4. RNA damage repair

4.1 Why repair damaged RNA?

The emergence of an RNA cleavage-ligation pathway in the wake of a host-parasite encounter (Fig. 1) brought to the fore the rather overlooked subject of RNA damage repair. RNA is susceptible to the same agents that threaten DNA. Radiation and chemicals that break the DNA backbone and modify its bases have similar effects on RNA and its precursors. RNA is also attacked by stress responsive RNases (Thompson & Parker, 2009) and various secreted ribotoxins (Wool *et al*, 1992;Masaki & Ogawa, 2002;Lu *et al*, 2005). What is more, its backbone is more sensitive to spontaneous hydrolysis than DNA's. Yet, the repair of damaged RNA seems necessary only in cases where its replenishment by re-synthesis is not possible, e.g., when a DNA template to transcribe from is missing.

Thus, it is conceivable that RNA repair tools played a critical role in sustaining the genomes of the hypothetical RNA and RNA/Protein Worlds (Cech, 2009). One may further speculate that some of these tools could have evolved into extant devices with similar RNA repair tasks or expatiated roles in other RNA transactions (Abelson *et al*, 1998;Sidrauski *et al*, 1996) or even in DNA repair (Aas *et al*, 2003;Tell *et al*, 2010).

RNA repair can be the only option also in extant situations, especially when a DNA template to transcribe from is missing. A relevant example already given here is the reliance of phage T4 on its tRNA repair proteins as a means to overcome the disruption of tRNA^{Lys} by the host's ACNase PrrC (section 2.1). Another relevant example is the AlkB RNA demethylase of certain single stranded plant RNA viruses. The intended role of this demethylase is probably the removal of toxic methyl groups from the viral genomic RNA (van den *et al*, 2008). Homologous bacterial and human RNA-specific AlkB methylases could save the resources and/or time needed to re-synthesize damaged RNAs. In fact, these enzymes have been found able to resuscitate damaged RNA models while distinguishing between natural base modifications and toxic ones. However, the biological relevance of these findings remains uncertain (Aas *et al*, 2003;Ougland *et al*, 2004). Another DNA repair protein with possible roots in RNA metabolism is the abasic DNA endonuclease APE1 (Tell *et al*, 2010). Below we focus on recently discovered cellular tools able to repair nicked RNA, as do the phage T4-encoded proteins Pnk and Rnl1 that counteract PrrC and are frustrated perhaps by RloC.

4.2 Cellular RNA nick repair systems

The RNA phosphodiester linkage is vulnerable to nucleophilic attack. Deprotonation of its adjacent 2' oxygen, subsequent formation of a pentameric phosphate intermediate and 5'-O protonation disrupt it, yielding 2', 3' cyclic phosphate and 5'-OH cleavage ends. This reaction occurs spontaneously and nonspecifically under physiological conditions but is also catalyzed at critical target sites by stress-responsive tRNases (Thompson & Parker, 2009) and secreted ribotoxins (Wool *et al*, 1992;Masaki & Ogawa, 2002;Lu *et al*, 2005;Jablonowski *et al*, 2006;Klassen *et al*, 2008). Some of the small self-cleaving ribozymes that catalyze it also catalyze the reverse reaction, converting 2',3'-cyclic-P and 5'-OH ends into a 3'-5' phosphodiester linkage (Ferre-D'Amare & Scott, 2010). A similar RNA ligase activity involved in tRNA splicing was detected early on in HeLa cell extracts (Filipowicz & Shatkin, 1983) and later in an archaeon (Gomes & Gupta, 1997). The protein catalyzing it termed RtcB has been recently identified in an archaeon, human cells and bacteria (Englert *et al*, 2011;Popow *et al*, 2011;Tanaka & Shuman, 2011). The archaeal and human proteins join 5' and 3' exons of tRNAs and the human possibly also those of the mRNA of an unfolded-protein-response factor (Englert *et al*, 2011;Popow *et al*, 2011). A role for the bacterial RtcB has not been assigned yet. However, its possible participation in an RNA-nick-repair pathway is suggested by the operon RtcB shares with the RNA 3'-P cyclase RtcA. RtcA turns the 3'-P end into 2', 3'-P> through an adenylated intermediate, analogous to the manner in which RNA and DNA ligases activate 5'-P termini (Genschik *et al*, 1998). Thus, combined, RtcAB could convert a 3'-P and 5'-OH pair into a 3'-5' phosphodiester linkage. Unlike RtcA, the RtcB mediated transesterification reaction does not require an energy source although it may be allosterically directed by bound GTP (Tanaka & Shuman, 2011). Given their ability to repair such RNA nicks, RtcAB or RtcB alone could mend accidentally broken RNAs, restore RNAs temporarily inactivated by stress-responsive RNases (Neubauer *et al*, 2009;Zhang *et al*, 2005) or counteract ribotoxins secreted by rival cells (Masaki & Ogawa, 2002). Moreover, the existence of both RtcA and RtcB in all three domains of life (Tanaka & Shuman, 2011;Englert *et al*, 2011;Popow *et al*, 2011) suggests that their cooperation could be rather widespread.

A more intricate RNA-nick-repair pathway is catalyzed by the bacterial proteins PnkP and Hen1. PnkP and Hen1 share the same operon and form a tetrameric P_2H_2 complex (Martins & Shuman, 2005;Chan *et al*, 2009b). The reactions catalyzed by the PnkP component of the complex resemble those mediated by phage T4 Pnk and Rnl1 (section 1) and the yeast and plant tRNA splicing ligase (Abelson *et al*, 1998). What makes this repair system unique is its ability to render the restored phosphodiester linkage immune to re-cleavage by virtue of the 2'-O methylase activity of Hen1 (Chan *et al*, 2009b). PnkP comprises an N-terminal kinase domain, a central metallophosphoesterase domain and a C-terminal ligase domain. Thus, it comprises functions similar to those of the yeast tRNA splicing ligase but differs in domain order and different origin of the phosphoesterase domain (Apostol *et al*, 1991;Martins & Shuman, 2005). Interestingly, by itself the bacterial PnkP heals 2', 3'-cyclic P and 5'-OH termini pairs and undergoes the first step in the RNA ligase reaction, its auto-adenylation, but does not proceed to activate the 5'-P end and generate the phosphodiester linkage (Martins & Shuman, 2005). This deficiency is corrected by expressing PnkP with the 2'-O methylase Hen1. Within the resultant PnkP/Hen1 complex PnkP heals and seals the cleavage termini while Hen1 2'-O methylates the dephosphorylated 3'-end prior to the ligation step. This modification renders the restored ligation junction immune to re-cleavage (Chan *et al*, 2009b). The bacterial Hen1 is so named because it resembles in sequence and structure the methylase domain of eukaryal miRNA methyltransferase Hen1 (Chan *et al*, 2009a). The eukaryal Hen1 protects the 3'-terminal ribose of miRNA from exonucleolytic degradation or utilization as replication primer (Chen, 2005).

As with bacterial RtcAB, the biological role of the PnkP/Hen1 is not known. Noteworthy in this regard is that PnkP/Hen1 is most abundant among *Actinobacteria*. In contrast, RtcAB is more prevalent among *Proteobacteria* and has not been detected yet in *Actinobacteria*. This coincidence raises the possibility that the two systems provide similar benefits to their respective hosts. In theory, PnkP/Hen1 complexes could defend their host cells from secreted ribotoxins more efficiently than RtcAB due to the ability to prevent re-cleavage of the susceptible RNA. It is noteworthy though that colicin-like ribotoxins that target rRNA (Bowman *et al*, 1971;Senior & Holland, 1971) or tRNA anticodon loops (Masaki & Ogawa, 2002) have not been identified yet in bacteria likely to accommodate PnkP/Hen1.

If PnkP/Hen1 were to counteract an ACNase that cleaves its substrate 3' to the wobble base like colicin E5 (Ogawa *et al*, 1999), then the repaired tRNA would contain a 2'-O methylated wobble nucleotide. Such a protective modification need not impair the tRNA's function since it exists in some natural bacterial tRNAs (Juhling *et al*, 2009). However, it cannot be excluded that PnkP/Hen1 plays additional or other roles and may be exploited differently in different bacterial hosts. One example of such a different role is hinted at by the juxtaposition of the PnkP/Hen1 and CRISPR-Cas loci of *Microscilla marina*. The CRISPR-Cas system confers adaptive immunity against foreign nucleic acids. During its antiviral interference activity specific RNA portions of the CRISPR transcript are used to target a Cas protein to cleave the invasive nucleic acid (Deveau *et al*, 2010). Hence, it may be asked if *M. marina* PnkP/Hen1 catalyze some RNA processing and/or modification steps during CRISPR RNA maturation. Finally, in a reversal of roles, one could envisage PnkP/Hen1 encoding phage able to prevent re-cleavage of a tRNA by the ACNase they counteract.

4.3 An essential eukaryal DNA repair protein is related to T4 Pnk

There are a number of examples of DNA repair devices that could have originated from RNA-specific progenitors, some of them already alluded to above. Here it will suffice to describe just one of them, related to the phage T4-encoded end healing protein Pnk. This conserved eukaryal protein termed interchangeably PNKP and Pnk1 contains 5'-kinase and 3'-phosphatase domains resembling those of T4 Pnk but arranged in the reverse order, the phosphoesterase domain preceding the kinase domain. The mammalian PNKP is also endowed with an N-terminal FHA (Fork Head Associated) phosphopeptide binding domain that links PNKP to the scaffold proteins XRCC1 and XRCC4 (Bernstein *et al*, 2009). The latter recruit PNKP to exercise its functions in base excision repair (Hegde *et al*, 2008) or NHEJ (Lieber, 2008). PNKP's essential role in these ssDNA and DSB repair pathways is to convert 3'-P and 5'-OH DNA termini into 3'-OH and 5'-P pairs that are ligatable or fit for gap-filling by a DNA polymerase. A wide DNA binding cleft accounts for the ability of this protein to prefer nicked duplexes and recessed 5'-termini over ssDNA substrates and distinguishes it from the RNA end healing phage counterpart. The 3'-P and 5'-OH DNA termini are caused by ionizing radiation, genotoxic chemicals and enzymatic reactions. Specific examples include excision of abasic sites (Hazra *et al*, 2002), DSB generated by DNase II (Evans & Aguilera, 2003) and release of camptothecin- trapped topoisomerase I-DNA adducts by a tyrosine-DNA specific phosphodiesterase (Pouliot *et al*, 1999). Failure to repair such lesions underlies several inborn neural disorders. Conversely, PNKP can render cancer cells resistant to certain genotoxic drugs and, therefore, is considered itself a potential therapeutic target (Weinfeld *et al*, 2011).

5. Conclusions

In this chapter we addressed the possible biological role of the conserved bacterial anticodon nuclease RloC that combines two seemingly conflicting properties. One, predicted by resemblance of its regulatory region to the universal DNA-damage-checkpoint/DNA repair protein Rad50/SbcC is monitoring DNA insults. The second, predicted by its tRNase activity is disabling the translation apparatus. The co-existence of such functions in the same molecule and the regulation of one by the other suggests that RloC is designed to block translation in response to DNA damage. Such a response is suicidal since it prevents recovery from DNA damage. Hence, it must be executed only under special circumstances where cell death is advantageous. One possibility considered here is that RloC benefits its host cell by acting as an antiviral contingency during recovery from DNA damage. Under these conditions bacterial cells may shut off their primary antiviral defense, i.e., their DNA restriction activity. RloC's suicidal activity would not rescue the infected cell but would prevent the spread of the infection to other vulnerable members of the population recovering from DNA damage.

Another unique property, which could make RloC particularly suited to thwart phage infection, is the ability of this ACNase to excise its substrate's wobble nucleotide. In this regard RloC differs from its distant homologue the ACNase PrrC, which only incises its tRNA substrate and is counteracted by phage tRNA repair enzymes. Therefore, it seems conceivable that the harsher lesion inflicted by RloC will encumber such phage reversal. The possibility that RloC is a more efficient antiviral device than PrrC is also hinted at by its ~3-fold more frequent occurrence among bacteria.

While these notions are supported by some demonstrated properties of RloC, testing them and identifying RloC's true call requires studying this protein under physiological conditions; ideally, using a natural host encoding it and cognate phages endowed with tRNA repair enzymes.

The RNA repair pathway instigated by PrrC and possibly avoided by RloC brings to the fore the rather overlooked issue of RNA-damage-repair. Such repair would seem necessary only under circumstances such as the absence of a DNA template to transcribe from. Nonetheless, recent discoveries of various cellular RNA repair devices distributed in the three domains of life suggest that RNA damage repair is more prevalent, exercised perhaps also during responses to nutritional, pathogenic and other forms of stress. RNA repair is also of interest because many of its devices seem to have evolved to serve in other RNA transactions and even in DNA repair. Conversely, the vast repertoire of DNA repair, RNA splicing and RNA editing reactions may be exploited by investigators to discover novel RNA repair phenomena.

Work in G.K's laboratory was supported by grants from the Israeli Science Foundation, Jerusalem, the United States-Israel Bi-national Science Foundation and the Israeli Ministry of Science. G.K. is an incumbent of the Louise and Nahum Barag Chair in Cancer Molecular Genetics.

6. References

Aas,P.A., Otterlei,M., Falnes,P.O., Vagbo,C.B., Skorpen,F., Akbari,M., Sundheim,O., Bjoras,M., Slupphaug,G., Seeberg,E., & Krokan,H.E. (2003) Human and bacterial oxidative demethylases repair alkylation damage in both RNA and DNA. *Nature*, 421, 859-863.

Abelson,J., Trotta,C.R., & Li,H. (1998) tRNA splicing. *Journal of Biological Chemistry*, 273, 12685-12688.

Ambudkar,S.V., Kim,I.W., Xia,D., & Sauna,Z.E. (2006) The A-loop, a novel conserved aromatic acid subdomain upstream of the Walker A motif in ABC transporters, is critical for ATP binding. *FEBS Lett.*, 580, 1049-1055.

Amitsur,M., Benjamin,S., Rosner,R., Chapman-Shimshoni,D., Meidler,R., Blanga,S., & Kaufmann,G. (2003) Bacteriophage T4-encoded Stp can be replaced as activator of anticodon nuclease by a normal host cell metabolite. *Mol.Microbiol.*, 50, 129-143.

Amitsur,M., Levitz,R., & Kaufmann,G. (1987) Bacteriophage T4 anticodon nuclease, polynucleotide kinase and RNA ligase reprocess the host lysine tRNA. *The EMBO Journal*, 6, 2499-2503.

Amitsur,M., Morad,I., Chapman-Shimshoni,D., & Kaufmann,G. (1992) HSD restriction-modification proteins partake in latent anticodon nuclease. *The EMBO Journal*, 11, 3129-3134.

Apostol,B.L., Westaway,S.K., Abelson,J., & Greer,C.L. (1991) Deletion analysis of a multifunctional yeast tRNA ligase polypeptide. Identification of essential and dispensable functional domains. *Journal of Biological Chemistry*, 266, 7445-7455.

Becker,A. & Hurwitz,J. (1967) The enzymatic cleavage of phosphate termini from polynucleotides. *Journal of Biological Chemistry*, 242, 936-950.

Belogurov,A.A., Delver,E.P., Agafonova,O.V., Belogurova,N.G., Lee,L.Y., & Kado,C.I. (2000) Antirestriction protein Ard (Type C) encoded by IncW plasmid pSa has a high similarity to the "protein transport" domain of TraC1 primase of promiscuous plasmid RP4. *Journal of Molecular Biology*, 296, 969-977.

Bernstein,N.K., Hammel,M., Mani,R.S., Weinfeld,M., Pelikan,M., Tainer,J.A., & Glover,J.N. (2009) Mechanism of DNA substrate recognition by the mammalian DNA repair enzyme, Polynucleotide Kinase. *Nucleic Acids Res.*, 37, 6161-6173.

Blakely,G.W. & Murray,N.E. (2006) Control of the endonuclease activity of type I restriction-modification systems is required to maintain chromosome integrity following homologous recombination. *Mol.Microbiol.*, 60, 883-893.

Blanga-Kanfi,S., Amitsur,M., Azem,A., & Kaufmann,G. (2006) PrrC-anticodon nuclease: functional organization of a prototypical bacterial restriction RNase. *Nucleic Acids Res.*, 34, 3209-3219.

Blondal,T., Hjorleifsdottir,S., Aevarsson,A., Fridjonsson,O.H., Skirnisdottir,S., Wheat,J.O., Hermannsdottir,A.G., Hreggvidsson,G.O., Smith,A.V., & Kristjansson,J.K. (2005) Characterization of a 5'-polynucleotide kinase/3'-phosphatase from bacteriophage RM378. *J Biol.Chem.*, 280, 5188-5194.

Blondal,T., Hjorleifsdottir,S.H., Fridjonsson,O.F., Aevarsson,A., Skirnisdottir,S., Hermannsdottir,A.G., Hreggvidsson,G.O., Smith,A.V., & Kristjansson,J.K. (2003) Discovery and characterization of a thermostable bacteriophage RNA ligase homologous to T4 RNA ligase 1. *Nucleic Acids Res.*, 31, 7247-7254.

Bowman,C.M., Dahlberg,J.E., Ikemura,T., Konisky,J., & Nomura,M. (1971) Specific inactivation of 16S ribosomal RNA induced by colicin E3 in vivo. *Proc.Natl.Acad.Sci.U.S.A*, 68, 964-968.

Cameron,V. & Uhlenbeck,O.C. (1977) 3' Phosphatase activity in T4-polynucleotide kinase. *Biochemistry*, 16, 5120-5126.

Cech,T.R. (2009) Evolution of biological catalysis: ribozyme to RNP enzyme. *Cold Spring Harb.Symp.Quant.Biol.*, 74, 11-16.

Chan,C.M., Zhou,C., Brunzelle,J.S., & Huang,R.H. (2009a) Structural and biochemical insights into 2'-O-methylation at the 3'-terminal nucleotide of RNA by Hen1. *Proc.Natl.Acad.Sci.U.S.A.*

Chan,C.M., Zhou,C., & Huang,R.H. (2009b) Reconstituting Bacterial RNA Repair and Modification in Vitro. *Science*, 326, 247.

Chapman,D., Morad,I., Kaufmann,G., Gait,M.J., Jorissen,L., & Snyder,L. (1988) Nucleotide and deduced amino acid sequence of *stp*: the bacteriophage T4 anticodon nuclease gene. *Journal of Molecular Biology*, 199, 373-377.

Chen,J., Lu,G., Lin,J., Davidson,A.L., & Quiocho,F.A. (2003) A tweezers-like motion of the ATP-binding cassette dimer in an ABC transport cycle. *Mol.Cell*, 12, 651-661.

Chen,X. (2005) MicroRNA biogenesis and function in plants. *FEBS Lett.*, 579, 5923-5931.

Connelly,J.C., Kirkham,L.A., & Leach,D.R. (1998) The SbcCD nuclease of Escherichia coli is a structural maintenance of chromosomes (SMC) family protein that cleaves hairpin DNA. *Proc.Natl.Acad.Sci.U.S.A*, 95, 7969-7974.

Cromie,G.A., Connelly,J.C., & Leach,D.R. (2001) Recombination at double-strand breaks and DNA ends: conserved mechanisms from phage to humans. *Mol.Cell*, 8, 1163-1174.

Darmon,E., Eykelenboom,J.K., Lincker,F., Jones,L.H., White,M., Okely,E., Blackwood,J.K., & Leach,D.R. (2010) E. coli SbcCD and RecA control chromosomal rearrangement induced by an interrupted palindrome. *Mol.Cell*, 39, 59-70.

Darmon,E., Lopez-Vernaza,M.A., Helness,A.C., Borking,A., Wilson,E., Thacker,Z., Wardrope,L., & Leach,D.R. (2007) SbcCD regulation and localization in Escherichia coli. *Journal of Bacteriology*, 189, 6686-6694.

David,M., Borasio,G.D., & Kaufmann,G. (1982) Bacteriophage T4-induced anticodon-loop nuclease detected in a host strain restrictive to RNA ligase mutants. *Proceedings of the National Academy of Sciences U.S.A.*, 79, 7097-7101.

Davidov,E. & Kaufmann,G. (2008) RloC: a wobble nucleotide-excising and zinc-responsive bacterial tRNase. *Mol.Microbiol.*, 69, 1560-1574.

Depew,R.E. & Cozzarelli,N.R. (1974) Genetics and physiology of bacteriophage T4 3'-phosphatase: evidence for the involvement of the enzyme in T4 DNA metabolism. *Journal of Virology*, 13, 888-897.

Depew,R.E., Snopek,T.J., & Cozzarelli,N.R. (1975) Characterization of a new class of deletions of the D region of the bacteriophage T4 genome. *Virology*, 64, 144-145.

Deveau,H., Garneau,J.E., & Moineau,S. (2010) CRISPR/Cas system and its role in phage-bacteria interactions. *Annu.Rev.Microbiol.*, 64, 475-493.

Dryden,D.T., Murray,N.E., & Rao,D.N. (2001) Nucleoside triphosphate-dependent restriction enzymes. *Nucleic Acids Res.*, 29, 3728-3741.

El Omari K., Ren,J., Bird,L.E., Bona,M.K., Klarmann,G., LeGrice,S.F., & Stammers,D.K. (2006) Molecular architecture and ligand recognition determinants for T4 RNA ligase. *Journal of Biological Chemistry*, 281, 1573-1579.

Englert,M., Sheppard,K., Aslanian,A., Yates,J.R., III, & Soll,D. (2011) Archaeal 3'-phosphate RNA splicing ligase characterization identifies the missing component in tRNA maturation. *Proc.Natl.Acad.Sci.U.S.A*, 108, 1290-1295.

Evans,C.J. & Aguilera,R.J. (2003) DNase II: genes, enzymes and function. *Gene*, 322, 1-15.

Fernandez De Henestrosa,A.R., Ogi,T., Aoyagi,S., Chafin,D., Hayes,J.J., Ohmori,H., & Woodgate,R. (2000) Identification of additional genes belonging to the LexA regulon in Escherichia coli. *Mol.Microbiol.*, 35, 1560-1572.

Ferre-D'Amare,A.R. & Scott,W.G. (2010) Small self-cleaving ribozymes. *Cold Spring Harb.Perspect.Biol.*, 2, a003574.

Filipowicz,W. & Shatkin,A.J. (1983) Origin of splice junction phosphate in tRNAs processed by HeLa cell extract. *Cell*, 32, 547-557.

Friedberg,E.C., Walker,G.C., Siede,W., Wood,R.D., & Ellenberger,T. (2006) *DNA Repair And Mutagenesis*, Second edn, p. 1. ASM Press, Washington, DC.

Galburt,E.A., Pelletier,J., Wilson,G., & Stoddard,B.L. (2002) Structure of a tRNA Repair Enzyme and Molecular Biology Workhorse: T4 Polynucleotide Kinase. *Structure*, 10, 1249-1260.

Genschik,P., Drabikowski,K., & Filipowicz,W. (1998) Characterization of the Escherichia coli RNA 3'-terminal phosphate cyclase and its sigma54-regulated operon. *Journal of Biological Chemistry*, 273, 25516-25526.

Gerlt,J.A. (1993) Mechanistic principles of enzyme-catalyzed cleavage of phosphodiester bonds. Nucleases (ed. by S. M. Linn, R. S. Lloyd, & R. J. Roberts), pp. 1-34. Cold Spring Harbor Laboratory Press, Plainview NY.

Gomes,I. & Gupta,R. (1997) RNA splicing ligase activity in the archaeon Haloferax volcanii. *Biochem.Biophys.Res.Commun.*, 237, 588-594.

Guo,X., Chen,X., Weber,I.T., Harrison,R.W., & Tai,P.C. (2006) Molecular basis for differential nucleotide binding of the nucleotide-binding domain of ABC-transporter CvaB. *Biochemistry*, 45, 14473-14480.

Hazra,T.K., Izumi,T., Boldogh,I., Imhoff,B., Kow,Y.W., Jaruga,P., Dizdaroglu,M., & Mitra,S. (2002) Identification and characterization of a human DNA glycosylase for repair of modified bases in oxidatively damaged DNA. *Proc.Natl.Acad.Sci.U.S.A*, 99, 3523-3528.

Hegde,M.L., Hazra,T.K., & Mitra,S. (2008) Early steps in the DNA base excision/single-strand interruption repair pathway in mammalian cells. *Cell Res.*, 18, 27-47.

Hirano,T. (2006) At the heart of the chromosome: SMC proteins in action. *Nat.Rev.Mol.Cell Biol.*, 7, 311-322.

Hirano,T. (2005) SMC proteins and chromosome mechanics: from bacteria to humans. *Philos.Trans.R.Soc.Lond B Biol.Sci.*, 360, 507-514.

Ho,C.K. & Shuman,S. (2002) Bacteriophage T4 RNA ligase 2 (gp24.1) exemplifies a family of RNA ligases found in all phylogenetic domains. *Proceedings National Academy of Sciences USA*, 99, 12709-12714.

Hopfner,K.P., Craig,L., Moncalian,G., Zinkel,R.A., Usui,T., Owen,B.A., Karcher,A., Henderson,B., Bodmer,J.L., McMurray,C.T., Carney,J.P., Petrini,J.H., & Tainer,J.A. (2002) The Rad50 zinc-hook is a structure joining Mre11 complexes in DNA recombination and repair. *Nature*, 418, 562-566.

Hopfner,K.P., Karcher,A., Shin,D.S., Craig,L., Arthur,L.M., Carney,J.P., & Tainer,J.A. (2000) Structural biology of Rad50 ATPase: ATP-driven conformational control in DNA double-strand break repair and the ABC-ATPase superfamily. *Cell*, 101, 789-800.

Hopfner,K.P. & Tainer,J.A. (2003) Rad50/SMC proteins and ABC transporters: unifying concepts from high-resolution structures. *Curr.Opin.Struct.Biol.*, 13, 249-255.

Jabbar,M.A. & Snyder,L. (1984) Genetic and physiological studies of an Escherichia coli locus that restricts polynucleotide kinase- and RNA ligase-deficient mutants of bacteriophage T4. *Journal of Virology*, 51, 522-529.

Jablonowski,D., Zink,S., Mehlgarten,C., Daum,G., & Schaffrath,R. (2006) tRNAGlu wobble uridine methylation by Trm9 identifies Elongator's key role for zymocin-induced cell death in yeast. *Mol.Microbiol.*, 59, 677-688.

Janscak,P., Sandmeier,U., Szczelkun,M.D., & Bickle,T.A. (2001) Subunit assembly and mode of DNA cleavage of the type III restriction endonucleases EcoP1I and EcoP15I. *Journal of Molecular Biology*, 306, 417-431.

Jiang,Y., Blanga,S., Amitsur,M., Meidler,R., Krivosheyev,E., Sundaram,M., Bajji,A.C., Davis,D.R., & Kaufmann,G. (2002) Structural features of tRNALys favored by anticodon nuclease as inferred from reactivities of anticodon stem and loop substrate analogs. *Journal of Biological Chemistry*, 277, 3836-3841.

Jiang,Y., Meidler,R., Amitsur,M., & Kaufmann,G. (2001) Specific Interaction between Anticodon Nuclease and the tRNA(Lys) Wobble Base. *Journal of Molecular Biology*, 305, 377-388.

Juhling,F., Morl,M., Hartmann,R.K., Sprinzl,M., Stadler,P.F., & Putz,J. (2009) tRNAdb 2009: compilation of tRNA sequences and tRNA genes. *Nucleic Acids Res.*, 37, D159-D162.

Kaufmann,G. & Kallenbach,N.R. (1975) Determination of recognition sites of T4 RNA ligase on the 3'-OH and 5'-P termini of polyribonucleotide chains. *Nature*, 254, 452-454.

Klaiman,D., Amitsur,M., Blanga-Kanfi,S., Chai,M., Davis,D.R., & Kaufmann,G. (2007) Parallel dimerization of a PrrC-anticodon nuclease region implicated in tRNALys recognition. *Nucleic Acids Res.*, 35, 4704-4714.

Klaiman,D. & Kaufmann,G. (2011) Phage T4-induced dTTP accretion bolsters a tRNase-based host defense. *Virology*, 414, 97-101.

Klassen,R., Paluszynski,J.P., Wemhoff,S., Pfeiffer,A., Fricke,J., & Meinhardt,F. (2008) The primary target of the killer toxin from Pichia acaciae is tRNA(Gln). *Mol.Microbiol.*

Levitz,R., Chapman,D., Amitsur,M., Green,R., Snyder,L., & Kaufmann,G. (1990) The optional *E. coli prr* locus encodes a latent form of phage T4-induced anticodon nuclease. *The EMBO Journal*, 9, 1383-1389.

Lieber,M.R. (2008) The mechanism of human nonhomologous DNA end joining. *Journal of Biological Chemistry*, 283, 1-5.

Linder,P., Doelz,R., Gubler,M., & Bickle,T.A. (1990) An anticodon nuclease gene inserted into a *hsd* region encoding a type I DNA restriction system. *Nucleic Acids Res.*, 18, 7170.

Lu,J., Huang,B., Esberg,A., Johansson,M.J., & Bystrom,A.S. (2005) The Kluyveromyces lactis gamma-toxin targets tRNA anticodons. *RNA.*, 11, 1648-1654.

Makovets,S., Powell,L.M., Titheradge,A.J., Blakely,G.W., & Murray,N.E. (2004) Is modification sufficient to protect a bacterial chromosome from a resident restriction endonuclease? *Mol.Microbiol.*, 51, 135-147.

Martins,A. & Shuman,S. (2005) An end-healing enzyme from Clostridium thermocellum with 5' kinase, 2',3' phosphatase, and adenylyltransferase activities. *RNA.*, 11, 1271-1280.

Masaki,H. & Ogawa,T. (2002) The modes of action of colicins E5 and D, and related cytotoxic tRNases. *Biochimie*, 84, 433-438.

Mascarenhas,J., Sanchez,H., Tadesse,S., Kidane,D., Krisnamurthy,M., Alonso,J.C., & Graumann,P.L. (2006) Bacillus subtilis SbcC protein plays an important role in DNA inter-strand cross-link repair. *BMC.Mol.Biol.*, 7, 20.

Mathews,C.K. (1994) An overview of the T4 develpmental program. Molecular Biology of bacteriophage T4 (ed. by J. D. Karam, J. W. Drake, K. N. Kreuzer, G. Mosig, D. W. Hall, F. A. Eiserling, L. W. Black, E. K. Spicer, E. Kutter, K. Carlson, & E. S. Miller), pp. 1-8. American Society for Microbiology, Washignton, DC.

Meidler,R., Morad,I., Amitsur,M., Inokuchi,H., & Kaufmann,G. (1999) Detection of Anticodon Nuclease Residues Involved in tRNALys Cleavage Specificity. *Journal of Molecular Biology*, 287, 499-510.

Meineke,B., Schwer,B., Schaffrath,R., & Shuman,S. (2010) Determinants of eukaryal cell killing by the bacterial ribotoxin PrrC. *Nucleic Acids Res.*

Miller,E.S., Heidelberg,J.F., Eisen,J.A., Nelson,W.C., Durkin,A.S., Ciecko,A., Feldblyum,T.V., White,O., Paulsen,I.T., Nierman,W.C., Lee,J., Szczypinski,B., & Fraser,C.M. (2003a) Complete genome sequence of the broad-host-range vibriophage KVP40: comparative genomics of a T4-related bacteriophage. *J Bacteriol.*, 185, 5220-5233.

Miller,E.S., Kutter,E., Mosig,G., Arisaka,F., Kunisawa,T., & Ruger,W. (2003b) Bacteriophage T4 Genome. *Microbiology and Molecular Biology Reviews*, 67, 86-156.

Miller,W.G., Pearson,B.M., Wells,J.M., Parker,C.T., Kapitonov,V.V., & Mandrell,R.E. (2005) Diversity within the Campylobacter jejuni type I restriction-modification loci. *Microbiology*, 151, 337-351.

Moody,J.E. & Thomas,P.J. (2005) Nucleotide binding domain interactions during the mechanochemical reaction cycle of ATP-binding cassette transporters. *J.Bioenerg.Biomembr.*, 37, 475-479.

Moreno-Herrero,F., de,J.M., Dekker,N.H., Kanaar,R., Wyman,C., & Dekker,C. (2005) Mesoscale conformational changes in the DNA-repair complex Rad50/Mre11/Nbs1 upon binding DNA. *Nature*, 437, 440-443.

Morera,S., Imberty,A., schke-Sonnenborn,U., Ruger,W., & Freemont,P.S. (1999) T4 phage beta-glucosyltransferase: substrate binding and proposed catalytic mechanism. *Journal of Molecular Biology*, 292, 717-730.

Murray,N.E. (2000) Type I restriction systems: sophisticated molecular machines (a legacy of Bertani and Weigle). *Microbiology and Molecular Biology Reviews*, 64, 412-434.

Neubauer,C., Gao,Y.G., Andersen,K.R., Dunham,C.M., Kelley,A.C., Hentschel,J., Gerdes,K., Ramakrishnan,V., & Brodersen,D.E. (2009) The structural basis for mRNA recognition and cleavage by the ribosome-dependent endonuclease RelE. *Cell*, 139, 1084-1095.

Novogrodsky,A., Tal,M., Traub,A., & Hurwitz,J. (1966) The enzymatic phosphorylation of ribonucleic acid and deoxyribonucleic acid. II. Further properties of the 5'-hydroxyl polynucleotide kinase. *Journal of Biological Chemistry*, 241, 2933-2943.

Ogawa,T., Tomita,K., Ueda,T., Watanabe,K., Uozumi,T., & Masaki,H. (1999) A Cytotoxic Ribonuclease Targeting Specific Transfer RNA Anticodons. *Science*, 283, 2097-2100.

Ougland,R., Zhang,C.M., Liiv,A., Johansen,R.F., Seeberg,E., Hou,Y.M., Remme,J., & Falnes,P.O. (2004) AlkB restores the biological function of mRNA and tRNA inactivated by chemical methylation. *Mol.Cell*, 16, 107-116.

Paull,T.T. (2010) Making the best of the loose ends: Mre11/Rad50 complexes and Sae2 promote DNA double-strand break resection. *DNA Repair (Amst)*, 9, 1283-1291.

Penner,M., Morad,I., Snyder,L., & Kaufmann,G. (1995) Phage T4-coded Stp: double-edged effector of coupled DNA and tRNA-restriction systems. *Journal of Molecular Biology*, 249, 857-868.

Popow,J., Englert,M., Weitzer,S., Schleiffer,A., Mierzwa,B., Mechtler,K., Trowitzsch,S., Will,C.L., Luhrmann,R., Soll,D., & Martinez,J. (2011) HSPC117 is the essential subunit of a human tRNA splicing ligase complex. *Science*, 331, 760-764.

Pouliot,J.J., Yao,K.C., Robertson,C.A., & Nash,H.A. (1999) Yeast gene for a Tyr-DNA phosphodiesterase that repairs topoisomerase I complexes. *Science*, 286, 552-555.

Richardson,C.C. (1965) Phosphorylation of nucleic acid by an enzyme from T4 bacteriophage infected E. coli. *Proceedings National Academy of Sciences USA*, 54, 158-165.

Runnels,J., Soltis,D., Hey,T., & Snyder,L. (1982) Genetic and physiological studies of the role of RNA ligase of bacteriophage T4. *Journal of Molecular Biology*, 154, 273-286.

Sargent,R.G. & Mathews,C.K. (1987) Imbalanced deoxyribonucleoside triphosphate pools and spontaneous mutation rates determined during dCMP deaminase-defective bacteriophage T4 infections. *Journal of Biological Chemistry*, 262, 5546-5553.

Schirmer,T. & Evans,P.R. (1990) Structural basis of the allosteric behaviour of phosphofructokinase. *Nature*, 343, 140-145.

Schmidt,F.J. & Apirion,D. (1983) T4 transfer RNAs: Pardigmatic System for the Study of RNA Processing. Bacteriophage T4 (ed. by C. K. Mathews, E. M. Kutter, G. Mosig, & P. B. Berget), pp. 193-217. American Society for Microbiology, Washington.

Senior,B.W. & Holland,I.B. (1971) Effect of colicin E3 upon the 30S ribosomal subunit of Escherichia coli. *Proc.Natl.Acad.Sci.U.S.A*, 68, 959-963.

Sidrauski,C., Cox,J.S., & Walter,P. (1996) tRNA ligase is required for regulated mRNA splicing in the unfolded protein response. *Cell*, 87, 405-413.

Silber,R., Malathi,V.G., & Hurwitz,J. (1972) Purification and properties of bacteriophage T4-induced RNA ligase. *Proceedings National Academy of Sciences USA*, 69, 3009-3013.

Sirotkin,K., Cooley,W., Runnels,J., & Snyder,L. (1978) A role in true-late gene expression for the T4 bacteriophage 5'-polynucleotide kinase 3'-phosphatase. *Journal of Molecular Biology*, 123, 221-233.

Stanley,L.K., Seidel,R., van der,S.C., Dekker,N.H., Szczelkun,M.D., & Dekker,C. (2006) When a helicase is not a helicase: dsDNA tracking by the motor protein EcoR124I. *The EMBO Journal*, 25, 2230-2239.

Steyaert,J. (1997) A decade of protein engineering on ribonuclease T1--atomic dissection of the enzyme-substrate interactions. *Eur.J Biochem.*, 247, 1-11.

Storvik,K.A. & Foster,P.L. (2011) The SMC-like protein complex SbcCD enhances DNA polymerase IV-dependent spontaneous mutation in Escherichia coli. *Journal of Bacteriology*, 193, 660-669.

Stracker,T.H. & Petrini,J.H. (2011) The MRE11 complex: starting from the ends. *Nat.Rev.Mol.Cell Biol.*, 12, 90-103.

Takami,H., Nishi,S., Lu,J., Shimamura,S., & Takaki,Y. (2004) Genomic characterization of thermophilic Geobacillus species isolated from the deepest sea mud of the Mariana Trench. *Extremophiles.*, 8, 351-356.

Tanaka,N. & Shuman,S. (2011) RtcB is the RNA ligase component of an Escherichia coli RNA repair operon. *Journal of Biological Chemistry*.

Tell,G., Wilson,D.M., III, & Lee,C.H. (2010) Intrusion of a DNA repair protein in the RNome world: is this the beginning of a new era? *Mol.Cell Biol.*, 30, 366-371.

Thompson,D.M. & Parker,R. (2009) Stressing out over tRNA cleavage. *Cell*, 138, 215-219.

Thoms,B. & Wackernagel,W. (1984) Genetic control of damage-inducible restriction alleviation in Escherichia coli K12: an SOS function not repressed by lexA. *Mol.Gen.Genet.*, 197, 297-303.

Tyndall,C., Meister,J., & Bickle,T.A. (1994) The Escherichia coli prr region encodes a functional type IC DNA restriction system closely integrated with an anticodon nuclease gene. *Journal of Molecular Biology*, 237, 266-274.

van den,B.E., Omelchenko,M.V., Bekkelund,A., Leihne,V., Koonin,E.V., Dolja,V.V., & Falnes,P.O. (2008) Viral AlkB proteins repair RNA damage by oxidative demethylation. *Nucleic Acids Res.*, 36, 5451-5461.

van Noort,J., van der Heijden,T., Dutta,C.F., Firman,K., & Dekker,C. (2004) Initiation of translocation by Type I restriction-modification enzymes is associated with a short DNA extrusion. *Nucleic Acids Res.*, 32, 6540-6547.

van,N.J., van Der,H.T., de,J.M., Wyman,C., Kanaar,R., & Dekker,C. (2003) The coiled-coil of the human Rad50 DNA repair protein contains specific segments of increased flexibility. *Proc.Natl.Acad.Sci.U.S.A*, 100, 7581-7586.

Weinfeld,M., Mani,R.S., Abdou,I., Aceytuno,R.D., & Glover,J.N. (2011) Tidying up loose ends: the role of polynucleotide kinase/phosphatase in DNA strand break repair. *Trends Biochem.Sci.*

Wheeler,L.J., Ray,N.B., Ungermann,C., Hendricks,S.P., Bernard,M.A., Hanson,E.S., & Mathews,C.K. (1996) T4 phage gene 32 protein as a candidate organizing factor for the deoxyribonucleoside triphosphate synthetase complex. *Journal of Biological Chemistry*, 271, 11156-11162.

White,M.A., Eykelenboom,J.K., Lopez-Vernaza,M.A., Wilson,E., & Leach,D.R. (2008) Non-random segregation of sister chromosomes in Escherichia coli. *Nature*, 455, 1248-1250.

Williams,G.J., Lees-Miller,S.P., & Tainer,J.A. (2010) Mre11-Rad50-Nbs1 conformations and the control of sensing, signaling, and effector responses at DNA double-strand breaks. *DNA Repair (Amst)*, 9, 1299-1306.

Williams,G.J., Williams,R.S., Williams,J.S., Moncalian,G., Arvai,A.S., Limbo,O., Guenther,G., Sildas,S., Hammel,M., Russell,P., & Tainer,J.A. (2011) ABC ATPase signature helices in Rad50 link nucleotide state to Mre11 interface for DNA repair. *Nat.Struct.Mol.Biol.*, 18, 423-431.

Williams,R.S., Williams,J.S., & Tainer,J.A. (2007) Mre11-Rad50-Nbs1 is a keystone complex connecting DNA repair machinery, double-strand break signaling, and the chromatin template. *Biochem.Cell Biol.*, 85, 509-520.

Wool,I.G., Gluck,A., & Endo,Y. (1992) Ribotoxin recognition of ribosomal RNA and a proposal for the mechanism of translocation. *Trends Biochem.Sci.*, 17, 266-269.

Yamagami,H. & Endo,H. (1969) Loss of lysis inhibition in filamentous Escherichia coli infected with wild-type bacteriophage T4. *Journal of Virology*, 3, 343-349.

Youell,J. & Firman,K. (2008) EcoR124I: from plasmid-encoded restriction-modification system to nanodevice. *Microbiology and Molecular Biology Reviews*, 72, 365-77, table.

Zhang,Y., Zhang,J., Hara,H., Kato,I., & Inouye,M. (2005) Insights into the mRNA cleavage mechanism by MazF, an mRNA interferase. *J Biol.Chem.*, 280, 3143-3150.

4

The Potential Roles of DNA-Repair Proteins in Centrosome Maintenance

Mikio Shimada, Akihiro Kato and Junya Kobayashi
Radiation Biology Center, Kyoto University
Japan

1. Introduction

The centrosome, an organelle that regulates microtubules, is necessary for proper cell division in mammalian cells (Doxsey, 2001; Nigg, 2002, 2007). The existence of centrosomes was first reported 100 years ago by Theodor Boveri (Boveri, 2008). A centrosome is composed of two centrioles and is surrounded by pericentriolar material (PCM), which provides a binding site for the γ-tubulin ring complex (γ-TuRC). The γ-TuRC acts as a microtubule nucleation template, and it attaches to the PCM to form microtubules (Fig. 1). The number of centrosomes is precisely regulated, and the duplication cycle is synchronized to the cell cycle. Centrosomes duplicate once in the S phase and mature in the G2 phase, and in the M phase, centrosomes are divided into daughter cells (Fig. 2). The number of centrosomes and their functions are regulated by many proteins including centrosome proteins, cell-cycle proteins, and DNA-repair proteins, and recently, the role of DNA-repair proteins in centrosome maintenance has been clarified. In this chapter, we introduce recent findings about the roles of DNA-repair proteins in centrosome maintenance.

2. Centrosomes and aneuploidy

Many cancer cells possess extra centrosomes, which is called centrosome amplification and means overduplication of centrosomes. Extra centrosomes can lead to multipolar cell divisions, subsequent aneuploidy, and cell death (Kwon et al., 2008). Although almost all multipolar cell division results in cell death via mitotic catastrophe (Ganem et al., 2009), some multipolar cells divide into daughter cells to maintain aneuploidy. Aneuploidic cells are believed to potentially cause tumorigenesis. Recent studies suggest that aneuploidic cells are produced by a clustering of extra centrosomes, which accumulate at the two poles, and microtubules from each of the extra centrosomes attach to the chromosomes prior to mitosis (Kwon et al., 2008) (Fig. 2). The tension created by the extra centrosomes leads to improper chromosome segregation (Godinho et al., 2009).

Several environmental factors and chemicals, or carcinogens, including ionizing radiation and benzopyrene, can induce extra centrosomes (Sato et al., 2000). Thus, failure of the centrosome duplication cycle could cause tumorigenesis via chromosome aneuploidy.

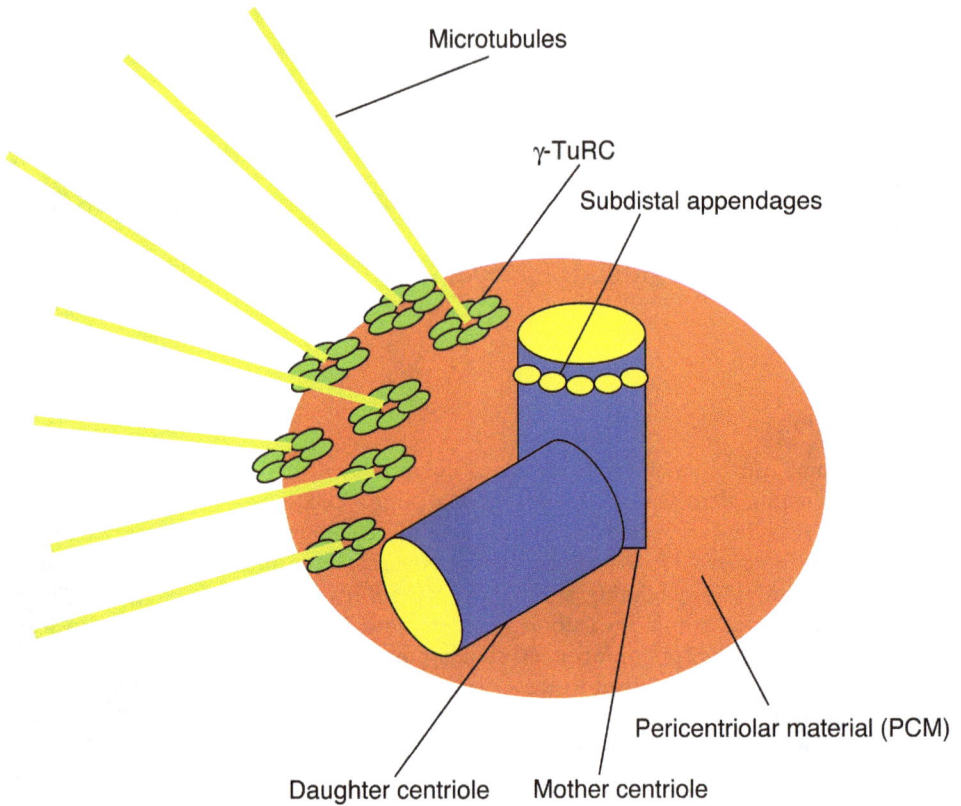

Centrosomes are located at the periphery of the nucleus and consist of a mother centriole and a daughter centriole, surrounded by the pericentriolar material (PCM). The γ-tubulin ring complex (γ-TuRC) binds to the PCM to form microtubules.

Fig. 1. Centrosome structure.

3. Centrosomes and the cell cycle

Centrosome duplication is controlled by several cell-cycle regulators (Fukasawa, 2007). The cyclin E/CDK2 complex is responsible for initiating DNA synthesis and regulates cell-cycle progression (Matsumoto et al., 1999). This complex also contributes to centrosome duplication (Fig. 3). Cyclin E contains the centrosome localization signal (CLS), and overexpression of mutated cyclin E through a CLS deletion results in failed centrosome duplication (Matsumoto and Maller, 2002). The CDKN1A product, p21, is a negative regulator of CDK2. As the expression of p21 is regulated by p53-dependent transcription, the absence of p53 abrogates p21-dependent repression of CDK2 and subsequently leads to centrosome duplication. The DNA synthesis inhibitor, hydroxyurea (HU), induces cell-cycle arrest at the G1/S phase. Cells possessing wild-type p53 prevent HU-induced overduplication of centrosomes by inhibiting CDK2 through p53/p21. In contrast, the absence of functional p53 abolishes the p21-dependent repression of CDK2, leading to centrosome amplification. p53

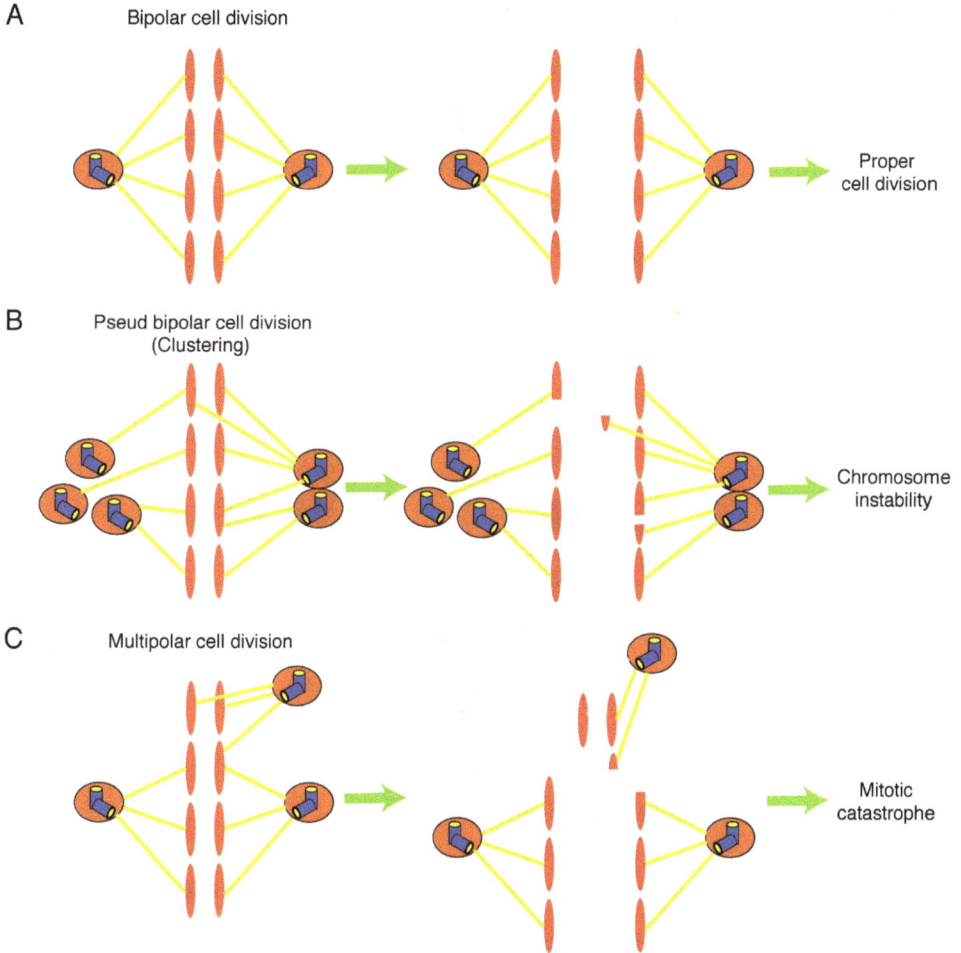

(A) Two centrosomes separate at the two poles, and normal cell division progresses. (B) Overduplicated centrosomes accumulate into the two poles and form a pseudo-bipolar spindle, leading to improper cell division and chromosome instability. (C) Overduplicated centrosomes form a multipolar spindle, leading to a failure of cytokinesis and mitotic catastrophe.

Fig. 2. Cell division during mitosis with normal and abnormal number of centrosomes.

also contributes to abrogation of the linkage between the cell cycle and the centrosome duplication cycle because the p53-dependent G2/M checkpoint is activated in an ataxia telangiectasia mutated (ATM)/ATM- and Rad3-related (ATR)-dependent manner after DNA damage such as from irradiation.

4. Centrosomes and DNA-repair proteins

DNA-repair-related proteins, including ATM, ATR, checkpoint kinase 1 (CHK1), CHK2, PARP1, Nijmegen breakage syndrome (NBS1), BRCA1, BRCA2, RAD51, RAD51 paralogs, and TOPBP1 localize at centrosomes, and defects in these proteins cause several functional aberrations in centrosomes (Fig. 4).

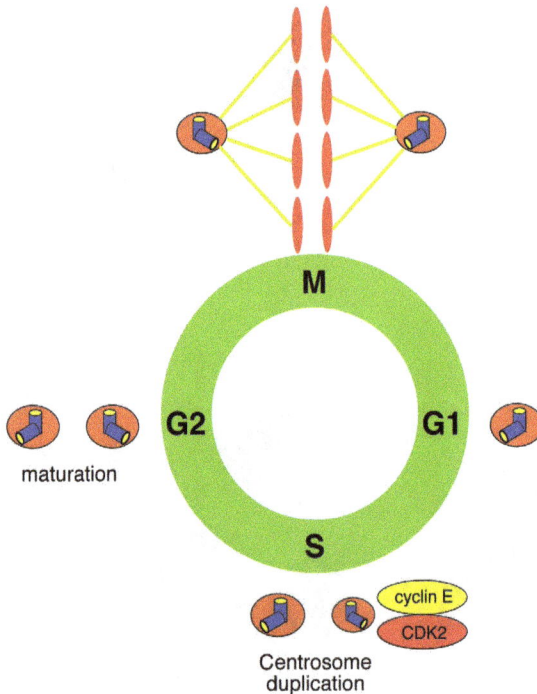

Fig. 3. Centrosome duplication and the cell cycle.

Centrosomes are duplicated once in S phase and mature in G2 phase. In M phase, centrosomes divide into daughter cells. Cyclin E/CDK2 activity is important for centrosome duplication.

4.1 ATM and ATR

ATM and ATR, central protein kinases in the DNA damage response (Bensimon et al., 2010), phosphorylate CEP63, a centrosomal protein, leading to proper control of spindle assembly after DNA damage (Smith et al., 2009). Rad51-deficient chicken DT40 cell lines show centrosome amplification, but a Rad51/Atm-double knockout DT40 cell line revealed a decrease in centrosome amplification compared to Rad51-single knockout cell lines (Dodson et al., 2004). Furthermore, treating Rad51-deficient cells with wortmannin or caffeine, inhibitors of ATM and ATR, results in a decrease in centrosome amplification. These results suggest that ATM could contribute to centrosome amplification in Rad51-deficient cells by regulating the G2/M checkpoint or an unknown function in the centrosome duplication pathway. ATR is mutated in some individuals with Seckel Syndrome (ATR Seckel), which is an autosomal recessive disorder that includes intrauterine growth retardation and microcephaly. Seckel syndrome patient cells have aberrant centrosome and checkpoint regulation (Alderton et al., 2004). *Pericentrin* is a mutated gene in PCNT Seckel syndrome

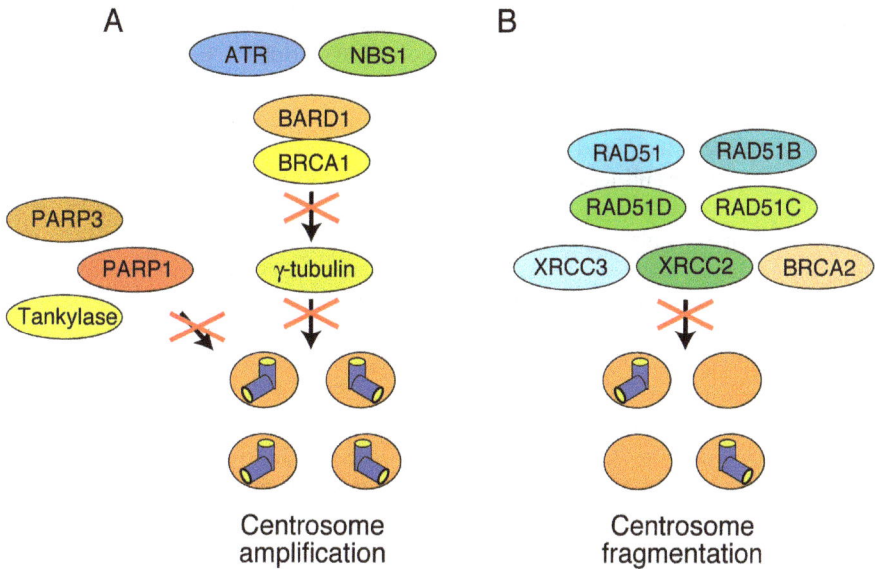

(A) Ataxia telangiectasia mutated (ATM) and Rad3-related (ATR), Nijmegen breakage syndrome (NBS)1, and BRCA1-BARD1 complex-dependent γ-tubulin monoubiquitination is important for centrosome duplication. Defects in these proteins result in centrosome amplification. A defect in PARP-1, PARP-3, or tankylase also leads to centrosome amplification. (B) In contrast, a defect in RAD51, RAD51 paralogs, or BRCA2 results in centrosome fragmentation.

Fig. 4. DNA-repair proteins and centrosome maintenance.

(Griffith et al., 2008; Rauch et al., 2008) that is involved in the ATR-dependent DNA damage signaling pathway. Exposure to UV light or HU induces the activation of ATR, and activated ATR phosphorylates CHK1. Phosphorylated CHK1 accumulates at the centrosome and its localization causes inhibition of Cdc25 activity, which prevents activation of cyclin B/CDK1. Hence, the ATR-dependent G2/M checkpoint may contribute to centrosome amplification.

4.2 CHK1, CHK2

CHK1 is an essential gene for mammalian cells and functions in the cell-cycle checkpoint (Shimada et al., 2008). Loss of functional Chk1 in human or chicken cell lines causes a G2/M checkpoint deficiency and increased sensitivity to DNA damage treatment (Bourke et al., 2007). CHK1 localizes at the centrosome. Chk1-deficient chicken cell lines abolish irradiation-induced centrosome amplification. CHK1 interacts with MCPH1 and pericentrin in the centrosome, and MCPH1 knockdown decreases the accumulation of Chk1 and pericentrin in centrosomes (Tibelius et al., 2009). These results suggest that CHK1 accumulation in the centrosome is dependent on MCPH1. Thus, Chk1 participates in the regulation of centrosome number through checkpoint control or phosphorylation of unknown substrates by Chk1.

Chk2 is another important cell-cycle checkpoint kinase that is activated in response to DNA damage (Tsvetkov et al., 2003; Golan et al., 2010). Chk2 and Plk1, which are mitotic kinases, co-localize at the centrosome. This interaction may be important for the DNA mitotic damage-dependent checkpoint, although the details remain unknown.

4.3 BRCA1 and BRCA2

About 10% of women diagnosed with breast cancer have inherited mutations in *BRCA1* or *BRCA2* (Irminger-Finger and Jefford, 2006). Both *BRCA1* and *BRCA2*, products of the familial breast cancer susceptibility gene, are involved in several cellular functions, such as DNA repair, transcriptional regulation, cell-cycle checkpoints, and centrosome maintenance. BRCA1 forms a heterodimer complex with BRCA1-associated RING domain (BARD1), which functions as an E3 ubiquitin ligase. Both BRCA1 and BARD1 contain a RING domain, which mediates DNA–protein and protein–protein interactions, a nuclear export signal sequence at their N-terminus, and tandem BRCT (BRCA1 carboxy-terminal) domains. The BRCA1–BARD1 complex monoubiquitylates γ-tubulin at Lysine 48 and Lysine 344, and overexpression of mutated γ-tubulin at the K48 ubiquitination site results in centrosome amplification and aberration of microtubule nucleation (Starita et al., 2005; Simons et al., 2006). Overexpression of mutated γ-tubulin (K344R) results in an aberration of microtubule nucleation only, suggesting that BRCA1 controls centrosome function by monoubiquitination of γ-tubulin. The BRCA1–BARD1 complex also ubiquitylates the nucleolar phosphoprotein nucleophosmin (NPM also known as B23), which functions in nucleolar organization, cell-cycle regulation, and centrosome duplication. The BRCA1–BARD1 complex polyubiquitinates NPM, leading to its degradation. Aurora A, which localizes at the centrosome and is an important factor for mitotic progression, phosphorylates BRCA1, which contributes to regulation of centrosome duplication. Furthermore, the BRCA1–BARD1 complex regulates microtubule organization through a Ran-dependent import pathway.

A BRCA2 mutation is involved in approximately 50% of hereditary breast cancers (Yoshida and Miki, 2004). BRCA2 has no sequential or structural similarity with either BRCA1 or BARD1 and localizes at the centrosome. Interaction of BRCA2 with plectin, a cytoskeletal cross-linker protein, is necessary for centrosome anchoring to the nucleus (Niwa et al., 2009). BRCA2 also forms a complex at the centrosome with NPM and ROCK2, an effector of Rho small GTPase. A definite BRCA2 deletion can abrogate the association of BRCA2 with NPM, and cells expressing this deletion mutant show centrosome amplification (Wang et al., 2011), suggesting that the BRCA2–NPM complex maintains centrosome duplication and controls cell division.

4.4 NBS1

NBS, which is caused by an *NBS1* gene mutation, is characterized by growth retardation, a birdlike face, immunodeficiency, predisposition to malignancy, and microcephaly (Matsuura et al., 1998). NBS patient cells have a defect in the cell-cycle checkpoint and hyper-radiosensitivity. NBS1 is a multifunctional protein that participates in homologous recombination repair, DNA replication, the cell-cycle checkpoint, and apoptosis (Tauchi et al., 2002). NBS1 forms a complex with MRE11 and RAD50 (MRN complex), and this complex is required for recruitment of ATM to DNA damage sites and for efficient phosphorylation of ATM substrates (Iijima et al., 2008). NBS1 contains a forkhead-associated (FHA) domain and a BRCT domain at the N-terminus, the binding motif for MRE11, ATM, and RNF20, which is a E3 ubiquitin ligase for H2B, at the C-terminus (Nakamura et al., 2011). The NBS1 FHA domain is required for ATR interaction (Shimada et al., 2009). NBS1 knockdown by siRNA in human or mouse cells causes centrosome amplification and decreases BRCA1-dependent monoubiquitination of γ-tubulin. Furthermore, the NBS1 N-terminus, which interacts with ATR, is indispensable for the monoubiquitination of γ-tubulin. NBS1 potentially plays a role in genome integrity via centrosome and nucleus volume control (Shimada and Komatsu, 2009; Shimada et al., 2010).

4.5 PARP family

PARP1 catalyzes the formation of long branched polyADP-ribosylation covalently attached to target proteins using NAD$^+$ as a substrate. Many proteins are poly(ADP-ribosyl)ated by PARP1, and this modification may be involved in transcriptional regulation and DNA repair (Miwa and Masutani, 2007). PARP1$^{-/-}$ mouse cell lines show centrosome amplification (Kanai et al., 2003). Other PARP family proteins, such as PARP3 and tankylase (also known as PARP5a), localize at the centrosome (Smith and de Lange, 1999; Augustin et al., 2003). These reports suggest that PARP family proteins are involved in the control of centrosome duplication.

4.6 RAD51 paralogs

RAD51 and five paralogs, RAD51B (RAD51L1), RAD51C (RAD51L2), RAD51D (RAD51L3), XRCC2, and XRCC3, play important roles in homologous recombination (HR) repair (Date et al., 2006; Renglin Lindh et al., 2007; Cappelli et al., 2011). These proteins have a consensus domain including Walker A and B ATPase domains and are necessary for chromosome stability and the control of chromosome segregation. In mammalian cells, XRCC2 forms a complex with RAD51B and RAD51C, and XRCC3 forms a complex with RAD51C. The XRCC2 complex is involved in the RAD51 loading step to ssDNA in HR repair. The XRCC3 complex is involved in Holliday junction resolution. Loss of RAD51, RAD51B, RAD51C, RAD51D, XRCC2, or XRCC3 leads to centrosome amplification and chromosome instability. RAD51C, XRCC2, or XRCC3-deficient cell lines show centrosome amplification in the M phase, but only XRCC2-deficient cell lines show centrosome amplification at interphase (Renglin Lindh et al., 2007), suggesting that RAD51C and XRCC3, but not XRCC2, may be involved in the same centrosome duplication pathway.

4.7 Nonhomologous end-joining repair proteins

Nonhomologous end-joining (NHEJ) repair proteins such as DNA-PKcs also localize at centrosomes (Zhang et al., 2007). Our previous reports showed that DNA-PKcs-deficient cell lines (SCID) have a slightly increased centrosome number compared to wild-type cell lines. Moreover, another NHEJ factor, Ku70, found in a Ku70-deficient cell line also has a slight increase in centrosome number compared to complementary cell lines (Shimada et al., 2010), indicating that NHEJ factors may be involved in centrosome functions different from HR factors.

4.8 Other DNA-repair-related proteins

TopBP1, a sensor protein involved in the DNA damage response, localizes at the centrosome during mitosis but not at interphase (Bang et al., 2011). TopBP1 interacts with the centrosome through its C-terminus and eliminates TopBP1 localization, resulting in a delay in mitotic progression. SMC1, a condensin protein important during chromosome condensation, also localizes at the centrosome but its role in centrosome maintenance is unclear.

5. Conclusion

DNA-repair proteins are necessary for genome integrity. Their main functions are to control DNA repair and control the cell-cycle checkpoint. Recent studies have not clarified the role of DNA damage repair proteins in centrosome maintenance, although interactions between DNA-repair proteins and centrosomal proteins may have an important role in centrosome maintenance and microtubule regulation such as ATM/ATR-dependent CEP63 phosphorylation. How these interactions contribute to centrosome maintenance and microtubule regulation is unclear, so investigating the relationship between DNA-repair proteins and centrosomal proteins is important. Furthermore, the linkage between centrosome amplification and tumorigenesis is key to developing clinical targets. Inhibitors of the DNA-repair protein PARP-1 and the centrosomal protein Aurora A could be a focus for anticancer drugs. Investigations into the molecular signaling pathway of DNA-repair proteins during centrosome maintenance may contribute to advanced options for clinical therapeutics.

6. Acknowledgments

This work was supported by grants from the Ministry of Education, Culture, Sports, Science, and Technology, Japan. We thank M. Takado for critical reading of the manuscript.

7. References

Alderton, G.K., H. Joenje, R. Varon, A.D. Borglum, P.A. Jeggo, and M. O'Driscoll. 2004. Seckel syndrome exhibits cellular features demonstrating defects in the ATR-signalling pathway. *Hum Mol Genet*. 13:3127–3138.

Augustin, A., C. Spenlehauer, H. Dumond, J. Menissier-De Murcia, M. Piel, A.C. Schmit, F. Apiou, J.L. Vonesch, M. Kock, M. Bornens, and G. De Murcia. 2003. PARP-3 localizes preferentially to the daughter centriole and interferes with the G1/S cell cycle progression. *J Cell Sci*. 116:1551–1562.

Bang, S.W., M.J. Ko, S. Kang, G.S. Kim, D. Kang, J. Lee, and D.S. Hwang. 2011. Human TopBP1 localization to the mitotic centrosome mediates mitotic progression. *Exp Cell Res*. 317:994–1004.

Bensimon, A., A. Schmidt, Y. Ziv, R. Elkon, S.Y. Wang, D.J. Chen, R. Aebersold, and Y. Shiloh. 2010. ATM-dependent and -independent dynamics of the nuclear phosphoproteome after DNA damage. *Sci Signal*. 3:rs3.

Bourke, E., H. Dodson, A. Merdes, L. Cuffe, G. Zachos, M. Walker, D. Gillespie, and C.G. Morrison. 2007. DNA damage induces Chk1-dependent centrosome amplification. *EMBO Rep*. 8:603–609.

Boveri, T. 2008. Concerning the origin of malignant tumours by Theodor Boveri. Translated and annotated by Henry Harris. *J Cell Sci*. 121 Suppl 1:1–84.

Cappelli, E., S. Townsend, C. Griffin, and J. Thacker. 2011. Homologous recombination proteins are associated with centrosomes and are required for mitotic stability. *Exp Cell Res*. 317:1203–1213.

Date, O., M. Katsura, M. Ishida, T. Yoshihara, A. Kinomura, T. Sueda, and K. Miyagawa. 2006. Haploinsufficiency of RA-51B causes centrosome fragmentation and aneuploidy in human cells. *Cancer Res*. 66:6018–6024.

Dodson, H., E. Bourke, L.J. Jeffers, P. Vagnarelli, E. Sonoda, S. Takeda, W.C. Earnshaw, A. Merdes, and C. Morrison. 2004. Centrosome amplification induced by DNA damage occurs during a prolonged G2 phase and involves ATM. *EMBO J*. 23:3864–3873.

Doxsey, S. 2001. Re-evaluating centrosome function. *Nat Rev Mol Cell Biol.* 2:688–698.

Fukasawa, K. 2007. Oncogenes and tumour suppressors take on centrosomes. *Nat Rev Cancer.* 7:911–924.

Ganem, N.J., S.A. Godinho, and D. Pellman. 2009. A mechanism linking extra centrosomes to chromosomal instability. *Nature.* 460:278–282.

Godinho, S.A., M. Kwon, and D. Pellman. 2009. Centrosomes and cancer: how cancer cells divide with too many centrosomes. *Cancer Metastasis Rev.* 28:85–98.

Golan, A., E. Pick, L. Tsvetkov, Y. Nadler, H. Kluger, and D.F. Stern. 2010. Centrosomal Chk2 in DNA damage responses and cell cycle progession. *Cell Cycle.* 9:2645–2654.

Griffith, E., S. Walker, C.A. Martin, P. Vagnarelli, T. Stiff, B. Vernay, N. Al Sanna, A. Saggar, B. Hamel, W.C. Earnshaw, P.A. Jeggo, A.P. Jackson, and M. O'Driscoll. 2008. Mutations in pericentrin cause Seckel syndrome with defective ATR-dependent DNA damage signaling. *Nat Genet.* 40:232–236.

Iijima, K., M. Ohara, R. Seki, and H. Tauchi. 2008. Dancing on damaged chromatin: functions of ATM and the RAD50/MRE11/NBS1 complex in cellular responses to DNA damage. *J Radiat Res (Tokyo).* 49:451–464.

Irminger-Finger, I., and C.E. Jefford. 2006. Is there more to BARD1 than BRCA1? *Nat Rev Cancer.* 6:382–391.

Kanai, M., W.M. Tong, E. Sugihara, Z.Q. Wang, K. Fukasawa, and M. Miwa. 2003. Involvement of poly(ADP-Ribose) polymerase 1 and poly(ADP-Ribosyl)ation in regulation of centrosome function. *Mol Cell Biol.* 23:2451–2462.

Kwon, M., S.A. Godinho, N.S. Chandhok, N.J. Ganem, A. Azioune, M. Thery, and D. Pellman. 2008. Mechanisms to suppress multipolar divisions in cancer cells with extra centrosomes. *Genes Dev.* 22:2189–2203.

Matsumoto, Y., and J.L. Maller. 2002. Calcium, calmodulin, and CaMKII requirement for initiation of centrosome duplication in *Xenopus* egg extracts. *Science.* 295:499–502.

Matsumoto, Y., K. Hayashi, and E. Nishida. 1999. Cyclin-dependent kinase 2 (Cdk2) is required for centrosome duplication in mammalian cells. *Curr Biol.* 9:429–432.

Matsuura, S., H. Tauchi, A. Nakamura, N. Kondo, S. Sakamoto, S. Endo, D. Smeets, B. Solder, B.H. Belohradsky, V.M. Der Kaloustian, M. Oshimura, M. Isomura, Y. Nakamura, and K. Komatsu. 1998. Positional cloning of the gene for Nijmegen breakage syndrome. *Nat Genet.* 19:179–181.

Miwa, M., and M. Masutani. 2007. PolyADP-ribosylation and cancer. *Cancer Sci.* 98:1528–1535.

Nakamura, K., A. Kato, J. Kobayashi, H. Yanagihara, S. Sakamoto, D.V. Oliveira, M. Shimada, H. Tauchi, H. Suzuki, S. Tashiro, L. Zou, and K. Komatsu. 2011. Regulation of homologous recombination by RNF20-dependent H2B ubiquitination. *Mol Cell.* 41:515–528.

Nigg, E.A. 2002. Centrosome aberrations: cause or consequence of cancer progression? *Nat Rev Cancer.* 2:815–825.

Nigg, E.A. 2007. Centrosome duplication: of rules and licenses. *Trends Cell Biol.* 17:215–221.

Niwa, T., H. Saito, S. Imajoh-ohmi, M. Kaminishi, Y. Seto, Y. Miki, and A. Nakanishi. 2009. BRCA2 interacts with the cytoskeletal linker protein plectin to form a complex controlling centrosome localization. *Cancer Sci.* 100:2115–2125.

Rauch, A., C.T. Thiel, D. Schindler, U. Wick, Y.J. Crow, A.B. Ekici, A.J. van Essen, T.O. Goecke, L. Al-Gazali, K.H. Chrzanowska, C. Zweier, H.G. Brunner, K. Becker, C.J. Curry, B. Dallapiccola, K. Devriendt, A. Dorfler, E. Kinning, A. Megarbane, P. Meinecke, R.K. Semple, S. Spranger, A. Toutain, R.C. Trembath, E. Voss, L. Wilson,

R. Hennekam, F. de Zegher, H.G. Dorr, and A. Reis. 2008. Mutations in the pericentrin (*PCNT*) gene cause primordial dwarfism. *Science*. 319:816–819.

Renglin Lindh, A., N. Schultz, N. Saleh-Gohari, and T. Helleday. 2007. RAD51C (RAD51L2) is involved in maintaining centrosome number in mitosis. *Cytogenet Genome Res*. 116:38–45.

Sato, N., K. Mizumoto, M. Nakamura, H. Ueno, Y.A. Minamishima, J.L. Farber, and M. Tanaka. 2000. A possible role for centrosome overduplication in radiation-induced cell death. *Oncogene*. 19:5281–5290.

Shimada, M., and K. Komatsu. 2009. Emerging connection between centrosome and DNA repair machinery. *J Radiat Res (Tokyo)*. 50:295–301.

Shimada, M., H. Niida, D.H. Zineldeen, H. Tagami, M. Tanaka, H. Saito, and M. Nakanishi. 2008. Chk1 is a histone H3 threonine 11 kinase that regulates DNA damage-induced transcriptional repression. *Cell*. 132:221–232.

Shimada, M., R. Sagae, J. Kobayashi, T. Habu, and K. Komatsu. 2009. Inactivation of the Nijmegen breakage syndrome gene leads to excess centrosome duplication via the ATR/BRCA1 pathway. *Cancer Res*. 69:1768–1775.

Shimada, M., J. Kobayashi, R. Hirayama, and K. Komatsu. 2010. Differential role of repair proteins, BRCA1/NBS1 and Ku70/DNA-PKcs, in radiation-induced centrosome overduplication. *Cancer Sci*. 101:2531–2537.

Simons, A.M., A.A. Horwitz, L.M. Starita, K. Griffin, R.S. Williams, J.N. Glover, and J.D. Parvin. 2006. BRCA1 DNA-binding activity is stimulated by BARD1. *Cancer Res*. 66:2012–2018.

Smith, E., D. Dejsuphong, A. Balestrini, M. Hampel, C. Lenz, S. Takeda, A. Vindigni, and V. Costanzo. 2009. An ATM- and ATR-dependent checkpoint inactivates spindle assembly by targeting CEP63. *Nat Cell Biol*. 11:278–285.

Smith, S., and T. de Lange. 1999. Cell cycle dependent localization of the telomeric PARP, tankyrase, to nuclear pore complexes and centrosomes. *J Cell Sci*. 112 (Pt 21):3649–3656.

Starita, L.M., A.A. Horwitz, M.C. Keogh, C. Ishioka, J.D. Parvin, and N. Chiba. 2005. BRCA1/BARD1 ubiquitinate phosphorylated RNA polymerase II. *J Biol Chem*. 280:24498–24505.

Tauchi, H., J. Kobayashi, K. Morishima, D.C. van Gent, T. Shiraishi, N.S. Verkaik, D. vanHeems, E. Ito, A. Nakamura, E. Sonoda, M. Takata, S. Takeda, S. Matsuura, and K. Komatsu. 2002. Nbs1 is essential for DNA repair by homologous recombination in higher vertebrate cells. *Nature*. 420:93–98.

Tibelius, A., J. Marhold, H. Zentgraf, C.E. Heilig, H. Neitzel, B. Ducommun, A. Rauch, A.D. Ho, J. Bartek, and A. Kramer. 2009. Microcephalin and pericentrin regulate mitotic entry via centrosome-associated Chk1. *J Cell Biol*. 185:1149–1157.

Tsvetkov, L., X. Xu, J. Li, and D.F. Stern. 2003. Polo-like kinase 1 and Chk2 interact and co-localize to centrosomes and the midbody. *J Biol Chem*. 278:8468–8475.

Wang, H.F., K. Takenaka, A. Nakanishi, and Y. Miki. 2011. BRCA2 and nucleophosmin coregulate centrosome amplification and form a complex with the Rho effector kinase ROCK2. *Cancer Res*. 71:68–77.

Yoshida, K., and Y. Miki. 2004. Role of BRCA1 and BRCA2 as regulators of DNA repair, transcription, and cell cycle in response to DNA damage. *Cancer Sci*. 95:866–871.

Zhang, S., P. Hemmerich, and F. Grosse. 2007. Werner syndrome helicase (WRN), nuclear DNA helicase II (NDH II) and histone gammaH2AX are localized to the centrosome. *Cell Biol Int*. 31:1109–1121.

5

Mitochondrial DNA Damage: Role of Ogg1 and Aconitase

Gang Liu[1] and David W. Kamp[2]
[1]Clinical Medicine Research Center, Affiliated Hospital of Guangdong Medical College
[2]Department of Medicine, Northwestern University Feinberg School of Medicine and Jesse
Brown VA Medical Center
[1]PR China
[2]USA

1. Introduction

Mitochondria have a vital role in respiration-coupled energy production, amino acid and fatty acid metabolism, Fe^{2+}/Ca^{2+} homeostasis and the integration of apoptotic signals that regulate cellular life and death (Babcock et al., 1997; Loeb et al., 2005; Taylor & Turnbull, 2005; Kroemer et al., 2007). Given the importance of these cellular functions regulated by the mitochondria with implications for aging, degenerative diseases and carcinogenesis, it is not surprising that this organelle has been the subject of intensive investigation for decades and continues to challenge investigators. Mitochondria produce nearly 90% of all the energy made in the body by oxidative phosphorylation that occurs via the electron transport chain (ETC). Mitochondria are the major cellular site of reactive oxygen species (ROS) production. It is estimated that 1–5% of the oxygen consumed in the mitochondrial ETC is converted to ROS (Kroemer et al., 2007). Mammalian mitochondria have a covalently closed round mitochondrial DNA (mtDNA) that is replicated and expressed within the mitochondria in close proximity to the ETC and potentially damaging ROS (Clayton 1982; Clayton 1984; Kroemer et al., 2007). Mammalian mtDNA contains 37 genes that encode 13 proteins (all of which are involved in the ETC), 22 tRNAs, and 2 rRNAs (Anderson et al., 1981). The remaining mitochondrial ETC proteins, the metabolic enzymes, the DNA and RNA polymerases and the ribosomal proteins are all encoded by nuclear genome.

Oxidative stress-induced mtDNA damage is implicated in a wide range of pathologic processes including carcinogenesis, aging and degenerative diseases of various organs and tissues (Bohr et al., 2002; Van Houten et al., 2006; Kroemer et al., 2007; Gredilla et al., 2010). In this review, we summarize the evidence that mtDNA damage augments mitochondria-regulated (intrinsic) apoptosis; an event that underlies the pathophysiologic mechanisms of diverse diseases. We focus our attention on one form of oxidative stress, exposure to asbestos fibers, which are well known to cause pulmonary fibrosis (asbestosis) and malignancies (e.g. mesothelioma and lung cancer). Specifically, we examine the role of a mitochondrial oxidative DNA repair enzyme (8-oxoguanine DNA glycosylase; Ogg1) and a recently described novel mechanism whereby mitochondrial Ogg1 acts as a mitochondrial aconitase chaperone protein to prevent oxidant-induced alveolar epithelial cell (AEC) mitochondrial dysfunction and intrinsic apoptosis. We discuss studies showing that mitochondrial aconitase, a crucial Kreb cycle enzyme, also functions in mtDNA maintenance and are a mitochondrial redox-sensor that is susceptible to oxidative degradation. Finally, we review accumulating evidence for important crosstalk between p53, which is a crucial DNA damage response protein, Ogg1, mtDNA damage and apoptosis.

2. MtDNA damage: Role of mitochondrial ROS

Individual cells contain several thousand copies of mtDNA, and in normal individuals, almost all of the mtDNA is similar. However, in some cases, especially in mitochondrial diseases, wild-type and variant mtDNAs coexist. The mutation rate of mtDNA is several folds higher in mtDNA than nuclear DNA (Bohr et al 2002; Van Houten et al 2006; Gredilla et al., 2010). There are three reasons for the high mutation rate in mtDNA. The first is that mtDNA, which is located along the mitochondrial inner membrane, is vulnerable to ROS-induced damage due to its close proximity to high levels of ROS produced from the ETC (Nass 1969; Albring et al., 1977; Chance et al., 1979; Shigenaga et al., 1994; Gredilla et al., 2010). The second reason is that mtDNA has no histone-containing protein shield as does the nuclear genome, so that mtDNA is uniquely susceptible to ROS-induced stress. Finally, mitochondria have a limited DNA repair systems as compared to what is present in the nucleus (see for review: Gredilla at al 2010). Collectively, these conditions cause mtDNA to accumulate various somatic mtDNA mutations in mitotic (Michikawa et al., 1999) and post-mitotic tissues (Soong et al., 1992; Corral-Debrinski et al., 1992; Liu et al., 1998). Mitochondrial DNA mutations and insertions/deletions have been observed in many types of human cancer (Bohr et al 2002). Mitochondrial functional defects have also been observed due to abnormal expression of mtDNA encoded proteins and defective oxidative phosphorylation (Kroemer et al 2007). Mitochondrial dysfunction and mtDNA mutations are also implicated in the development and complications of diabetic cardiomyopathy as well as directly associated with different types of neurodegenerative diseases (Medikayala et al., 2011). An emerging regulatory role for mitochondrial topoisomerases appears important for mtDNA integrity in the myocardium (Medikayala et al., 2011).

The most frequently formed mitochondrial ROS are hydrogen peroxide (H_2O_2), superoxide anion ($O_2\bullet^-$), singlet oxygen, and hydroxyl radicals ($OH\bullet$). Nearly 1-5% of the total molecular oxygen utilized by mammalian mitochondria is converted into ROS (Boveris & Chance 1977). Not surprisingly, mitochondria are one of the main cellular targets of oxidative damage resulting in relatively high levels of oxidized proteins, lipids and nucleic acids in mammalian mitochondria under normal metabolic conditions (see for reviews Raha & Robinson, 2000; Kroemer et al 2007). Generation of ROS produce a variety of lesions in cellular DNA, such as single or double strand breaks, intra- and inter-strand cross-linking and base damage (see for reviews Upadhyay & Kamp, 2003; Gredilla et al., 2010). Persistent DNA damage can cause cell cycle arrest, induction of transcription, induction of signal transduction pathways, replication errors, and genomic instability. Mitochondrial ROS can induce oxidative mitochondrial as well as nuclear DNA damage that results in apoptosis, if cells survive, promotes DNA mutations. For example, DNA damage is an early event in asbestos-exposed cells that can trigger apoptosis by inducing mitochondrial ROS production that may in part account for its malignant potential (see for reviews Kamp et al., 1992; Hardy & Aust, 1995; Jaurand 1997; Shukla et al., 2003; Liu et al., 2010).

Mitochondrial-associated gene expression, which is significantly different in cancer cells as compared to normal cells, identifies the changes in mitochondrial function emerging in developing cancer cells (see for review Ralph et al., 2010). Cancer cell development is dependent on the interactions of key oncogenes and tumor suppressor genes and their encoded products (see for review Janicke et al., 2008; Ralph et al., 2010). Studies in yeast show that mtDNA mutations can either reduce or extend life span depending upon the severity, context, and developmental stage of mtDNA damage (Powell et al., 2000). Unexpectedly, complete absence of mtDNA in yeast is associated with increased life span (Powell et al., 2000). Mice with a homozygous mutation in the exonuclease domain of mtDNA polymerase gamma (POLG) have been used as a model of mitochondrial dysfunction and aging. These mice possess an mtDNA mutator phenotype, accumulating lot of deletions and point mutations in mtDNA. These mice do not display signs of elevated ROS generation, but instead exhibit increased apoptosis, a number of age-related phenotypes, and a shortened life span (Kujoth et al., 2005; Trifunovic et al., 2004). As recently reviewed elsewhere (Kamp et al., 2011), chronic inflammation can promote all stages of tumorigenesis including mtDNA damage important in regulating mitochondrial function that coordinates life and death signaling pathways. Lung mesothelial cell mtDNA damage is evident following exposure to a four-fold lower concentration of crocidolite asbestos than required for causing nuclear DNA damage (Shukla et al., 2003). Several lines of evidence implicate mtDNA oxidative injury as a key trigger of apoptosis that can result in inflammation-associated cancer including: (1) cell death is often associated with mtDNA oxidative lesions, (2) mtDNA damage result in ATP depletion and mitochondrial dysfunction, (3) enhancing mtDNA repair can prevent cell death, and (4) defective mtDNA repair enhances cell death (see for review Kamp et al., 2011).

Apoptosis, or programmed cell death, is an important mechanism by which cells with extensive DNA damage, including mtDNA damage, are eliminated without inciting an inflammatory response. Notably, cell-sorting experiments established that persistent mtDNA damage is necessary for triggering mitochondrial dysfunction and apoptosis (Santos et al., 2003). Although much is known about the complex molecular pathways regulating apoptosis, the precise mechanisms involved remain incompletely understood (see for reviews: Kroemer et al., 2007; Kim et al., 2008; Youle & Strasser, 2008; Franco et al., 2009). The two major pathways regulating apoptosis include the mitochondria (intrinsic) death pathway and the death receptor (extrinsic) pathway. The intrinsic death pathway is activated by various stimuli, such as ROS, DNA damage, and calcium, that result in permeabilization of the outer mitochondrial membrane (OMM), a reduction in mitochondrial membrane potential change ($\triangle\psi m$) and the release of apoptogenic proteins, including cytochrome c that activate caspase-9 and, ultimately caspase-3. Notably, mtDNA damage that occurs following oxidative stress or mutations in mitochondrial DNA polymerase are implicated in premature aging as well as tumor metastasis (Trifunovic et al., 2004; Ishikawa et al., 2008)

ROS and DNA damage, including that caused by asbestos, trigger intrinsic apoptosis that can be blocked by antioxidants and iron chelators (Kroemer et al., 2007; Youle & Strasser, 2008; Franco et al., 2009; Kamp et al., 1995; Aljandali 2001; Panduri 2003; Panduri 2004). Herein we focus on asbestos-induced apoptosis to lung cells. Accumulating evidence over the past decade convincingly demonstrate that all forms of asbestos fibers, as opposed to inert particulates (e.g. titanium dioxide [TiO_2]), cause apoptosis in AEC as well as mesothelial cells via the mitochondria-regulated death pathway (reviewed in Kamp et al., 2011). Our group used human A549 cell and rat primary cells isolated alveolar type II to show that asbestos causes both a dose- and time-dependent reduction in $\triangle\psi$m that was associated with release of cytochrome c from the mitochondria to the cytoplasm as well as activation of caspase-9 (Panduri et al., 2003). In this study, both an iron chelator (phytic acid) and a free radical scavenger (sodium benzoate) blocked asbestos-induced reductions in $\triangle\psi$m and caspase-9, implying the importance of both iron-derived ROS and the mitochondrial death pathway. Furthermore, asbestos-induced apoptosis in A549 cells that stably overexpress Bcl-xl, an anti-apoptotic Bcl-2 family member, was significantly attenuated as compared to wild-type cells as evidenced by preservation of the OMM integrity and reduced DNA fragmentation (Panduri et al., 2003). Using confocal microscopy and Western blotting of mitochondrial proteins, we showed that asbestos stimulates mitochondrial translocation of pro-apoptotic Bax and that these effects are blocked by phytic acid (Panduri et al., 2006). Notably, using A549-ρ0 cells that lack mtDNA and a functional electron transport chain necessary for mitochondrial ROS generation, asbestos-induced ROS production, caspase-9 activation, and intrinsic apoptosis were all completely blocked (Panduri et al., 2006). These findings establish an important role for mitochondrial ROS in mediating asbestos-induced AEC apoptosis.

3. Ogg1 and mitochondrial base excision repair

Oxidative stress can induce many types of DNA base damage including two of the most abundant lesions, 8-hydroxyguanine (8-oxoG) and thymine glycol (TG) (Demple & Harrison, 1994; Dizdaroglu 1992; Bohr et al., 2002; Gredilla et al., 2010). Further, 8-oxoG is more susceptible to oxidative attack than guanine itself, resulting in the formation of oxidation products such as guanidinohydantoin and spiroiminodihydantoin (Bjelland & Seeberg, 2003; Hailer et al., 2005). The 8-oxoG residue exists predominantly in its keto form at physiological pH, resulting in the normal anti conformation around the N-glycosylic bond, and forming a common Watson-Crick base pair with cytosine. 8-oxoG adopts a *syn* conformation and base pairs with adenine leading to transversion mutations in replicating cells (Shibutani 1991), which may play a role in the development of cancer and the process of aging (Ames 1989; Lindahl 1993). In contrast, TG strongly blocks DNA replication (Ide et al., 1985; Clark & Beardsley, 1987) and transcription (Hatahet et al., 1994; Htun & Johnston, 1992) and must be efficiently removed and repaired to maintain genetic stability. Therefore, inefficient repair of oxidative mtDNA damage augments the accumulation of mtDNA damage and mutations that can lead to mitochondrial dysfunction and apoptosis. In this section we focus attention on repair of 8-oxoG by mitochondrial 8-oxoguanine DNA glycosylase 1 (mt-Ogg1) since it is among the best characterized mitochondrial base excision repair (BER) proteins.

The BER pathway accounts for the repair of the majority of spontaneously formed oxidized bases in mtDNA important for preserving the genome stability required for long-term cell survival (Barnes & Lindahl, 2004; Gredilla et al. 2010). All mitochondrial DNA repair enzymes, including those involved in BER, are encoded in the nucleus and imported into the mitochondria (Gredilla et al. 2010). The BER pathway removes small covalent modifications, which do not distort the DNA helix, such as the base modifications generated by ROS and single-strand breaks. The BER pathway in mitochondria and nucleus is highly conserved in all cellular organisms, from bacteria to man. BER is carried out in four sequential enzymatic steps catalyzed by the enzymes DNA glycosylase, AP-endonuclease, DNA polymerase and DNA ligase (Dianova et al., 2001; Gredilla et al., 2010). The initial steps in the BER pathway are recognition and removal of the aberrant base by a DNA glycosylase. Most DNA glycosylases remove several structurally different damaged bases, and some of them have overlapping substrate specificities, which may indicate that they serve as back-up systems for each other (Dianovet al., 2001). The mammalian DNA glycosylase, Ogg1, recognizes and removes 8-oxoG that is base-paired with cytosine in DNA (Aburatani et al., 1997; Radicella et al., 1997). Ogg1 is a bifunctional DNA glycosylase, with an associated AP-lyase activity, cleaving DNA at abasic sites through a β-elimination mechanism (Bjoras et al., 1997). The human OGG1 gene is located on chromosome 3p26.2. Studies of mice that are deficient in Ogg1 demonstrate that this enzyme is responsible for most of the BER activity that is initiated at 8-oxoG in mammalian cells (Klungland et al., 1999). Interestingly, using fluorometric techniques to identify the site of Ogg1 DNA repair activity following exposure to oxidative stress, the mitochondria, rather than the nucleus, was primary site of Ogg1 DNA repair activity (Mirbahai et al., 2010). In Ogg1 knockout mice, the mitochondrial genome contains almost nine times more 8-oxoguanine than control animals, whereas in the nuclear DNA the level of 8-oxoguanine is increased only twofold (Souza-Pinto et al., 2001). OGG1 gene mutations or polymorphisms increase the risk of various malignancies including lung, kidney, gastric, and colorectal cancer, as well as leukemia (Chevillard et al., 1998; Shinmura et al., 1998; Audebert et al., 2000; Bohr et al., 2002; Elahi et al., 2002; Fortini et al., 2003; Russo et al., 2004; Mambo et al., 2005). Furthermore, reduced Ogg1 activity is a risk factor in lung and head and neck cancer (Paz-Elizur et al., 2008).

Several groups have demonstrated that overexpression of mitochondria-targeted Ogg1 prevents mtDNA damage and intrinsic apoptosis caused by ROS-exposed vascular endothelial and asbestos-exposed cells (Dobson et al., 2002; Ruchko et al., 2005; Rachek et al., 2006; Harrison et al., 2007; Panduri et al., 2009; Ruchko et al., 2010). This suggests a prominent role of mt-Ogg1 in regulating intrinsic apoptosis in diverse settings of oxidative stress. Alternative splicing of the OGG1 transcript results in two isoforms: α-Ogg1 and β-Ogg1 (Gredilla et al., 2010). β-Ogg1 levels in the mitochondria are 20-fold greater than α-Ogg1 levels yet, curiously, β-Ogg1 lacks 8-oxoG DNA glycosylase activity (Hashiguchi et al., 2004). This finding suggests a role for Ogg1 that is independent of DNA repair. Our group recently reported that overexpression of mitochondrial α-Ogg1 mutants lacking 8-oxoG DNA repair activity were as effective as wild type mt-Ogg1 in preventing oxidant-induced caspase-9 activation and intrinsic apoptosis. Mitochondria-targeted Ogg1 did not alter the levels of mitochondrial ROS produced but, interestingly, preserved mitochondrial aconitase suggesting a novel role for Ogg1 as discussed further below (Panduri et al., 2009).

4. Aconitase and mitochondrial DNA

Aconitase, an enzyme that is vital for carbohydrate and energy metabolism, is responsible for the interconversion of citrate and isocitrate in the tricarboxylic acid (TCA) cycle (Emptage et al., 1983). The importance of mitochondrial aconitase is suggested by the observation that citrate levels in the human prostate appear important for promoting oncogenic conditions. Normal citrate-producing prostate epithelial cells can develop into citrate-oxidizing malignant cells that result in a net increase of 22 ATP/mol glucose that affords energy for malignant-associated activities (Costello & Franklin, 1994). It has been suggested that mitochondrial aconitase is a key enzyme associated with this bioenergy transformation since loss of its activity reduces cellular survival (Singh et al., 2006).

Mitochondrial aconitase is an iron-sulfur protein that is vulnerable to oxidative inactivation and is implicated as a mitochondrial redox-sensor (Gardner et al., 1994; Bulteau et al., 2003). Aconitase inactivation can further promote oxidant generation by releasing redox-active Fe from the $(4Fe–4S)^{2+}$ center following exposure to oxidants such as $O_2 \bullet^-$ (Gardner et al., 2000) or deficiency of mitochondrial manganese superoxide dismutase (MnSOD) (Williams et al., 1998). Oxidative-inactivation of aconitase is associated with decreased Drosophila lifespan (Yan et al., 1997). Reduced aconitase activity has also been described in a number of neurodegenerative diseases, including progressive supranuclear palsy (Park 2001), Friedreich's ataxia (Bradley 2000), and Huntington's disease (Tabrizi 1999).

Collectively, the above findings suggested a key role for mitochondrial aconitase beyond the TCA cycle. In this regard, a provocative finding in yeast showed that mitochondrial aconitase preserves mtDNA independent of aconitase's catalytic activity (Chen et al., 2005). This was the first suggestion of a dual role for aconitase as a mitochondrial TCA enzyme as well as in mtDNA maintenance, mitochondrial aconitase co-precipitates with frataxin, which is an iron chaperone protein that prevents aconitase oxidative inactivation and/or augments aconitase reactivation (Bulteau et al., 2004). This study suggested that prevention of oxidative inactivation of mitochondrial aconitase may be important for the pathogenesis of a degenerative disease (e.g. Friedrich's ataxia). Further evidence for this possibility was our recent finding that mt-Ogg1 overexpression completely blocks oxidant induced decreases in AEC mitochondrial aconitase activity and protein expression (Panduri et al., 2009). Moreover, using immunoprecipitation to explore the possible interactive effects between mitochondrial Ogg1 and aconitase, mitochondrial aconitase coprecipitated with both wild-type and mutant mt-Ogg1. Notably, overexpression of mitochondrial aconitase eliminated oxidant induced AEC apoptosis whereas Ogg1 underexpression using shRNA techniques reduced basal mitochondrial aconitase levels and augmented oxidant-induced AEC apoptosis (Panduri et al., 2009). These latter findings are in accord with several recent studies showing that Ogg1 deficiency increases oxidant-induced apoptosis (Youn et al., 2007; Bacsi et al., 2007; Xie et al., 2008). Collectively, these results suggest a novel interaction between an mtDNA repair enzyme (mt-Ogg1) and aconitase in preventing intrinsic AEC apoptosis following exposure to oxidative stress (e.g. asbestos or H_2O_2).

The underlying mechanisms that account for the interactive protective effect of mt-Ogg1 and aconitase require further study but there are at least two possibilities, which are not mutually exclusive. First, mt-Ogg1 may block key oxidative modification sites on mitochondrial aconitase responsible for triggering degradation by mitochondrial Lon protease (Bota & Davies, 2002; Bota et al., 2005). Lon protease selectively degrades oxidatively modified aconitase at a much higher rate than unexposed aconitase; a finding that may be important in defending the mitochondria against the accumulation of oxidized proteins as well as ensuring that such cells will undergo intrinsic apoptosis (Wallace, 1999; Bota et al., 2005; Bota & Davies, 2002; Panduri et al. 2009). Support for this possibility is our finding that MG132, a protease inhibitor that blocks mitochondrial Lon protease (Granot et al., 2007), attenuates asbestos-induced reductions in mitochondrial aconitase activity (Panduri et al., 2009). Second, overexpression of mt-Ogg1 or aconitase may preserve mtDNA levels necessary to prevent activation of intrinsic apoptosis. Future studies are required to clarify these possibilities as well as to determine precisely how mt-hOgg1 interacts with aconitase and whether other mtDNA repair proteins act similarly.

5. p53 and mitochondrial DNA repair

p53 functions as the "gatekeeper" of the genome by integrating various signals and initiating appropriate biological responses including cell cycle arrest, differentiation, apoptosis, senescence, and anti-angiogenesis (see for reviews Levine 1997; Vogelstein et al., 2000; Vousden et al., 2009). Previous studies have shown that the functions of p53 are mediated by transcriptional activation that regulates expression of downstream target genes (El-Deiry 1998). Expression of some cellular genes, including WAF1, CIP1, p21, IGF-BP3, mdm2, cyclin G, PCNA, and GADD45, are directly regulated by p53-mediated transactivation (Ko & Prives, 1996). p53 is also a redox-sensitive transcription factor whose function is integrally connected to ROS production as well as mediating the down-stream cellular effects following oxidative stress including the induction of apoptotic cell death (reviewed in Sablina et al., 2005; Janicke et al., 2008; Vaseva et al., 2009; Liu et al., 2010). ROS can induce p53 expression whereas p53 stabilization can augment further ROS production, often via effects on the mitochondria (Janicke et al., 2008; Liu et al., 2010). The mitochondria are an important target of transcription-dependent and -independent actions of p53 required to trigger apoptosis. By regulating thousands of genes, either directly or indirectly, p53 is implicated in numerous key cellular roles, including a recently described role for mtDNA maintenance (El-Deiry et al., 1992; Janicke et al., 2008, Bakhanashvili et al., 2008; Lebedeva et al., 2009).

The mechanism by which p53 regulates cellular responses following exposure to oxidative stress generally depends on the levels of ROS. A biphasic response is seen in which low basal p53 expression promotes ROS homeostasis and cell survival by augmenting anti-oxidant defenses as one of its tumor-suppressing mechanisms while higher levels of ROS induce persistent p53 expression that blocks the cell cycle enabling time for DNA repair and, if repair is insufficient, triggers apoptosis (Bensaad et al., 2005; Janicke et al., 2008; Vousden et al., 2009). Notably, p53 also enhances Ogg1 activity for 8-oxoG removal suggesting a link between Ogg1, p53 and mtDNA (Achanta & Huang , 2004). A recently described role for p53 in mtDNA maintenance following exposure to mitochondrial ROS is evidenced by its involvement in maintaining mtDNA copy number and mtDNA synthesis (Bakhanashvili et al., 2008; Lebedeva et al., 2009). Cells that are p53-depleted exhibit significant disruption of cellular ROS homeostasis that are characterized by reduced mitochondrial biogenesis and increased H_2O_2 production (Lebedeva et al., 2009). In contrast, thymic lymphomas derived from the p53$^{-/-}$ mouse (a common model of carcinogenesis) have highly significant upregulation of mitochondrial biogenesis, mitochondrial protein translation, mtDNA copy number, ROS levels, anti-oxidant defenses, proton transport, ATP synthesis, hypoxia response, and glycolysis, indicating important mitochondrial bioenergetic profile changes of cells occurs during the process of malignant transformation (Samper et al., 2009). Hypoxia stimulates mitochondrial ROS production, which activates p53 stabilization and localization to the mitochondria where p53 has many effects including inhibiting MnSOD thereby promoting $O_2{}^{\bullet-}$ formation and greater oxidative damage (Ralph et al., 2010) as well as regulating mtDNA repair and replication as noted above. Taken together, the emerging evidence strongly implicate that p53 is a key regulator of mitochondrial function, including ROS production and associated mtDNA repair following oxidative damage, as well as mtDNA replication and mitochondrial biogenesis (Ralph et al., 2010).

It is known well that most human tumors contain mutations in one or more p53 gene family members (see for reviews Janicke et al., 2008; Vousden & Prives, 2009). In this section we focus on the role of p53 in the lungs exposed to asbestos fibers. Altered p53 expression has been implicated in the pathophysiology of pulmonary fibrosis, including that due to asbestos, as well as asbestos-associated malignancies, especially bronchogenic lung cancer (Nelson et al., 2001; Mishra et al., 1997; Burmeister et al., 2004; Plataki et al., 2005). Asbestos activates p53 and p21 expression in lung epithelial and mesothelial cells that result in cell cycle arrest (Levresse et al., 1997; Matsuoka et al., 2003; Kopnin et al., 2004). Furthermore, increased p53 levels are detected in lung cancers of patients with asbestosis (Nuorva et al., 1994) and p53 point mutations are present in the lung epithelium of smokers and asbestos-exposed individuals (Husgafvel-Pursiainen et al., 1997). Crocidolite asbestos promotes p53 gene mutations predominantly in axons 9 through 11 in BALB/c-3 T3 cells (Lin et al., 2000). Finally, studies in lung epithelial and mesothelial cells using gene expression microarray techniques have established that induction of p53 gene expression following asbestos fiber exposure is an important event (Nymark et al., 2007; Hevel et al., 2008). Thus, p53 has a crucial role regulating lung cellular DNA damage response following exposure to oxidative stress as occurs with asbestos and tobacco smoke.

The mechanisms by which p53 regulate apoptosis are complex and incompletely understood. One established pathway involves intrinsic apoptosis via p53 crosstalk with the mitochondria by increasing transcription of pro-apoptotic stimuli (e.g. Bax and BH3-only proteins) while inhibiting gene expression of anti-apoptotic Bcl-2 family members (Miyashita et al., 1995; Oda et al., 2000; Nakano et al., 2001; Janicke et al., 2008; Vousden & Prives, 2009). There is considerable evidence that p53 phosphorylation at the Ser15 position following exposure to DNA damaging agents, including asbestos, is in part responsible for p53 stabilization and its subsequent mitochondrial translocation. Several different proteins have been implicated in the phosphorylation of p53 at Ser15, including members of the phosphatidylinostitol 3-kinase-related kinase (PI3K) family such as DNA-activated protein kinase (DNA-PK) and ataxia-telangiectasia mutated (ATM) kinase, as well as members of the mitogen-activated protein kinase (MAPK). In one study, suppression of DNA-PK coupled with a mutated form of ATM inhibited asbestos-induced Ser15 phosphorylation and accumulation of p53 (Matsuoka et al., 2003). Considerable evidence has established that p53 is a crucial regulator of mitochondrial function, including ROS generation and mtDNA repair following oxidative damage as well as mitochondrial biogenesis and mtDNA replication (see for review Liu et al., 2010). For example, p53 mediates asbestos-induced mitochondria-regulated apoptosis in lung epithelial cells and this is blocked in cells incapable of producing mitochondrial ROS (Panduri et al., 2006). Notably, loss of p53 results in mtDNA depletion, altered mitochondrial function and increased H_2O_2 production (Lebedeva et al., 2009).

The above data are providing insights into the molecular mechanisms by which p53 regulates the cellular response to DNA damage caused by exposure to oxidative stress that is likely important in the pathogenesis of inflammation-associated cancer (see for review: Kamp et al., 2011). An important link between p53 and Ogg1 is suggested by the finding that Ogg1 is under transcriptional regulation by p53 in colon and renal epithelial cells (Youn et al., 2007). In this study, the expression and activity of Ogg1 were decreased in HCT116p53-/- cells. Further, gel-shift assays showed that p53 binds to the putative cis-elements within the OGG1 promoter while supplementing p53 in HCT116p53-/- cells enhanced OGG1 transcription. In renal epithelial cells, tuberin also regulates OGG1 expression since transcriptional activity of the OGG1 promoter is decreased in tuberin-null cells; an effect that in part is mediated by the transcription factor NF-YA (Habib et al., 2008).

p53 modulates cellular metabolism by enhancing aerobic respiration and blocking glycolysis in most cell types; findings that are likely important in cellular malignant transformation (Bensaad et al., 2006; Bensaad et al., 2007). Interestingly, there is some evidence that p53 impacts mitochondrial aconitase levels since thymoquinone, a p53-dependent antineoplastic drug, reduces aconitase enzyme activity in isolated rat liver mitochondria (Roepke et al., 2007). Also, mitochondrial aconitase gene expression in prostate carcinoma cells is inhibited by both endogenous p53 induction by camptothecin treatment and exogenous p53 induction by transient overexpression of p53 (Tsui et al., 2011). Further, these investigators showed that mitochondrial aconitase is a p53-downregulated gene. Camptothecin did not affect mitochondrial aconitase reporter activity in p53-null PC-3 cells suggesting that the decrease in mitochondrial aconitase gene expression by camptothecin occurs via p53 activation. The relevance of these findings to other cell types as well as the *in vivo* significance requires further study.

6. Conclusion

In this review we have summarized emerging evidence demonstrating an important interactive effect between mitochondrial Ogg1, mitochondrial aconitase, and p53 in mtDNA repair and oxidant-induced intrinsic apoptosis. Although we focused on the role of oxidative stress caused by exposure to asbestos fibers, it is likely that many of the described interactive effects between mt-Ogg1, aconitase, p53 and intrinsic apoptosis will have broader implications but this awaits future investigations. Additional studies are necessary to further characterize the role of mitochondrial Ogg1 and aconitase in preventing mtDNA damaging (including following asbestos exposure), p53 activation and intrinsic apoptosis. It will also be of considerable interest to better understand the molecular mechanisms by which mitochondrial Ogg1 binds aconitase. Finally, and perhaps most importantly, we reason that the asbestos paradigm will continue to provide insights into the molecular mechanisms underlying the interactive effects between mt-Ogg1, aconitase, p53 and intrinsic apoptosis that should shed light into the pathogenesis of other more common diseases, such as lung cancer and idiopathic pulmonary fibrosis, for which more effective management regimens are urgently required. Strategies aimed at augmenting mtDNA integrity by increasing mt-Ogg1 and/or aconitase levels to mitigate the deleterious effects of oxidative stress may prove useful for developing novel therapeutic treatments for tumors and degenerative diseases as well as modulating the effects of aging.

7. Abbreviations

Electron transport chain (ETC)
outer mitochondrial membrane (OMM)
alveolar epithelial cell (AEC)
reactive oxygen species (ROS)
mitochondrial human 8-oxoguanine-DNA glycosylase 1 (mt-hOgg1)
alveolar type II (AT2) cells
hydrogen peroxide (H_2O_2),
superoxide anion (O_2-)
hydroxyl radical (HO•)
8-hydroxydeoxyguanosine (8OHdG)

mitochondrial DNA (mtDNA)
tricarboxcylic acid (TCA)
mitochondrial membrane potential ($\triangle\psi m$)
titanium dioxide (TiO_2)
thymine glycol (TG)
base excision repair (BER)
8-hydroxyguanine (8-oxoG)
manganese superoxide dismutase (MnSOD)

8. References

Aburatani, H.; Hippo, Y.; Ishida, T.; Takashima, R., Matsuba, C.; Kodama, T. et al. (1997). Cancer Res 57: 2151–2156.

Achanta, G.; Huang, P.;(2004). "Role of p53 in sensing oxidative DNA damage in response to reactive oxygen species-generating agents." Cancer Res 64(17):6233-9.

Albring, M.; Griffith, J.; Attardi, G. (1977). "Association of a protein structure of probable membrane derivation with HeLa cell mitochondrial DNA near its origin of replication." Proc Natl Acad Sci U S A 74: 1348–1352.

Aljandali, A.; Pollack, H.; Yeldandi, A.; Li, Y.; Weitzman, S. A.; Kamp, D. W. (2001). "Asbestos causes apoptosis in alveolar epithelial cells: role of iron-induced free radicals." J Lab Clin Med 137 (5): 330–339.

Ames, B. N. (1989). "Endogenous oxidative DNA damage aging and cancer." Free Radic. Res. Commun. 7: 121–128.

Anderson, S.; Bankier, A. T.; Barrell, B. G.; de Bruijn, M. H.; Coulson, A. R.; Drouin, J.; Eperon, I. C.; Nierlich, D. P.; Roe, B. A.; Sanger, F.; Schreier, P. H.; Smith, A. J.; Staden, R.; Young, I. G. (1981). "Sequence and organization of the human mitochondrial genome." Nature 290: 457–465.

Audebert, M.; Chevillard, S.; Levalois, C.; Gyapay, G.; Vieillefond,A.; Klijanienko J et al. (2000). Cancer Res 60: 4740–4744.

Babcock, D. F.; Herrington, J.; Goodwin, P. C.; Park, Y. B. and Hille, B. (1997). "Mitochondrial participation in the intracellular Ca2. Network." J Cell Biol 136: 833–844.

Bacsi, A.; Chodaczek, G.; Hazra, T. K.; Konkel, D.; Boldogh, I. (2007). "Increased ROS generation in subsets of OGG1 knockout fibroblast cells." Mech. Ageing Dev. 128(11–12): 637–649.

Bakhanashvili,M.; Grinberg, S.; Bonda, E.; Simon, A.J.; Moshitch-Moshkovitz, S.; Rahav, G.(2008). "p53 in mitochondria enhances the accuracy of DNA synthesis." Cell Death Differ 15(12):1865-74.

Barnes, D. E.; Lindahl, T. (2004). "Repair and genetic consequences of endogenous DNA base damage in mammalian cells." Annu. Rev. Genet. 38: 445–476.

Bensaad, K.; Tsuruta, A.; Selak, M. A.; Vidal, M. N. C.; Nakano, K.; Bartrons, R.; Gottlieb, E.; Vousden, K. H. (2006). "TIGAR, a p53-inducible regulator of glycolysis and apoptosis." Cell 126: 107–120.

Bensaad, K.; Vousden, K. H. (2005). "Savior and slayer: the two faces of p53." Nat. Med. 11: 1278–1279.

Bensaad, K.; Vousden, K. H. (2007). "p53:New roles in metabolism." Trends Cell Biol 17: 286–291.

Bjelland, S.; Seeberg, E. (2003). "Mutagenicity, toxicity and repair of DNA base damage induced by oxidation." Mutat. Res. 531: 37–80.

Bjoras, M.; Luna, L.; Johnsen, B.; Hoff, E.; Haug, T.; Rognes, T.; Seeberg, E. (1997). "Opposite base-dependent reactions of a human base excision repair enzyme on DNA containing 7,8-dihydro-8-oxoguanine and abasic sites." EMBO J. 16: 6314–6322.

Bohr, V. A.; Stevnsner, T.; de Souza-Pinto, N. C. (2002). "Mitochondrial DNA repair of oxidative damage in mammalian cells." Gene 286: 127–134.

Bota, D. A.; and Davies, K. J. A. (2002). "Lon protease preferentially degrades oxidized mitochondrial aconitase by an ATP-stimulated mechanism." Nat. Cell Biol. 4 (9): 674–680

Bota, D.A.; Ngo, J.K.; Davies, K.J. (2005). "Downregulation of the human Lon protease impairs mitochondrial structure and function and causes cell death." Free Radic Biol Med 38(5):665-77.

Bota, D. A.; Van Remmen, H.; Davies, K. J. (2002). "Modulation of Lon protease activity and aconitase turnover during aging and oxidative stress." FEBS Lett. 532 (1-2): 103–106.

Boveris, A.; Chance, B. (1977). "The mitochondrial generation of hydrogen peroxide." General properties and effect of hyperbaric oxygen. Biochem. J. 134: 707–716.

Bradley, J.L.; Blake, J.C.; Chamberlain, S.; Thomas, P.K.; Cooper, J.M.; Schapira,A.H.(2000). "Clinical, biochemical and molecular genetic correlations in Friedreich's ataxia." Hum Mol Genet 9(2):275-82.

Bulteau, A. L.; Ikeda-Saito, M.; Szweda, L. I. (2003). "Redox-dependent modulation of aconitase activity in intact mitochondria." Biochemistry 42 (50): 14846–14855.

Bulteau, A. L.; O'Neill, H. A.; Kennedy, M. C.; Ikeda-Saito, M.; Isaya, G.; Szweda, L. I. (2004). "Frataxin acts as an iron chaperone protein to modulate mitochondrial aconitase activity." Science 305 (5681): 242–245.

Burmeister, B.; Schwerdtle, T.; Poser, I.; Hoffmann, E.; Hartwig, A.; Muller, W. U.; Rettenmeier, A. W.; Seemayer, N. H.; Dopp, E. (2004). "Effects of asbestos on initiation of DNA damage, induction of DNA-strand breaks, P53-expression and apoptosis in primary, SV40-transformed and malignanthumanmesothelial cells." Mutat. Res. 558 (1-2): 81–92.

Chance, B.; Sies, H.; Boveris, A. (1979). "Hydroperoxide metabolism in mammalian organs." Physiol Rev 59: 527–605.

Chen, X. J.; Wang, X.; Kaufman, B. A.; Butow, R. A. (2005). "Aconitase couples metabolic regulation to mitochondrial DNA maintenance." Science 307 (5710):714–717.

Chevillard, S.; Radicella, J. P.; Levalois, C.; Lebeau, J.; Poupon, M. F.; Oudard, S.; Dutrillaux, B.; Boiteux, S. (1998). "Mutations in OGG1, a gene involved in the repair of oxidative DNA damage, are found in human lung and kidney tumors." Oncogene 16:3083–3086.

Clark, J. M.; Beardsley, G. P. (1987). "Functional effects of cis-thymine glycol lesions on DNA synthesis in vitro." Biochemistry 26: 5398–5403.

Corral-Debrinski, M.; Horton, T.; Lott, M. T.; Shoffner, J. M.; Beal, M. F.; Wallace, D. C. (1992). "Mitochondrial DNA deletions in human brain: regional variability and increase with advanced age." Nat Genet 2: 324–329.

Costello, L. C.; Franklin, R. B. (1994). "The bioenergetic theory of prostate malignancy." Prostate 25:162–166.

Clayton, D. A. (1982). "Replication of animal mitochondrial DNA." Cell 28: 693–705.

Clayton, D. A. (1984). "Transcription of the mammalian mitochondrial genome." Annu Rev Biochem 53:573–594.

Demple, B.; Harrison, L. (1994). "Repair of oxidative damage to DNA: enzymology and biology."Annu. Rev. Biochem. 63: 915–948.

De Souza-Pinto, N. C.; Eide, L.; Hogue, B. A.; Thybo, T.; Stevnsner, T.; Seeberg, E.; Klungland, A.; Bohr, V. A. (2001). "Repair of 8-oxodeoxyguanosine lesions inmitochondrial 498 DNA depends on the oxoguanine DNA glycosylase (OGG1) gene and 8-499 oxoguanine accumulates in the mitochondrial DNA of OGG1-defective mice." Cancer Res. 61: 5378–5381.

Dianov, G.L.;Souza-Pinto,N.; Nyaga,S.G.; Thybo,T.; Stevnsner,T.;Bohr,V.A.(2001)."Base excision repair in nuclear and mitochondrial DNA." Prog Nucleic Acid Res Mol Biol 68:285-97.

Dianova I.I.; Bohr, V. A.; and Dianov, G. L. (2001). "Interaction of human AP endonuclease 1 with flap endonuclease 1 and proliferating cell nuclear antigen involved in long-patch base excision repair." Biochem 40: 12639-12644.

Dizdaroglu, M. (1992). "Oxidative damage to DNA in mammalian chromatin." Mutat. Res. 275: 331–342.

Dobson, A. W.; Grishko, V. S.; LeDoux, P.; Kelley, M. R.; Wilson, G. L.; Gillespie, M. N. (2002) "Enhanced mtDNA repair capacity protects pulmonary artery endothelial cells from oxidant-mediated death." Am. J. Physiol. Lung Cell Mol. Physiol. 283 (1): 205–210.

Dobson, A. W.; Xu, Y.; Kelley, M. R.; LeDoux, S. P.; Wilson, G. L. (2000). "Enhanced mitochondrial DNA repair and cellular survival after oxidative stress by targeting the human 8-oxoguanine glycosylase repair enzyme to mitochondria." J Biol Chem 275: 37518–37523.

Elahi,A.; Zheng, Z.;Park, J.;Eyring, K.; McCaffrey,T.;Lazarus,P.(2002). "The human OGG1 DNA repair enzyme and its association with orolaryngeal cancer risk." Carcinogenesis 23(7):1229-34.

El-Deiry, W. S.; Kern, S. E.; Pietenpol, J. A.; Kinzler, K. W.; Vogelstein, B. (1992). "Definition of a consensus binding site for p53, Nat." Genet. 1: 45–49.

El-Deiry, W. S. (1998). "Regulation of p53 downstream genes, Semin." Cancer Biol. 8:345–357.

Emptage, M. H.; Kent, T. A.; Kennedy, M. C.; Beinert, H.; Munck, E. (1983). "Mossbauer and EPR studies of activated aconitases: Development of a localized valence state at a subsite of the [4Fe-4S] cluster on binding of citrate." Proc Natl Acad Sci U S A 80: 4674–4678.

Fortini, P.; Pascucci, B.; Parlanti, E.; D'Errico, M.; Simonelli, V.; Dogliotti, E. (2003). "8-Oxoguanine DNA damage: at the crossroad of alternative repair pathways." Mutat Res 531:127–139.

Franco, R.; Sanchez-Olea, R.; Reyes-Reyes, E. M.; Panayiotidis, M. I. (2009). "Environmental toxicity, oxidative stress and apoptosis: menage a trois." Mutat. Res. 674 (1–2): 3–22.

Gardner, P. R. (1997). "Superoxide-driven aconitase Fe–S center cycling." Biosci Rep 17: 33–42.

Gardner, P. R.; Nguyen, D. D.; White, C. W. (1994). "Aconitase is a sensitive and critical target of oxygen poisoning in cultured mammalian cells and in rat lungs." Proc Natl Acad Sci U S A 91(25): 12248-12252.

Granot,Z.;Kobiler,O.;Melamed-Book,N.;Eimerl,S.;Bahat,A.;Lu,B.;Braun,S.;Maurizi,M.R.;Suzuki, C.K.;Oppenheim,A.B.;Orly,J.(2007). "Turnover of mitochondrial steroidogenic acute regulatory (StAR) protein by Lon protease: the unexpected effect of proteasome inhibitors." Mol Endocrinol 21(9):2164-77.

Gredilla, R.; Bohr, V.A.; Stevnsner, T.(2010)" Mitochondrial DNA repair and association with aging--an update." Exp Gerontol. 45(7-8):478-88.

Habib, S. L.; Riley, D. J.; Mahimainathan, L.; Bhandari, B.; Choudhury, G. G.; Abboud, H. E. (2008). "Tuberin regulates the DNA repair enzyme OGG1." Am. J. Physiol. Renal Physiol. 294 (1): F281–290.

Hailer, M. K.; Slade, P. G.; Martin, B. D.; Rosenquist, T. A.; Sugden, K. D. (2005). "Recognition of the oxidized lesions spiroiminodihydantoin and guanidinohydantoin in DNA by the mammalian base excision repair glycosylases NEIL1 and NEIL2." DNA Repair 4: 41–50.

Hardy, J. A.; Aust, A. E. (1995). "The effect of iron binding on the ability of crocidolite asbestos to catalyze DNA single-strand breaks." Carcinogenesis 16 (2): 319–325.

Harrison, J. F.; Rinne, M. L.; Kelley, M. R.; Druzhyna, N. M.; Wilson, G. L.; Ledoux, S.P. (2007). "Altering DNA base excision repair: use of nuclear and mitochondrial-targeted N-methylpurine DNA glycosylase to sensitize astroglia to chemotherapeutic agents." Glia. 55: 1416–1425.

Hashiguchi, K.; Stuart, J. A.; de Souza-Pinto, N. C.; Bohr, V. A. (2004). "The C-terminal α-O helix of human Ogg1 is essential for 8-oxoguanine DNA glycosylase activity: the mitochondrial β-Ogg1 lacks this domain and does not have glycosylase activity." Nucleic Acids Res. 32: 5596–5608.

Hatahet, Z.; Purmal, A. A.; Wallace, S. S. (1994). "Oxidative DNA lesions as blocks to in vitro transcription by phage T7 RNA polymerase." Ann. NY Acad. Sci. 726: 346–348.

Hevel, J.M.; Olson-Buelow, L.C.; Ganesan, B.; Stevens, J.R.; Hardman, J.P.; Aust, A.E.(2008). "Novel functional view of the crocidolite asbestos-treated A549 human lung epithelial transcriptome reveals an intricate network of pathways with opposing functions." BMC Genomics 9:376.

Htun, H.; Johnston, B. H. (1992). "Mapping adducts of DNA structural probes using transcription and primer extension approaches." Methods Enzymol. 212: 272–294.

Husgafvel-Pursiainen, K.; Kannio, A.; Oksa, P.; Suitiala, T.; Koskinen, H.; Partanen, R.; Hemminki, K.; Smith, S.; Rosenstock-Leibu, R.; Brandt-Rauf, P. W. (1997). "Mutations, tissue accumulations, and serum levels of p53 in patients with occupational cancers from asbestos and silica exposure." Environ. Mol. Mutagen. 30 (2): 224–230.

Ide, H.; Kow, Y. W.; Wallace, S. S. (1985). "Thymine glycols and urea residues in M13 DNA constitute replicative blocks in vitro." Nucleic Acids Res. 13: 8035–8052.

Ishikawa,K.; Takenaga,K.; Akimoto,M.; Koshikawa,N.; Yamaguchi,A.; Imanishi,H.; Nakada,K.; Honma,Y.; Hayashi,J.(2008). "ROS-generating mitochondrial DNA mutations can regulate tumor cell metastasis." Science. 320(5876):661-4.

Janicke, R. U.; Sohn, D.; Schulze-Osthoff, K. (2008). "The dark side of a tumor suppressor: anti-apoptotic p53." Cell Death Differ. 15 (6): 959–976.

Jaurand, M. C. (1997). "Mechanisms of fiber-induced genotoxicity." Environ Health Perspect 105 (Suppl. 5): 1073–1084.

Kamp, D. W.; Graceffa, P.; Pryor, W. A.; Weitzman, S. A. (1992). "The role of free radicals in asbestos-induced diseases". Free Radic. Biol. Med. 12 (4): 293–315.

Kamp, D. W.; Israbian, V. A.; Preusen, S. E.; Zhang, C. X.; Weitzman, S. A. (1995). "Asbestos causes DNA strand breaks in cultured pulmonary epithelial cells: role of iron-catalyzed free radicals." Am. J. Physiol. 268 (3 Pt 1): L471–480.

Kamp, D.W., Shacter E, Weitzman SA. Chronic inflammation and cancer: The role of the mitochondria. Oncology 2011; 25:400-13.

Ke-Hung Tsui, Tsui-Hsia Feng, Yu-Fen Lin, Phei-Lang Chang, Horng-Heng Juang. (2011). "p53 Downregulates the Gene Expression of Mitochondrial Aconitase in Human Prostate Carcinoma Cells." The Prostate 71: 62-70.

Kim, I.; Xu, W.; Reed, J. C. (2008). "Cell death and endoplasmic reticulum stress: disease relevance and therapeutic opportunities." Nat. Rev. Drug Discov. 7 (12): 1013–1030.

Klungland, A.; Rosewell, I.; Hollenbach, S.; Larsen, E.; Daly, G.; Epe, B. et al. (1999). Proc Natl Acad Sci USA 96: 13300–13305.

Ko, L. J.; Prives, C. (1996). " p53: puzzle and paradigm." Genes Dev. 10: 1054-1072.

Kopnin, P. B.; Kravchenko, I. V.; Furalyov, V. A.; Pylev, L. N.; Kopnin, B. P. (2004). "Cell typespecific effects of asbestos on intracellular ROS levels, DNA oxidation and G1 cell cycle checkpoint." Oncogene 23 (54): 8834–8840.

Kroemer, G.; Galluzzi, L.; Brenner, C. (2007). "Mitochondrial membrane permeabilization in cell death." Physiol. Rev 87 (1): 99–163.

Kujoth, G. C.; Hiona, A.; Pugh, T. D.; Someya, S.; Panzer, K.; Wohlgemuth, S. E.; Hofer, T.; Seo, A. Y.; Sullivan, R.; Jobling, W. A. et al. (2005). "Mitochondrial DNA mutations, oxidative stress, and apoptosis in mammalian aging." Science 309: 481–484.

Lebedeva, M. A.; Eaton, J. S.; Shadel, G. S. (2009). "Loss of p53 causes mitochondrial DNA depletion and altered mitochondrial reactive oxygen species homeostasis." Biochim. Biophys. Acta. May 1787 (5): 328-34.

Levine, A. J. (1997). "p53, the cellular gatekeeper for growth and division." Cell 88: 323-331.

Levresse, V.; Renier, A.; Fleury-Feith, J.; Levy, F.; Moritz, S.; Vivo, C.; Pilatte, Y.; Aurand, M. C. (1997). "Analysis of cell cycle disruptions in cultures of rat pleural mesothelial cells exposed to asbestos fibers." Am. J. Respir. Cell Mol. Biol. 17 (6): 660–671.

Lindahl, T. (1993). "Instability and decay of the primary structure of DNA." Nature 362: 709–715.

Lin, F.; Liu, Y.; Keshava, N.; Li, S. (2000). "Crocidolite induces cell transformation and p53 gene mutation in BALB/c-3T3 cells." Teratog. Carcinog. Mutagen. 20 (5): 273–281.

Liu, V. W.; Zhang, C.; Nagley, P. (1998). "Mutations in mitochondrial DNA accumulate differentially in three different human tissues during ageing." Nucleic. Acids Res. 26: 1268–1275.

Liu, G.; Beri, R.; Mueller, A.; Kamp, D.W. "Molecular mechanisms of asbestos-induced lung epithelial cell apoptosis." Chemico-Biol Interactions 2010; 188:309-18.

Loeb, L. A.; Wallace, D. C. and Martin, G. M. (2005). "The mitochondrial theory of aging and its relationship to reactive oxygen species damage and somatic mtDNA mutations." Proc. Natl. Acad. Sci. U S A 102: 18769–18770.

Mambo, E.; Chatterjee, A.; de Souza-Pinto, N. C.; Mayard, S.; Hogue, B. A.; Hoque, M. O.; Dizdaroglu, M.; Bohr, V. A.; Sidransky, D. (2005). "Oxidized guanine lesions and hOgg1 activity in lung cancer." Oncogene 24: 4496–4508.

Lebedeva, M. A.; Eaton, J. S.; Shadel, G. S.(2009). "Loss of p53 causes mitochondrial DNA depletion and altered mitochondrial reactive oxygen species homeostasis." Biochimica et Biophysica Acta 1787:328–334.

Matsuoka, M.; Igisu, H.; Morimoto, Y. (2003). "Phosphorylation of p53 protein in A549 human pulmonary epithelial cells exposed to asbestos fibers." Environ. Health Perspect. 111 (4): 509–512.

Medikayala, S.; Piteo, B.; Zhao, X. and Edwards, J. G. (2011). "Chronically elevated glucose compromises myocardial mitochondrial DNA integrity by alteration of mitochondrial topoisomerase function." Am. J. Physiol. Cell Physiol. 300: C338–C348.

Michikawa, Y.; Mazzucchelli, F.; Bresolin, N.; Scarlato, G.; Attardi, G. (1999). "Aging-dependent large accumulation of point mutations in the human mtDNA control region for replication." Science 286: 774–779.

Mirbahai,L.; Kershaw,R.M.; Green,R.M.; Hayden,R.E.; Meldrum,R.A.; Hodges,N.J.(2010). "Use of a molecular beacon to track the activity of base excision repair protein OGG1 in live cells." DNA Repair (Amst) 9(2):144-52.

Mishra, A.; Liu, J. Y.; Brody, A. R.; Morris, G. F. (1997). "Inhaled asbestos fibers induce p53 expression in the rat lung." Am. J. Respir. Cell Mol. Biol. 16 (4): 479–485.

Miyashita, T.; Reed, J. C. (1995). "Tumor suppressor p53 is a direct transcriptional activator of the human bax gene." Cell, 80 (2): 293–299.

Nakano, K.; Vousden, K. H. (2001). "PUMA, a novel proapoptotic gene, is induced by p53." Mol. Cell 7 (3):683–694.

Nass, M. M. (1969). "Mitochondrial DNA. I. Intramitochondrial distribution and structural relations of single- and double-length circular DNA." J. Mol. Biol. 42: 521–528.

Nelson, A.; Mendoza, T.; Hoyle, G. W.; Brody, A. R.; Fermin, C.; Morris, G. F. (2001). "Enhancement of fibrogenesis by the p53 tumor suppressor protein in asbestos-exposed rodents." Chest 120 (1 Suppl): 33S–34S.

Nuorva, K.; Makitaro, R.; Huhti, E.; Kamel, D.; Vahakangas, K.; Bloigu, R.; Soini, Y.; Paakko, P. (1994). "p53 protein accumulation in lung carcinomas of patients exposed to asbestos and tobacco smoke." Am. J. Respir. Crit. Care Med. 150 (2): 528–533.

Nymark, P.; Lindholm, P.M.; Korpela, M.V.; Lahti, L.; Ruosaari, S.; Kaski, S.; Hollmen, J.; Anttila,S.; Kinnula,V.L.; Knuutila,S. (2007). "Gene expression profiles in asbestos-exposed epithelial and mesothelial lung cell lines." BMC Genomics 8:62.

Oda, E.; Ohki, R.; Murasawa, H.; Nemoto, J.; Shibue, T.; Yamashita, T.; Tokino, T.; Taniguchi, T.; Tanaka, N. (2000). "Noxa, a BH3-only member of the Bcl-2 family and candidate mediator of p53-induced apoptosis." Science 288 (5468): 1053–1058.

Panduri, V.; Liu, G.; Surapureddi, S.; Kondapalli, J.; Soberanes, S.; de Souza-Pinto, N. C.; Bohr, V. A.; Budinger, G. R.; Schumacker, P. T.; Weitzman, S. A.; Kamp, D. W. (2009). "Role of mitochondrial hOGG1 and aconitase in oxidant-induced lung epithelial cell apoptosis." Free Radic. Biol. Med. 47 (6): 750–759.

Panduri, V.; Weitzman, S. A.; Chandel, N.; Kamp, D. W. (2003). "The mitochondria regulated death pathway mediates asbestos-induced alveolar epithelial cell apoptosis." Am. J. Respir. Cell Mol. Biol. 28 (2): 241–248.

Panduri, V.; Weitzman, S. A.; Chandel, N. S.; Kamp, D. W. (2004). "Mitochondrial-derived free radicals mediate asbestos-induced alveolar epithelial cell apoptosis." Am. J. Physiol. Lung Cell Mol. Physiol. 286 (6): L1220–L1227.

Panduri, V. S.; Surapureddi, S.; Soberanes, S. A.; Weitzman, N.; Chandel, D.; Kamp, W. (2006). "P53 mediates amosite asbestos-induced alveolar epithelial cell mitochondria-regulated apoptosis." Am. J. Respir. Cell Mol. Biol. 34 (4) : 443–452.

Paradies, G.; Petrosillo, G.; Paradies, V.; Ruggiero, F. M. (2010). "Oxidative stress, mitochondrial bioenergetics, and cardiolipin in aging." Free Radic. Biol. Med. 48: 1286–1295.

Park, L. C. (2001). J. Neurosci. Res. 66 : 1028-1034.

Plataki, M.; Koutsopoulos, A. V.; Darivianaki, K.; Delides, G.; Siafakas, N. M.; Bouros, D. (2005). "Expression of apoptotic and antiapoptotic markers in epithelial cells in idiopathic pulmonary fibrosis." Chest. 127 (1): 266–274.

Powell, C. D.; Quain, D. E. and Smart, K. A. (2000). "he impact of media composition and petite mutation on the longevity of a polyploid brewing yeast strain." Lett. Appl. Microbiol. 31: 46–51.

Rachek, L. I.; Grishko, V. I.; Ledoux, S. P.; Wilson, G. L. (2006). "Role of nitric oxide-induced mtDNA damage in mitochondrial dysfunction and apoptosis." Free Radic. Biol. Med. 40 (5): 754–762.

Rachek, L. I.; Thornley, N. P.; Grishko, V. I.; LeDoux, S. P.; Wilson, G. L. (2006). "Protection of INS-1 cells from free fatty acid-induced apoptosis by targeting hOGG1 to mitochondria." Diabetes 55: 1022–1028.

Radicella, J. P.; Dherin, C.; Desmaze, C.; Fox, M. S.; Boiteux, S. (1997). Proc. Natl. Acad. Sci. U S A 94: 8010–8015.

Raha, S. ; Robinson, B. H. (2000). "Mitochondria, oxygen free radicals, diseases and ageing." Trends Biochem. Sci. 25: 502–508.

Ralph, S. J.; Rodríguez-Enríquez, S.; Neuzil, J.; Saavedra, E.; Moreno-Sánchez, R. (2010). "The causes of cancer revisited: "Mitochondrial malignancy" and ROS-induced oncogenic transformation - Why mitochondria are targets for cancer therapy." Molecular Aspects of Medicine 31: 145–170.

Roepke, M.; Diestel, A.; Bajbouj, K.; Walluscheck, D.; Schonfeld, P.; Roessner, A.; Schneider-Stock, R.; Gali-Muhtasib, H. (2007). "Lack of p53 augments thymoquinone-induced apoptosis and caspase activation in human osteosarcoma cells." Cancer Biol. Ther. 6: 160–169.

Ruchko, M.; Gorodnya, O.; LeDoux, S. P.; Alexeyev, M. F.; Al-Mehdi, A. B.; Gillespie, M. N. (2005). "Mitochondrial DNA damage triggers mitochondrial dysfunction and apoptosis in oxidant-challenged lung endothelial cells." Am. J. Physiol. Lung Cell. Mol. Physiol. 288: L530–L535.

Ruchko, M. V.; Gorodnya, O. M.; Zuleta, A.; Pastukh, V. M.; Gillespie, M. N. (2010). "The DNA glycosylase Ogg1 defends against oxidant-induced mtDNA damage and apoptosis in pulmonary artery endothelial cells." Free Radic Biol Med 50(9):1107-13.

Russo, M. T.; De Luca, G.; Degan, P.; Parlanti, E.; Dogliotti, E.; Barnes, D. E.; Lindahl, T.; Yang, H.; Miller, J. H.; Bignami, M. (2004). "Accumulation of the oxidative base lesion 8-hydroxyguanine in DNA of tumor-prone mice defective in both the Myh and Ogg1 DNA glycosylases." Cancer Res. 64: 4411–4414.

Sablina, A. A.; Budanov, A.V.; Ilyinskaya, G. V.; Agapova, L. S.; Kravchenko, J. E.; Chumakov, P. M. (2005). "The antioxidant function of the p53 tumor suppressor." Nat. Med. 11:1306–1313.

Samper, E.; Morgado, L.; Estrada, J. C.; Bernad, A.; Hubbard, A.; Cadenas, S.; Melov, S. (2009). "Increase in mitochondrial biogenesis, oxidative stress, and glycolysis in murine lymphomas." Free Radic. Biol. Med. 46: 387–396.

Santos, J. H.; Hunakova, L.; Chen,Y.; Bortner,Carl.; Van Houten, B. (2003). "Cell sorting experiments link persistent mitochondrial DNA damage with loss of mitochondrial membrane potential and apoptotic cell death." J. Biol. Chem 278(3):1728-1734.

Shibutani, S.; Takeshita, M.; Grollman, A. P. (1991). "Insertion of specific bases during DNA synthesis past the oxidation-damaged base 8-oxodG." Nature 349: 431–434.

Shigenaga, M. K.; Hagen, T. M.; Ames, B. N. (1994). "Oxidative damages and mitochondrial decay in aging." Proc. Natl. Acad. Sci. U S A 91: 10771–10778.

Shinmura,K.;Kohno,T.;Kasai,H.;Koda,K.;Sugimura,H.;Yokota,J.(1998).   "Infrequent mutations of the hOGG1 gene, that is involved in the excision of 8-hydroxyguanine in damaged DNA, in human gastric cancer." Jpn J Cancer Res 89(8):825-8.

Shukla, A.; Gulumian, M.; Hei, T.; Kamp, D. W.; Rahman, Q.; Aust, A. E. (2003). Mossman, B.T. "Multiple roles of oxidants in the pathogenesis of asbestos-induced diseases." Free Radic Biol Med 34 (9): 1117–1129.

Soong, N. W.; Hinton, D. R.; Cortopassi, G.; Arnheim, N. (1992). "Mosaicism for a specific somatic mitochondrial DNA mutation in adult human brain." Nature Genet. 2: 318–323.

Shukla, A.; Jung, M.; Stern, M.; Fukagawa, N. K.; Taatjes, D. J.; Sawyer, D. (2003). "Asbestos induces mitochondrial DNA damage and dysfunction linked to the development of apoptosis." Am. J. Physiol. Lung Cell Mol. Physiol. 285(5): L1018-1025.

Singh, K. K.; Desouki, M. M.; Franklin, R. B.; Costello, L. C. (2006). "Mitochondrial aconitase and citrate metabolism in malignant and nonmalignant human prostate tissues." Mol. Cancer 5:14.

Tabrizi,S.J.;Cleeter,M.W.;Xuereb,J.;Taanman,J.W.;Cooper,J.M.;Schapira,A.H.(1999). "Biochemical abnormalities and excitotoxicity in Huntington's disease brain." Ann Neurol 45(1):25-32.

Tamar, P. E.; Ziv, S.; Yael, L. D.; Dalia, E.; Laila, C. (2008). "Roisman, Zvi Livneh DNA repair of oxidative DNA damage in human carcinogenesis: Potential application for cancer risk assessment and prevention." Cancer Letters 266: 60–72.

Taylor, R. W. and Turnbull, D. M. (2005). "Mitochondrial DNA mutations in human disease. " Nat. Rev. Genet. 6: 389 –402.

Trifunovic, A.; Wredenberg, A.; Falkenberg, M.; Spelbrink, J. N.; Rovio, A. T.; Bruder, C. E.; Bohlooly-Y, M.; Oldfors, A.; Wibom, R. (2004). "Premature ageing in mice expressing defective mitochondrial DNA polymerase." Nature 429: 417–423.

Trifunovic, A.; Hansson, A.; Wredenberg, A.; Rovio,A.T.; Dufour, E.; Khvorostov, I.; Spelbrink, J.N.; Wibom, R.; Jacobs, H.T.; Larsson, N.G.(2005). "Somatic mtDNA mutations cause aging phenotypes without affecting reactive oxygen species production." Proc Natl Acad Sci U S A 102(50):17993-8.

Upadhyay, D.; Kamp, D. W. (2003). "Asbestos-induced pulmonary toxicity: role of DNA damage and apoptosis." Exp Biol Med 228 (6): 650–659.

Van, H. B.; Woshner, V.; Santos,J.H.(2006). "Role of mitochondrial DNA in toxic responses to oxidative stress." DNA Repair (Amst). 5(2):145-52.

Vaseva, A. V.; Moll, U. M. (2009). "The mitochondrial p53 pathway." Biochim. Biophys. Acta 1787: 414–420.

Vásquez-Vivar, J.; Kalyanaraman, B.; Kennedy, M. C. (2000). "Mitochondrial aconitase is a source of hydroxyl radical: an electron spin resonance investigation." J. Biol. Chem. 275: 14064–14069.

Vogelstein, B.; Lane, D. ; Levine, A. J. (2000). "Surfing the p53 network." Nature 408: 307–310.

Vousden, K. H.; Prives, C. (2009). "Blinded by the light: the growing complexity of p53." Cell 137 (3):413–431.

Wallace DC.(1999). "Mitochondrial diseases in man and mouse." Science 283(5407):1482-8.

Williams, M. D.; Van Remmen, H.; Conrad, C. C.; Huang, T. T.; Epstein, C. J. and Richardson, A. (1998) J. Biol. Chem. 273: 28510-28515.

Xie, Y.; Yang, H.; Miller, J. H.; Shih, D. M.; Hicks, G. G.; Xie, J.; Shiu, R. P. (2008). "Cells deficient in oxidative DNA damage repair genes My hand Ogg1 are sensitive to oxidants with increased G2/Marrest and multinucleation." Carcinogenesis 29 (4): 722–728.

Yan,L. J.; Levine, R. L.; Sohal, R. S. (1997). "Oxidative damage during aging targets mitochondrial aconitase." Proc Natl Acad Sci U S A 94(21):11168-72.

Youn, C. K.; Song, P. I.; Kim, M. H.; Kim, J. S.; Hyun, J. W.; Choi, S. J.; Yoon, S. P.; Chung, M. H.; Chang, I. Y.; You, H. J. (2007) "Human 8-oxoguanine DNA glycosylase suppresses the oxidative stress induced apoptosis through a p53-mediated signaling pathway in human fibroblasts." Mol. Cancer Res. 5(10): 1083-1098.

Youle, R. J. ; Strasser, A. (2008). "The BCL-2 protein family: opposing activities that mediate cell death." Nat. Rev. Mol. Cell Biol. 9 (1): 47–59.

6

Shared Regulatory Motifs in Promoters of Human DNA Repair Genes

Lonnie R. Welch[1], Laura M. Koehly[2] and Laura Elnitski[3]

[1]School of Electrical Engineering and Computer Science, Ohio University, Athens, Ohio
[2]Social and Behavioral Research Branch,
National Human Genome Research Institute, NIH, Bethesda, Maryland
[3]Genome Technology Branch
National Human Genome Research Institute, NIH, Bethesda, Maryland
USA

1. Introduction

This manuscript presents methods used to test, and resulting evidence to support the hypothesis that specialized transcription factor binding sites coordinate the expression of DNA repair genes. Building on the seminal work of the Elnitski laboratory (Yang et al. 2007), which identified the most complete set of human transcripts under the control of bidirectional promoters and identified the first putative regulatory networks that make use of the bidirectional promoter structure, the authors present additional details of these regulatory networks.

Much of the work regarding the regulation of DNA repair proteins is aimed at the level of protein-protein interactions and post-translational processing events (Hurley et al. 2007, Jensen et al. 2011, Shibata et al. 2010). However, transcriptional activation of DNA repair genes is likely to utilize shared factors, especially in cases of induced activation, which have not been thoroughly evaluated. Yang, Koehly and Elnitski reported the discovery and characterization of 5,653 bidirectional promoters in the human genome (Yang et al. 2007). Prior to that date, bidirectional promoters were annotated only for protein-coding genes, and only 1,352 examples had been reported in the human genome. The work of Yang et al. included evidence from all noncoding-RNA genes, as well. Each bidirectional promoter regulates the expression of two genes, oriented in opposite directions with transcription start sites within 1000 bp of one another. The authors developed a novel approach to map all bidirectional promoters by analyzing the public expressed-sequence-tag (EST) data. The prevalence of this promoter structure led the authors to explore the hypothesis that it plays a role in regulation of certain classes of genes. They discovered that many more DNA repair genes have bidirectional promoters than previously reported and that many genes with somatic mutations in cancer have bidirectional promoters. The relevance of DNA repair genes to cancers (Kinsella et al. 2009, Liang et al. 2009, Smith et al. 2010, Kelley et al. 2008, Li et al. 2009, Bellizii et al. 2009, Naccarati et al. 2007, Berwick et al. 2000)) and the association of bidirectional promoters with DNA repair genes suggested that bidirectional promoters might indicate a higher-order type of regulatory structure that could be detected through common features at the DNA sequence level. If true, these features should discriminate bidirectional promoters and unidirectional promoters of genes with DNA repair functions.

Thus, this chapter presents additional evidence of these regulatory networks. Specifically, this chapter provides evidence that there are distinct regulatory signatures for (1) genes involved in certain types of cancers, (2) bidirectional versus unidirectional promoters and (3) specific DNA repair pathways. The authors have identified transcription factor binding sites in bidirectional promoters of genes implicated in breast and ovarian (B/O) cancers. Additionally, they have discovered novel transcription factor binding sites that may serve as regulatory elements to distinguish DNA repair genes with bidirectional promoters from DNA repair genes with unidirectional promoters. Applications of this work extend to a collection of novel transcription factor binding sites shared among genes acting as checkpoint factors of DNA repair pathways. These findings have important implications – as evidence of novel regulatory mechanisms, and new insights into cancer biology (i.e., genomic elements relevant to transcriptional regulation) are gained.

2. Regulatory features of genes implicated in breast and ovarian cancers

This section provides evidence to support the hypothesis that there are distinct regulatory control systems among bidirectional and unidirectional promoters. Additionally, this section presents transcription factor binding sites discovered in bidirectional promoters of genes implicated in breast and ovarian cancers.

As reported in Yang et al. 2007, we identified transcription factor binding sites for known factors in genes implicated in B/O cancers. The enrichment of bidirectional promoters in several cancer genes, and in additional genes having functions in DNA repair, suggests common mechanisms of regulation. We used expression clustering and enrichment of genes with bidirectional promoters to group the cancer genes into expression groups from the full genome to address features common among the clusters that might indicate the presence of regulatory networks. The cancer-related genes that were identified and studied are listed below, along with their descriptions from GeneCards (Safran et al. 2010). The Elnitski group was the first to report that this set of genes has bidirectional promoters.

All genes were assessed for the top most related gene expression profiles in the genome using the gene sorter tool at the UCSC Genome Browser and expression data from the Novartis GNF Atlas2 (containing expression profiles for 96 tissues). Each cluster was then compared to all the others to identify intersection points (by gene names) among the lists of co-expressed genes. Using a process of multidimensional scaling, the gene lists were compared and a putative regulatory network was generated (Figure 1). The *MLH1* gene appeared in several co-expression clusters and therefore occupied a central location with connections to 7 other genes (*BARD1, FANCA, BRCA1, CHK2, BRCA2, TP53* and *FANCF*). Two additional genes co-occupied the central position with *MLH1*. *COMMD3* (an uncharacterized protein) and *ITGB3BP*, a regulator of apoptosis in breast cancer cells.

2.1 Network visualization

The bidirectional promoters that are associated with the breast and ovarian cancer genes were considered an affiliation network or a bipartite graph. In this example nodes represent the genes in the co-expression clusters and edges connect the genes appearing in more than one list. The higher the number of appearances of any gene from the ten co-expression lists, the more central its position in the network. Geodesic distances between genes were computed (e.g. length of the shortest path between genes through promoters, and the geodesic distance matrix was scaled using a metric multidimensional scaling (MDS)

Gene	Description from GeneCards (Safran 2010)
BARD1	This gene encodes a protein which interacts with the N-terminal region of BRCA1.
BRCA1	This gene encodes a nuclear phosphoprotein that plays a role in maintaining genomic stability, and it also acts as a tumor suppressor.
BRCA2	Inherited mutations in BRCA1 and this gene, BRCA2, confer increased lifetime risk of developing breast or ovarian cancer.
CHK2	In response to DNA damage and replication blocks, cell cycle progression is halted through the control of critical cell cycle regulators. The protein encoded by this gene is a cell cycle checkpoint regulator and putative tumor suppressor.
ERBB2	This gene encodes a member of the epidermal growth factor (EGF) receptor family of receptor tyrosine kinases.
TP53	This gene encodes tumor protein p53, which responds to diverse cellular stresses to regulate target genes that induce cell cycle arrest, apoptosis, senescence, DNA repair, or changes in metabolism.
FANCA	DNA repair protein that may operate in a post-replication repair or a cell cycle checkpoint function. May be involved in inter-strand DNA cross-link repair and in the maintenance of normal chromosome stability.
FANCB	DNA repair protein required for FANCD2 ubiquitination.
FANCD2	Required for maintenance of chromosomal stability. Promotes accurate and efficient pairing of homologs during meiosis. Involved in the repair of DNA double-strand breaks, both by homologous recombination and single-strand annealing. May participate in S phase and G2 phase checkpoint activation upon DNA damage. Promotes BRCA2/FANCD1 loading onto damaged chromatin.
FANCF	DNA repair protein that may operate in a postreplication repair or a cell cycle checkpoint function. May be implicated in interstrand DNA cross-link repair and in the maintenance of normal chromosome stability.

Table 1. The B/O cancer-related genes that were studied.

algorithm (in UCINET 6; Borgatti et al., 2002). The distance between the 10 B/O cancer genes represents their similarity based on the number of shared genes found in the co-expression clusters. Genes in the center of the network were present in the largest number of gene clusters, seven out of 10, indicating that co-expression clusters intersect through common regulatory nodes.

2.2 Transcription factor binding site analysis

A systematic search of transcription factor binding sites in the list of bidirectional promoters was used to assess regulatory connections at the DNA level, and revealed several in common (using a motif finding algorithm we searched for the motifs reported in (Xie et al. 2005)). Notably, identical ELK1 binding sites were located at the same distance from ERBB2, FANCD2, and BRCA2 transcription start sites (Yang et al. 2007). ETS factor binding sites were present as a trio with SP1 and PAX4/RXR binding sites in the majority of the promoters. The transcription factors for which binding motifs were found in all of the promoters along with their descriptions from GeneCards (Safran et al. 2010) are reported in Table 2.

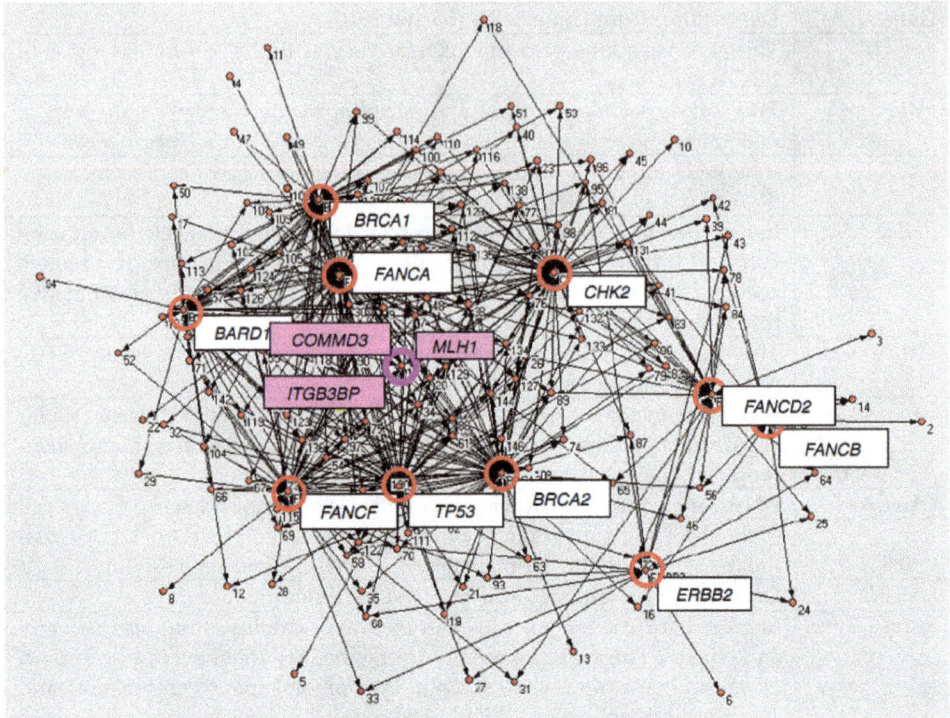

Fig. 1. Co-expression clustering analysis of 10 DNA repair genes finds intersecting nodes.

Transcription Factor	Description from GeneCards
Sp1	Transcription factor that can activate or repress transcription in response to physiological and pathological stimuli. Regulates the expression of a large number of genes involved in a variety of processes such as cell growth, apoptosis, differentiation and immune responses. May have a role in modulating the cellular response to DNA damage.
NFAT	The nuclear factor of activated T-cells family of transcription factors.
EGR-1	The protein encoded by this gene belongs to the EGR family of C2H2-type zinc-finger proteins. It is a nuclear protein and functions as a transcriptional regulator. Studies suggest this is a cancer suppressor gene.
PAX4	This gene is a member of the paired box (PAX) family of transcription factors. These genes play critical roles during fetal development and cancer growth.
ELK1	ELK1 is a member of ETS oncogene family. The protein encoded by this gene is a nuclear target for the ras-raf-MAPK signaling cascade.

Table 2. Transcription factor binding sites in the promoters of the B/O cancer genes.

3. Unbiased assessment of transcription factor binding sites in two subgroups of genes from DNA repair pathways

The research reported in (Yang et al. 2007) provides strong evidence that a unique set of regulatory proteins control genes that contain bidirectional promoters by comparing co-expression clusters of genes enriched for bidirectional promoters versus those depleted for bidirectional promoters. This section reports on a study that identified transcription factor binding sites that are specific to genes in DNA repair pathways (Lichtenberg et al. 2009). The promoters of genes from the DNA repair pathways were partitioned into two groups, those that are bidirectional (32 promoters) and those that are unidirectional (42 promoters).

3.1 Assessment of individual sites

Each group of promoters was analyzed to discover putative transcription factor binding sites. The analysis was performed with WordSeeker motif discovery software (Lichtenberg et al. 2010), which employs high performance supercomputer-based algorithms to perform motif enumeration and to construct Markov models. Our analysis revealed that the average nucleotide G+C content of the bidirectional promoters was slightly higher than the unidirectional promoters, 59.87% versus 50.84%, respectively. These differences were rigorously controlled by the use of the Markov model, which examines background frequencies of each nucleotide in the collection of sequences. Unique sets of binding sites were identified for each group, some of which represent novel binding sites.

A statistical analysis of the promoters of the DNA repair genes revealed a number of significant DNA binding site motifs. Some of the discovered motifs correspond to recognition sequences of known proteins. These are listed in Table 3, along with their p-values and the corresponding transcription factors known to bind to the motifs (as determined by the TRANSFAC database (Wingender et al. 2000) and the JASPAR database (Bryne et al. 2008)). In addition, novel motifs, representing uncharacterized transcription factor binding sites, were discovered in the bidirectional and unidirectional promoters from DNA repair pathway genes (see Table 4 for the motifs and their p-values).

Motif (bidirectional promoters)	p-Value	Transcription Factor	Motif (unidirectional promoters)	p-Value	Transcription Factor
AGGGCCGT	0.04142	*MYB*	ACCCGCCT	0.00656	*SP1*
CAGGGGCC	0.02841	*V$WT1_Q6*	AGGAAACA	0.03295	*NFAT*
CGTGGGGG	0.04701	*E2F*	ATTAAAAT	0.05372	*OCT1*
GGCCCGCC	0.06682	*SP1*	CGGAAACC	0.04210	*AREB6*
TCCCGGCT	0.05408	*ELK1*	GCAGGGCG	0.07134	*PF0096*
TCCCGGGA	0.06861	*STAT5A*	GGGGAGTA	0.03321	*FOXC1*
TCGCGCCA	0.01539	*PF0112*	GGGGCTGC	0.06212	*LRF*
TCTGAGGA	0.01350	*TFIIA*	TGGGCGGA	0.06334	*GC*

Table 3. Enriched motifs matching characterized transcription factor binding sites discovered in the bidirectional promoters (columns 1 and 2) and in the unidirectional promoters (columns 3 and 4).

Motif (bidirectional promoters)	P-Value	Motif (unidirectional promoters)	P-Value
ACTCCAGC	0.06212	AGCCGGCT	0.05007
AGAAAAGA	0.02756	ATTCCCAG	0.05599
AGGGAGGG	0.07159	CCTCTTTA	0.03381
CAGCAGCC	0.10540	CGCCCCTT	0.11386
CGACTCCG	0.02756	CGGCGGCG	0.04742
CGCGGCCG	0.03377	CTCCCGCT	0.05998
CGGGCCGA	0.06548	CTTCTTTC	0.03773
GCCCCTCC	0.07021	GCGCCGCG	0.09760
GCCGGCGA	0.03662	GGGCGCCC	0.08390
GGCAGGGA	0.10334	GTGCGTTT	0.06286
GGGCCAGG	0.09632	TCCGCCGG	0.05794
GGGGCCGG	0.05265	TCTCCCCT	0.07881
TCTGGGAT	0.01466	TCTTCTTC	0.04649
TGAAGCCA	0.05699	TGCGCCGA	0.04148
TGCCCGCG	0.08277	TTGGTCTC	0.08543
TGCGGAAT	0.02132	TTTCTCCA	0.06840
TGCTGAGA	0.03377	TTTTTTGA	0.04742

Table 4. Uncharacterized motifs discovered in the promoters of DNA repair genes. Words are ordered alphabetically.

3.2 Assessment of paired binding sites

To identify putative regulatory modules (co-acting regulatory elements), we identified statistically overrepresented pairs of DNA motifs in each set of promoters. Motif pairs are shown in Table 5. The motif pair scores are computed as the product of (1) the number sequences, S, in which the pair occurs and (2) the natural log of the ratio of S and the expected value of S, E_s; i.e., the score is $S \cdot \ln(S/E_s)$. The genomic signatures (significant DNA motifs and motif pairs) of the bidirectional promoters were virtually non-overlapping with the signatures of unidirectional promoters. This provides strong support for the hypothesis that the regulatory mechanisms of bidirectional promoters are unique. Additionally, this work contributes a significant enhancement to the available knowledge about transcriptional regulation of genes involved in DNA repair pathways, and implicates the presence of a regulatory network.

4. Unbiased assessment of transcription factor binding sites of checkpoint factor genes from DNA repair pathways

We have performed a focused, detailed characterization of the checkpoint factors in DNA repair pathways (Elnitski et al. 2010). The checkpoint factors (Kanehisa et al. 2008, Wood 2005, Helleday et al. 2008) are activated upon detection of DNA damage, resulting in halting the cell cycle so that subsequent DNA repair pathways can mend the damage. In addition to examining the most recognized promoter in each gene (the 5' end of the full-length transcription unit), we assessed alternative start sites for each checkpoint factor gene as independent regulatory units, to discover putative transcription factor binding sites. In this

Co-Occurring Motif Pair (Bidirectional Promoters)		Score	Co-Occurring Motif Pair (Unidirectional Promoters)		Score
TCTGAGGA	TCGCGCCA	12.1158	GTTCATTC	TCCGCCGG	11.2184
ACTCCAGC	TCGCGCCA	11.8387	CTGTGTGC	TGCGCCGA	11.1966
GCCCAGCC	TCCGCCGC	11.1827	TGACGCGA	CTCCCGCT	10.9997
GCCCAGCC	CGGAGCGC	10.8711	AGCCGGCT	GGGGAGTA	10.0590
TGCCCGCG	TCCCGGGA	10.7404	ATTGCAGG	ATTCTCTC	9.5459
GGCAGGGA	GGGCCAGG	9.8609	GGGGAGTA	AGGAAACA	9.3177
TCCCGGGA	TCGCGCCA	9.8112	CTGGGAGC	GTTCATTC	9.0337
AGCCTGTC	TCCCGGGA	9.7646	CCTTCCGA	CTGGGAGC	8.8439
GGAGGCTG	TCGCGCCA	9.7250	TGGGCGGA	ACCCGCCT	8.7895
TCCGCCGC	GCCCCTCC	9.6830	TTTCTCCA	CGGAAACC	8.6446
AGAAAAGA	TCGCGCCA	9.4042	CCCCCGCG	ACCCGCCT	8.5339
GCCCAGCC	GCCCCTCC	9.2808	TCCGCCGG	GGGGCTGC	7.7522
TGCCAAAA	GCCGGCGA	9.2604	AGCTGGCT	CCAGGCTG	7.7192
CAGCAGCC	TGCGGAAT	9.1297	TTGGTCTC	AGGAAACA	7.6068
AGGGCCGT	TCCCGGCT	9.1249	CTGGGAGC	TCCGCCGG	7.3021

Table 5. Putative transcription factor binding modules discovered in promoters of DNA repair genes.

section we report the DNA motifs that were discovered, along with several clusters of related genes and promoters. We hypothesize that these similar components implicate regulatory networks responsible for co-regulation of the checkpoint factor genes.

We studied fourteen checkpoint factor genes, which are listed in Table 6. The number of alternative promoters per gene, shown in parentheses, varied for each gene. Because most of the genes have alternative promoters, we analyzed a total of thirty promoters. The complete set of alternative promotes is shown in Table 7. Alternative promoters were identified using annotations of genes in the UCSC Human Genome Browser. Transcription start sites of transcript isoforms served as the coordinates around which 900 bp upstream and 100 bp downstream were defined as the putative promoter region. Alternative promoters with significant overlap were truncated or removed from the analysis. DNA sequences were obtained for the forward and reverse strands of the genome to ensure coverage of words that might have biased nucleotide content and be subject to omission during the Markov model analysis stage.

Gene	Description from GeneCards (Safran 2010)
ATM (5)	The protein encoded by this gene (ataxia telangiectasia mutated) belongs to the PI3/PI4-kinase family. This protein functions as a regulator of a wide variety of downstream proteins, including *p53*, *BRCA1*, *CHK2*, *RAD17*, *RAD9*, and *NBS1*. This protein and the closely related kinase ATR are thought to be master controllers of cell cycle checkpoint signaling pathways, required for cell response to DNA damage and for genome stability.
ATR (2)	The protein encoded by this gene (ataxia telangiectasia and Rad3 related) belongs the PI3/PI4-kinase family, and is most closely related to ATM. Both proteins share similarity with

	Schizosaccharomyces pombe rad3, a cell cycle checkpoint gene required for cell cycle arrest and DNA damage repair in response to DNA damage. This kinase has been shown to phosphorylate *CHK1*, *RAD17*, and *RAD9* and *BRCA1*. Transcript variants utilizing alternative polyA sites exist.
ATRIP (1)	The product of this gene (ATR interacting protein) is an essential component of the DNA damage checkpoint, and binds to single-stranded DNA coated with replication protein A that accumulates at sites of DNA damage. The encoded protein interacts with the ataxia telangiectasia and Rad3 related protein, a checkpoint kinase, resulting in accumulation of the kinase at intranuclear foci induced by DNA damage. Multiple transcript variants encoding different isoforms have been found for this gene.
CHEK1 (3)	Required for checkpoint mediated cell cycle arrest in response to DNA damage or the presence of unreplicated DNA. May also negatively regulate cell cycle progression during unperturbed cell cycles. Binds to and phosphorylates *CDC25A*, *CDC25B* and *CDC25C*. Binds to and phosphorylates *RAD51*. Binds to and phosphorylates *TLK1*. May also phosphorylate multiple sites within the C-terminus of *TP53*, which promotes activation of *TP53* by acetylation and enhances suppression of cellular proliferation.
CHEK2 (2)	The protein encoded by this gene is a cell cycle checkpoint regulator and putative tumor suppressor. It contains a forkhead-associated protein interaction domain essential for activation in response to DNA damage and is rapidly phosphorylated in response to replication blocks and DNA damage. This protein interacts with and phosphorylates *BRCA1*, allowing *BRCA1* to restore survival after DNA damage. Three transcript variants encoding different isoforms have been found for this gene.
CLK2 (2)	This gene encodes a member of the *CLK* family of dual specificity protein kinases. *CLK* family members have been shown to interact with, and phosphorylate, serine- and arginine-rich (SR) proteins of the spliceosomal complex, which is a part of the regulatory mechanism that enables the SR proteins to control RNA splicing.
HUS1 (1)	The protein encoded by this gene is a component of an evolutionarily conserved, genotoxin-activated checkpoint complex that is involved in the cell cycle arrest in response to DNA damage. This protein forms a heterotrimeric complex with checkpoint proteins *RAD9* and *RAD1*. DNA damage induced chromatin binding has been shown to depend on the activation of the checkpoint kinase ATM, and is thought to be an early checkpoint signaling event.
MDC1 (2)	The protein encoded by this gene (mediator of DNA-damage checkpoint) is required to activate the intra-S phase and G2/M phase cell cycle checkpoints in response to DNA damage. This nuclear protein interacts with phosphorylated histone H2AX near sites of DNA double-strand breaks through its *BRCT* motifs, and facilitates

	recruitment of the ATM kinase and meiotic recombination 11 protein complex to DNA damage foci.
NBS1 (1)	The encoded protein is a member of the *MRE11/RAD50* double-strand break repair complex which consists of 5 proteins. This gene product is thought to be involved in DNA double-strand break repair and DNA damage-induced checkpoint activation.
P53/TP53 (3)	This gene encodes tumor protein *p53*, which responds to diverse cellular stresses to regulate target genes that induce cell cycle arrest, apoptosis, senescence, DNA repair, or changes in metabolism.
PER1 (1)	This gene is a member of the Period family of genes and is expressed in a circadian pattern in the suprachiasmatic nucleus, the primary circadian pacemaker in the mammalian brain. Genes in this family encode components of the circadian rhythms of locomotor activity, metabolism, and behavior. The specific function of this gene is not yet known. Alternative splicing has been observed in this gene; however, these variants have not been fully described.
RAD1 (2)	This gene encodes a component of a heterotrimeric cell cycle checkpoint complex, known as the 9-1-1 complex, that is activated to stop cell cycle progression in response to DNA damage or incomplete DNA replication. The 9-1-1 complex is recruited by *RAD17* to affected sites where it may attract specialized DNA polymerases and other DNA repair effectors. Alternatively spliced transcript variants of this gene have been described.
RAD17 (3)	The protein encoded by this gene is highly similar to the gene product of Schizosaccharomyces pombe rad17, a cell cycle checkpoint gene required for cell cycle arrest and DNA damage repair in response to DNA damage. This protein recruits the *RAD1-RAD9-HUS1* checkpoint protein complex onto chromatin after DNA damage,. The phosphorylation of this protein is required for the DNA-damage-induced cell cycle G2 arrest, and is thought to be a critical early event during checkpoint signaling in DNA-damaged cells. Eight alternatively spliced transcript variants of this gene, which encode four distinct proteins, have been reported.
RAD9A (2)	This gene product is highly similar to Schizosaccharomyces pombe rad9, a cell cycle checkpoint protein required for cell cycle arrest and DNA damage repair in response to DNA damage. This protein is found to possess 3' to 5' exonuclease activity, which may contribute to its role in sensing and repairing DNA damage. It forms a checkpoint protein complex with *RAD1* and *HUS1*. This complex is recruited by checkpoint protein *RAD17* to the sites of DNA damage, which is thought to be important for triggering the checkpoint-signaling cascade. Use of alternative polyA sites has been noted for this gene.

Table 6. The checkpoint factors genes that were studied. The number of alternative promoters is shown in parentheses next to each gene name.

Checkpoint Factors	Alternative promoters (hg18 coordinates)
ATM	(ATM5) chr11:107662328-107663378_+ (ATM5) chr11:107662328-107663378_- (ATM2) chr11:107643346-107644396_+ (ATM2) chr11:107643346-107644396_- (ATM3) chr11:107597768-107598818_+ (ATM3) chr11:107597768-107598818_- (ATM4) chr11:107671910-107672960_+ (ATM4) chr11:107671910-107672960_- (ATM5) chr11:107679611-107680661_+ (ATM5) chr11:107679611-107680661_-
ATR	(ATR1) chr3:143780308-143781358_+ (ATR1) chr3:143780308-143781358_- (ATR2) chr3:143671051-143672101_+ (ATR2) chr3:143671051-143672101_-
ATRIP	chr3:48462221-48463271_+ chr3:48462221-48463271_-
CHEK1	(CHEK13) chr11:125000333-125001383_+ (CHEK13) chr11:125000333-125001383_- (CHEK12) chr11:125018185-125019235_+ (CHEK12) chr11:125018185-125019235_- (CHEK13) chr11:124999245-125000295_+ (CHEK13) chr11:124999245-125000295_-
CHEK2	(CHEK22) chr22:27467772-27468822_+ (CHEK22) chr22:27467772-27468822_- (CHEK22) chr22:27460665-27461715_+ (CHEK22) chr22:27460665-27461715_-
CLK2	(CLK22) chr1:153509855-153510905_+ (CLK22) chr1:153509855-153510905_- (CLK22) chr1:153514075-153515125_+ (CLK22) chr1:153514075-153515125_-
HUS1	chr7:47985721-47986771_+ chr7:47985721-47986771_-

MDC1	(MDC1$_2$) chr6:30792781-30793831_+ (MDC1$_2$) chr6:30792781-30793831_- (MDC1$_2$) chr6:30789060-30790110_+ (MDC1$_2$) chr6:30789060-30790110_-
NBS1	chr8:91066025-91067075_+ chr8:91066025-91067075_-
P53 (TP53)	(TP53$_3$) chr17:7519486-7520536_+ (TP53$_3$) chr17:7519486-7520536_- (TP53$_2$) chr17:7531538-7532588_+ (TP53$_2$) chr17:7531538-7532588_- (TP53$_3$) chr17:7520612-7521662_+ (TP53$_3$) chr17:7520612-7521662_-
PER1	chr17:7996377-7997427_+ chr17:7996377-7997427_-
RAD1	(RAD1$_2$) chr5:34954089-34955139_+ (RAD1$_2$) chr5:34954089-34955139_- (RAD1$_2$) chr5:34951438-34952488_+ (RAD1$_2$) chr5:34951438-34952488_-
RAD17	(RAD17$_3$) chr5:68699879-68700929_+ (RAD17$_3$) chr5:68699879-68700929_- (RAD17$_2$) chr5:68723716-68724766_+ (RAD17$_2$) chr5:68723716-68724766_- (RAD17$_3$) chr5:68701287-68702337_+ (RAD17$_3$) chr5:68701287-68702337_-
RAD9A	(RAD9A$_2$) chr11:66918716-66919766_+ (RAD9A$_2$) chr11:66918716-66919766_- (RAD9A$_2$) chr11:66914998-66916048_+ (RAD9A$_2$) chr11:66914998-66916048_-

Table 7. Alternative promoters, indicated by their genomic coordinates, of genes involved in cell-cycle checkpoint factor pathways.

Statistical analysis of thirty promoters found several interesting DNA words, which predict DNA elements that participate in the regulation of the DNA repair checkpoint factors. The most significant words discovered are listed in Table 8. Words that are shared among the gene sets identify regulatory relationships. Reverse complement words are reported separately, as internal verification on the process. Words without a reverse complement example indicate a particular bias in the nucleotide content.

Word	Promoters	Sln(S/Es)
ACAGCCAT	ATM_2	5.41
	$CHEK2_2$	
	$CLK2_1$	
ATGGCTGT	ATM_2	5.41
	$CHECK2_2$	
	$CLK2_1$	
GCCTGGGA	ATR_1	5.40
	$CHEK2_1$	
	$CLK2_2$	
	$MDC1_1$	
	$MDC1_2$	
	$RAD1_2$	
TCCCAGGC	ATR_1	5.40
	$CHEK2_1$	
	$CLK2_2$	
	$MDC1_1$	
	$MDC1_2$	
	$RAD1_2$	
ACTCCCTA	ATM_3	5.29
	$CHEK2_1$	
	$RAD17_2$	
TAGGGAGT	ATM_3	5.29
	$CHEK2_1$	
	$RAD17_2$	
AGCGGCCA	ATR_1	5.24
	ATR_2	
	$CHEK1_1$	
TGGCCGCT	ATR_1	5.24
	ATR_2	
	$CHEK1_1$	
GAAATGAA	ATM_2	5.24
	ATM_3	
	ATR_2	
	$CLK2_2$	
	$HUS1$	
	$MDC1_1$	
TTCATTTC	ATM_2	5.24
	ATM_3	
	ATR_2	
	$CLK2_2$	
	$HUS1$	
	$MDC1_1$	
AATGCAGG	$RAD1_1$	4.97
	$TP53_1$	
	$TP53_2$	
	$TP53_3$	

CCTGCATT	$RAD1_1$	4.97
	$TP53_1$	
	$TP53_2$	
	$TP53_3$	
ATCCCTGA	$ATRIP$	4.73
	$CHEK1_3$	
	$RAD1_2$	
	$RAD17_1$	
TCAGGGAT	$ATRIP$	4.73
	$CHEK1_3$	
	$RAD1_2$	
	$RAD17_1$	
GTATTTTA	ATM_4	4.58
	$CHEK1_2$	
	$NBS1$	
	$RAD17_1$	
	ATR_2	
	$TP53_1$	
	$HUS1$	
	$RAD17_1$	

Table 8. Top 15 enumerated DNA words, based on the $S \cdot \ln(S/E_S)$ overrepresentation score, and the alternative promoters, identified by subscript.

5. Visualization and interpretation of data

Shared words among the checkpoint factor genes suggested the presence of regulatory networks. We assessed the relationships by generating network depictions in the form of interaction networks (Figure 2) and a circos diagram (Figure 3) constructed from the summary data in Table 9. To derive Figure 2, a metric MDS was conducted on the affiliation network defined in Table 9. The resulting graph was then spring-embedded, with node repulsion, to facilitate visualization (Borgatti, 2002). The interaction network depicts the distribution of the DNA words among the genes (note that each gene appears once, representing all alternative promoters as a single node). Genes are denoted by blue squares and words are represented with red circles. Bold lines indicate multiple occurrences of a word. Reverse complement words are shown independently.

The circos diagram represents the information in a closed circular space, wherein connections between words on one side of the diagram extend to genes on the other side. The putative nodes of the regulatory networks are defined by multiple edges, representing a characterized transcription factor or a novel DNA binding site, or a checkpoint factor gene.

Some of the discovered words correspond to known binding sites for transcription factors, reported in the JASPAR and TRANSFAC databases of transcription factors (see Table 10). The relationships between the top fifteen words and the transcription factors are depicted in the circos diagram in Figure 4. Note that multiple binding site motifs were discovered for many of the transcription factors, and that several of the sites match the binding patterns of more than one transcription factor.

Row segments

Column segments

	ATM	ATR	ATRIP	CHEK1	CHEK2	CLK2	HUS1	MDC1	NBS1	P53	PER1	RAD1	RAD17	RAD9A
ACAGCCAT	1	0	0	0	1	1	0	0	0	0	0	0	0	0
ATGGCTGT	1	0	0	0	1	1	0	0	0	0	0	0	0	0
GCCTGGGA	0	1	0	0	1	1	0	2	0	0	0	1	0	0
TCCCAGGC	0	1	0	0	1	1	0	2	0	0	0	1	0	0
ACTCCCTA	1	0	0	0	1	0	0	0	0	0	0	0	1	0
TAGGGAGT	1	0	0	0	1	0	0	0	0	0	0	0	1	0
AGCGGCCA	0	2	0	1	0	0	0	0	0	0	0	0	0	0
TGGCCGCT	0	2	0	1	0	0	0	0	0	0	0	0	0	0
GAAATGAA	2	1	0	0	0	1	1	1	0	0	0	0	0	0
TTCATTTC	2	1	0	0	0	1	1	1	0	0	0	0	0	0
AATGCAGG	0	0	0	0	0	0	0	0	0	3	0	1	0	0
CCTGCATT	0	0	0	0	0	0	0	0	0	3	0	1	0	0
ATCCCTGA	0	0	1	1	0		0	0	0	0	0	0	1	1
TCAGGGAT	0	0	1	1	0	0	0	0	0	0	0	0	1	1
GTATTTTA	1	0	0	1	0	0	0	0	1	0	0	0	1	0

Cell value

Table 9. The top ranked words (rows of the table), based on statistical significance ($S \cdot \ln(S/E_s)$), and the number of occurrences of each word in the promoter regions of genes (columns).

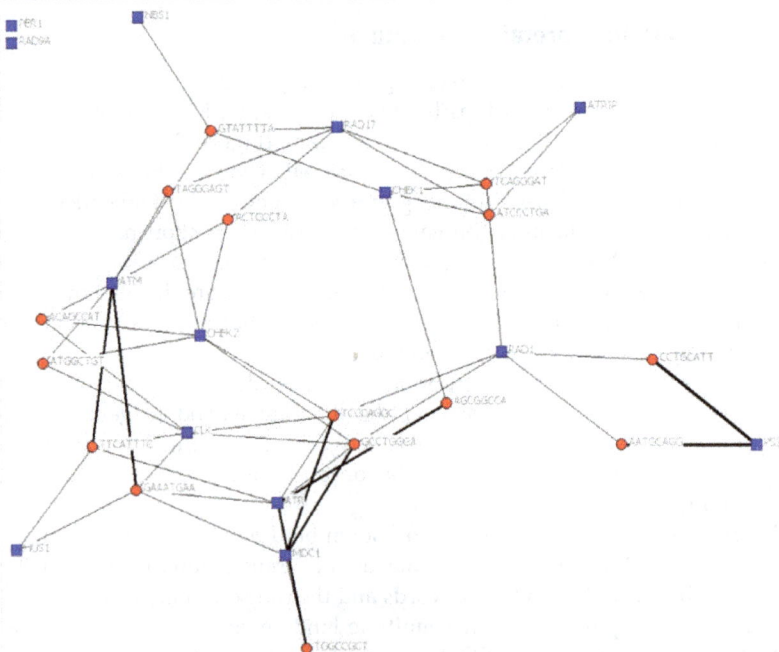

Fig. 2. Model of the checkpoint regulatory network using multidimensional scaling.

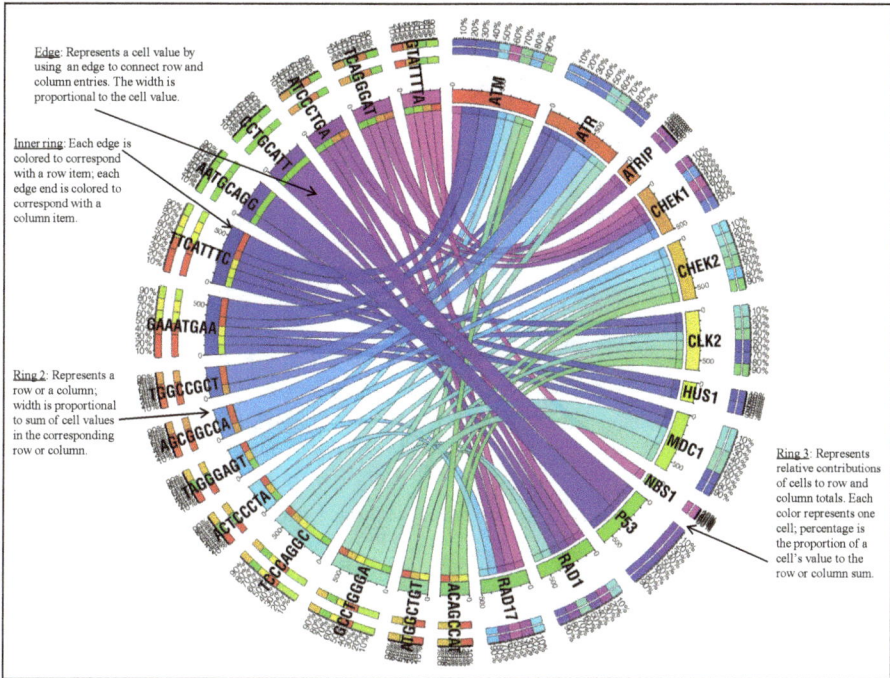

Fig. 3. Circos 2diagram of the top 15 words, based on statistical significance, and their occurrences in gene promoter regions.

TFBS	$S \ln(S/E_s)$	TF
ACCCCCAC	3.76	PF0091, Pax-4
ACTCCCTA	4.67	Helios A, p300
ATGGCTGT	5.42	Cap
ATTAAAGA	3.72	Pax-2
CGGAGCCC	3.95	LF-A1
CTGAAATT	3.80	STAT1, STAT6
CTTTTGAA	3.83	TCF-4
GAAAAATT	3.76	CIZ
GCACCTGC	3.68	PF0035, AP-4, cap, Lmo2 complex
GTGGCTGC	3.64	cap
TACTTTTT	3.82	FOXC, CIZ, RUSH-1alpha1
TATATTTA	3.82	FOXL1, PF0028, PF0054
TCCTTTCT	3.70	Pax-2
TTTTTATA	3.64	FOXL1

Table 10. Known transcription factor binding sites (with significance scores and corresponding transcription factor) discovered in the promoters of the checkpoint factors genes.

Additional insight into the regulatory network for the checkpoint factors can be seen in Figure 5, which replaces the DNA binding site motifs with the names of implicated transcription factors for each DNA repair gene. The diagram indicates the discovery of specific transcription factors involved in the control of each gene and shared among multiple genes. Up to seven transcription factors were discovered for each gene.

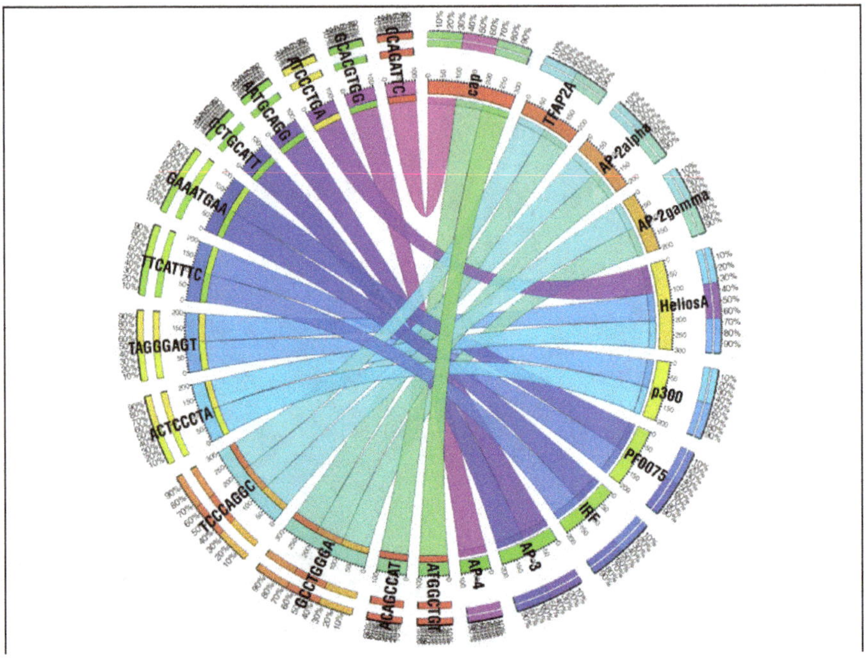

Fig. 4. Circos diagram showing the top 15 DNA motifs found in promoters of checkpoint factor genes and their related transcription factors (number of occurrences are multiplied by 100).

6. Conclusions

This chapter provides a summary of research into transcriptional regulatory networks controlling DNA repair pathways, bidirectional versus unidirectional promoters of DNA repair genes, and bidirectional promoters of breast and ovarian cancer genes. DNA words are shared among these promoters, and these words represent both known and unknown binding sites for transcription factors. When possible, we report the highest scoring assignment of transcription factor to DNA word. Our research represents a novel approach to identifying factors involved in transcriptional regulation of DNA repair genes. Many of these proteins have dual roles in transcription and DNA repair. Although many of the regulatory relationships are characterized at the level of protein-protein interactions, little research is available on the transcriptional regulatory networks that control DNA repair gene expression. We present evidence that regulatory networks exist among these genes, and support the claim that bidirectional promoters (implicated in B/O cancers) have a distinct network from unidirectional promoters. The identification of putative binding sites provides the first step in the elucidation of higher-order interdependencies among DNA repair genes in the cell. We also report preliminary findings on pairs of binding sites that represent regulatory modules. Furthermore, we show that there is much overlap among promoters of DNA repair genes, and that shared DNA binding motifs can be distributed among a collection of alternative promoters, each having distinct combinations of regulatory elements. The complex nature of the data can be simplified for visual interpretation using visualization techniques such as network modeling and circos diagrams.

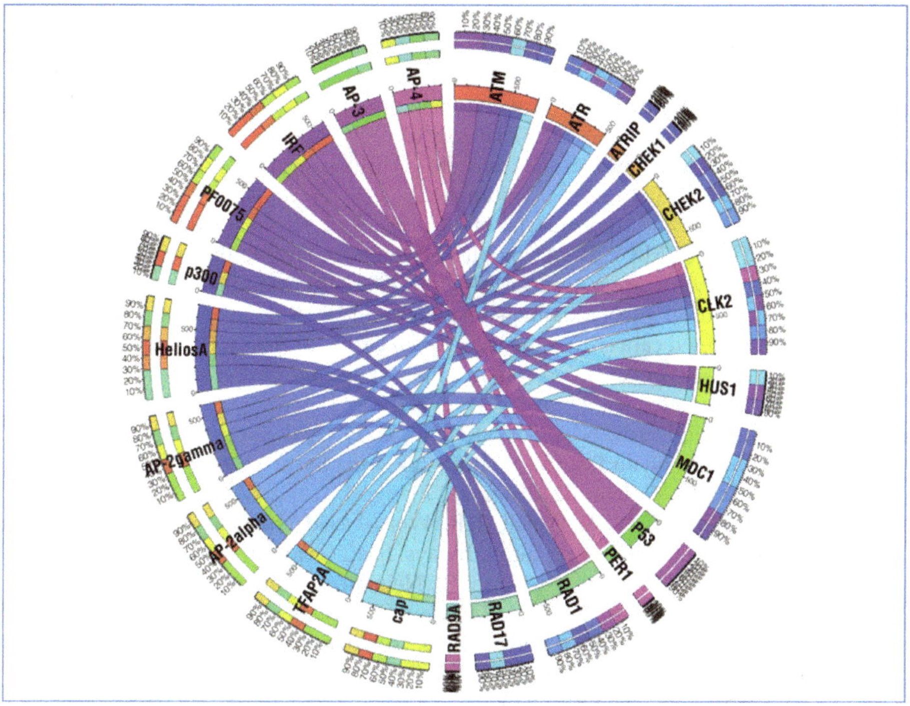

Fig. 5. Relationships between genes and transcription factors.

7. Acknowledgments

This work was supported by the Intramural Research Program of NHGRI (LE and LK) and by the Ohio Plant Biotechnology Consortium, the Choose Ohio First Program of the University System of Ohio, and the Ohio University Graduate Research and Education Board (LW).

8. References

Bellizzi, A.M., & Frankel, W.L. Colorectal cancer due to deficiency in DNA mismatch repair function: a review. *Adv Anat Pathol.* 2009 Nov;16(6):405-17.

Berwick, M. & Vineis, P. Markers of DNA repair and susceptibility to cancer in humans: an epidemiologic review. *J Natl Cancer Inst.* 2000 Jun 7;92(11):874-97.

Borgatti, S.P., Everet, M.G., and Freeman, L.C. 2002. Ucinet 6 for Windows: Software for Social Network Analysis. Harvard: Analytic Technologies.

Bryne, J.C., & Valen, E., Tang, M.H., Marstrand, T., & Winther, O. JASPAR, the open access database of transcription factor-binding profiles: new content and tools in the 2008 update. *Nucleic Acids Res* 2008, 36:D102-106.

Elnitski, L., Lichtenberg, J., & Welch, L.R.Regulatory network nodes of checkpoint factors in DNA repair pathways. *BCB '10: Proceedings of the First ACM International Conference on Bioinformatics and Computational Biology,* 529-536. 2010.

Helleday, T., Petermann, E., Lundin, C., Hodgson B., & Sharma, R.A. DNA repair pathways as targets for cancer therapy, *Nature Reviews Cancer* (2008) 8 (3): 193-204.

Hurley, P.J. & Bunz, F. ATM and ATR: components of an integrated circuit. *Cell Cycle*. 2007 Feb 15;6(4):414-7.

Jensen, N.M., Dalsgaard, T., Jakobsen, M., Nielsen, R.R., Sørensen, C.B., Bolund, L., & Jensen, T.G. An update on targeted gene repair in mammalian cells: methods and mechanisms. *J Biomed Sci*. 2011 Feb 2;18:10.

Kanehisa, M., Araki, M., Goto, S., Hattori, M., Hirakawa, M., Itoh, M., Katayama, T., Kawashima, S., Okuda, S., Tokimatsu, T., & Yamanishi, Y. KEGG for linking genomes to life and the environment. *Nucleic Acids Research* (2008) 36:D480-D484.

Kelley, M.R. & Fishel, M.L. DNA repair proteins as molecular targets for cancer therapeutics. *Anticancer Agents Med Chem*. 2008 May;8(4):417-25.

Kinsella, T.J. Understanding DNA damage response and DNA repair pathways: applications to more targeted cancer therapeutics. *Semin Oncol*. 2009 Apr;36(2 Suppl 1):S42-51.

Li, C., Wang, L.E., & Wei, Q. DNA repair phenotype and cancer susceptibility--a mini review. *Int J Cancer*. 2009 Mar 1;124(5):999-1007.

Liang, Y., Lin, S.Y., Brunicardi, F.C., Goss, J., & Li, K. DNA damage response pathways in tumor suppression and cancer treatment. *World J Surg*. 2009 Apr;33(4):661-6.

Lichtenberg, J., Jacox, E., Welch, J.D., Kurz, K., Liang, X., Yang, M.Q., Drews, F., Ecker, K., Lee, S.S., Elnitski, L., & Welch, L.R. Word-based characterization of promoters involved in human DNA repair pathways. *BMC Genomics* 10(Suppl 1):S18. 2009.

Lichtenberg, J., Kurz, K., Liang, X., Al-ouran, R., Neiman, L. Nau, L.J., Welch, J.D., Jacox, E., Bitterman, T., Ecker, K., Elnitski, L., Drews, F., Lee, S.S., & Welch, L.R. WordSeeker: Concurrent bioinformatics software for discovering genome-wide patterns and word-based genomic signatures. *BMC Bioinformatics*, 2010, 11(Suppl 12):S6.

Naccarati, A., Pardini, B., Hemminki, K., &Vodicka, P. Sporadic colorectal cancer and individual susceptibility: a review of the association studies investigating the role of DNA repair genetic polymorphisms. *Mutat Res*. 2007 May-Jun;635(2-3):118-45.

Safran, M., Dalah, I., Alexander, J., Rosen, N., Iny Stein, T., Shmoish, M., Nativ, N., Bahir, I., Doniger, T., Krug, H., Sirota-Madi, A., Olender, T., Golan, Y., Stelzer, G., Harel, A., & Lancet, D. GeneCards Version 3: the human gene integrator. *Database* (Oxford). 2010 Aug 5;2010:baq020.

Shibata, A., Barton, O., Noon, A. T., Dahm, K., Deckbar D., Goodarzi, A.A., Löbrich, M., & Jeggo, P.A. Role of ATM and the damage response mediator proteins 53BP1 and MDC1 in the maintenance of G(2)/M checkpoint arrest. *Mol Cell Biol*. 2010 Jul;30(13):3371-83.

Smith, J., Tho, L.M., Xu, N., & Gillespie, D.A. The ATM-Chk2 and ATR-Chk1 pathways in DNA damage signaling and cancer. *Adv Cancer Res*. 2010;108:73-112.

Wingender, E., Chen, X., Hehl, R., Karas, H., & Liebich, I. TRANSFAC: an integrated system for gene expression regulation. *Nucleic Acids Res* 2000, 28:316-319.

Wood R.D., Mitchell, M., & T. Lindahl, Human DNA repair genes, 2005, *Mutation Res* (2005) 577 (1-2): 275-83.

Xie, X., Lu J., Kulbokas, E.J., Golub, T.R., Mootha, V., Lindblad-Toh, K., Lander, E.S., & Kellis, M. Systematic discovery of regulatory motifs in human promoters and 3' UTRs by comparison of several mammals. *Nature*. 2005 Mar 17;434(7031):338-45.

Yang, M.Q., Koehly, L.M., & Elnitski, L.L. Comprehensive annotation of bidirectional promoters identifies co-regulation among breast and ovarian cancer genes. *PLoS Comput Biol*. 2007 Apr 20;3(4):e72.

7

Structure-Function Relationship of DNA Repair Proteins: Lessons from BRCA1 and RAD51 Studies

Effrossyni Boutou[1,2], Vassiliki Pappa[3],
Horst-Werner Stuerzbecher[4] and Constantinos E. Vorgias[1]
[1]Department of Biochemistry & Molecular Biology
Faculty of Biology, School of Sciences Athens University, Athens
[2]Prenatal Diagnosis Lab, Laiko Hospital, Athens
[3]2nd Propaedeutic Pathology Clinic, Medical School, Athens University, Athens
[4]Molecular Cancer Biology Group, Institute of Pathology, UK-SH, Luebeck
[1,2,3]Greece
[4]Germany

1. Introduction

Accurate transfer of genetic information is vital for all living organisms in order to guarantee species survival. DNA damage occurs spontaneously during a cell's life due to either endogenous causes such as Reactive Oxygen Species (ROS) produced during metabolism or due to exogenous insults such as Ionizing Radiation (IR) or genotoxic agents in food / water and environment, to which an organism is exposed. Endogenous damage, due to intrinsic instability of chemical bonds in DNA structure, occurs spontaneously under normal physiologic conditions and is calculated to be approximately 10^4 events per cell, per day (Lindahl, 1993). Moreover, during DNA replication base adducts can cause collapse of replication forks and DNA double strand breaks (DSBs) are introduced in order to reinitiate genome duplication process.

As the genome carries all necessary information for life and evidently preservation of genome integrity is critical for cell survival, a number of mechanisms have evolved over time to ensure the most effective performance of the genome repair procedure. DNA repair mechanisms are capable of repairing practically all different types of chromosomal lesions (single and double strand breaks, base modifications, etc.) ensuring that genetic information is accurately transferred to the next generation. The cell's response to DNA damage (DNA Damage Response, DDR) encompasses a complex network of proteins, consisting of DNA damage recognition, signal transduction, transcriptional regulation, cell cycle control, DNA repair and verification of the repair efficiency, depending on the type of lesion, the replication status of the genome as well as the cell cycle stage. (scheme 1). Many excellent recent reviews as well as other chapters in the current volume extensively cover this topic (Rogakou, 1999; Lisby & Rothstein 2005; Murphy & Moynahan, 2010).

Defects in repair efficiency are the consequence of dysfunction of either upstream damage signalling or the central repair process. The current chapter covers topics referring to

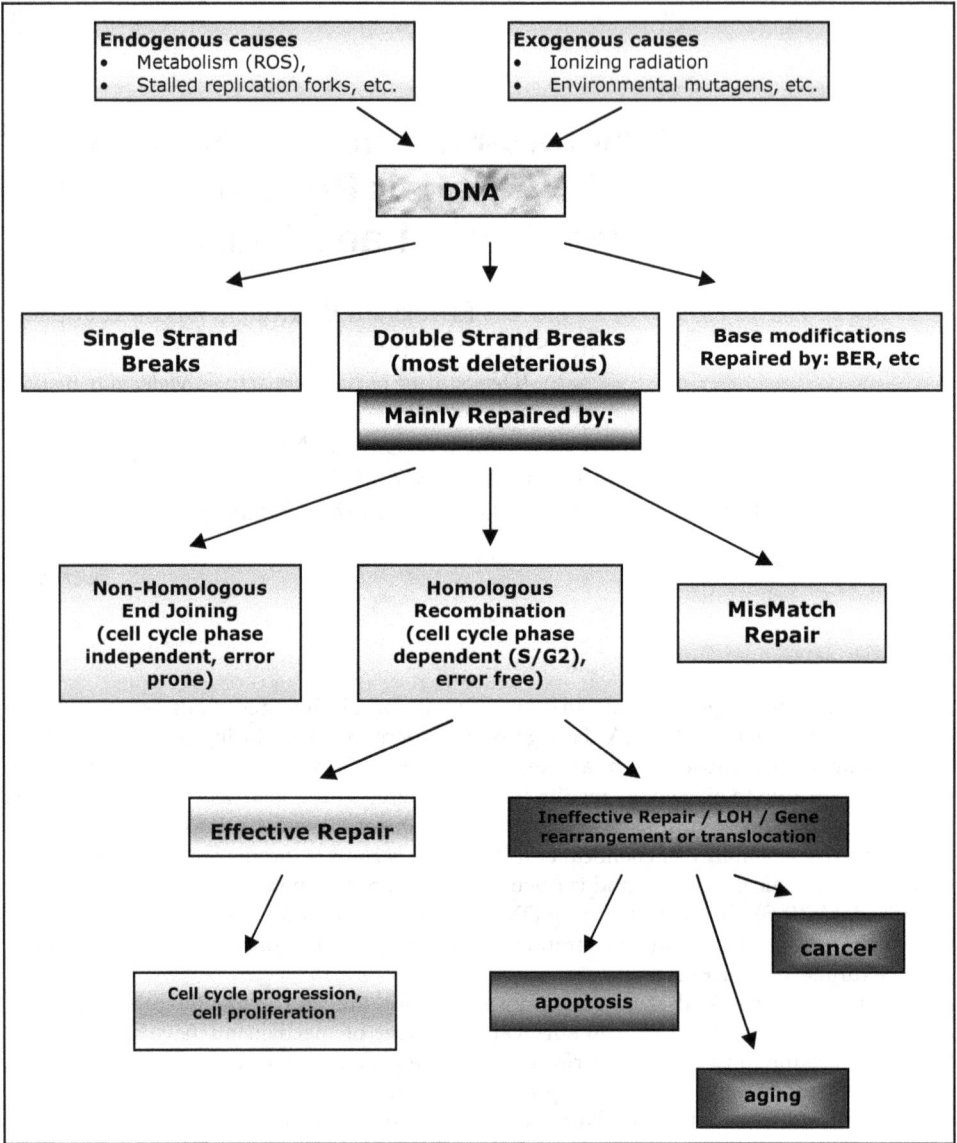

Scheme 1. Simplified diagram of DNA damage response network

factors/events influencing the structure - function relation of key molecules involved in each of the two processes, namely the BRCA1 and RAD51/Rad51 proteins, respectively. The breast cancer susceptibility gene 1 (brca1), isolated by reverse genetics in 1994, encodes for a large multifunctional protein (BRCA1) whose function is regulated by multiple post-translational modification events, driving the multi tasks performed, by which BRCA1 conducts almost all steps of DDR. The important anti-tumorigenic role of BRCA1 is strongly supported by its correlation with increased breast & ovarian familial cancer susceptibility in individuals carrying BRCA1 mutations. On the other hand, RAD51 is a relatively small and rigid protein playing a basic role (homology search strand and strand exchange) in the high fidelity DNA repair mechanism of Homologous Recombination (HR). RAD51 appears absolutely vital for cell survival, as its depletion results in embryonic lethality, it is highly conserved throughout evolution and up to now there is not a single mutation in the amino acid sequence detected in any type of cancer, although there is a strong correlation between its expression levels and both cancer development and cancer progression.

2. Consequences of genomic instability

Loss or insufficiency of DDR and genome repair can lead to an increased susceptibility to cancer due to the consequential genomic instability. Ineffective repair may result in subsequent mutations of genes required for cellular replication and division. The genome repair pathways also communicate with processes involved in induction of senescence and apoptosis when the damage cannot be repaired. Carefully balanced signalling cascades and regulatory systems are implicated in the maintenance of healthy cell survival in order to unfavour tumorigenesis and maintain stem and progenitor cells for renewal (anti-ageing) (Seviour & Lin 2010). Therefore, an effectively repaired genome is crucial not only for cancer prevention but also for lifespan extension. This notion is even more enhanced by the emerging benefit of the response of HR defective tumors to double strand break (DSB) producing therapies a promising and continuously evolving field. A clearer understanding of the biochemical, structural and genetic processes in conjuction with clinical data will lead to the development of more effective treatment strategies for both cancer and ageing processes.

2.1 Genomic instability and cancer

It is generally accepted that tumors are derived from a single genetically unstable cell, and that the unstable cell population as a whole continues to acquire further chromosomal abnormalities over time, although the precise mechanisms of acquisition of these abnormalities still remain unclear. Hereditary cancers are often characterized by the presence of a specific type of genomic instability, termed chromosomal instability. In these cancers, chromosomal instability can often be attributed to mutations in DNA repair genes, suggesting that the driving force behind tumor development is an increase in spontaneous genetic mutations resulting from lack of appropriate management of DNA damage. A second form of genomic instability, termed microsatellite instability, is also associated with defects in DNA repair, namely the mismatch repair system. However, in non-hereditary sporadic tumors, the picture is less clear. It should be emphasized that cancer is an extremely complex set of diseases, and that cancer cells develop many different mechanisms to achieve a similar phenotype of independent and uncontrolled growth (Hanahan & Weinberg, 2000; Luo et al., 2009 as reviewed by Schild & Wiese 2010).

Many of the DDR components including BRCA1 are known to be lost or mutated in human tumors. While loss of BRCA1 has been shown to lead to the development of mammary tumors in mouse models, the genetic diversity within those tumors suggests that the loss of BRCA1 may not directly be responsible for tumorigenesis. It is more likely, therefore, that the role of BRCA1 in the initiation of cancer is a result of its effects on DNA repair and the maintenance of genomic integrity. BRCA1 -/- tumors are shown to display numerous chromosomal aberrations. Analysis of BRCA1 -/- mouse models, coupled with the study of human BRCA1 -/- tumors, has revealed prevalence for p53 mutations in these tumors, which is likely to be caused by the decrease in genomic stability associated with the defects in DNA repair. Overall, these and many other data suggest that the loss of cell cycle checkpoints confers a selection advantage to cells with DNA repair defects, thereby triggering tumorigenesis in genetically unstable cells. Moreover, an increase in genomic instability is significantly correlated with the metastatic potential of the tumor. Further studies are required to determine whether this involvement in metastasis is a result of acquired genetic mutations resulting from DNA repair defects, or whether other mechanisms are required for this process (Murphy & Moynahan, 2010).

DNA repair by the high fidelity mechanism of homologous recombination, termed as Homologous Recombinational Repair (HRR) is practically the only 'error free' repair mechanism of the cell and as it requires a sister chromatid, normally is active in late S and G2 phases of the cell cycle. HRR involves a compex network of recombination mediators and co-mediators. Defects in recombination mediators and co-mediators, leading to impaired HRR, are indicated as major contributors in carcinogenesis and particularly in breast cancer (reviewed by Pierce et al., 2001; Henning & Stuerzbecher, 2003; Murphy & Moynahan, 2010). Nevertheless, up to now not a single mutation in the coding region of RAD51, the central recombinase in the HRR pathway, has been found in many tumor types examined. However, many primary tumor cells and cancer cell lines express significantly modified levels of RAD51 (Maacke et al., 2000; Henning & Stuerzbecher 2003; Klein 2008) and at least partly, this misregulation in protein expression levels is correlated with the polymorphism G->C in the 5'untranslated region of rad51 mRNA, as shown in some cases of hereditary breast tumors with BRCA2 mutation. As extensively discussed in the excellent and comprehensive review of Schild & Wiese 2010, RAD51 overexpression presumably complements initial HRR defects, thereby limiting genomic instability during carcinogenic progression and may explain the high frequency of *TP53* mutations in human cancers, as wild-type p53 represses RAD51 expression. Notably, both positive and negative regulations of HRR are required to maintain genomic stability by precise repair and suppression of deleterious rearrangements.

2.2 Genomic instability and ageing

DNA damage is a prominent cause of cancer in frequently dividing cells since cell proliferation is a prerequisite for the manifestation of genetic changes as permanent mutations. In contrast, DNA damage in infrequently dividing cells is likely a prominent cause of ageing (Best, 2009). Therefore, in addition to its role in the maintenance of genomic integrity, the DDR has been hypothesized to play a critical role in organismal ageing. Supporting to this hypothesis is the observation that DNA repair disorders such as Werner's syndrome, Bloom's syndrome and Ataxia telangiectasia, syndromes also characterised by premature ageing and / or retarded growth, are often called "segmental progerias" ("accelerated ageing diseases"). Individuals suffering from such diseases appear elderly and suffer from ageing-related diseases at an abnormally young age, while not manifesting all the symptoms of old age.

Ageing, resulting from the accumulation of damage to molecules, cells, organs and tissues over time, is believed to be caused by two cellular processes: senescence and apoptosis.

2.2.1 Senescence and DNA damage

Senescence, a phenomenon describing the irreversible cease of cell division, was initially described by Hayflick and Moorhead in 1961 and includes replicative senescence and oncogene-induced senescence, both of which involve aspects of the DDR.

Replicative senescence results from progressive shortening of telomeres with repeated rounds of cell replication. p53–dependent senescence serves as a tumor suppressor mechanism and is activated by the uncapping of critically shortened telomeres which are recognised as damaged DNA (Feldser and Greider, 2007). Recent studies argue that p53 can either activate or suppress senescence in cells, depending on their specific transcriptional activities and its interaction with partner molecules. As described by Vigneron & Vousden, 2010, the role of p53 in cell fate determination is even more complex as it involves epigenetic modifications of chromosomal DNA and relates chronic DDR signalling with increased levels of p53 acetylation. In addition to p53 other DNA damage response proteins like ATM have been associated with replicative senescence. ATM depletion in mice results in an increase in both chromosomal end-to-end fusion events and cell cycle-dependent telomere loss. These mice exhibit a premature ageing phenotype as defined by increased hair graying, alopecia and marked weight loss. Expression of mutant BRCA1 in mice also results in premature ageing, accompanied by an increase in cellular senescence and an increased susceptibility to certain cancers. The enhanced senescence observed in these mice may interfere with the fact that senescent cells have been noted to modify their tissue microenvironment. This phenomenon is thought to synergize with the accumulation of DNA damage over time to encourage cancer growth.

Oncogene-induced senescence can be induced by the overexpression of oncogenes by among others the induction of DNA damage resulting from both the generation of reactive oxygen species (ROS) and the hyper-replication of DNA. Both of these mechanisms activate the DNA damage response, which result in senescence by similar processes that induce replicative senescence.

2.2.2 Apoptosis and DNA damage

The accumulation of DNA damage can also lead to apoptosis. Activation of p53 by DNA damage and its role in the regulation of expression of pro-apoptotic proteins has been well documented. This is further supported by the fact that functional p53 is not detected to the majority of tumors. In cases of decline of the immune system an increase in p53 mediated apoptosis has been observed, linking ageing with apoptotic function. Moreover, constitutively activated p53 in mice also showed that, while high levels of p53 protect against cancer, it also accelerates the ageing process by reducing the mass of various tissues. The human condition Ataxia telangiectasia, which results from mutations in ATM, is associated with substantial neuro-degeneration. This has been shown in a mouse model to result from an accumulation of neurons harboring genomic damage, due to the inability of the mutant ATM protein to stimulate the p53 apoptotic cascade. Chk2 has also been shown to regulate apoptosis in a p53-dependent manner *in vitro* and *in vivo* in response to DNA damage (Seviour & Lin, 2010). Notably, the major recombinase of HR, Rad51, seems also to interact with p53, possibly serving as a tool for monitoring the extension as well the effectiveness of DNA repair processes (Henning & Sturzbecher Toxicology, 2003; Morita et al., 2010).

Consequently, impaired DDR appears to have dramatic effects on both tumorigenesis and premature ageing. At the molecular level, DDR impairment could be attributed to irregular interactions between the complexes involved in each process due to structural changes of DDR components resulting from either mutagenesis, modified post-translational modifications of altered protein levels, guiding equilibrium in favour of abnormal decisions.

3. Structural & functional integrity is essential for protein interactions with partner molecules

The efficient performance of DNA repair processes requires the coordinated actions of many players and mainly depends on proper interactions between the protein components of each pathway involved which in turn mainly relies on their functional structure (structure – function relationship). At least three major mechanisms influence protein function due to or independently of its tertiary structure:

a. Missense mutations in the coding region can modify the primary structure of a protein resulting in dys-functional folding of the protein, and / or instability in the cellular environment.

b. Post-translational modifications (phosphorylation, ubiquitination, ribosylation and acetylation) regulating distinct interactions with partner molecules driving the various pathways in which the protein in question is implicated.

c. Regulation of the protein levels, availability through regulation of the quantity of the protein, which can drive cell decisions in improper pathways, leading to abnormal cell cycle progression, cell division and possibly malignant transformation or aggressive tumor progression.

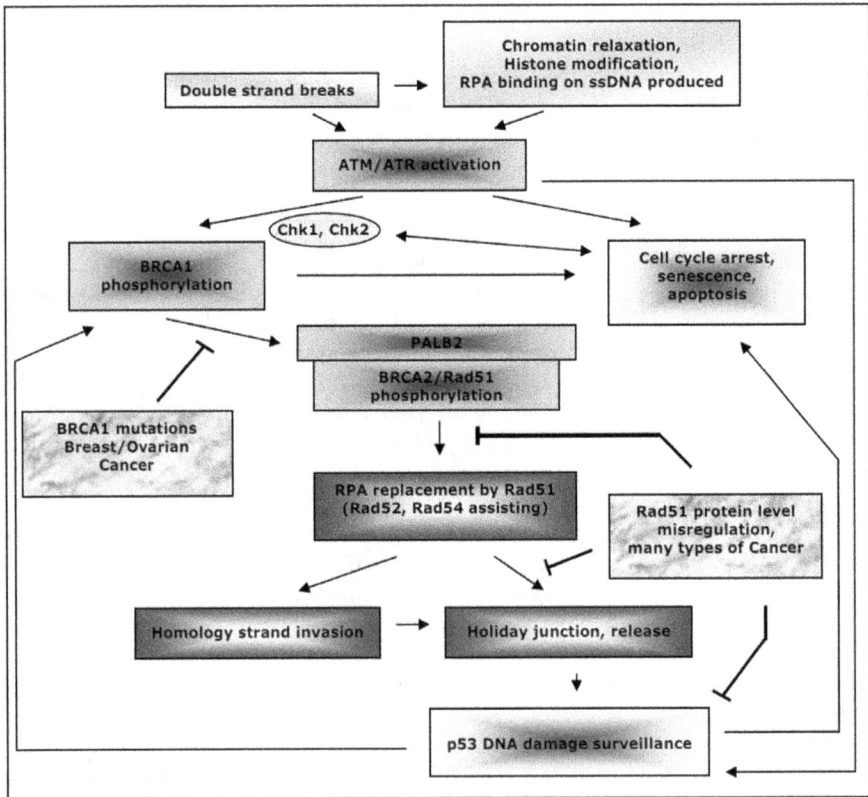

Scheme 2. Simplified Scheme of major steps of DNA Double strand break repair by Homologous Recombination.

The fundamental proteins involved in the HRR pathways are highly conserved in almost all organisms ranging from bacteria to human. The significance of this repair system is also indicated by the fact that defects in HRR cause human hereditary cancers as well as sporadic tumors. In many cases the dysfunction of proteins observed in many tumors helps to elucidate all three categories of the mechanisms mentioned above and to clarify different aspects of DDR pathways (Murphy & Moynahan, 2010).

Herein we will focus on current information regarding the structure – function relationship of two key players in regulation and performance of DSB repair – the most deleterious reported lesion of the genome – BRCA1 and RAD51. BRCA1 is a core component of many multi-molecular complexes involved in DNA damage detection, HR regulation, cell cycle regulation and genome transcription. RAD51, a key factor of HRR, replaces RPA on the produced single strands of damaged DNA and performs the search for homologous DNA strand and exchange in order to restore the damaged DNA sequence according to its sister chromatid. Moreover, RAD51 is implicated in telomere maintenance via ALT pathway and is also involved in mitochondrial DNA repair. Via at least its indirect interaction with BRCA1 as well as with direct p53 complex formation, RAD51 seems to be an interplayer responsible for communication between DNA repair effectiveness, cell replication, apoptosis or senescence decisions.

3.1 The BRCA1 structure – function relation paradigm

The Breast Cancer Susceptibility Gene 1 protein (BRCA1) is a multifunctional nuclear phosphoprotein of 1863 residues (220-240 kDa). BRCA1 was attributed the role of a tumor suppressor involved in multiple cellular functions (Starita & Parvin, 2003). Most of BRCA1 is located in the cell nucleus and is phosphorylated in a cell cycle-dependent manner by a number of kinases (reviewed by Ouchi, 2006). Depending on the position and the number of phosphorylated residues, BRCA1 participates in different multiprotein complexes performing diverse tasks. Therefore, BRCA1 has been implicated in a variety of functions required for the maintenance of genomic stability (Rowling et al., 2010).

Regarding DDR, BRCA1 has been attributed many roles in regulation of genome integrity including DNA replication, cell cycle checkpoint control, apoptosis, regulation of transcription, chromatin unfolding and protein ubiquitination. The ascribed functions are exerted through an extensive number of protein interactions reported (Jasin M. 2002 as cited in Murphy & Moynahan 2010). In brief, upon detection of chromatid relaxation due to breakage of both strands of the double helix of DNA, BRCA1 – being activated by ATR kinase – is recruited to the damage breakpoint assisting assembly of the BRCA2 – RAD51 complex in order to replace RPA and restore damage by the high fidelity process of HRR. In parallel, BRCA1 interaction with Fanconi Anemia (FA) and other complexes regulates G1/S and G2/M checkpoints. BRCA1 implication in cell cycle regulation is assisted by complex formation with BRCA1 interacting protein C-terminal helicase (BRIP1) and CtIP which are activated in S-phase by post-translational modifications. A graphical representation of the BRCA1 protein, including sites of both post-translational modifications and regions involved in protein-protein interactions, is depicted in Fig 1.

The amino-terminal region of BRCA1 contains a distinct ~100aa RING finger motif involved in ubiquitin ligase activity and enables BRCA1 to mono- or poly-ubiquitinate cellular proteins. BARD1 (another RING and BRCT domain-containing protein) is the 'permanent' partner of BRCA1 in the formation of the ubiquitine ligase complex. Phosphorylation of specific residues of BRCA1 appears to regulate its participation in transcription regulation and ubiquitination of substrate proteins. As many different BRCA1 species are produced by alternative splicing of its mRNA, the phosphorylated residues each form contains may regulate different functions. Moreover, the balance between full length and spliced forms of BRCA1 may play an important role in tumor suppression (Ouchi, 2006).

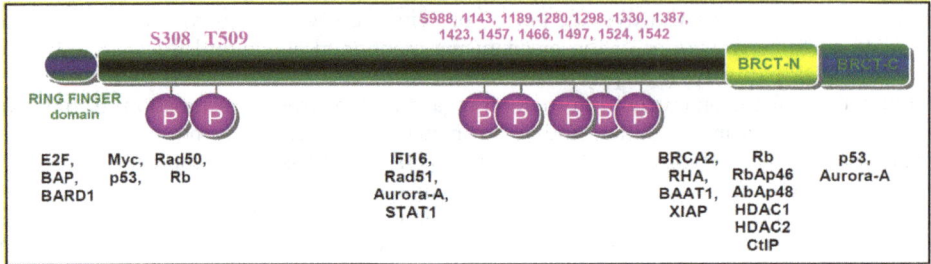

Fig. 1. Primary structure of BRCA1 Tumor Suppressor Protein. The phosphorylation sites and and its binding partners are indicated.

The carboxy-terminal domain of BRCA1 contains two structurally identical BRCT (**BRCA1** **C-terminal**) tandem repeats each containing ~90 amino acid residues. BRCT domains are found in proteins involved in DNA repair and maintenance of genomic stability, and more recently, the BRCT repeat has been recognized as a phosphopeptide-binding domain. The structure of each repeat consists of a parallel four-stranded β-sheet located at the central part of the domain surrounded by three α-helices (Fig. 2). The two BRCT repeats fold together in a specific head-to-tail manner, giving rise to the formation of a conserved, almost all-hydrophobic, inter-repeat interface, forming a phosphopeptide binding pocket. BRCT like domains have also been found in BRCA1 interacting proteins such as 53BP1 and BARD1. BRCT repeats are a family of phosphopeptide binding domains implicated in DNA damage response. Therefore, BRCTs are considered as protein-docking modules involved in eukaryotic DNA repair. Although BRCTs are characterized by low sequence homology they retain a generally well-conserved structure organization.

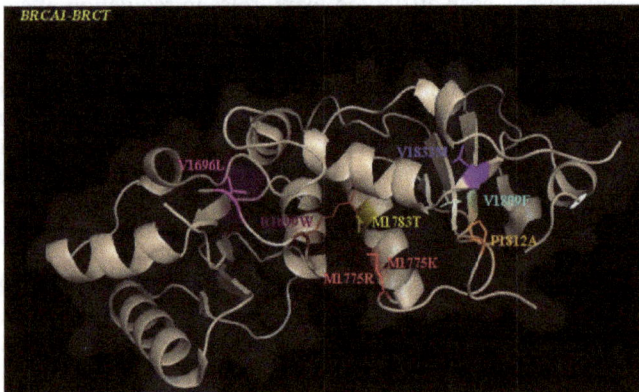

Fig. 2. Ribbon representation of the BRCA1-BRCTstructure. Positions of selected cancer–related mutations are indicated. M1775K and M1783T are located at the inter-BRCT-repeat interface where the BRCA1-BRCT binding groove for Phe 13 is also located. The exposed V1696L is located at the N-terminal BRCTstructural repeat. V1809F, P1812A are found at the C-terminal BRCT repeat. The positions of missense mutations from previously published studies are also depicted. (From Drikos et al., 2009).

Analysis of the BRCA1 mutational database (BIC, http://research.nhgri.nih.gov/bic/) indicates that both RING and BRCT repeats are most frequently mutated in women at risk of cancer, and have been further studied. Many research groups have used structural and biochemical methods to probe the function of BRCA1 and characterize the plethora of unclassified variants identified in breast cancer patients found in BIC database. Among the hundreds of distinct mutations uncovered in BRCA1, for the vast majority, there is insufficient genetic linkage data to determine the cancer risk associated with them (Glover, 2006).

The central region of BRCA1, between the two terminal domains, bears relatively low sequence identity between mammalian BRCA1 homologs, and attempts to define structured domains within this region indicate that this part of BRCA1 is largely unstructured (Glover, 2006; Mark, 2005). This region is extensively phosphorylated by DNA damage-associated kinases like ATM and may serve as a phosphorylation-dependent docking site for other proteins involved in the DNA damage response, or even for damaged DNA itself (Paul et al., 2001; Mark, 2005; Ouchi, 2006).

3.2 BRCA1 structure modifications found in cancer

Mutations in brca1 and brca2 genes have been found in 30-50% of hereditary breast and ovarian cancers. Women carrying BRCA1 mutations are particularly susceptible to the development of breast or ovarian cancer at an age earlier than 35-40 years old with a probability rate of 45-60% and 20-40%, respectively.

Most cancer-associated BRCA1 mutations identified so far, result in the premature translational termination of the protein and influence BRCA1 integrity and function. A large number of missense mutations is located in BRCT tandem repeats of BRCA1, while only few of them may cause loss of the protein's function, abolition of protein interactions and protein miss-localization. Therefore, it seems that the BRCT repeats in BRCA1 are essential for the tumor suppressing function of the protein as protein truncation and missense variants within the BRCT domain have been shown to be associated with human breast and ovarian cancers.

Variants that result in large truncations are deleterious to function and therefore can be classified as disease-associated. In contrast, missense mutations typically remain unclassified. Thus, the BIC database currently contains more than 108 missense mutations in the BRCT domains of BRCA1, but only 7% of them have been classified. These missense mutations may be either polymorphisms or mutations predisposing the carrier to cancer progression. The variants D1692Y, C1697R, R1699W, A1708E, S1715R, P1749R and M1775R all appear to be associated with an increased risk of breast cancer, while M1625I appears to be a benign polymorphism (Williams et al., 2003). Unfortunately, most of the missense mutations could not been assessed for disease association. An attempt to classify these variants by measuring the thermodynamic stability of the BRCA1 BRCT domains resulted in investigation of the effects of 36 missense mutations (Rowling et al., 2010). The mutations show a range of effects. Some do not change the stability, whereas others destabilize the protein by as much as 6 kcal mol−1; one-third of the mutants were considered to destabilize the protein by an even greater amount, as they could not be expressed in soluble form in *Escherichia coli*. Several computer algorithms were used in an attempt to predict the mutant effects. According to these results the variants were grouped into two classes (destabilizing by less than or more than 2.2 kcal mol−1). Importantly, with the exception of the few mutants located in the binding site, none showed a significant reduction in affinity for phosphorylated substrate. These results indicate that despite very large losses in stability, the integrity of the structure is not compromised by the mutations. Thus, the majority of mutations seem to cause loss of function by reducing the proportion of BRCA1 molecules that are in the folded state and increasing the proportion of molecules that are unfolded. The authors predict that small molecule stabilization of the structure could be a generally applicable preventative therapeutic strategy for rescuing many BRCA1 mutations. Another recent approach by Lee et al., 2010, extended in 117 variants, comprehensively shows how functional and structural information can be useful in the development of models to assess cancer risk.

Cancer-associated mutations in the BRCT domain of BRCA1 (BRCA1-BRCT) abolish its tumor suppressor function by disrupting interactions with other proteins such as BACH1. Many cancer-related mutations do not cause sufficient destabilization to lead to global unfolding under physiological conditions, and thus abrogation of function probably is due to localized structural changes. Molecular dynamics simulations on three cancer-associated mutants, A1708E, M1775R, and Y1853ter, and on the wild type and benign M1652I mutant, followed by comparison of the structures and fluctuations showed that only the cancer-associated mutants exhibited significant backbone structure differences from the wild-type crystal structure in BACH1-binding regions, some of which are far from the mutation sites. These BACH1-binding regions of the cancer-associated mutants also exhibited increases in their fluctuation magnitudes compared with the same regions in the wild type and M1562I mutant, as quantified by quasiharmonic analysis. The increased fluctuations in the disease-related mutants suggest an increase in vibrational entropy in the unliganded state that could result in a larger entropy loss in the disease-related mutants upon binding BACH1 than in the wild type. Vibrational entropies of the A1708E and wild type in the free state and bound to a BACH1-derived phosphopeptide, calculated using quasiharmonic analysis, determined the binding entropy difference DeltaDeltaS between the A1708E mutant and the wild type. In overall such biophysical/biochemical studies supported by suited algorithms showed that the observed differences in structure, flexibility, and entropy of binding are likely to be responsible for abolition of BACH1 binding, and illustrate that many disease-related mutations could have very long-range effects. Such methods have potential for identifying correlated motions responsible for other long-range effects of deleterious mutations. (Gough et al., 2007)

The C-terminal BRCT domains are also evidenced to mediate the transcriptional activity of BRCA1. Most of the published mutations within the BRCT domains have been reported to affect BRCA1 nuclear functions including DNA repair and transcriptional activity. The biochemical and biophysical studies of our group have already demonstrated that mutations of the BRCT domain: (i) affected the folding of the domain to a varying degree depending on the induced destabilization and (ii) altered and abolished the affinity of BRCT domain to synthetic phosphopeptides corresponding to BRCT interacting regions of pBACH1/ BRIP1 and pCtIP, by affecting the structural integrity of the BRCT active sites.

BRCA1 is a nuclear-cytoplasmic shuttling protein and its nuclear localization is regulated by the combined action of nuclear localization (NLS) and nuclear export signals (NES). In most cases, however, cellular and ectopically expressed BRCA1 are primarily nuclear due to nuclear import mediated by the two NLSs and interaction with the RING domain binding protein, BARD1, which can carry BRCA1 into the nucleus and trap it there by masking its nuclear export signal.

Despite the structural studies of BRCA1-BRCT protein mutants, the influence of these mutations at protein localization in cellular level has not yet been adequately addressed. Only few of them have already determined to present protein mislocalization. BRCA1 mutations of the BRCT domain altered BRCA1 localization, causing the protein to be excluded from the nucleus. Two of the C-terminal mutations (M1775R and Y1853X) that restricted nuclear localization are identical to mutations that disrupt BRCA1 C-terminal folding, suggesting that the conformational changes they elicit might be deleterious to BRCA1 nuclear transport. This nuclear exclusion was not due to increased nuclear export, but to reduced nuclear import. Similar findings were observed for both the overexpressed and endogenous forms of the BRCT mutant, BRCA1 (5382insC). Also, Chen et al., 1995 have published controversial findings, which claimed that BRCA1 was detected almost exclusively in the cytoplasm in breast cancer tissues, but remained nuclear in normal tissue and in other cancer cell types.

In our laboratory more than fifteen BRCA1-BRCT proteins mutants have already been studied for structural and functional alterations in protein's integrity. The most destabilizing protein mutants such as M1775K, V1809F (Fig3) were collected in order to be examined in cellular level about their impact in BRCA1 subcellular compartmentalization. M1775K is a rare breast cancer-linked mutation and it has been identified only in two unrelated families of European ancestry with a history of breast cancer. Met1775 is strongly involved in the phosphopeptide-binding pocket of the BRCT domain. The mutation of Met1775, namely the mutation M1775R, is much more frequent worldwide among patients with hereditary breast/ovarian cancer, its association with the disease is epidemiologically established and was the first characterized to be linked to cancer. The M1775R mutation has already shown to change the intracellular localization of BRCA1 protein which is less focused into the nucleus. The M1775K missense variant according to our *in vitro* experiments fails to bind to synthetic peptides such pBACH1/BRIP1 or pCtIP. Structural analysis of the interatomic interactions of Lys1775 show a direct clash of its side chain with Phe 13 of either phosphopeptide, a result arising from the disruption of the BRCT-phosphopeptide binding pocket.

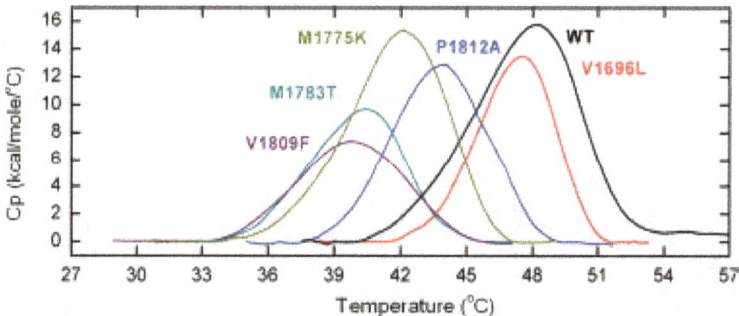

Fig. 3. DSC profiles for the thermally induced denaturation of BRCT-wt and of five missense variants V1809F, M1783T, M1775K, P1812A, and V1696L. (from Drikos et al., 2009).

V1809F is a rare mutation linked to hereditary breast/ ovarian cancer. Only a few cases of the mutation have been submitted to the BIC database with loss of function reported by *in vitro* experiments, regarding the interactions with synthetic phosphopeptides pBACH1/BRIP1 and pCtIP. The residue Val1809 is conserved among species. Val1809 and Met1775 are crucial for the integrity of phosphopeptide binding pocket of the BRCA1 protein and exhibit no binding to either pBACH1/BRIP1 or pCtIP synthetic phosphopeptides. These results, in combination with the fact that V1809F resembles structural destabilization of the native fold similar to M1775R, strongly supports the classification of V1809F as pathogenic.

Additionally, the variant M1652I is located at the first tandem of the BRCT domain and seems to have neutral influence on breast cancer pathogenesis. Based on preliminary structural studies by our laboratory, this variant is less involved in structural alteration of BRCT but further analysis is required. M1652I is classified as low risk mutation. Therefore we decided to include it to our study in order to compare it with more destabilizing mutants such as V1809F and M1775K.

In order to assess how the selected BRCA1-BRCT mutants influence the subcellular localization of BRCA1, we produced BRCA1-GFP fusion proteins with the corresponding mutations introduced at the BRCT domain. The GFP-BRCA1-BRCT mutated proteins were inserted into MCF-7 cells and their subcellular localization was assessed by fluorescent microscopy.

According to our results, destabilizing mutations of the carboxyl terminal region of BRCA1 seem to influence protein localization and presumably DDR. As shown in Fig 4, BRCA1-V1809F-GFP and M17775K are restricted to the cytoplasm in contrast to the nuclear-cytoplasmic localization of BRCA1wt and M1652I. As EGFP-BRCA1-M1652I shows similar subcellular distribution to the BRCA1wt–GFP protein (detected both in the nucleus and the cytoplasm), it is presumed that the structural change caused by replacement of M1652 to Ile has a minor effect of BRCA1 nuclear transport. UV irradiation of cells expressing wt or the mutants mentioned failed to drive M1175K and V1809F to the cell nucleus in contrast to both wt and M1652I which were then detected exclusively in the nucleus and shown to at least in part colocalize with Rad51 foci (data not shown from Drikos et al., submitted). Mutations such as M1775K and V1809F which disrupt BRCA1 C-terminal folding, appears that result to conformational and functional changes which might be restrictive to BRCA1 nuclear transport in contrast to more mild missense variants such as M1652I. These results suggest that structural integrity modifications of the BRCA1-BRCT domain can be reflected to the protein's subcellular localization and therefore can serve for further characterization and classification of the variant, in combination with the structural data (table 1). M1175K and V1809F are located near to the binding site of the inter-repeat region and affect through hydrophobic interactions the structural and functional integrity of the domain.

	Effect of mutation	Cancer Risk of the mutation	Structural Stability	Functional Activity with synthetic peptides	Subcellular localization (-UV)	Subcellular localization (+UV)
GFP-BRCA1-V1809F	Potential hydrophobic	Deleterious	Destabilizing	Alter binding affinity	Cytoplasmic	Cytoplasmic
GFP-BRCA1-M1775K	Potential hydrophobic	Deleterious	Destabilizing	Alter binding affinity	Cytoplasmic	Cytoplasmic
GFP-BRCA1-M1652I	No effect	Neutral	No effect	Unknown	Nuclear-Cytoplasmic	Nuclear
GFP-BRCA1wt					Nuclear-Cytoplasmic	Nuclear

Table 1. Summary of the impact of BRCA1-BRCT mutants on the structural, functional and cellular levels. Mutations such as M1775K and V1809F, which disrupt BRCA1 C-terminal folding, induce also alterations of the integrity of the BRCA1-BRCT domain and the proteins subcellular localization.

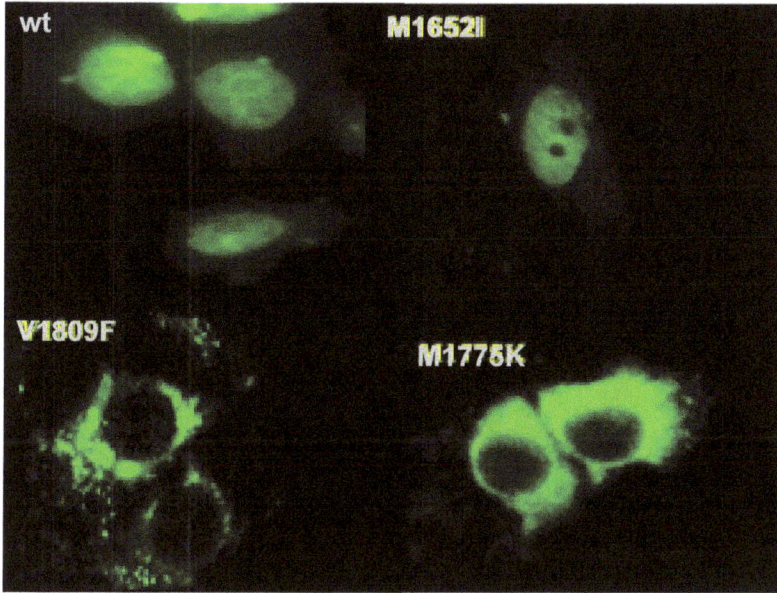

Fig. 4. Subcellular mis-localization of cancer linked GFP-BRCA1-BRCT mutations, M1775K and V1809F in the cytoplasm of MCF-7 cells in contrast to the wild type BRCA1 (wt) and the 'neutral' mutation of BRCT, M1652I which are detected in both the nucleus and the cytoplasm.

A living cell is a dynamic unit with flexible equilibrium between different processes which drive cell fate and determination decisions. The various pathways involved are either activated or suppressed as a result of qualitative and / or quantitative interactions between biomolecules. BRCA1 is an elegant paradigm of both kinds of interactions. Truncated or absent BRCA1 (abolishment of qualitative interactions) leads to impaired DNA repair, carcinogenesis and cancer progression. As indicated by the studies of various missense mutations there are cases where although the mutated BRCA1 seems to function properly, a significant proportion of BRCA1 molecules adopts an unfolding state and only few molecules are found in a given time in proper and functional structure. M1324K, R42573L mutations are good paradigms where biophysical studies of these BRCA1 mutants indicated that although interactions with phosphopeptides were attained, the majority of mutant molecules were detected in improper folding state. Moreover, as many missense variants remain to be characterized, combination of biophysical with cell/molecular biology studies, as in the case of M1775K and V1809F mutations, is expected to substantially contribute in their classification regarding to cancer-relation.

3.3 The RAD51 structure – function relation paradigm

DNA double strand breaks (DSBs), produced by either exogenous causes or in order to restore stalled replication forks during genome replication, are detrimental to cell survival and are mainly restored by either Homologous Recombination Repair (HRR) or Non-Homologous End Joining (NHEJ). The two pathways compete for each other while a number of factors such as cell cycle stage and availability of duplicated DNA regulate the final choice. HRR is the prominent (high fidelity) DSB repair pathway, requiring an homologous DNA sequence present (the identical sister chromatid located in proper distance) and therefore is mainly active during S/G2 phases of cell cycle, while the error prone NHEJ pathway is mainly active during G1 and early S phase, although in certain cases can work in all cell cycle phases. HRR compete NHEJ pathway through a complicated manner, where initial DNA damage signalling factors play important roles. Damage processing and effectiveness of repair are incessantly checked by sensor molecules which through a series of distinct but interconnected pathways prolong cell cycle arrest, induce senescence or apoptosis depending on the information flow and the signals produced (Freeman & Monteiro 2010).

An important early step in HRR is the conversion of double to single stranded DNA in the area of the double strand break, which in turn is coated by the Replication Protein A (RPA) and can be extended up to 4 – 5 Kb on both sides along the break point. Displacement of RPA by RAD51, assembled as a nucleoprotein filament on the ssDNA, is the initial step towards HRR and is highly regulated through interactions with a variety of accessory proteins referred to as the 'recombination mediators' (Essers et al., 2002; Henning & Sturzbecher 2003; Schild & Wiese, 2010; Forget & Kowalczykowski, 2010; West, 2003; San Filippo et al., 2008; Li & Heyer, 2008). The central event in HRR is the synapsis of the single-stranded (ss)DNA molecule - produced along the double strand break point - with homologous duplex DNA. The strand invasion is mediated by the strand transferase RAD51 oligomerized on ssDNA as an active nucleoprotein filament (and the corresponding co-factors needed for filament assembly and function), which initiates the strand exchange that leads to recombination. RAD51, a recombinase essential for cell viability, is one of the most conserved molecules known. RAD51 mediates strand exchange via distinct reactions grouped into the presynaptic, synaptic, and postsynaptic phases (2) (Heyer, 2007; Shivji et al., 2009). The major steps of HRR process are schematically illustrated in Scheme 3.

RAD51 assembly on ssDNA and subsequent catalysis of homology dependent strand invasion is mainly driven by the tumor suppressor protein BRCA2 while during the different phases of HRR RAD51 interacts sequentially with other molecules involved in HRR, cell cycle control and cell fate decisions. RAD51, as part of dynamic structures called DNA damage foci, seems to be a stably associated core component, whereas other co-factors such as Rad52 and Rad54 rapidly and reversibly interact with the structure (Essers et al., 2002). RAD51 function depends on protein re-localization and is mainly regulated by various post-translational modifications, mainly phosphorylation (Slupianek et al., 2001; Venkitaraman, 2001), as well as non-covalent interaction with SUMO (Ouyang et al., 2009). Along evolution recombinase molecules are highly conserved, starting from the prokaryotic orthologue RecA to mammalian RAD51. The fact the RecA seems not to be an ancestor of RAD51 but these two molecules are considered to have evolved by converging evolution, suggests that the structure obtained is crucial for the specific recombination function and cannot afford modifications. This notion is further supported by the absence of RAD51 coding region mutants in any cancer type, while cells or animals that do not express RAD51 eventually are not viable (Tsuzuki et al., 1996; Sonoda et al., 1998). Despite the non-detection of RAD51 mutants itself it is clear that mutations in recombination mediators and co-mediators, which control RAD51 activity and availability, are highly related to cancer susceptibility and particularly breast cancer (Venkitaraman, 2009; Rahan et al., 2007; Seviour & Lin, 2010). Additionally, the 'guardian of the genome' p53, found mutated in more than 50% of cancers, also directly interacts with RAD51 presumably connecting HRR efficiency to cell cycle control and apoptosis (Henning & Sturzbecher, 2003; Gatz & Wiesmuller, 2006; Lazao-Trueba & Silva, 2006).

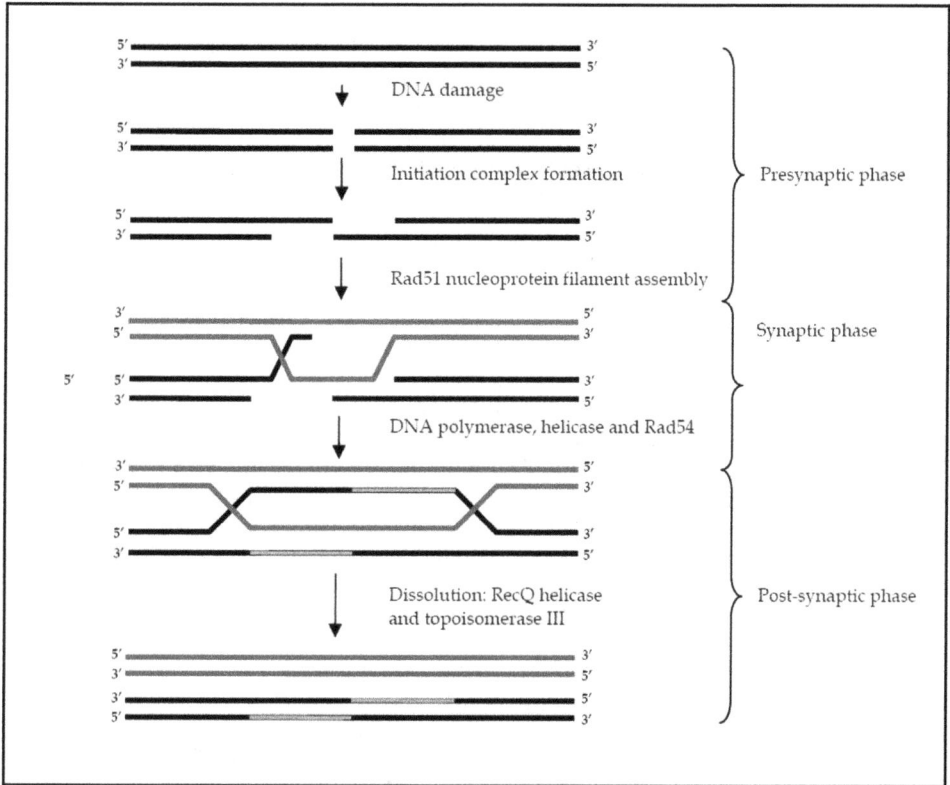

Scheme 3. Simplified description of HRR major steps. The presynaptic phase involves resection of the 5' terminated strand at the DNA double strand break point (black line) and the formation of the RAD51 active nucleoprotein filament on 3'ssDNA tails. During the synaptic phase RAD51 traces the homologous strand (grey line)(usually the sister chromatid) and performs the strand exchange. After DNA heteroduplex extension and branch migration (newly synthesized DNA is shown as framed grey line) the Holiday junctions produced are separated resulting in two intact homologous DNA molecules. (Resolution or dissolution of holiday junction may also involve crossing over resulting in chimeric but still homologous DNA molecules).

RAD51 function is mainly controlled by the breast cancer susceptibility gene 2 product BRCA2 which acts as a recombination mediator (scheme 4). Briefly, BRCA2 targets RAD51 to ssDNA for assembly into a nucleoprotein filament, stabilizes the ATP-bound form of RAD51 and inhibits RAD51 assembly on dsDNA (Shivji et al., 2009). BRCA2 is an extremely large protein of 3418 residues and essentially contributes to RAD51-mediated HRR through several regions. RAD51 interacts with 8 copies of ~35 residues repeated motifs (BRC repeats) located at exon 11 (Yu et al., 2003), as well as with an unrelated carboxyl-terminal motif in exon 27 (Esashi et al., 2007). The BRC repeats sequence, unlike the C-terminal motif, is evolutionarily conserved. RAD51 replaces RPA in ssDNA, a process regulated by the DNA-binding domain of BRCA2, in cooperation with the BRC repeats and the contribution of other RAD52 epistasis group members as Rad52 and Rad54. BRC repeats of BRCA2 bind to the core of RAD51 by mimicking the structure of an adjacent Rad51 monomer (Pellegrini at al., 2002). RAD51 loading on ssDNA is promoted by the BRCA2[BRC1–8] region while RAD51 assembly on dsDNA is at the same time suppressed. This way the efficiency of RAD51-mediated HRR is further enhanced. RAD51 function can either be stimulated or suppressed by activities of the BRC repeats, depending on the experimental conditions used and the BRC: RAD51 molar ratio used (Galkin et al., 2005; Shivji et al., 2009; Carreira et al., 2009; Rajendra & Venkitaraman, 2010). BRC4 also blocks nucleation of RAD51 onto dsDNA while not disassembling Rad51-dsDNA filaments. (Carreira A, et al., 2009). At lower molar ratios BRC3 or BRC4 actually bind and form stable complexes with RAD51-DNA nucleoprotein filaments. Only at high concentrations of the BRC repeats are filaments disrupted. The specific protein-protein contacts occur in the RAD51 filament by means of the N-terminal domain of RAD51 for BRC3 and the nucleotide-binding core of RAD51 for BRC4 (Galkin et al., 2005; Rajendra & Venkitaraman 2009). These observations show that the BRC repeats bind distinct regions of RAD51 and are nonequivalent in their mode of interaction. These results might explain how disruption of a single RAD51 interaction site in BRCA2 might modulate the ability of RAD51 to promote recombinational repair and lead to an increased risk of breast cancer. Moreover, the dysregulated molar ratio present in a cell may drive hyper-recombination effects leading to abnormal outcome and in part may explain why mutations in BRCA2 predispose individuals to breast cancer, a consequence of the role of BRCA2 in DNA repair.

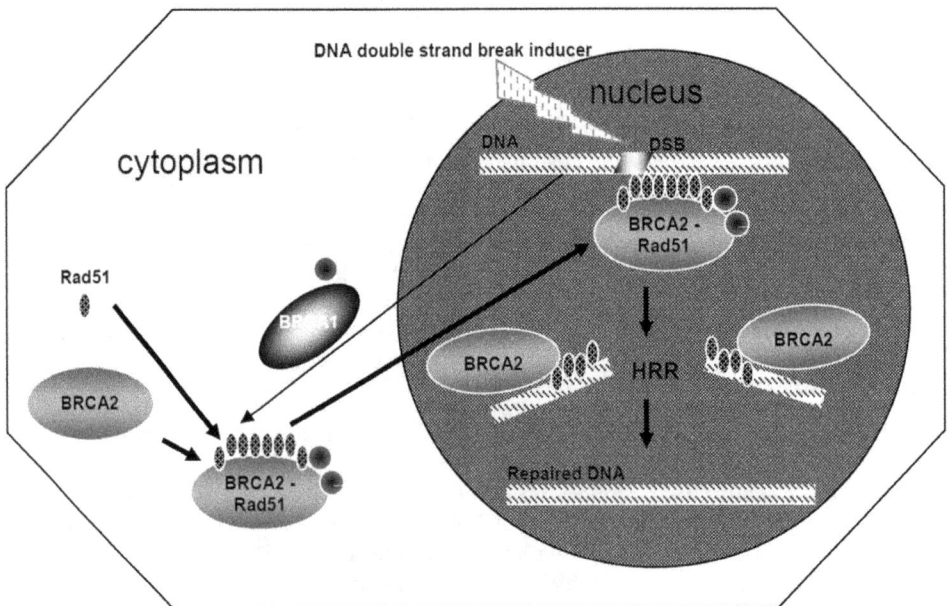

Scheme 4. Simplistic cartoon of the HRR process depicting Rad51 functions.

RAD51 filaments are further stabilized by direct interaction of the BRCA2 C terminus to the interface created by two adjacent RAD51 protomers. This way filaments cannot be dissociated by the BRC repeats. Interaction of the BRCA2 C terminus with the RAD51 filament causes a large movement of the flexible RAD51 N-terminal domain that is important in regulating filament dynamics. RAD51 interaction with the BRCA2 C-terminal region may facilitate efficient nucleation of RAD51 multimers on DNA and thereby stimulate recombination-mediated repair. (Esashi et al., 2007). Data from studies the Caenorhabditis elegans BRCA2 homolog CeBRC-2 support a model where an interaction with RAD-51 alone is likely involved in filament nucleation, whereas a second independent interaction is involved in *in situ* stabilization of RAD51 filaments by BRCA2 and provide further insight into why mutations in many different positions within BRCA2 lead to loss of genomic stability (Petalcorin et al., 2007).

RAD51, the central homology strand search and strand exchange effector in HRR can serve as a nice example to show how unregulated protein levels can abolish normal cell fate decisions and result in premature ageing or malignancies, depending on the mechanisms involved. RAD51 is one of the most conserved proteins known and essential for cell survival (Henning & Sturzbecher, 2003; Sonoda et al., 1998). While no mutations have ever been detected in human cancers, in many tumors significantly up- or downregulated levels of RAD51 have been observed (Maacke et al., 2000; Henning & Sturzbecher 2003; Klein, 2008). Moreover, high-level expression of RAD51 is an independent prognostic marker of survival in non-small-cell lung cancer patients (Qiao et al., 2005). In addition, haematopoietic progenitor cells, when Rad51 is overexpressed showed elevated levels of chromosomal alterations, similar to those observed in tumors of the hematopoietic system (Francis and Richardson, 2007). Notably, both positive and negative regulation of HRR is required in order to maintain genomic stability with precise repair and suppression of deleterious rearrangements. The only tumorigenesis-related variation found in the *rad51* gene is a G->C change in the 5' untranslated rad51 mRNA region. This variation has been correlated to higher risk for breast cancer in BRCA2 mutation carriers and is possibly involved in mRNA modified translation capability resulting in abnormal RAD51 protein levels (Antoniou et al., 2007).

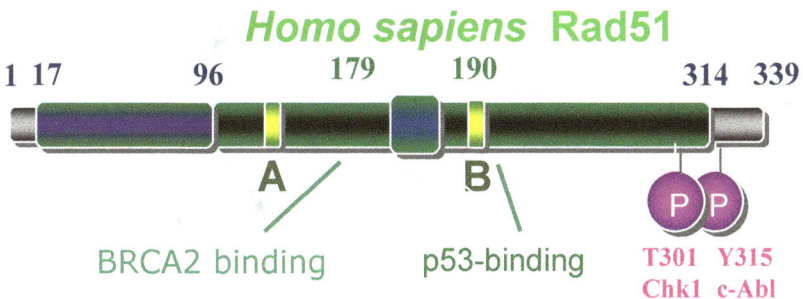

Scheme 5. Scematic representation of human Rad51 protein. The areas responsible for interaction with p53 and BRCA2 are indicated.

Since Rad51 overexpression can compensate for loss of function of other key molecules of DDR, including BRCA1 and BRCA2, experimental evidences from various research groups support two models: 1. Rad51 abnormal levels lead to genomic instability early in cancer development, thereby placing Rad51 modified expression as a leading cause of transformation and 2. Rad51 overexpression can protect cancer cells from DNA damage as more effective repair occurs further stabilizing the neoplastic clone and render it more aggressive and metastatic (Schild & Wiese, 2010). As cancer is an extremely complex set of diseases and can develop by different aetiologies while achieving similar phenotype of independent and uncontrolled growth the two models presented by Schild and Wiese can each adequately explain the neoplastic procedure of different cancer types.

The exact causes of Rad51 overexpression are still poorly explored but there is a number of data indicating both transcriptional regulation and protein stability and turnover modification. p53, the tumor suppressor that is implicated in DNA repair control, is involved in transcriptional regulation of rad51 (Arias-Lopez et al., 2006). p53 is mutated in about half of human cancers resulting in loss of suppression of rad51 transcriptional regulation. Notably, as p53 directly interacts with Rad51 (Stürzbecher et al., 1996), in cases of p53 mutations inhibition of Rad51 activity could be abolished. Moreover, in cases of either TP53 deletion or some TP53 point mutations Rad51 expression up-regulation is detected. The Transcriptional activator protein 2 (AP2), in combination with p53 suppresses rad51 transcription (Hannay et al., 2007). Abl kinase phosphorylates Rad51 in Tyr315 and in cases of the presence of the oncogenic constituvely active BCR/Abl fusion tyrosine kinase (i.e. in Ph+ leukaemias) Rad 51 expression is increased (Slupianek et al., 2001).

Aiming in further clarifying aspects of structure-function relationship of RAD51, we produced several RAD51 mutants by altering amino acid residue candidates to be involved in RAD51-BRCA2 or RAD51-p53 interaction (fig 5). Exogenous expression the RAD51m6 mutant, fused to EYFP, altered their subcellular localization compared to the wt protein.

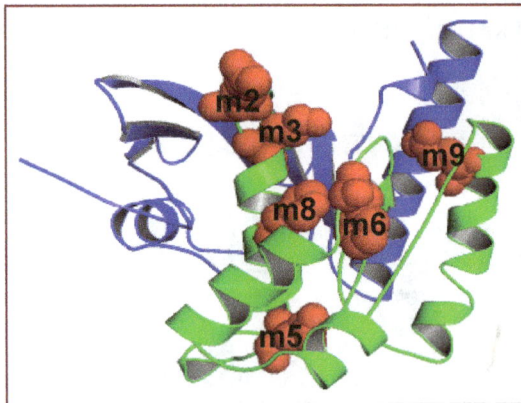

Fig. 5. Solved structure of Rad51-BRC4 complex (adapted from PDB: 1NOW, (Pellegrini et al., 2002)) where candidate residues presumed to alter Rad51-BRCA2/p53 complex interaction are indicated. In vitro Site-Directed mutagenesis was employed in order to alter each indicated residue to Ala. Mutant form positions are indicated.

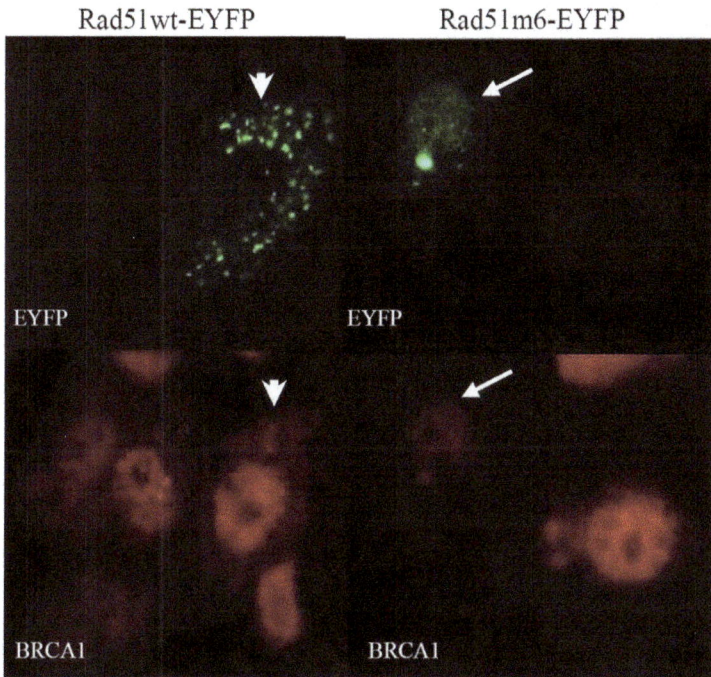

Fig. 6. MCF7 cells expressing exogenous Rad51wt-EYFP and Rad51m6-EYFP as indicated. Endogenous BRCA1 expression, as detected by immunofluoresence, shows a significant reduction in all cells expressing the Rad51m6-EYFP in comparison to both the Rad51wt-EYFP expressing as well as to non-transfected cells (Boutou et al., unpublished data).

Moreover cells expressing RAD51m6 showed a modification in their cell cycle progression (data not shown) accompanied by modifications in expression of BRCA1 (fig 6), p53 and p21^{wafl} (data not shown). Notably, RAD51m6 electively kill certain cancer cell lines as HeLa cells, but do not affect the Caspase 3 defective MCF-7 cells.

Double strand breaks (DSBs) of DNA is the most deleterious damage of the genome since if not repaired accurately can result in ICL, translocations, chromatin rearrangements, LOH and mutation accumulation. HRR restores DNA damage in mitotic cells by gene conversion, where the broken sequence is converted to the sequence of the repair template (original sequence), which remains unaltered. In case of HRR misregulation other templates can be used including homologous chromosomes and repetitive elements on heterologous chromosomes. Such data indicate that single amino acid residue alterations of Rad51 are capable to modify the behaviour of the entire protein, presumably through structural modifications. These results, combined with the fact that RAD51 protein in nature is not mutated, suggest that its proper function is strongly dependent on its high degree of structural conservation throughout evolution. Moreover, the combination of the active peptide of such mutants under the promoter of human RAD51 which in the absence of the N-terminal region of Rad51 enhances expression up to 10 fold in cancer cell lines, could serve as a potential anti-cancer agent, selectively targeting malignant cells.

4. Conclusions

In conclusion maintenance of genome integrity depends on structure-function relationship of the protein molecules involved. Proper response to DNA damage mainly relies on functional components of DDR driving their appropriate complex formation with partner proteins. These processes are regulated by a number of post-translational modifications, distinct protein isoforms and protein availability (stability / degradation). In case these interactions are deregulated due to genetic / epigenetic causes a balanced cell cycle progression and cell fate determination are abolished in favour of cancer/ageing. Structural / biophysical data accompanied by functional experiments of key DNA repair molecules are significant for: (a) elucidating which residues or structural elements are really necessary for proper function at the molecular level, (b) asses/classify variants identified in individuals, (c) enriching diagnostic markers in cancer and (d) designing effective small molecules to target protein molecules essential for cell survival and genome integrity.

Co-operation of various disciplines is a fundamental prerequisite for fulfilling such a vision, and numerous attempts worldwide work on this subject with promising results.

5. References

Arias-Lopez C., Lazaro-Trueba I., Kerr P., Lord C.J., Dexter T., Iravani M., Ashworth A. and Silva A. p53 modulates homologous recombination by transcriptional regulation of the RAD51 gene. *EMBO Rep.* 2006 7: 219-224

Best B.P. Nuclear DNA damage as a direct cause of aging. *Rejuvenation Research* 2009, 12: 199–208.

Buchhop S., Gibson M.K., Wang X.W., Wagner P., Sturzbecher H-W. and Harris C.C. Interaction of p53 with the human Rad51 protein. *Nucleis Acids Res.* 1997 25: 3868-3874.

Carreira A, Hilario J, Amitani I, Baskin RJ, Shivji MK, Venkitaraman AR, Kowalczykowski SC. The BRC repeats of BRCA2 modulate the DNA binding selectivity of RAD51. Cell 2009 136:1032–1043.

Deng C.-X. BRCA1: cell cycle checkpoint, genetic instability, DNA damage response and cancer evolution. *Nucleic Acids Res.*, 2006, 34: 1416–1426

Drikos I. Nounesis G. and Vorgias C.E. Characterization of cancer-linked BRCA1-BRCT missense variants and their interaction with phosphoprotein targets. *Proteins.* 2009 77: 464-76.

Esashi F, Galkin VE, Yu X, Egelman EH, West SC. Stabilization of RAD51 nucleoprotein filaments by the C-terminal region of BRCA2. *Nat Struct Mol Biol.* 2007 14:468-74.

Essers J., Houtsmuller A., van Vielen L., Paulusma C., Nigg A., Pastink A., Vermeulen W., Hoeijmakers J. and Kanaar R. Nuclear dynamics of RAD52 group homologous recombination proteins in response to DNA damage. *EMBO J* 2002 21 2030-37

Feldser D.M., Greider C.W. Short telomeres limit tumor progression in vivo by inducing senescence. *Cancer Cell.* 2007 11:461-9.

Forget A. & Kowalczykowski S.C. Single-molecule imaging brings Rad51 nucleoprotein filaments into focus. *TICS* 2010 20: 269-76

Francis R, Richardson C. Multipotent hematopoietic cells susceptible to alternative double-strand break repair pathways that promote genome rearrangements. *Genes Dev.* 2007 21:1064-74.

Freeman A. and Monteiro A. Phosphatases in the cellular response to DNA damage. *Cell Comm.Signaling* 2010, 8:27-38

Friedberg EC, Walker GC, Siede W, Wood RD, Schultz RA, Ellenberger T. (2006). *DNA Repair and Mutagenesis, part 3.* ASM Press. 2nd edition

Galkin V. E., Esashi F., Yu X. , Yang S., West S. C., and Egelman E. H. BRCA2 BRC motifs bind RAD51–DNA filaments. *PNAS* 2005, 102: 8537–8542

Gatz S.A & Wiesmuller L p53 in recombination and repair. *Cell Death Differ.* 2006 13: 1003-1016

Glover JN, Insights into the molecular basis of human hereditary breast cancer from studies of theBRCA1 BRCT domain. *Familial Cancer* 2006 5: 89-93

Gough CA, Gojobori T, Imanishi T. Cancer-related mutations in BRCA1-BRCT cause long-range structural changes in protein-protein binding sites: a molecular dynamics study. *Proteins,* 2007 66:69-86.

Hannay J.A., Liu J., Zhu Q.S., Bolshakov S.V., Li L., Pisters P.W., Lazar A.J., Yu D., Pollock R.E. and Lev D. Rad51 overexpression contributes to chemoresistance in human soft tissue sarcoma cells: a role for p53/activator protein 2 transcriptional regulation. *Mol. Cancer Ther.* 2007 6: 1650-1660.

Henning W, Sturzbecher HW. 'Homologous recombination and cell cycle checkpoints: Rad51 in tumour progression and therapy resistance'. *Toxicology.* 2003, 193: 91-109.

Heyer W.-D. Biochemistry of eukaryotic homologous recombination. *Top Curr Genet.* 2007, 17: 95–133

Klein H. The consequences of Rad51 overexpression for normal and tumor cells. *DNA Repair (Amst)* 2008 7: 686-693.

Lazao-Trueba I. Arias. C. & Silva A. Double bolt regulation of Rad51 by p53: a role for transcriptional repression. *Cell Cycle* 2006 5: 1062-1065

Lee MS, Green R, Marsillac SM, Coquelle N, Williams RS, Yeung T, Foo D, Hau DD, Hui B, Monteiro AN, Glover JN. Comprehensive analysis of missense variations in the BRCT domain of BRCA1 by structural and functional assays. *Cancer Res.* 2010 70:4880-90.

Li X & Heyer WD. *Cell Res* 2008 Homologous recombination in DNA repair and DNA damage tolerance. 18: 99-113

Lindahl T. Instability and decay of the primary structure of DNA. *Nature.* 1993, 362: 709-715

Lisby M & Rothstein R. Localization of checkpoint and repair proteins in eukaryotes. Biochimie 2005, 87: 579-89.

Lodish H, Berk A, Matsudaira P, Kaiser CA, Krieger M, Scott MP, Zipursky SL, Darnell J. (2004). *Molecular Biology of the Cell,* p963. WH Freeman: New York, NY. 5th ed.

Maacke H, Opitz S, Jost K, Hamdorf W, Henning W, Kruger S, Feller AC, Lopens A, Diedrich K, Schwinger E, Sturzbecher HW. 'Over-expression of wild-type Rad51 correlates with histological grading of invasive ductal breast cancer'. *Int J Cancer.* 2000, 88: 907-13.

Mark W.Y. Liao J.C, Lu Y., Ayed A, Laister R, Szymczyna B, Chakrabartty A, Arrowsmith CH. Characterization of segments from the central region of BRCA1: an intrinsically disordered scaffold for multiple protein-protein and protein-DNA interactions? *J. Mol.Biol.* 2005, 345: 275-87

Morita R, Nakane S, Shimada A, Inoue M, Iino H, Wakamatsu T, Fukui K, Nakagawa N, Masui R, Kuramitsu S. Molecular mechanisms of the whole DNA repair system: a comparison of bacterial and eukaryotic systems. *J Nucleic Acids.* 2010:179594.

Murphy CG, Moynahan ME. BRCA gene structure and function in tumor suppression: a repair-centric perspective. *Cancer J.* 2010 16: 39-47.

Ouchi T. BRCA1 Phosphorylation. Biological Consequences. *Cancer Biol. Ther.* 2006, 5: 470-475

Ouyang KJ, Woo LL, Zhu J, Huo D, Matunis MJ, Ellis NA. 2009 SUMO modification regulates BLM and RAD51 interaction at damaged replication forks. *PLoS Biol.* 7 e1000252

Paull TT, Cortez D, Bowers B, Elledge SJ, Gellert M. 2001, Direct DNA binding by Brca1. *Proc Natl Acad Sci U S A.* 98 6086-91

Pellegrini L, Yu DS, Lo T, Anand S, Lee M, Blundell TL, Venkitaraman AR. Insights into DNA recombination from the structure of a RAD51-BRCA2 complex. *Nature.* 2002 420: 287-93.

Petalcorin MI, Galkin VE, Yu X, Egelman EH, Boulton SJ. 2007 Stabilization of RAD-51-DNA filaments via an interaction domain in Caenorhabditis elegans BRCA2. *Proc Natl Acad Sci U S A*. 104: 8299–8304

Qiao GB, Wu YL, Yang XN, Zhong WZ, Xie D, Guan XY, Fischer D, Kolberg HC, Kruger S, Stuerzbecher HW. High-level expression of Rad51 is an independent prognostic marker of survival in non-small-cell lung cancer patients. *Br J Cancer*. 2005, 93:137-43.

Rahan R et al., Links between DNA double strand break repair and breast cancer: accumulating evidence from both familiar and nonfamilial cases. *Cancer Lett*. 2007 248: 1-17

Rajendra E. and Venkitaraman A. R. Two modules in the BRC repeats of BRCA2 mediate structural and functional interactions with the RAD51 recombinase. *Nucleic Acids Res*. 2010, 38: 82–96

Rogakou E. Megabase chromatin domains involved in DNA double strand breaks *in vivo*. J. Cell Biol. 1999, 146 905-916

Rowling PJ, Cook R, Itzhaki LS. 2010 Toward classification of BRCA1 missense variants using a biophysical approach. *J Biol Chem*. JBC 285(26): 20080–20087

San Filippo J, Sung P, Klein H. Mechanism of eukaryotic homologous recombination. *Annu Rev Biochem*. 2008 77 229

Schild D, Wiese C. Overexpression of RAD51 suppresses recombination defects: a possible mechanism to reverse genomic instability. *Nucleic Acids Res*. 2010 38 1061-70

Seviour E & Lin S-Y 2010 Aging 2 900-907 The DNA damage response: Balancing the scale between cancer and ageing;

Shivji MK, Mukund SR, Rajendra E, Chen S, Short JM, Savill J, Klenerman D, Venkitaraman AR. The BRC repeats of human BRCA2 differentially regulate RAD51 binding on single- versus double-stranded DNA to stimulate strand exchange. *Proc Natl Acad Sci U S A*. 2009 106 13254–13259

Slupianek A. Schmutte C., Tombline G., Nieborowska-Skorska M., Hoser G., Nowicki M.O. Pierce A.J. Fishel R. and Skorski T. BCR/ABL regulates mammalian RecA homologs, resulting in drug resistance. *Mol. Cell* 2001 8: 795-806.

Sonoda E, Sasaki MS, Buerstedde JM, Bezzubova O, Shinohara A, Ogawa H, Takata M, Yamaguchi-Iwai Y, Takeda S. Rad51-deficient vertebrate cells accumulate chromosomal breaks prior to cell death. 1998 EMBO J 17 598-608

Starita LM, Parvin JD. The multiple nuclear functions of BRCA1: transcription, ubiquitination and DNA repair. *Curr Opin Cell Biol* 2003, 15: 345-50

Stürzbecher HW, Donzelmann B, Henning W, Knippschild U, Buchhop S. p53 is linked directly to homologous recombination processes via RAD51/RecA protein interaction. *EMBO J*. 1996 **15**:1992-2002.

Tsuzuki T, Fujii Y, Sakumi K, Tominaga Y, Nakao K, Sekiguchi M, Matsushiro A, Yoshimura Y, MoritaT. Targeted disruption of the Rad51 gene leads to lethality in embryonic mice. *Proc Natl Acad Sci U S A*. 1996 93: 6236-6240

Venkitaraman A.R. Linking the cellular functions of BRCA genes to cancer pathogenesis and treatment. *Ann.Rev. Pathol*. 2009 4: 461-487

Vigneron A, Vousden KH. p53, ROS and senescence in the control of aging. *Aging (Albany NY)*, 2010 2:471-4.

West SC. Molecular views of recombination proteins and their control. *Nat Rev Mol Cell Biol*. 2003 4:435-45.

Williams RS, Chasman DI, Hau DD, Hui B, Lau AY, Glover JN. Detection of protein folding defects caused by BRCA1-BRCT truncation and missense mutations. *J. Biol. Chem*. 2003, 278: 53007-16

8

The Involvement of E2F1 in the Regulation of XRCC1-Dependent Base Excision DNA Repair

Yulin Zhang and Dexi Chen

Department of infectious diseases, Beijing You'an Hospital
Beijing liver disease research institute, Capital Medical University, Beijing
China

1. Introduction

Reactive oxygen species, ionizing radiation and alkylating agents can attack on DNA resulting in single or double strand breaks, generation of abasic sites, base and sugar lesions [1]. Double-strand breaks (ds breaks) are repaired by two different types of mechanism. One type takes advantage of proteins that promote homologous recombination (HR) to obtain instructions from the sister or homologous chromosome for proper repair of breaks. The other type permits joining of ends even if there is no sequence similarity between them. The latter process is called non-homologous end joining (NHEJ).The process by which complex single-strand breaks (those that cannot be directly religated) are repaired (SSBR) in some ways resembles NHEJ. Here we shall mainly discuss the mechanism of base excision repair (BER) of SSBR.

2. Base excision DNA repair

The major pathway to remove damaged DNA bases is Base Excision Repair (BER, *Fig. 1*). BER can be divided into five steps: (i) excision of damaged base by the specific DNA glycosylase and formation of apurinic/apyrimidinic (AP) site; (ii) cleavage of phosphodiester bond at AP site by AP-endonuclease or AP-lyase; (iii) removal of chemical groups interfering with gap filling and ligation; (iv) gap filling; (v) ligation [2].

The first step of the BER pathway is recognition of damaged base by the specific DNA glycosylase, which cleaves N-glycosidic bond leaving behind a free base and an AP site. In humans about 10 DNA glycosylases of different, but partially overlapping substrate specificities are known [3]. Some of them are bifunctional enzymes, which have endowed AP-lyase activity and cleave phosphodiester bonds at 3' side of AP site either by ß- or ß/δ - elimination. *E. coli* endonuclease III (Nth), its human homolog, hNTH1 and human 8-oxoG DNA glycosylase (OGG1) catalyse reaction of ß- elimination, which creates alpha/ß-unsaturated aldehyde (3'dRP) at the 3' end of cleaved DNA strand. Bacterial formamidopyrimidine DNA glycosylase (Fpg), endonuclease VIII (Nei) and two human homologs of the latter, NEIL1 and NEIL2 catalyse ß/ δ-elimination and remove deoxyribose residue leaving phosphate at the 3' end of cleaved DNA strand. Monofunctional DNA glycosylases need the assistance of AP-endonucleases, which hydrolyse phosphodiester bond at the 5' end of the AP site. This yields DNA single strand break (SSB) with the 5'end bearing baseless deoxyribose (5′dRP) and the 3′ end with the free hydroxyl group. Both AP sites and SSBs can be formed due to spontaneous hydrolysis of purines, as well as upon DNA damaging agents, like ionizing radiation or oxidation.

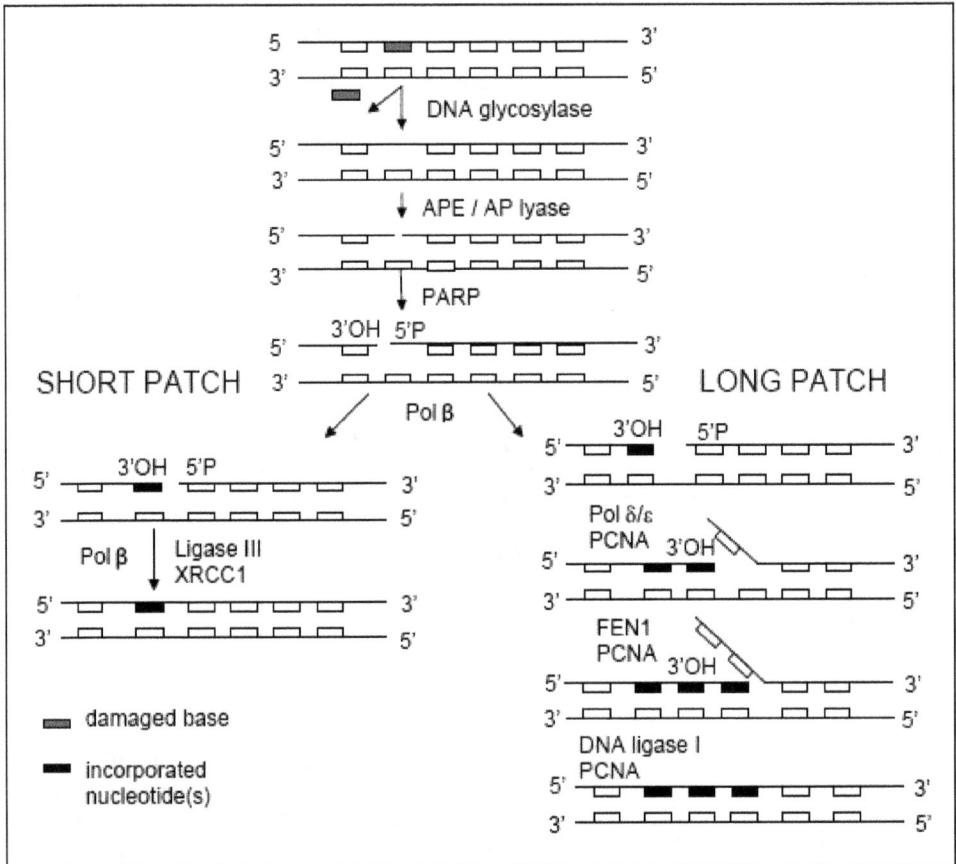

Fig. 1. Base Excision Repair pathway

Before filling the gap by DNA polymerases possible additional chemical groups present on 3′OH end, which may block polymerisation, must be removed. Bacterial enzymes Xth (exonuclease III) and Nfo (Endonuclease IV), besides of cleaving phosphodiester bonds at 5′ AP-site, have as well 3′ phosphatase and 3′ phosphodiesterase activities and remove phosphates and phosphoglycolates from 3′ hydroxyl group of cleaved DNA strand[4]. In contrast, the major mammalian AP-endonuclease, APE1 effciently removes 3′ phosphoglycolate groups, but has a very weak 3′ phosphatase activity [5]. Phosphate groups left e.g. by NEIL1 glycosylase at 3′hydroxyls are most probably removed by polynucleotide kinase[6]. After cleavage of phosphodiester bond, repair may be continued on two alternative pathways (Fig. 1): short-patch BER (SP-BER) or long patch BER (LP-BER). During SP-BER in mammals, only one missing nucleotide is incorporated by DNA polymerase ß (pol ß), which has also endowed 5′dRPase activity and can remove baseless sugar from the

5'site of DNA break. In LP-BER a longer fragment ranging from 2 to 12 nucleotides is excised and re-synthesized [2]. Initially DNA polymerase elongates 3' end by a few nucleotides and moves aside a DNA fragment which contains 5' deoxyribophosphate. Subsequently, such flap structure is cleaved out by specific flap endonuclease, FEN1. It is believed that in LP-BER the first nucleotide is incorporated by DNA polymerase ß, while next ones by DNA polymerases δ or ε [2]. LP-BER demands also other assisting proteins, PCNA (*proliferating cell nuclear antigen*) and RPC (*replication protein C*).

The last stage of BER is ligation of repaired DNA fragments by DNA ligase. Different DNA ligases(LIG) are operating in short and long patch BER, LIG1 in LP-BER and LIG3alpha in SP-BER. LIG3alpha remains in complex with XRCC1 (x-ray repair cross-complementing group 1) protein, which activates ligation of DNA ends by LIG3alpha.

3. The role of XRCC1 protein in base excision DNA repair

X-ray cross-complementing group 1 (XRCC1) is a 70- kDa protein comprising three functional domains; an N-terminal DNA binding domain, a centrally located BRCT I and a C-terminal BRCT II domain. It has no known enzymatic activity. Since it specifically interacts with nicked and gapped DNA *in vitro*[7-9], and rapidly and transiently responds to DNA damage in cells, it may serve as a strand-break sensor [10, 11].

DNA single-strand breaks (SSBs) are one of the most frequent types of DNA damage in cells [12]. SSBs can lead to the accumulation of mutations or can be converted from single to cytotoxic double-strand breaks. Thus, SSBs pose a critical threat to the genetic stability and survival of cells[13]. Various proteins have been identified that are part of the repair machinery for SSBs, including XRCC1 protein. XRCC1 has been shown to be critically involved in DNA SSB repair in studies using XRCC1-mutant cells and XRCC1 knockout mice[14], which have increased sensitivity to alkylating agents, ultraviolet and ionizing radiation [15], as well as elevated levels of sister chromatic exchange. Since XRCC1 interacts with many proteins known to be involved in BER and SSBR, it has been proposed that XRCC1 functions as a scaffold protein able to coordinate and facilitate the steps of various DNA repair pathways[11, 16]. For example, XRCC1 interacts with several DNA glycosylases involved in repair of both oxidative and alkylated base lesions, and stimulates their activity[17, 18]. This protein interacts with DNA ligase III, polymerase beta and poly (ADP-ribose) polymerase to participate in the base excision repair pathway. It is recruited to the site of DNA damage by several DNA glycosylases, e.g. OGG1 or NTH1 and remains at the site of repair till the last stage of ligation (*Fig. 2*), regulating and coordinating the whole process. XRCC1 facilitates exchange of DNA glycosylase with AP-endonuclease at the damaged substrate, which increases the excision rate of modified base, regulates pol ß interactions with APE1, and finally activates ligation step [17]. Binding of XRCC1 to Polynucleotide Kinase (PNK) enhances its capacity for damage discrimination, and binding of XRCC1 to DNA enables displacement of PNK from the phosphorylated product [19] thus accelerating SSBR of damaged DNA[20]. XRCC1 associates with Tyrosyl-DNA phosphodiesterase1(Tdp1) and enhances its activity required for repair of Top1-associated SSBs. It may act to recruit Tdp1 to these damaged sites[21]. Biochemical and nuclear magnetic resonance (NMR) experiments have demonstrated protein-protein interaction between the N-terminal domain of XRCC1 and the polymerase domain of pol β[22-25]. Additionally, stabilization of DNA ligase IIIα is dependent on its interaction with the BRCT II domain of XRCC1[26]. Aprataxin also interacts with XRCC1 and functions to maintain

XRCC1 stability, thus further linking the neurological degeneration associated with ataxia to an inefficiency of SSBR[27-29].

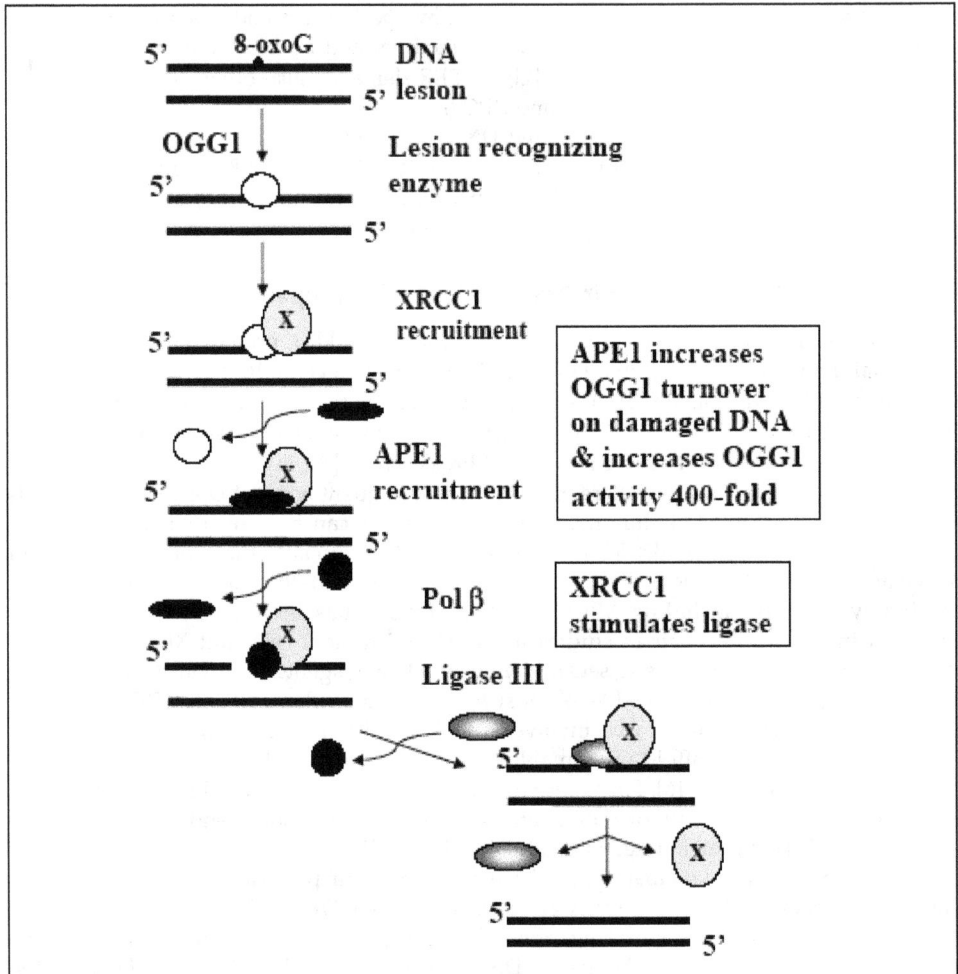

Fig. 2. Coordinative role of XRCC1 protein in BER

Several additional proteins participate in BER and play regulative and coordinative role. The most important proteins are: PARP1 (polyADP ribose polymerase, which binds to free DNA ends and protects them against degradation, participates in chromatin relaxation and probably modulates binding of repair proteins to the site of damage by interaction with poly(ADP-ribose) chains [22, 30, 31], PCNA (proliferating cells nuclear antigen, DNA polymerase processivity subunit in LP-BER), RFC (replication factor C, loading PCNA on DNA), WRN (helicase deficient in Werner syndrome, a premature aging disease) or CSB (helicase deficient in Cockayne syndrome, neurodevelopmental and premature aging disease).

4. The role of E2F1 in XRCC1 associated base excision DNA repair

E2F1 is a member of E2F family of transcription factors which plays an important role in promoting both cellular proliferation and cell death. E2F1 is important for regulating S-phase specific genes as well as promoting apoptosis, just as other "activating" E2F family members [32, 33]. Simultaneously, E2F1 regulates DNA repair through interaction with other factors including RB family proteins, p53 and X-ray repair cross-complementing group 1 (XRCC1) protein.

4.1 E2F family
The E2F transcription factor family consists of at least seven distinct genes divided into two groups. E2F1, E2F2, E2F3, E2F4, and E2F5 constitute one group, while the related DP1 and DP2 genes constitute the other group. Several forms of the DP2 (also referred to as DP3) protein can be produced as the result of alternative splicing, thus providing additional complexity to the E2F family. A functional E2F transcription factor consists of a heterodimer containing an E2F polypeptide and a DP polypeptide. Each of the five E2F polypeptides can heterodimerize with either DP1 or DP2 (DP3). Furthermore, each of these E2F/DP heterodimers (referred to as E2F factors hereafter) can bind consensus E2F sites *in vitro* and stimulate transcription when overexpressed[34].

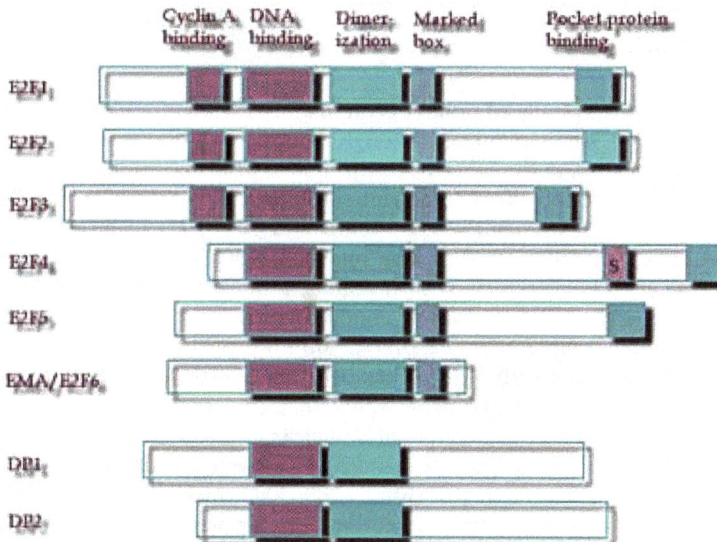

Fig. 3. the members of E2F

All of the E2F subgroup proteins have a similar structure although E2F1, E2F2, and E2F3 are more closely related to each other than to E2F4 and E2F5 (*Fig. 3*). The DNA-binding domain found in the amino terminus represents the area of greatest homology between the five E2F species. Adjacent to the DNA-binding domain is the DP dimerization domain, which contains within it a leucine heptad repeat. The carboxy termini of the five E2F polypeptides contain the defined transcriptional activation domains, which are characterized by an abundance of acidic residues. Embedded within the transactivation domain of each E2F is a region of homology involved in binding to the pocket proteins (Rb, p107, and p130). An additional region of homology, termed the Marked box, lies between the DP dimerization and transcriptional activation domains. Although this Marked box motif is highly conserved between the different E2Fs, its function is unknown. The amino termini of E2F1, E2F2, and E2F3 contain an additional region of homology not found in E2F4 or E2F5. This region has been demonstrated to have several functions, including binding to the cyclin A protein. The E2F4 protein contains a stretch of consecutive serine residues between the Marked box and the pocket protein binding domain not found in other E2F family members. DP1 and DP2 polypeptides contain DNA-binding and dimerization domains related to the E2F proteins but do not contain transcriptional activation domains or regions homologous to the pocket protein binding or Marked box domains.

An additional E2F family member has recently been isolated and termed EMA (E2F-binding site modulating activity) or E2F6[35, 36]. EMA/E2F6 shares homology with the E2F polypeptides in the DNA-binding domain, the DP dimerization domain and the Marked box, but lacks the pocket protein binding domain and acidic transcriptional activation domain found in the carboxy terminus of the other E2F species (figure 1). Like the other E2F polypeptides, EMA/E2F6 dimerizes with DP1 or DP2 and, in conjunction with a DP partner, binds E2F DNA-binding sites with preference for a subset of sites with the core sequence TCCCGCC. EMA/E2F6 appears to function as a repressor of E2F site-dependent transcription independent of pocket protein binding. The mechanism of repression is either through competitive inhibition with other E2F species or through an active transcriptional repression domain located in the amino terminus of EMA/E2F6.

4.2 E2F factors and Rb family of pocket proteins

The activity of E2F factors is regulated through association with the retinoblastoma tumor suppressor protein (Rb) and the other pocket proteins, p107 and p130. Binding of Rb, p107 or p130 converts E2F factors from transcriptional activators to transcriptional repressors. The interplay among G1 cyclins (D-type cyclins and cyclin E), cyclin-dependent kinases (cdk4, 6, and 2), cdk inhibitors, and protein phosphatases determines the phosphorylation state of the pocket proteins which in turn regulates the ability of the pocket proteins to complex with E2F. E2F activity is further regulated through direct interactions with other factors, such cyclin A, Sp1, p53 and the ubiquitin-proteasome pathway. Deregulated expression of E2F family member genes has been shown to induce both inappropriate S phase entry and apoptosis. Experiments show that dimerization between E2F1 and its partner DP1 is stable and that E2F1 stimulates nuclear localization of DP1[37]. E2F1/DP1 is acetylated by the three acetyltransferases P300/CBP-associated factor (PCAF), cAMP-response element-binding protein (CREBBP) and p300 which stabilizes E2F1 protein[38]. The acetylated complex is able to bind to PCAF to form an active dimer. The complex ability to bind to DNA on the promoter sites of its target genes along with its transcriptional activity are increased at the G1/S transition. During G2, the complex is phosphorylated by CycA2/CDK2[39]. The affinity between E2F1 and DP1 is then diminished leading to the dissociation of the complex and the release of PCAF[40]. The proteins undergo further modifications before degradation: E2F1 is deacetylated by histone deacetylase1(HDAC1) [41], dephosphorylated and phosphorylated *de novo* during S phase by Transcription factor II H (TFIIH) kinase for rapid degradation[42].

Upon DNA damage, the complex PCAF/E2F1/DP1 can be phosphorylated and stabilized either by Checkpoint kinases (CHEK1 and CHEK2) through phosphorylation at Ser-364, or by ataxia telangiectasia mutated(ATM) and ATR (ATM and Rad3-related) [43, 44], preventing E2F1 ubiquitination[45]. E2F1 mediates the transcription of many genes involved in apoptosis. However, E2F1 transcriptional activity can also be inhibited when bound to the topoisomerase TopBP1 in order to give time to the cell to repair the damage [46]. Mutations of the RB gene represent the most frequent molecular defect in Osteosarcoma. Studies in animal models and in human cancers have shown that deregulated E2F1 overexpression possesses either "oncogenic" or "oncosuppressor" properties, depending on the cellular context. High E2F1 levels exerted a growth-suppressing effect that relied on the integrity of the DNA damage response network. Surprisingly, induction of p73, an established E2F1 target, was also DNA damage response-dependent. Furthermore, a global proteome analysis associated with bioinformatics revealed novel E2F1-regulated genes and potential E2F1-driven signaling networks that could provide useful targets in challenging this aggressive neoplasm by innovative therapies[47]. Similarly, deregulation of the Rb/E2F pathway in human fibroblasts results in an E2F1-mediated apoptosis dependent on ATM, Nijmegen breakage syndrome 1 (NBS1), CHEK2 and p53. E2F1 expression results in MRN(Mre11-Rad50-Nbs10 foci formation, which is independent of the Nbs1 interacting region and the DNA-binding domain of E2F1. E2F1-induced MRN foci are similar to irradiation-induced foci (IRIF) that result from double-strand DNA breaks because they correlate with 53BP1 and gammaH2AX foci, do not form in NBS cells, do form in AT cells and do not correlate with cell cycle entry. In fact, in human fibroblasts, deregulated E2F1 causes a G1 arrest, blocking serum-induced cell cycle progression, in part through an Nbs1/53BP1/p53/p21(WAF1/CIP1) checkpoint pathway. This checkpoint protects against apoptosis because depletion of 53BP1 or p21(WAF1/CIP1) increases both the rate and extent of apoptosis. Nbs1 and p53 contribute to both checkpoint and apoptosis pathways. These results suggest that E2F1-induced foci generate a cell cycle checkpoint that, with sustained E2F1 activity, eventually yields to apoptosis. Uncontrolled proliferation due to Rb/E2F deregulation as well as inactivation of both checkpoint and apoptosis programs would then be required for transformation of normal cells to tumor cells[48]. ZBRK1 is a zinc finger-containing transcriptional repressor that can modulate the expression of GADD45A, a DNA damage response gene, to induce cell cycle arrest in response to DNA damage. Liao et al found that the ZBRK1 promoter contains an authentic E2F-recognition sequence that specifically binds E2F1, but not E2F4 or E2F6, together with chromatin remodeling proteins CtIP and CtBP to form a repression complex that suppresses zinc finger protein (ZBRK1) transcription. Furthermore, loss of RB-mediated transcriptional repression led to an increase in ZBRK1 transcript levels, correlating with increased sensitivity to ultraviolet (UV) and methyl methanesulfonate-induced DNA damage. Thus, the RB.CtIP (CtBP interacting protein)/CtBP (C terminus-binding protein) /E2F1 complex plays a critical role in ZBRK1 transcriptional repression, and loss of this repression may contribute to cellular sensitivity of DNA damage, ultimately leading to carcinogenesis[49]. One study suggested that E2F1 is also a transcriptional regulator of Xeroderma pigmentosum group C(XPC) and Rb/E2F1 tumor suppressor pathway is involved in the regulation of the DNA lesion recognition step of nucleotide excision repair[50]. Disruption of pRB-E2F interactions by E1A is a key event in the adenoviral life cycle that drives expression of early viral transcription and induces cell cycle progression. This function of E1A is complicated by E2F1. pRB-E2F1 interactions are resistant to E1A-mediated disruption. Using mutant forms of pRB that selectively force E2F1 to bind through only one of the two binding sites on pRB, E1A is unable to disrupt E2F1's unique interaction with pRB. Furthermore, analysis of pRB-E2F complexes during adenoviral infection reveals the selective maintenance of pRB-E2F1 interactions despite the presence of E1A[51].

4.3 E2F1 factors and DNA repair

The E2F1 transcription factor is post-translationally modified and stabilized in response to various forms of DNA damage to regulate the expression of cell cycle and pro-apoptotic genes. E2F1 also forms foci at DNA double-strand breaks (DSBs). The absence of E2F1 leads to spontaneous DNA breaks and impaired recovery following exposure to ionizing radiation. E2F1 deficiency results in defective NBS1 phosphorylation and foci formation in response to DSBs but does not affect NBS1 expression levels. Moreover, an increased association between NBS1 and E2F1 is observed in response to DNA damage, suggesting that E2F1 may promote NBS1 foci formation through a direct or indirect interaction at sites of DNA breaks. E2F1 deficiency also impairs RPA and Rad51 foci formation indicating that E2F1 is important for DNA end resection and the formation of single-stranded DNA at DSBs. These findings establish new roles for E2F1 in the DNA damage response, which may directly contribute to DNA repair and genome maintenance[52]. Chromatin structure is known to be a barrier to DNA repair and a large number of studies have now identified various factors that modify histones and remodel nucleosomes to facilitate repair. In response to ultraviolet (UV) radiation several histones are acetylated and this enhances the repair of DNA photoproducts by the nucleotide excision repair (NER) pathway. The E2F1 transcription factor accumulates at sites of UV-induced DNA damage and directly stimulates NER through a non-transcriptional mechanism. E2F1 associates with the general control nonderepressible(GCN5) acetyltransferase in response to UV radiation and recruits GCN5 to sites of damage. UV radiation induces the acetylation of histone H3 lysine 9 (H3K9) and this requires both GCN5 and E2F1. Moreover, as previously observed for E2F1, knock down of GCN5 results in impaired recruitment of NER factors to sites of damage and inefficient DNA repair. These findings demonstrate a direct role for GCN5 and E2F1 in NER involving H3K9 acetylation and increased accessibility to the NER machinery[53].

Mice lacking E2F1 have increased levels of epidermal apoptosis compared to wild-type mice following exposure to ultraviolet B (UVB) radiation. Moreover, transgenic overexpression of E2F1 in basal layer keratinocytes suppresses apoptosis induced by UVB. Inhibition of UVB-induced apoptosis by E2F1 is unexpected given that most studies have demonstrated a proapoptotic function for E2F1. E2F1-mediated suppression of apoptosis does not involve alterations in mitogen-activated protein kinase activation or B-cell lymphoma (Bcl-2) downregulation in response to UVB and is independent of p53. Instead, inhibition of UVB-induced apoptosis by E2F1 correlates with a stimulation of DNA repair. Mice lacking E2F1 are impaired for the removal of DNA photoproducts, while E2F1 transgenic mice repair UVB-induced DNA damage at an accelerated rate compared to wild-type mice. These findings suggest that E2F1 participates in the response to UVB by promoting DNA repair and suppressing apoptosis[54]. One study showed that E2F1 has a direct, non-transcriptional role in DNA repair involving increased recruitment of NER factors to sites of damage[55].

4.4 The role of E2F1 in the regulation of XRCC1-dependent BER

The exact mechanism of E2F1 regulating XRCC1-dependent base excision DNA repair is still not completely clear. The E2F1 pathway is centrally involved in the highly complex networks coupling cellular proliferation and apoptosis. XRCC1, which plays a critical role in SSBR/BER [15], is a direct E2F1 target gene. E2F1 is upstream of XRCC1 significantly expands on prior observations that E2F plays a role in other repair pathways, such as MMR and NER [56-59]. The BER protein uracil-DNA glycosylase is also E2F-regulated[60]. Intriguingly, although E2F1 is best characterized as a transcription factor, E2F1 protein may have a direct role in DNA repair, as suggested by its localization to repair complexes[46, 61]. Thus, it is likely that multiple E2F-regulated mechanisms function in parallel with XRCC1 to stimulate repair. Chen found that enforced E2F expression stimulated XRCC1 levels and that(methylmethane sulfonate) MMS, which induces predominantly heat-labile DNA damage repaired by an XRCC1-mediated BER pathway[62, 63], causes an E2F1-dependent increase in XRCC1 expression. This is consistent with prior reports demonstrating that cellular stress increases endogenous XRCC1 levels [64-66], although this may be cell type-specific [67]. How MMS-induced stress activates the E2F1-XRCC1 axis remains unknown. Cellular sensitivity to MMS may involve an ATR-dependent pathway, and genetic evidence suggests that MMS-induced damage activates the yeast Rad53 (Chk2 human homologue) pathway [68, 69].

Given that the ATM/ATR and Chk2 pathways phosphorylate and activate E2F1 protein[70-73], it is possible that these kinases stimulate XRCC1 expression through E2F1 activation, although this remains to be demonstrated. Interestingly, Chk2-mediated stabilization of the FoxM1 transcription factor stimulates expression of DNA repair genes, including XRCC1 [74]. Given that XRCC1 function is complex, it is likely that its control involves multiple levels. Indeed, posttranslational mechanisms modulate XRCC1 function, as evidenced by the ability of DNA-dependent protein kinase to phosphorylate XRCC1[75] as well as the requirement of protein kinase CK2 to phosphorylate XRCC1 and enhance SSBR and genetic stability [76]. Consistent with the complex control of XRCC1, serum starvation followed by refeeding stimulated XRCC1 expression. This is consistent with cell conditions of high E2F activity but also suggests that serum/mitogenic factors may be important too. This could be a cell typespecific phenomenon, since density arrest and release does not alter XRCC1 levels in human T24 cells[77]. Nevertheless, the biological importance of E2F1 regulation of XRCC1 is suggested by the attenuated *in vivo* DNA repair in E2F1-/-*versus* E2F1+/+MEFs. Two different methods demonstrated reduced DNA repair after MMS-induced DNA damage, which correlates with the decreased XRCC1 levels observed in E2F1-/-cells. The repair of MMS-damaged DNA still occurs in E2F1-/- cells, suggesting that the E2F1-XRCC1 axis is not an absolute requirement in these systems. This is not surprising, given the complex and overlapping repair pathways involved. However, the significance of even a modestly reduced XRCC1-mediated repair function may have important implications for maintaining genomic stability and cell viability. Consistent with this notion of XRCC1 mediating E2F1 activity is the observation that loss of XRCC1 function resulted in an enhanced E2F1-induced apoptotic response in EM9 cells compared with AA8 cells.

Although E2F1 is a damage response protein, it also plays an important role in promoting the expression of a large number of genes required for replication and proliferation [57, 78-80]. Given the intimate relationship between proliferation and replication/repair, the control of XRCC1 by E2F in undamaged cells further integrates SSBR with cell cycle progression as might be expected if enhanced SSBR were necessary to repair SSBs at replication forks [15, 81-83]. Whether and in what context the other E2F family members play a role, as well as what specific SSBR pathways are utilized (*e.g.* long patch BER), remains to be explored.

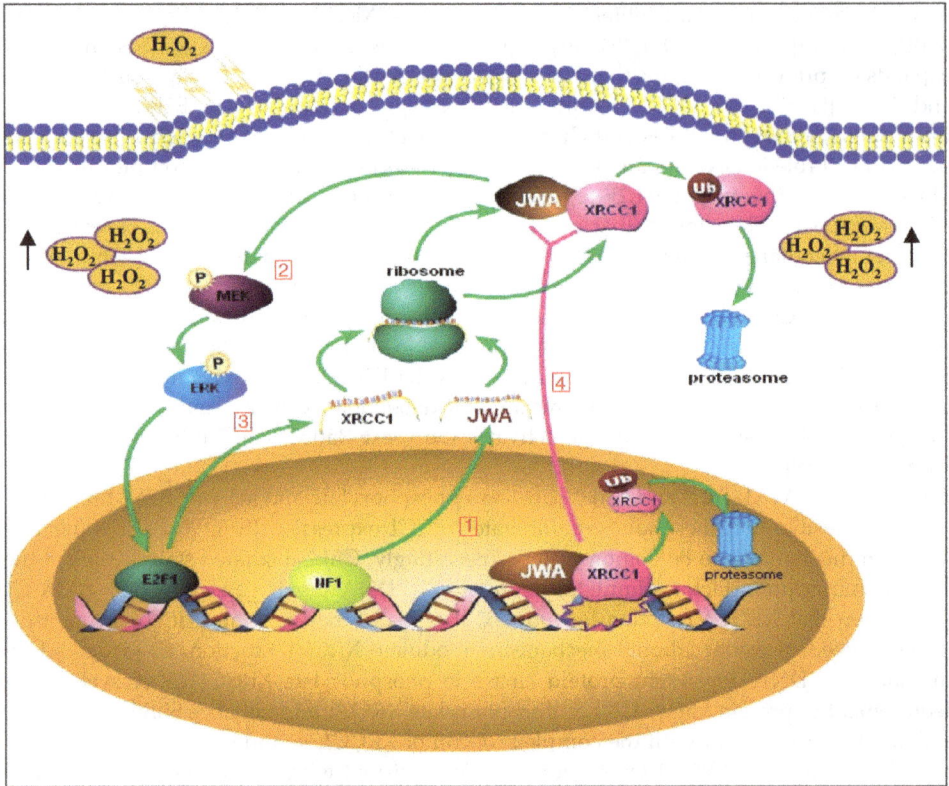

Fig. 4. JWA-E2F1-XRCC1 regulation network in base excision repair

The rapid response of XRCC1-dependent SSBR, especially in S/G2 phase, has been reported [81], and an increased co-localization of XRCC1 with proliferating cell nuclear antigen (PCNA) was observed at sites of replication during S-phase[84]. These results indicate the importance of XRCC1-dependent SSBR and its regulation during the cell cycle. Phosphorylation of E2F1 at serine-31 (S31) in response to DNA damage is required for the activation of ATM-, ATR-and ChK2-dependent DNA damage response pathways [44, 45, 85, 86]. E2F1 has also been suggested to play a potential role in nucleotide excision repair pathway (NER)[54]. Recent reports have shown that XRCC1 is a direct target of E2F1 that is involved in the enhancement of SSBR and BER, which maintain genomic stability and contribute to cell survival[63]. Over-expression of E2F1 has been shown to induce quiescent cells to enter early S-phase and is capable of preventing cells from entering quiescence [87]. Recently, we showed that E2F1 regulates the expression of XRCC1 in response to activation of DNA repair processes, and the exact functional E2F1 binding sites in the XRCC1 promoter region were identified[63]. Certain BER proteins, such as the uracil-DNA glycosylase, have also been demonstrated to be regulated by E2F transcription factors[60]. The fact that enhanced E2F expression stimulates XRCC1-mediated activation of the BER pathway in response to MMS-induced, heat-labile DNA damage suggests it might also be able to promote the expression of a variety of genes involved in DNA replication and cell proliferation.

The p53-E2F network controls and integrates critical functions, such as proliferation, cell cycle checkpoints, apoptosis, and DNA repair [43, 73, 88, 89]. In particular, p53 can promote BER [90, 91], and our discovery that E2F1 may also promote BER expands our understanding of the p53-E2F1 network in regulating DNA repair [63]. Disruption of these cooperative pathways has profound implications for tumorigenesis, as evidenced by enhanced tumor formation in knock-out mouse models for both p53 and E2F1, although intriguingly, both oncogenic and tumor suppressor functions for E2F1 are suggested in compound p53-/- and E2F1-/- mice [73, 92, 93].

The JWA (ARL6IP5)-E2F1-XRCC1 network also plays crucial role in base excision repair [94]. Exposure to oxidative stress increases the generation of intracellular reactive oxygen species, which stimulates NF1 binding to the JWA promoter, enhancing JWA transcription and translation. Then JWA regulates the expression of E2F1, leading to increased transcription of XRCC1. Interactions between JWA and XRCC1 occur in both the cytoplasm and the nucleus when the cells are subjected to oxidative stress (*fig.4*)..

5. References

[1] Tudek, B., et al., *Modulation of oxidative DNA damage repair by the diet, inflammation and neoplastic transformation.* J Physiol Pharmacol, 2006. 57 Suppl 7: p. 33-49

[2] Fortini, P., et al., *The base excision repair: mechanisms and its relevance for cancer susceptibility.* Biochimie, 2003. 85(11): p. 1053-71

[3] Wood, R.D., M. Mitchell, and T. Lindahl, *Human DNA repair genes, 2005.* Mutat Res, 2005. 577(1-2): p. 275-83

[4] Doetsch, P.W. and R.P. Cunningham, *The enzymology of apurinic/apyrimidinic endonucleases.* Mutat Res, 1990. 236(2-3): p. 173-201

[5] Kelley, M.R. and S.H. Parsons, *Redox regulation of the DNA repair function of the human AP endonuclease Ape1/ref-1.* Antioxid Redox Signal, 2001. 3(4): p. 671-83

[6] Wiederhold, L., et al., *AP endonuclease-independent DNA base excision repair in human cells.* Mol Cell, 2004. 15(2): p. 209-20

[7] Marintchev, A., et al., *Solution structure of the single-strand break repair protein XRCC1 N-terminal domain.* Nat Struct Biol, 1999. 6(9): p. 884-93

[8] Mani, R.S., et al., *Biophysical characterization of human XRCC1 and its binding to damaged and undamaged DNA.* Biochemistry, 2004. 43(51): p. 16505-14

[9] Horton, J.K., et al., *XRCC1 and DNA polymerase beta in cellular protection against cytotoxic DNA single-strand breaks.* Cell Res, 2008. 18(1): p. 48-63

[10] Lan, L., et al., *In situ analysis of repair processes for oxidative DNA damage in mammalian cells.* Proc Natl Acad Sci U S A, 2004. 101(38): p. 13738-43

[11] Mortusewicz, O. and H. Leonhardt, *XRCC1 and PCNA are loading platforms with distinct kinetic properties and different capacities to respond to multiple DNA lesions.* BMC Mol Biol, 2007. 8: p. 81

[12] Lindahl, T., *Instability and decay of the primary structure of DNA.* Nature, 1993. 362(6422): p. 709-15

[13] Slupphaug, G., B. Kavli, and H.E. Krokan, *The interacting pathways for prevention and repair of oxidative DNA damage.* Mutat Res, 2003. 531(1-2): p. 231-51

[14] Tebbs, R.S., L.H. Thompson, and J.E. Cleaver, *Rescue of Xrcc1 knockout mouse embryo lethality by transgene-complementation.* DNA Repair (Amst), 2003. 2(12): p. 1405-17

[15] Caldecott, K.W., *XRCC1 and DNA strand break repair*. DNA Repair (Amst), 2003. 2(9): p. 955-69

[16] Wong, H.K. and D.M. Wilson, 3rd, *XRCC1 and DNA polymerase beta interaction contributes to cellular alkylating-agent resistance and single-strand break repair*. J Cell Biochem, 2005. 95(4): p. 794-804

[17] Marsin, S., et al., *Role of XRCC1 in the coordination and stimulation of oxidative DNA damage repair initiated by the DNA glycosylase hOGG1*. J Biol Chem, 2003. 278(45): p. 44068-74

[18] Campalans, A., et al., *XRCC1 interactions with multiple DNA glycosylases: a model for its recruitment to base excision repair*. DNA Repair (Amst), 2005. 4(7): p. 826-35

[19] Mani, R.S., et al., *XRCC1 stimulates polynucleotide kinase by enhancing its damage discrimination and displacement from DNA repair intermediates*. J Biol Chem, 2007. 282(38): p. 28004-13

[20] Whitehouse, C.J., et al., *XRCC1 stimulates human polynucleotide kinase activity at damaged DNA termini and accelerates DNA single-strand break repair*. Cell, 2001. 104(1): p. 107-17

[21] Plo, I., et al., *Association of XRCC1 and tyrosyl DNA phosphodiesterase (Tdp1) for the repair of topoisomerase I-mediated DNA lesions*. DNA Repair (Amst), 2003. 2(10): p. 1087-100

[22] Caldecott, K.W., et al., *XRCC1 polypeptide interacts with DNA polymerase beta and possibly poly (ADP-ribose) polymerase, and DNA ligase III is a novel molecular 'nick-sensor' in vitro*. Nucleic Acids Res, 1996. 24(22): p. 4387-94

[23] Kubota, Y., et al., *Reconstitution of DNA base excision-repair with purified human proteins: interaction between DNA polymerase beta and the XRCC1 protein*. EMBO J, 1996. 15(23): p. 6662-70

[24] Marintchev, A., et al., *Domain specific interaction in the XRCC1-DNA polymerase beta complex*. Nucleic Acids Res, 2000. 28(10): p. 2049-59

[25] Gryk, M.R., et al., *Mapping of the interaction interface of DNA polymerase beta with XRCC1*. Structure, 2002. 10(12): p. 1709-20

[26] Taylor, R.M., et al., *Role of a BRCT domain in the interaction of DNA ligase III-alpha with the DNA repair protein XRCC1*. Curr Biol, 1998. 8(15): p. 877-80

[27] Luo, H., et al., *A new XRCC1-containing complex and its role in cellular survival of methyl methanesulfonate treatment*. Mol Cell Biol, 2004. 24(19): p. 8356-65

[28] Date, H., et al., *The FHA domain of aprataxin interacts with the C-terminal region of XRCC1*. Biochem Biophys Res Commun, 2004. 325(4): p. 1279-85

[29] Clements, P.M., et al., *The ataxia-oculomotor apraxia 1 gene product has a role distinct from ATM and interacts with the DNA strand break repair proteins XRCC1 and XRCC4*. DNA Repair (Amst), 2004. 3(11): p. 1493-502

[30] Schreiber, V., et al., *Poly(ADP-ribose) polymerase-2 (PARP-2) is required for efficient base excision DNA repair in association with PARP-1 and XRCC1*. J Biol Chem, 2002. 277(25): p. 23028-36

[31] Leppard, J.B., et al., *Physical and functional interaction between DNA ligase IIIalpha and poly(ADP-Ribose) polymerase 1 in DNA single-strand break repair*. Mol Cell Biol, 2003. 23(16): p. 5919-27

[32] Sears, R.C. and J.R. Nevins, *Signaling networks that link cell proliferation and cell fate*. J Biol Chem, 2002. 277(14): p. 11617-20

[33] Ginsberg, D., *E2F1 pathways to apoptosis*. FEBS Lett, 2002. 529(1): p. 122-5

[34] Johnson, D.G. and R. Schneider-Broussard, *Role of E2F in cell cycle control and cancer*. Front Biosci, 1998. 3: p. d447-8

[35] Morkel, M., et al., *An E2F-like repressor of transcription*. Nature, 1997. 390(6660): p. 567-8

[36] Cartwright, P., et al., *E2F-6: a novel member of the E2F family is an inhibitor of E2F-dependent transcription*. Oncogene, 1998. 17(5): p. 611-23

[37] Magae, J., et al., *Nuclear localization of DP and E2F transcription factors by heterodimeric partners and retinoblastoma protein family members*. J Cell Sci, 1996. 109 (Pt 7): p. 1717-26

[38] Frolov, M.V. and N.J. Dyson, *Molecular mechanisms of E2F-dependent activation and pRB-mediated repression*. J Cell Sci, 2004. 117(Pt 11): p. 2173-81

[39] He, Y. and W.D. Cress, *E2F-3B is a physiological target of cyclin A*. J Biol Chem, 2002. 277(26): p. 23493-9

[40] Tsantoulis, P.K. and V.G. Gorgoulis, *Involvement of E2F transcription factor family in cancer*. Eur J Cancer, 2005. 41(16): p. 2403-14

[41] Martinez-Balbas, M.A., et al., *Regulation of E2F1 activity by acetylation*. EMBO J, 2000. 19(4): p. 662-71

[42] Ianari, A., et al., *Specific role for p300/CREB-binding protein-associated factor activity in E2F1 stabilization in response to DNA damage*. J Biol Chem, 2004. 279(29): p. 30830-5

[43] Dimova, D.K. and N.J. Dyson, *The E2F transcriptional network: old acquaintances with new faces*. Oncogene, 2005. 24(17): p. 2810-26

[44] Powers, J.T., et al., *E2F1 uses the ATM signaling pathway to induce p53 and Chk2 phosphorylation and apoptosis*. Mol Cancer Res, 2004. 2(4): p. 203-14

[45] Wang, B., et al., *A role for 14-3-3 tau in E2F1 stabilization and DNA damage-induced apoptosis*. J Biol Chem, 2004. 279(52): p. 54140-52

[46] Liu, K., et al., *Regulation of E2F1 by BRCT domain-containing protein TopBP1*. Mol Cell Biol, 2003. 23(9): p. 3287-304

[47] Liontos, M., et al., *Modulation of the E2F1-driven cancer cell fate by the DNA damage response machinery and potential novel E2F1 targets in osteosarcomas*. Am J Pathol, 2009. 175(1): p. 376-91

[48] Frame, F.M., et al., *E2F1 induces MRN foci formation and a cell cycle checkpoint response in human fibroblasts*. Oncogene, 2006. 25(23): p. 3258-66

[49] Liao, C.C., et al., *RB.E2F1 complex mediates DNA damage responses through transcriptional regulation of ZBRK1*. J Biol Chem, 2010. 285(43): p. 33134-43

[50] Lin, P.S., et al., *The role of the retinoblastoma/E2F1 tumor suppressor pathway in the lesion recognition step of nucleotide excision repair*. DNA Repair (Amst), 2009. 8(7): p. 795-802

[51] Seifried, L.A., et al., *pRB-E2F1 complexes are resistant to adenovirus E1A-mediated disruption*. J Virol, 2008. 82(9): p. 4511-20

[52] Chen, J., et al., *E2F1 promotes the recruitment of DNA repair factors to sites of DNA double-strand breaks*. Cell Cycle, 2011. 10(8): p. 1287-94

[53] Guo, R., et al., *GCN5 and E2F1 stimulate nucleotide excision repair by promoting H3K9 acetylation at sites of damage*. Nucleic Acids Res, 2011. 39(4): p. 1390-7

[54] Berton, T.R., et al., *Regulation of epidermal apoptosis and DNA repair by E2F1 in response to ultraviolet B radiation.* Oncogene, 2005. 24(15): p. 2449-60

[55] Guo, R., et al., *E2F1 localizes to sites of UV-induced DNA damage to enhance nucleotide excision repair.* J Biol Chem, 2010. 285(25): p. 19308-15

[56] Youn, C.K., et al., *Bcl-2 expression suppresses mismatch repair activity through inhibition of E2F transcriptional activity.* Nat Cell Biol, 2005. 7(2): p. 137-47

[57] Polager, S., et al., *E2Fs up-regulate expression of genes involved in DNA replication, DNA repair and mitosis.* Oncogene, 2002. 21(3): p. 437-46

[58] Bindra, R.S., et al., *Hypoxia-induced down-regulation of BRCA1 expression by E2Fs.* Cancer Res, 2005. 65(24): p. 11597-604

[59] Prost, S., et al., *E2F regulates DDB2: consequences for DNA repair in Rb-deficient cells.* Oncogene, 2007. 26(24): p. 3572-81

[60] Walsh, M.J., et al., *E2F-1 and a cyclin-like DNA repair enzyme, uracil-DNA glycosylase, provide evidence for an autoregulatory mechanism for transcription.* J Biol Chem, 1995. 270(10): p. 5289-98

[61] Maser, R.S., et al., *Mre11 complex and DNA replication: linkage to E2F and sites of DNA synthesis.* Mol Cell Biol, 2001. 21(17): p. 6006-16

[62] Lundin, C., et al., *Methyl methanesulfonate (MMS) produces heat-labile DNA damage but no detectable in vivo DNA double-strand breaks.* Nucleic Acids Res, 2005. 33(12): p. 3799-811

[63] Chen, D., et al., *E2F1 regulates the base excision repair gene XRCC1 and promotes DNA repair.* J Biol Chem, 2008. 283(22): p. 15381-9

[64] Yacoub, A., et al., *MAPK dependence of DNA damage repair: ionizing radiation and the induction of expression of the DNA repair genes XRCC1 and ERCC1 in DU145 human prostate carcinoma cells in a MEK1/2 dependent fashion.* Int J Radiat Biol, 2001. 77(10): p. 1067-78

[65] Yacoub, A., et al., *Epidermal growth factor and ionizing radiation up-regulate the DNA repair genes XRCC1 and ERCC1 in DU145 and LNCaP prostate carcinoma through MAPK signaling.* Radiat Res, 2003. 159(4): p. 439-52

[66] Shung, B., J. Miyakoshi, and H. Takebe, *X-ray-induced transcriptional activation of c-myc and XRCC1 genes in ataxia telangiectasia cells.* Mutat Res, 1994. 307(1): p. 43-51

[67] Thompson, L.H. and M.G. West, *XRCC1 keeps DNA from getting stranded.* Mutat Res, 2000. 459(1): p. 1-18

[68] Libutti, S.K., et al., *Parathyroid gland-specific deletion of the mouse Men1 gene results in parathyroid neoplasia and hypercalcemic hyperparathyroidism.* Cancer Res, 2003. 63(22): p. 8022-8

[69] Nakada, D., et al., *The ATM-related Tel1 protein of Saccharomyces cerevisiae controls a checkpoint response following phleomycin treatment.* Nucleic Acids Res, 2003. 31(6): p. 1715-24

[70] Lin, W.C., F.T. Lin, and J.R. Nevins, *Selective induction of E2F1 in response to DNA damage, mediated by ATM-dependent phosphorylation.* Genes Dev, 2001. 15(14): p. 1833-44

[71] Stevens, C., L. Smith, and N.B. La Thangue, *Chk2 activates E2F-1 in response to DNA damage.* Nat Cell Biol, 2003. 5(5): p. 401-9

[72] Blattner, C., A. Sparks, and D. Lane, *Transcription factor E2F-1 is upregulated in response to DNA damage in a manner analogous to that of p53.* Mol Cell Biol, 1999. 19(5): p. 3704-13

[73] Stevens, C. and N.B. La Thangue, *The emerging role of E2F-1 in the DNA damage response and checkpoint control.* DNA Repair (Amst), 2004. 3(8-9): p. 1071-9

[74] Tan, Y., P. Raychaudhuri, and R.H. Costa, *Chk2 mediates stabilization of the FoxM1 transcription factor to stimulate expression of DNA repair genes.* Mol Cell Biol, 2007. 27(3): p. 1007-16

[75] Levy, N., et al., *XRCC1 is phosphorylated by DNA-dependent protein kinase in response to DNA damage.* Nucleic Acids Res, 2006. 34(1): p. 32-41

[76] Loizou, J.I., et al., *The protein kinase CK2 facilitates repair of chromosomal DNA single-strand breaks.* Cell, 2004. 117(1): p. 17-28

[77] Dong, Z. and A.E. Tomkinson, *ATM mediates oxidative stress-induced dephosphorylation of DNA ligase IIIalpha.* Nucleic Acids Res, 2006. 34(20): p. 5721-279

[78] Ishida, S., et al., *Role for E2F in control of both DNA replication and mitotic functions as revealed from DNA microarray analysis.* Mol Cell Biol, 2001. 21(14): p. 4684-99

[79] Ren, B., et al., *E2F integrates cell cycle progression with DNA repair, replication, and G(2)/M checkpoints.* Genes Dev, 2002. 16(2): p. 245-56

[80] Weinmann, A.S., et al., *Isolating human transcription factor targets by coupling chromatin immunoprecipitation and CpG island microarray analysis.* Genes Dev, 2002. 16(2): p. 235-44

[81] Fan, J., et al., *XRCC1 co-localizes and physically interacts with PCNA.* Nucleic Acids Res, 2004. 32(7): p. 2193-201

[82] Moore, D.J., et al., *Mutation of a BRCT domain selectively disrupts DNA single-strand break repair in noncycling Chinese hamster ovary cells.* Proc Natl Acad Sci U S A, 2000. 97(25): p. 13649-54

[83] Taylor, R.M., A. Thistlethwaite, and K.W. Caldecott, *Central role for the XRCC1 BRCT I domain in mammalian DNA single-strand break repair.* Mol Cell Biol, 2002. 22(8): p. 2556-63

[84] Taylor, R.M., et al., *A cell cycle-specific requirement for the XRCC1 BRCT II domain during mammalian DNA strand break repair.* Mol Cell Biol, 2000. 20(2): p. 735-40

[85] Inoue, Y., M. Kitagawa, and Y. Taya, *Phosphorylation of pRB at Ser612 by Chk1/2 leads to a complex between pRB and E2F-1 after DNA damage.* EMBO J, 2007. 26(8): p. 2083-93

[86] Pommier, Y., et al., *Targeting chk2 kinase: molecular interaction maps and therapeutic rationale.* Curr Pharm Des, 2005. 11(22): p. 2855-72

[87] Kowalik, T.F., et al., *E2F1 overexpression in quiescent fibroblasts leads to induction of cellular DNA synthesis and apoptosis.* J Virol, 1995. 69(4): p. 2491-500

[88] Harris, S.L. and A.J. Levine, *The p53 pathway: positive and negative feedback loops.* Oncogene, 2005. 24(17): p. 2899-908

[89] Bell, L.A. and K.M. Ryan, *Life and death decisions by E2F-1.* Cell Death Differ, 2004. 11(2): p. 137-42

[90] Sengupta, S. and C.C. Harris, *p53: traffic cop at the crossroads of DNA repair and recombination.* Nat Rev Mol Cell Biol, 2005. 6(1): p. 44-55

[91] Adimoolam, S. and J.M. Ford, *p53 and regulation of DNA damage recognition during nucleotide excision repair.* DNA Repair (Amst), 2003. 2(9): p. 947-54

[92] Yamasaki, L., et al., *Tumor induction and tissue atrophy in mice lacking E2F-1.* Cell, 1996. 85(4): p. 537-48

[93] Field, S.J., et al., *E2F-1 functions in mice to promote apoptosis and suppress proliferation.* Cell, 1996. 85(4): p. 549-61

[94] Wang, S., et al., *JWA regulates XRCC1 and functions as a novel base excision repair protein in oxidative-stress-induced DNA single-strand breaks.* Nucleic Acids Res, 2009. 37(6): p. 1936-50

9

Post-Transcriptional Regulation of E2F Transcription Factors: Fine-Tuning DNA Repair, Cell Cycle Progression and Survival in Development & Disease

Lina Dagnino, Randeep Kaur Singh and David Judah
University of Western Ontario
Canada

1. Introduction

Cells are continually exposed to genotoxic stresses. Upon DNA damage, the cell activates a coordinated and complex series of responses (Levitt and Hickson, 2002). Multiple factors are implicated in each of these responses. Recently, it has become apparent that various transcription factors play important roles in cellular responses to genotoxic stress. In particular, E2F transcription factors are key for the activation of genes involved in these processes.

E2F family comprises two subfamilies, termed E2F and DP, and includes orthologs expressed across many species, from plants to higher vertebrates (McClellan and Slack, 2007). In mammals, multiple E2F (E2F-1 through -8) and DP (DP-1 through -4) genes have been identified. E2F-1, -2 and -3 are associated with DNA synthesis and cell cycle progression, and function as heterodimers with a DP member (McClellan and Slack, 2007). E2F-4 and -5 also require association with a DP protein, but often function to halt cell cycle progression associated with terminal differentiation or reversible entry into quiescence (McClellan and Slack, 2007). E2F-1 through -5 can mediate transcriptional activation when found as "free" E2F/DP dimers, but can also act as transcriptional repressors if they are associated with a member of the retinoblastoma (pRb) family of proteins (Hallstrom and Nevins, 2009). In contrast, E2F-6 lacks both transcriptional activation and pRb-binding domains, and functions as a constitutive transcriptional repressor. The most divergent members of the E2F family are E2F-7 and -8, which bind neither DP nor pRb-family proteins, and also function as transcriptional repressors to mediate cell cycle arrest (Lammens et al., 2009). To regulate gene expression, E2F factors bind GC-rich elements on proximal promoters, which can conform to either a consensus 5' -TTTC[CG]CGC-3' element, or to non-consensus sequences (Judah et al., 2010; Rabinovich et al., 2008). Considerable efforts have been directed to investigate whether different E2F proteins exhibit target selectivity. Genome-wide screens for E2F targets have revealed considerable overlap in the ability of individual E2F proteins to regulate their targets, although a few promoters activated by specific E2F forms have been identified (Cao et al., 2011).

In spite of the vast similarities in the activities of distinct E2F proteins and their ability to bind potential target Genes, to-date E2F1 is the principal E2F member shown to participate in cellular responses to DNA damage (Bracken et al., 2004). The role of E2F-1 upon DNA damage depends on cellular context. E2F-1 can either induce pro-apoptotic or anti-apoptotic outcomes. During the latter, E2F-1 can play roles to induce cell cycle arrest and upregulate DNA repair, by directing expression of multiple genes. These genes are involved in mismatch repair (MSH2, MLH1), nucleotide excision repair (DDB2, RPA), homologous recombination repair (RAD51, RAD54, RECQL), base excision repair (UNG, APE) & non-homologous end joining (Chang et al., 2006; Ishida et al., 2001; Polager and Ginsberg, 2008; Prost et al., 2007).

In humans, E2F-1 is a 437 amino acid protein, which shows constitutive and rapid nucleocytoplasmic shuttling in a variety of cells (Ivanova et al., 2007). E2F-1 stimulates cell proliferation by positively modulating transcription of genes necessary for DNA synthesis and cell cycle progression (Ivanova et al., 2005). In an apparently paradoxical manner, E2F-1 can also induce cell cycle arrest when associated with pRb, or apoptosis, by activating expression of pro-apoptotic genes (Polager and Ginsberg, 2008). The breadth of E2F-1 targets mediates the distinct biological activities of this transcription factor, which encompass both oncogenic and anti-oncogenic properties, as well as positive modulation of tissue regeneration after injury (D'Souza et al., 2002; Field et al., 1996).

2. E2F-1 and the DNA damage response

Genotoxic stress in cells activates the DNA damage response, and can occur as a result of a variety of insults. The latter include DNA double- strand breaks and single-strand damage. DNA damage can result from exogenous agents (e.g. radiation, exposure to reactive and mutagenic chemicals), or from endogenous products of cell metabolism (Shiloh, 2003). In response to DNA damage, cells activate multiple pathways that result in apoptosis or in DNA repair, cell cycle arrest, changes in gene expression, as well as in protein synthesis and degradation.

Cells require efficient response mechanisms to genotoxic stress, as this is a life-threatening event because it can significantly alter their genetic material. Multiple mechanisms have evolved to repair damage induced by genotoxic stress, including activation of a global signalling network termed the DNA damage response (DDR), which is capable of detecting distinct types of DNA damage, coordinating appropriate responses. The latter include transcriptional activation, cell cycle arrest, apoptosis, senescence and DNA repair (Shiloh, 2003). The DNA damage response plays a critical role in cell survival when damage occurs during DNA replication. In addition, there are specialized processes, including base-excision repair (BER), nucleotide-excision repair (NER) & nonhomologous end-joining, which recognize and repair specific types of lesions (Shiloh, 2003). Central to transduce signals that indicate DNA damage and initiate appropriate cellular responses are two related protein kinases, termed ATM (ataxia telangiectasia mutated) and ATR (ATM and Rad3-related). ATM can associate with its regulator, the MRN (Mre11-Rad50-NMS1) complex, when double-strand breaks (DSB) are generated (Levitt and Hickson, 2002). On the other hand, ATR forms complexes with its regulator ATRIP (ATR-interacting protein), which senses single-strand DNA (ssDNA) breaks generated by processing of double-strand breaks, as well as single-strand DNA which arises from stalled replication forks (Shiloh, 2003). These two kinases also phosphorylate E2F-1, thus initiating transcriptional activation of its target DNA repair genes.

2.1 Identification of E2F targets involved in DNA damage repair

Central to understanding the role of the E2F family of transcription factors in DNA repair has been the identification of a large number of putative and demonstrated E2F target genes. Although E2F proteins were originally characterized as important regulators of cell cycle progression, genome-wide screens have demonstrated much broader roles in a variety of primary and immortalized cell types. For example, E2F-1 and E2F-3 bind to the promoters of apurinic/apyrimidinic endonuclease (APE) and other repair enzymes in human primary epidermal keratinocytes, irrespective of their differentiation status (Chang et al., 2006). Similarly, in the GM06990 lymphoblastoid cell line, non-biased genome-wide screening has identified a large number of putative E2F-4 targets involved in responses to DNA damage (Lee et al., 2011). E2F targets important for DNA repair have also been identified in neoplastic cells following therapeutic intervention. For example, treatment of prostate cancer cells with histone deacetylase inhibitors reduces their ability to repair DNA damage induced by radio- and chemotherapy, thus reducing tumour mass (Kachhap et al., 2010). The impaired ability to repair DNA of treated cells was due, at least in part, to decreased recruitment to and activation by E2F-1 to the promoters of key DNA repair genes. Hence, the importance of E2F factors in DNA repair encompasses not only events during carcinogenesis, but also the potential impact of various therapies.

2.2 Role of E2F-1 in responses to DNA damage induced by UV radiation

UV radiation induces severe DNA damage, which is the principal cause of skin carcinogenesis in humans (Brash et al., 1996). UV-B radiation induces formation of cyclobutane pyrimidine dimers (CPD) and pyrimidine (6-4) pyrimidone photoproducts (6-4PP), which would result in loss of DNA integrity and genetic instability if left unrepaired. This type of damage to DNA triggers activation of the nucleotide-excision repair pathway, and can occur *via* one or more streams. Such DNA repair streams include (i) global genome repair (GGR), which repairs damage from the entire genome, (ii) transcription-coupled repair (TCR), which generally repairs damage on actively transcribed DNA strands & (iii) transcription domain-associated repair (DAR), which deals with repairing both strands of actively transcribed regions (Nouspikel, 2009).

Normal responses of the epidermis to UV damage are critically dependent on E2F-1 expression. Indeed, increased levels of epidermal apoptosis upon UV-B irradiation have been reported in E2F-1-null mouse epidermis, whereas repair of UV-B-induced DNA photoproducts is more efficient in keratinocytes that overexpress E2F-1 (Berton et al., 2005). UV-induced DNA damage results in stabilization of E2F-1 protein, which stimulates nucleotide excision repair (Berton et al., 2005; Pediconi et al., 2003; Wikonkal et al., 2003). The mechanisms involved include phosphorylation of E2F-1 on Ser31 by ATR and/or ATM kinases (Lin et al., 2001). This modification facilitates E2F-1 recruitment to sites of double-strand breaks or UV-induced DNA damage. Under these conditions, E2F-1 interacts with two key proteins involved in DNA repair: TopBP1 and GCN5 histone acetyltransferase (Guo et al., 2010a; Guo et al., 2010b). Formation of these E2F-1 complexes is necessary for efficient recruitment of factors involved in nucleotide excision repair. Importantly, the association of E2F-1 with TopBP1 and GCN5 occurs at the expense of the E2F-1-induced expression of pro-apoptotic p73, thus ensuring that DNA repair, rather than apoptosis, takes place (Berton et al., 2005; Pediconi et al., 2003; Wikonkal et al., 2003). In mouse embryo fibroblasts, UV-C irradiation results in the formation of both CPD and 6-4PP. In these cells, nucleotide excision repair is activated through pathways that involve activation of xeroderma pigmentosum

(XPC) gene expression by E2F-1 *via* increased binding to the XPC promoter (Lin et al., 2009). XPC is an essential mediator of DNA damage recognition during global genomic repair, and this phase of repair is actually more efficient in pRB-deficient cells, likely because lack of pRb increases E2F-1 activity.

The importance of E2F in repair of DNA damage induced by UV radiation is further demonstrated by the conservation of this pathway through evolution. For example, in *Arabidopsis* and in maize, MSH2 and MSH6, which are two genes that belong to the mismatch repair system, are targets of E2F transcriptional activation following DNA damage by UV-B radiation (Lario et al., 2011).

2.3 E2F is a key factor to maintain the balance between cell cycle arrest and expression of DNA repair genes following DNA damage

Given the key roles that pRb family proteins play in the regulation of E2F activity, it is not surprising that they also modulate the function of E2F factors following DNA damage. For example, the zinc finger-containing transcriptional repressor ZBRK1 is an important modulator of GADD45A transcription. The latter is involved in induction of cell cycle arrest in response to DNA damage (Siafakas and Richardson, 2009). E2F-1, but not other E2F proteins, binds to the ZBRK1 promoter, together with pRb, CtIP and CtBP, forming repressor complexes that interfere with ZBRK1 expression (Liao et al., 2010). In pRb-deficient cells, increased susceptibility to DNA damage induced by UV radiation or methylating agents occurs, partly as a result of abnormal cell cycle arrest and DNA repair. In a similar manner, E2F-1 is essential for normal expression of XRCC1 (x-ray repair cross-complementation group 1), which participates in the repair of single-strand breaks, thus ensuring efficient repair following DNA damage induced by methylating agents (Chen et al., 2008).

In contrast, loss of pRb can improve DNA repair in other circumstances, such as those involving activation of DDB2. Mutations in the *DDB2* gene, which encodes a protein involved in global genomic repair and repair of CPDs, gives rise to xeroderma pigmentosum, a disorder associated with increased risk of cutaneous and ocular tumours (Bennett and Itoh, 2008). DDB2 expression is positively regulated by E2F-1 and E2F-3. Further, deletion of pRb increases DDB2 mRNA and protein levels, together with ability of these cells to repair DNA damage. The latter is associated with more efficient CPD removal relative to that in pRb-expressing cells (Prost et al., 2007).

Solid tumours frequently exhibit hypoxic cores, which contribute to genetic instability within the tumour microenvironment (Bindra et al., 2005). This is partly due to decreased expression of DNA mismatch genes (MLH1 and MSH2), as well as repair genes (RAD51 and BRCA1). E2F factors can also be involved in the downregulation of some of these repair genes, in apparent contrast to their pro-repair roles in other circumstances. Specifically, hypoxic conditions result in the dephosphorylation of the pRb family member p130, which then associates with E2F-4 in the nucleus. This complex can efficiently bind to E2F sites on the RAD51 and BRCA1 promoters, thus interfering with their transcription (Bindra et al., 2005). Thus, E2F factors can positively or negatively regulate DNA repair, depending on cellular context. Given that E2F-4/p130 complexes are also important for cell cycle exit, a balance must exist between these two outcomes, which is essential to avoid increased genetic instability in transformed cells and their clonal expansion.

2.4 Role of E2F-1 in senescence-associated DNA damage

Senescence is defined as irreversible cell cycle arrest, which occurs both in cultured cells and *in vivo* (Lanigan et al., 2011). Senescence has been recognized as a key mechanism that acts as a barrier to tumour formation and progression. Thus, in spite of any DNA damage that may exist in a long-lived cell, if this cell is senescent it will not undergo clonal expansion to generate daughter cells with altered DNA. A number of molecular mechanisms control cellular senesce, and the E2F/pRb pathway is a key component (Lanigan et al., 2011). Under normal circumstances, the frequency of DNA mutations increases with age. DNA mismatch mutation repair is very efficient in mesenchymal cells from young individuals, as well as in embryonic fibroblasts (Chang et al., 2008). In contrast, these mechanisms are less efficient in senescent cells, in which MSH2 expression is decreased. Associated with these abnormalities is the inhibition of E2F-1 transcriptional activity, which leads to repression of MSH2 gene transcription. Thus, E2F-1 activity is essential to maintain normal capacity of cells to repair mismatch mutations. Whether the reduced activity of E2F-1 also increases the risk of transformation in senescent cells probably depends on cell context, extent of DNA damage, and presence of other oncogenic stimuli.

2.5 Role of E2F/DP interactions in DNA repair

The interactions between E2F-1 through -6 and their partner DP proteins are essential for normal transcriptional activity, and can also contribute to abnormal regulation of DNA repair factors. Again, depending on the exact context, E2F/DP interactions can positively or negatively modulate DNA repair. For example, following DNA damage by a variety of agents, including doxorubicin, etoposide and UV radiation, the abundance of DP-4 protein is substantially increased, replacing other DP proteins in E2F-1-containing complexes (Ingram et al., 2011). As a result, the capacity of E2F-1 to bind target promoters is strongly reduced, which can result in downregulation of cell cycle regulatory and/or DNA repair genes.

A positive modulatory role in DNA nucleotide excision repair through inhibition of repressor E2F complexes has been recently attributed to p14[Arf] (Dominguez-Brauer et al., 2009). Specifically, DNA damage induces p14[Arf] expression, which directly binds to DP-1, disrupting its interactions with E2F-4. As a result, repressive E2F-4/p130 complexes lose their ability to bind promoters of genes such as XPC, resulting in upregulation of their expression.

To-date, multiple mechanisms that regulate E2F-1 activity at the post-transcriptional level have been identified, although only a handful has been studied in the context of DNA repair. These forms of regulation of E2F-1 activity can have important consequences on its ability to modulate DNA damage responses, as discussed below.

3. Role of miRNAs in E2F regulation of cell growth and DNA repair

MicroRNAs (miRNAs) are short nucleotide sequences (~21-24nt) that pair with the 3'-untranslated regions of target mRNAs. They negatively regulate gene expression by mediating degradation of the target mRNA, or by inhibition of protein translation (Almeida et al., 2011). Small miRNAs regulate many cellular processes, such as apoptosis, differentiation, and proliferation. They are upregulated in many human disorders, including cancer and neurological diseases (Almeida et al., 2011). To-date, approximately 800 miRNAs have been identified in humans. A single miRNA can target multiple mRNAs (Griffiths-

Jones, 2004). Consistent with their role in cancer, miRNAs control cell proliferation by regulating E2F factors and, thereby, expression of genes that are important for cell cycle progression.

The E2F signalling pathway is regulated by many different types of miRNA clusters, including *miRNA-17-92, miRNA-106b-25, miRNA-34, miRNA330-3p, miRNA-128, miRNA-195, miRNA-37 and miRNA-193a,* as described below.

3.1 Growth-promoting miRNAs

O' Donnell et al. were the first to provide evidence that E2F is a target for miRNAs (O'Donnell et al., 2005). They showed that miRNA-17 and miRNA-20a decrease E2F-1 translation efficiency. This type of regulation prevents uncontrolled activation of E2F-1 during normal cell cycle progression. Disruption of miRNA-17 and miRNA-20a leads to improperly timed expression of E2F-1, resulting in the accumulation of DNA double strand breaks (Pickering et al., 2009).

An auto-regulatory loop between E2F-1 and E2F-3 and the miRNA-17-92 clusters has been demonstrated. E2F-1 and E2F-3 bind to and upregulate the transcription of the miRNA-17-92 cluster. In turn, the miRNA-17-92 cluster downregulates expression of these two transcription factors (Sylvestre et al., 2007; Woods et al., 2007). This negative feedback loop is important to prevent the accumulation of E2F-1 and E2F-3, thereby allowing proper progression of the cell cycle, preventing apoptosis. Another negative feedback loop has been observed between the miRNA-106b-25 clusters and E2F-1 (Petrocca et al., 2008). miRNA106b and miRNA93 downregulate E2F-1 expression. Reciprocally, transcription of these miRNAs is activated by E2F-1. In this manner, properly timed expression of E2F-1 during the G1/S transition is maintained, as the presence of these miRNAs prevents continuous E2F-1 expression throughout the cell cycle, which would induce apoptosis.

3.2 Tumor suppressor miRNAs

The E2F signalling pathway is also regulated by the miRNA-34 family of clusters (Tazawa et al., 2007). miRNA-34b decreases E2F-1 and E2F-3 transcript levels in a p53-dependent manner, inhibiting cell proliferation and inducing senescence in tumour cells. This demonstrates that miRNAs can function as tumor suppressors. A similar role has been suggested for miRNA-195 (Xu et al., 2009), miRNA-128 (Cui et al., 2010), miRNA-330-3p (Lee et al., 2009) and miRNA193a (Kozaki et al., 2008).

Overexpression of miRNA-195 causes cell cycle arrest at the G1/S boundary, by interfering with the expression of cell cycle regulatory proteins, such E2F-3, Cyclin D1 and cyclin-dependent kinase 6 (CDK6). As a result, pRb remains hypophosphorylated, allowing activation of E2F-dependent target genes (Xu et al., 2009). Exogenous expression of miRNA-127 in glioma cells represses E2F-3a translation, thereby decreasing cell proliferation (Cui et al., 2010). Similarly, in oral squamous cell carcinoma, miRNA193a significantly represses cell growth and down-regulates E2F-6 translation (Kozaki et al., 2008).

3.3 Role of miRNAs in modulation of DNA repair by E2F-1

Several miRNA clusters, including *mir17-92, mir-106a-92* and *mir106b-25,* are downregulated by p53 via E2F-dependent mechanisms. This leads to decreased proliferation and/or promotes senescence in normal and transformed cells (Brosh et al., 2008). In addition, in response to mitogenic stimulation, E2F-1 activates transcription of the miRNA clusters *let-*

7a-d, mir-15b-16-2 and *mir-106b-25* during the G1/S transition (Bueno et al., 2010). These miRNAs, in turn, regulate E2F-1 activity. In their absence, E2F-1 induces entry into S phase, but also DNA damage. Indeed, E2F-1 and other oncogenes can induce stalling and collapsing of DNA replication forks, leading to the formation of DNA double-strand breaks (Halazonetis et al., 2008). Thus, *let-7a-d, mir-15b-16-2* and *mir-106b-25* play key roles in prevention of DNA damage and replicative stress associated with abnormal regulation of E2F-1 (Zhang et al., 2011).

4. Regulation of E2F-1 by post-translational modifications

Another mode of E2F regulation that fine-tunes cell cycle progression and DNA repair occurs at the post-translational level. Post-translational modifications identified in E2F-1 include phosphorylation, acetylation, methylation & ubiquitination. These modifications can exert either activating or inhibitory effects on E2F-1 transcriptional activity.

4.1 Acetylation

E2F-1 is acetylated at three highly conserved lysine residues (K117, K120 and K125) by the p300/CREB-binding protein (CBP) or by p300/CBP-associated factor (P/CAF) acetyltransferase (Martinez-Balbas et al., 2000; Marzio et al., 2000). P/CAF directly interacts with E2F-1 through its adenosine deaminase 2 (ADA2) binding domain (Martinez-Balbas et al., 2000). Acetylation of E2F-1 allows for marked stabilization and significant increase in E2F-1 protein levels. This leads to an increase in transcriptional activation of E2F-1 target genes (Farhana et al., 2002; Martinez-Balbas et al., 2000).

Increases in E2F-1 protein levels upon DNA damage are partly due to cell type-specific acetylation (Blattner et al., 1999; Meng et al., 1999; Zhu et al., 1999). For example, adriamycin-mediated treatment induces E2F-1 acetylation in human glioblastoma T98G cells (Pediconi et al., 2003) , but not in HeLa cells (Ozaki et al., 2009). In response to DNA damage, E2F-1 switches to activate pro-apoptotic gene expression, rather than cell cycle progression. This change requires E2F-1 acetylation and recruitment to promoters of pro-apoptotic target genes, such as p73 (Pediconi et al., 2003). P/CAF, but not p300, is required for E2F-1 stabilization upon DNA damage by doxorubicin (Ianari et al., 2004). On the other hand, overexpression of p300 can be sufficient for acetylation and stabilization of E2F-1 in cells treated with camptothecin, a drug that causes double strand break during DNA replication (Galbiati et al., 2005). The distinct actions of these two acetyltransferase can thus determine the outcome of cellular responses by modulating cellular DNA damage checkpoints (p300) or apoptotic events (P/CAF). The stabilization of E2F-1 by acetylation could also allow it to directly interact with activating signal cointegrator-2 (ASC-2), a mitogenic transcription factor co-activator that regulates cellular proliferation and cell cycle progression (Kong et al., 2003).

4.2 Phosphorylation

E2F-1 is phosphorylated on several residues, giving rise to modifications that can alter different functional aspects. E2F-1 was first identified as a substrate for phosphorylation in a cell-free system (Bagchi et al., 1989). This post-translational modification interfered with E2F-1 DNA binding activity. Consistent with these observations, E2F-1 and E2F-3 showed decreased DNA binding capacity upon phosphorylation by cyclin A-activated cyclin-dependent kinase 2 (cdk2) (Dynlacht et al., 1997; Krek et al., 1995). Complexes containing cyclin A, cdk2, E2F-1, and DP-1 are formed during Late S-phase to terminate E2F-dependent DNA binding and transcription, and enable orderly S-phase progression (Krek et al., 1995). In the absence of cyclin A-cdk2 activity, there is decreased E2F-1 phosphorylation and increased DNA binding activity (Li et al., 1997). This results in S-phase delay and/or arrest, by mechanisms that involve transcriptional activation of E2F-dependent cell cycle checkpoint genes. Together, these data demonstrate that E2F-1 phosphorylation is essential for timely activation of E2F-1 function and orderly cell cycle progression and survival. A second proline-directed kinase, c-Jun N-terminal protein kinase (JNK1), can phosphorylate E2F-1 in response to stress stimuli mediated by tumor necrosis factor-alpha, decreasing its ability to bind DNA and activate target gene transcription (Kishore et al., 2003).

Following DNA damage, Chk2 and ATM phosphorylate E2F-1 on Ser364 and Ser31, respectively (Lin et al., 2001). E2F-1 phosphorylated on Ser31 subsequently interacts with 14-3-3τ (Wang et al., 2004). This interaction prevents E2F-1 association with the SKP1-Cullin-F-box/ S-phase kinase-associated protein 2 (SCF^{Skp2}) ubiquitin ligase. As a result, E2F-1 is not ubiquitinated and is protected from degradation. The net result of phosphorylation of E2F-1 at Ser31 and Ser364 after DNA damage is activation of the pro-apoptotic gene p73, as well as accumulation of p53 through upregulation of p19^{ARF} expression. The latter protein inhibits ubiquitination and degradation of p53, inducing apoptosis (Weber et al., 1999). In additon, the ATM and Chk2 promoters are activated by E2F-1, thereby forming a positive feedback pathway that promotes apoptosis (Berkovich and Ginsberg, 2003) .

The phosphorylation of E2F can also affect its ability to interact with other proteins. In *Drosophila melanogaster*, phosphorylation of E2F-1 and E2F-2 enhances their ability to interact with the SCF^{slmb} ubiquitin ligase complex, targetting it for degradation during S phase (Heriche et al., 2003). *In vitro*, E2F-1 is phosphorylated at Ser337 by complexes containing p34^{cdc2} and cyclin B (Dynlacht et al., 1997). The significance of this finding is not clear, as E2F-1 phosphorylation on these residues occurs during late G1 phase, and is mediated by cyclin D-cdk4 complexes (Mann and Jones, 1996). Phosphorylation of E2F-1 at Ser332 and Ser337 enhances E2F-1 interactions with the adenovirus E4 protein, simultaneously attenuating its ability to bind pRb (Fagan et al., 1994). Upon adenovirus infection, the enhanced interaction between E2F-1 and E4 increases the efficiency of E2A transcription, which is required for viral DNA replication (Hardy et al., 1989).

Changes in phosphorylation status also modulate the activity and subcellular localization of E2F-4 and E2F-5, although these changes are unlikely to be mediated by cyclin A-dependent cdk activity (Dynlacht et al., 1997). Regulation of E2F-4 and E2F-5 by phosphorylation is important during entry into quiescence associated with cell differentiation, but varies depending on the cell type. For example, hypophosphorylated forms of E3F-4 efficiently associate with p130 in the nucleus, forming transcriptional repressor complexes associated with growth arrest in muscle cells (Shin et al., 1995). In contrast, in human intestinal crypt cells, hypophosphorylated E2F-4 is imported into the nucleus in response to mitogenic stimuli or inhibition of p38 MAP kinase, where it activate genes necessary for S phase entry (Deschenes et al., 2004).

E2F-5 is phosphorylated by cyclin E/cdk2 complexes on Thr251 in the transcriptional activation domain, stimulating cell cycle progression (Morris et al., 2000). This modification stabilizes E2F-5 interaction with the co-activator p300/CBP, resulting in transcription of genes required for DNA synthesis. Significantly, phosphorylation of E2F-5 at Thr251 does not affect its DNA binding activity, intracellular localization or ability to interact with pRb family proteins.

In epidermal keratinocytes, E2F-1 is tightly regulated during normal proliferation and differentiation (Ivanova and Dagnino, 2007; Ivanova et al., 2009; Wong et al., 2003). E2F-1 is localized in the nucleus in undifferentiated keratinocytes, but differentiation induces its export to the cytoplasm, where it is degraded. The signaling pathways involved in E2F-1 turnover in differentiating keratinocytes involve activation by Ca^{2+} of protein kinase C eta and delta, followed by activation of p38β. The latter appears to phosphorylate E2F-1 at Ser403 and Thr433. Once E2F-1 is phosphorylated, it is exported from the nucleus in a CRM1-dependent fashion, and degraded in the proteasome. This sequence of events involving E2F-1 phosphorylation, ubiquitination, nuclear export and subsequent degradation is required for proper keratinocyte differentiation (Ivanova et al., 2006; Ivanova and Dagnino, 2007; Ivanova et al., 2009).

E2F-1 degradation subsequent to phosphorylation also occurs in HeLa cells. Specifically, phosphorylation of E2F-1 at Ser403 and Thr433 by TFIIH-cdk7 targets E2F-1 for degradation during S phase (Vandel and Kouzarides, 1999). Phosphorylation at Ser403 is also induced upon DNA damage (Real et al., 2010). Ser403 and Thr433 in E2F-1 are also phosphorylated by glycogen synthase kinase-3β (GSK3β) in HEK293T cells (Garcia-Alvarez et al., 2007). In U2OS osteosarcoma cells treated with doxorubicin, Ser403 is phosphorylated, but is not a substrate of either p38 MAP or GSK3β kinases (Real et al., 2010). Under these conditions, phosphorylation of Ser403 results in changes in E2F-1 target selectivity. Thus, the mechanisms and consequences of E2F-1 phosphorylation on Ser403 appear to be cell-type and context dependent (Ivanova et al., 2009).

4.3 Methylation

Lysine methylation plays critical regulatory roles for histones and non-histone proteins (Huang et al., 2008). The consequences of methylation on E2F-1 activity are controversial at present. E2F-1 is methylated by Set9, a histone H3 methyltransferase, at Lys185, both *in vitro* and in cultured cells (Kontaki et al., 2010, Xie et al., 2011). It has been reported that DNA damage in p53-deficent H1299 lung carcinoma cells is associated with loss of E2F-1 methylation by the lysine-specific demethylase 1(LSD1). Demethylation stabilizes E2F-1, allowing its upregulation of p73. Importantly, methylation of E2F-1 at Lys185 impairs its acetylation and phosphorylation on Ser364, targeting E2F-1 for ubiquitination and degradation in doxorubicin-treated cells (Kontaki et al., 2010). In stark contrast, methylation of E2F-1 at Lys185 by Set9 in U2OS and HCT116 cells treated with adriamycin resulted in E2F-1 stabilization and cell apoptosis (Xie et al., 2011). The reasons for these pronounced discrepancies are not clear.

4.4 Ubiquitination

Many studies have shown that the expression of E2F-1 is regulated by the ubiquitin proteasome pathway, and that E2F-1 is protected from degradation by binding to pRb (Campanero and Flemington, 1997; Hateboer et al., 1996; Hofmann et al., 1996). In mammalian and plant cells, E2F-1 is regulated at the S/G2 phases of the cell cycle through ubiquitination by the SCF[SKP2]-dependent pathway (del Pozo et al., 2002; Marti et al., 1999). *In vitro*, ROC-cullin ligase ubiquitinates E2F-1 in a Skp2-independent manner. Further, phosphorylation of E2F-1 by cyclin A/cdk complexes does not affect E2F-1 ubiquitination (Ohta and Xiong, 2001). Another E3 ubiquitin ligase complex, the anaphase-promoting complex or cyclosome (ACPC/C), also regulates E2F-1 stability during late S phase (Peart et al., 2010). The presence of multiple E3 ligases that interact with and mediate degradation of E2F-1 enables orderly control of E2F-1 expression under multiple circumstances.

5. Regulation of E2F activity by protein-protein interactions

The first type of protein-protein interactions shown to modulate E2F transcriptional activity included association with the retinoblastoma family of proteins (pRB, p107 and p130). pRB is a key regulator of E2F-1, -2 and -3 activity and G1/S-phase transition (Weintraub et al., 1995). The importance of pRb regulation of E2F is evidenced by the fact that a majority of human tumours exhibit inactivating alterations in the pRb pathway (Nevins, 2001). Subsequent studies have revealed thet E2F forms complexes with a multitude of additional proteins, underlining the levels of complexity of E2F regulation.

Protein-protein interactions also appear to assist or provide target specificity to E2F under certain conditions. This effects appear to involve cooperative interactions between E2F and other transcription factors, mediated by binding to neighbouring consensus sites on target promoters. Consensus binding sites for various transcription factors have been identified in the promoters of a subset of E2F target genes. These sites are generally adjacent to the E2F binding sites, and include recognition sequences for YY1, TFE3, and C/EBPα (Schlisio et al., 2002; van Ginkel et al., 1997). These sites possess biological significance, and assist E2F in binding to its consensus sequence. This determines the specific phase of the cell cycle in which E2F activates such promoters. In addition, as these other transcription factors do not interact equally well with all E2F members, they constitute a mechanism of activation of individual E2F factors (Giangrande et al., 2003; Schlisio et al., 2002).

5.1 Retinoblastoma family proteins

pRb binds predominantly to E2F-1, E2F-2, and E2F-3, blocking their transactivation domains (Flemington et al., 1993; Xiao et al., 2003). Under certain circumstances, such as during responses to transforming growth factor-beta in certain cell lines, pRb also binds E2F-4 and represses transcription (Yang, et al. 2008). The pRb family of proteins can also repress transcription of E2F target genes by recruiting other factors, such as histone deacetylases, thus creating transcriptional repressor complexes (Dick, 2007; Morrison et al., 2002; Herrera et al., 1996). pRb is, in turn, regulated by cyclin and cyclin-dependent kinases (Cdk),which deactivate pRb through phosphorylation. Specifically, Cyclin D/Cdk4 and Cyclin E/Cdk2 complexes phosphorylate pRb in the G1 phase of the cell cycle, allowing E2F-1, E2F-2 and E2F-3 to activate target genes (Connell-Crowley et al., 1997; Smith et al., 1996). The other pRb family proteins, p107 and p130, generally bind to E2F-4 and E2F-5, and function to modulate their nucleocytoplasmic shuttling during different periods of the cell cycle. Specifically, E2F-4 and E2F-5 translocate into the nucleus outside of the G1 and S-phases, and act as transcriptional repressors in complexes containing p107 and p130 (Ginsberg et al., 1994; Moberg et al., 1996) (Hijmans et al., 1995) (Guo et al., 2009).

5.2 DP proteins

Optimal binding of E2F to DNA requires cooperative interactions with a member of the other subfamily of E2F proteins, the DP (Dimerization Partner) family. In fact, with the exception of E2F-7 and -8, all functional E2F complexes identified contain a member of the E2F family associated with a DP protein. The DP family is composed of three known members, DP-1 (with isoforms DP-1α and DP-1β), DP-2 (and its mouse orthologue DP-3), and DP-4 (Helin and Harlow, 1994; Milton et al., 2006; Ormondroyd et al., 1995). Different DP proteins have distinct modulatory effects on E2F. For example, DP-1β can mediate E2F translocation to the nucleus, whereas DP-1α, which shows reduced affinity for E2F, participates in E2F nuclear export and translocation to the cytoplasm. In this manner, DP-1α indirectly represses the ability of E2F-1 to activate transcription (Ishida et al., 2005). DP-4 can mediate transcriptional repression as well (Milton et al., 2006). Furthermore, a growing body of evidence shows that other proteins that interact with DP factors, such as C/EBP, TRIP-Br and SOCS3, can modulate E2F activation of gene transcription (Masuhiro et al., 2008; Zaragoza et al., 2010).

5.3 C/EBP

CCAAT/Enhancer Binding Protein (C/EBP) factors are generally characterized as effectors of cellular growth arrest. Within the C/EBP family, C/EBPα has been shown to associate with and repress E2F-1 (Wang et al., 2007). This interaction has been demonstrated through co-immunoprecipitation assays and is independent of pRb family proteins. Rather, it requires the presence of DP-1 or DP-2 (Zaragoza et al., 2010).

The effect of C/EBP repression on E2F activity has been demonstrated in multiple tissues. In primary murine keratinocytes, C/EBPα and β are upregulated as these cells differentiate and move from the basal to the suprabasal layers of the epidermis. Further, the repression of E2F target genes via the action C/EBP is necessary for proper differentiation (Lopez et al., 2009). Interactions between C/EBP and E2F also play important roles during senescence. Indeed, C/EBPα and HDAC1 are recruited to hepatic DNA from older, but not young, mice (Wang et al., 2008). Recruitment of these two factors is accompanied by decreased transcription of E2F target genes.

In mouse 3T3-L1 preadipocytes, C/EBPα, but not C/EBPβ, disrupts E2F-p107 and induces E2F-p130 complexes, leading to decreased proliferation, likely involved in preadipocyte differentiation (Timchenko et al., 1999).

In mouse hepatocytes devoid of C/EBPβ, E2F target genes are repressed and DNA synthesis is severely impaired. In these cells, C/EBP β interacts with E2F-1, facilitating recruitment of CBP and p300 to E2F target genes. The recruitment of these multiprotein complexes results in upregulation of E2F targets involved in cell proliferation (Wang et al., 2007). C/EBPβ is also required for expression of E2F-3 and S-phase progression in uterine epithelial cells (Ramathal et al., 2010). In primary epidermal keratinocytes, C/EBPα interferes with DNA synthesis in response to DNA damage (Johnson, 2005). However, the mechanisms involved are not fully undertood. It has been proposed that C/EBPα functions with E2F/pRb complexes to repress transcription of S-phase genes. In neuroblastoma cells, C/EBP is involved in induction of apoptotic gene transcription by E2F-1 (Marabese et al., 2003).

5.4 SOCS3

The Suppressor of Cytokine Signaling (SOCS) family of proteins act as negative feedback regulators of the JAK-STAT pathway. Recently, SOCS factors have also been shown to associate with DP-1 and DP-3. SOCS3 inhibits transcriptional activation of E2F target genes and cell cycle progression. The mechanisms involved in this repression include SOCS3 inhibition of E2F/DP dimerization, thus preventing the formation of the E2F DNA-binding complexes (Masuhiro et al., 2008).

5.5 TRIP-BR1
The Transcriptional Regulator Interacting with the PHD zinc finger and/or the Bromodomain-1 (TRIP-Br1) protein (also known as p34) is a transcriptional modulator that directly interacts with DP-1, as well as with the co-activators p300/CBP and KRIP1(Hsu et al., 2001). As such, TRIP-Br co-activates E2F responsive genes, such as B-myb, in U2OS osteosarcoma cells, an ability potentiated by KRIP1. This effect is impaired by pRb. TRIP-Br1 also interferes with deactivation of Cyclin D/Cdk4 complex by p16^{INK4}, effectively activating E2F by inhibiting pRb (Sim et al., 2004).

5.6 p110 CUX1
Cut homeobox 1 (CUX1) proteins are transcription factors that can either activate or repress transcription. In particular, the CUX1 isoform p110 can stably interact with DNA and promote entry into the S-phase of the cell cycle (Truscott et al., 2008). P110 CUX1 interacts with E2F-1 or E2F-2, stimulating their recruitment to the DNA polymerase α gene promoter, in a manner that requires ability of E2F to bind DNA. Further, common targets for E2F and p110 CUX1 include genes involved in cell cycle progression, DNA repair and replication (Truscott et al., 2008).

5.7 YY1
The transcriptional repressor YY1 can bind to target sites adjacent to E2F binding elements in the promoters of genes such as *Cdc6* (Schlisio et al., 2002). In addition, the YY1 accessory protein Ring1 and YY1 binding protein (RYBP) can interact with E2F-2, -3 and -4 to synergistically enhance binding of E2F-2 and -3 (but not of E2F-1). In this manner, YY1 and RYBP not only enhance the binding and transcription of E2F to certain promoters, but also add specificity.

5.8 TFE3
Studies of the p68 promoter have shown that transcription factor E3 (TFE3) operates in a similar manner to YY1. Thus, the E Box bound by TFE3 and the E2F consensus sequence occur in close proximity in the *p68* promoter. TFE3 and E2F-3 bind to those sites cooperatively (Giangrande et al., 2003; Giangrande et al., 2004). This interaction requires E2F-3, but not TFE3 binding to DNA. Although a direct interaction between these two proteins was not be demonstrated, these two factors likely work together in a larger protein complex, or interact temporarily to recruit one another to the *p68* promoter.

6. Regulation of E2F activity by viral oncoproteins
Viruses work by hijacking the cellular machinery of their host cell, to facilitate their replication. Hence, it is not surprising that constitutive activation of E2F, which induces cell transition into a state of DNA replication (S-phase), is a critical step in the viral modification of infected cell functions.

6.1 Human papillomavirus protein E7
The human papillomaviruses (HPV) are commonly known oncoviruses. This notoriety is due to their ability to activate E2F proteins, causing rapid and unregulated progression through the cell cycle (Lee et al., 1998). HPV couples this action with deactivation of

pathways that act as fail-safe mechanisms for E2F activity, such as p53-mediated apoptosis (Moody and Laimins, 2010). The key HPV viral protein involved in activating E2Fs is E7. This protein carries an LXCXE domain characteristic of proteins that associate with pRb family proteins (Lee et al., 1998). In this manner, E7 proteins bind to pRb, p107 and p130, dissociating them from E2F factors. The mechanisms involved in this effect include blockade by E7 of the pRb-E2F binding domain (Lee et al., 1998). As a result, E2F species bind to and activate target genes without the possibility of repression. E7 also induces pRb proteasomal degradation, by increasing its ubiquitination (Moody and Laimins). Furthermore, there is evidence to indicate that E7 also binds to p300/CBP, allowing this acetyltransferase to facilitate and rapidly increase the transcription of E2F target genes (Bernat et al., 2003).

6.2 SV40 large-T antigen

The simian virus 40 (SV40) genome encodes a protein that shares some characteristics with HPV E7, termed large T-antigen. Similar to HPV E7, large-T antigen has a LXCXE domain, which can bind all three pRb family proteins, leading to release of free E2F and expression of its target genes (DeCaprio, 2009). In addition, large-T antigen binds preferentially to the hypophosphorylated form of pRb, present during the G1-phase of the cell cycle (Ludlow et al., 1989). The characterization of the interactions between large-T antigen and complexes containing p130 or p107 and E2F-4 has been central to understanding the mechanisms involved in deactivation of pRb family proteins by this viral factor (Sullivan et al., 2000). Dissociation of p107 or p130 from E2F also requires Large-T antigen interactions with the J type of chaperone protein Hsc70 and ATP.

Similar to HPV E7, large-T antigen binds to p300/CBP through its C-terminus (Eckner et al., 1996). This interaction is likely involved in histone acetylation and transcriptional activation of E2F target genes. Significantly, mutations in the C-termimus of large-T antigen impair its ability to bind p300/CBP, but are without effect on its capacity to disrupt pRb binding to E2F (Nemethova et al., 2004).

6.3 Adenovirus E1A

Adenovirus protein E1A functions in a similar manner to HPV E7 and SV40 large-T antigen. E1A interacts with multiple cellular proteins, including the pRb family and p300/CBP (Raychaudhuri, 1991; Liu, 2007). X-ray crystallographic characterization of E1A has revealed that its N-terminal domain competes with the transactivation domain of E2F for binding to pRb. This induces a decrease in E2F binding to pRb by competition (Liu, 2007). Similar to other viral oncoproteins, E1A also has an LXCXE domain that binds to pRb, p107 and p130 (Dyson, 1992). E1A also binds the 400-kDa protein p400, which mediates further interactions with TRRAP/PAF400, along with the DNA helicase TAP54α/β). Together, these proteins form a chromatin remodeling complex, which contributes to cell transformation and activation of E2F target genes that mediate viral DNA replication (Liu, 2007).

6.4 Human parvovirus NS1

Human parvovirus B19 (B19V) is the only pathogenic human parvovirus, and it targets cells of the erythroid lineage, expecially erythroid progenitors (Wan et al., 2010). The B19V protein NS1 (nonstructural protein 1) interacts with E2F-4 and E2F-5, inducing their nuclear accumulation and G2 arrest, necessary for viral replication (Wan et al., 2010). Simultaneously, NS1 also decreases expression of E2F-1, E2F-2, and E2F-3, resulting in transcriptional repression of genes necessary for the G2/M transition. Thus, B19V targets cells for arrest in the G2 phase by altering E2F activity, indicating the importance of this family of transcription factors in all phases of the cell cycle and multiple aspects of cell cycle progression and DNA replication and repair.

7. Conclusions

A large body of work has been focused on identifying the mechanisms that regulate E2F activity and its consequences on induction of DNA repair. As a result, it has become apparent that E2F activity is complex, and is regulated at multiple levels, including transcription, post-translational modifications and protein-protein interactions. However, understanding of how different post-translational modifications modulate E2F interactions with other proteins, allowing it to form transcriptional activator or repressor complexes is in its infancy.

The biological roles of the various modes of E2F modulation go well beyond normal development and cell differentiation, implicate mechanisms of DNA repair as a central function, and are involved in the genesis of multiple pathologies.

Although pRb family proteins form the central backbone of E2F regulation, they are only one component. Studies of HPV proteins have shown that, in addition to E7, the proteins E5 and E6 are critical for the functional transformation of a cell. In the case of HPV, these proteins serve to deactivate the p53 pathway, preventing the pro-apoptotic responses normally switched on with abnormal activation of E2F. Other viruses encode proteins that serve a similar function. The identification and study of these proteins may provide key insights into the function of these viruses and the pathways that regulate E2F during normal tissue development and homeostasis, and affect DNA repair mechanisms to ensure viral replication.

8. Acknowledgement

Work in the author's laboratory was supported with funds from the Canadian Institutes of Health Research.

9. References

Almeida, M.I., Reis, R.M., and Calin, G.A. (2011). MicroRNA history: Discovery, recent applications, and next frontiers. *Mutat Res*.

Bagchi, S., Raychaudhuri, P., and Nevins, J.R. (1989). Phosphorylation-dependent activation of the adenovirus-inducible E2F transcription factor in a cell-free system. *Proc Natl Acad Sci U S A* 86, 12.(4352-4356).

Bennett, D., and Itoh, T. (2008). The XPE gene of xeroderma pigmentosum, its product and biological roles. *Adv Exp Med Biol* 63757-64).

Berkovich, E., and Ginsberg, D. (2003). ATM is a target for positive regulation by E2F-1. *Oncogene* 22, 2.(161-167).

Bernat, A., Avvakumov, N., Mymryk, J.S., and Banks, L. (2003). Interaction between the HPV E7 oncoprotein and the transcriptional coactivator p300. *Oncogene* 22, 39.(7871-7881).

Berton, T.R., Mitchell, D.L., Guo, R., and Johnson, D.G. (2005). Regulation of epidermal apoptosis and DNA repair by E2F1 in response to ultraviolet B radiation. *Oncogene* 242449-2460).

Bindra, R.S., Gibson, S.L., Ment, A., Westermark, U., Jasin, M., Pierce, A.J., Bristow, R.G., Classon, M.K., and Glazer, P.M. (2005). Hypoxia-induced down regulation of BRCA1 expression by E2Fs. *Cancer Res* 6511597-11604).

Blattner, C., Sparks, A., and Lane, D. (1999). Transcription factor E2F-1 is upregulated in response to DNA damage in a manner analogous to that of p53. *Mol Cell Biol* 193704-3713).

Bracken, A.P., Ciro, M., Cocito, A., and Helin, K. (2004). E2F target genes: unraveling the biology. *Trends Biochem Sci* 29409-417).

Brash, D.E., Ziegler, A., Jonason, A.S., Simon, J.A., Kunala, S., and Leffell, D.J. (1996). Sunlight and sunburn in human skin cancer: p53, apoptosis and tumor promotion. *J Invest Dermatol Symp Proc* 1136-142).

Brosh, R., Shalgi, R., Liran, A., Landan, G., Korotayev, K., Nguyen, G.H., Enerly, E., Johnsen, H., Buganim, Y., Solomon, H., *et al.* (2008). p53-Repressed miRNAs are involved with E2F in a feed-forward loop promoting proliferation. *Mol Syst Biol* 4229).

Bueno, M.J., Gomez de Cedron, M., Laresgoiti, U., Fernandez-Piqueras, J., Zubiaga, A.M., and Malumbres, M. (2010). Multiple E2F-induced microRNAs prevent replicative stress in response to mitogenic signaling. *Mol Cell Biol* 30, 12.(2983-2995).

Campanero, M.R., and Flemington, E.K. (1997). Regulation of E2F through ubiquitin-proteasome-dependent degradation: stabilization by the pRB tumor suppressor protein. *Proc Natl Acad Sci U S A* 94, 6.(2221-2226).

Cao, A.R., Rabinovich, R., Xu, M., Xu, X., Jin, V.X., and Farnham, P.J. (2011). Genome-wide Analysis of Transcription Factor E2F1 Mutant Proteins Reveals That N- and C-terminal Protein Interaction Domains Do Not Participate in Targeting E2F1 to the Human Genome. *J Biol Chem* 286, 14.(11985-11996).

Chang, I.Y., Jin, M., Yoon, S.P., Youn, C.K., Yoon, Y., Moon, S.P., Hyun, J.W., Jun, J.Y., and You, H.J. (2008). Senescence-dependent MutS alpha dysfunction attenuates mismatch repair. *Mol Cancer Res* 6, 6.(978-989).

Chang, W.Y., Andrews, J., Carter, D.E., and Dagnino, L. (2006). Differentiation and injury-repair signals modulate the interaction of E2F and pRB proteins with novel target genes in keratinocytes. *Cell Cycle* 51872-1879).

Chang, W.Y., Bryce, D.M., D'Souza, S.J.A., and Dagnino, L. (2004). The DP-1 transcription factor is required for keratinocyte growth and epidermal stratification. *J Biol Chem* 27951343-51353).

Chen, D., Yu, Z., Zhu, Z., and Lopez, C.D. (2008). E2F1 regulates the base excision repair gene XRCC1 and promotes DNA repair. *J Biol Chem* 283, 22.(15381-15389).

Connell-Crowley, L., Harper, J.W., and Goodrich, D.W. (1997). Cyclin D1/Cdk4 regulates retinoblastoma protein-mediated cell cycle arrest by site-specific phosphorylation. *Mol Biol Cell* 8, 2.(287-301).

Cui, J.G., Zhao, Y., Sethi, P., Li, Y.Y., Mahta, A., Culicchia, F., and Lukiw, W.J. (2010). Micro-RNA-128 (miRNA-128) down-regulation in glioblastoma targets ARP5 (ANGPTL6), Bmi-1 and E2F-3a, key regulators of brain cell proliferation. *J Neurooncol* 98, 3.(297-304).

D'Souza, S.J.A., Vespa, A., Murkherjee, S., Maher, A., Pajak, S., and Dagnino, L. (2002). E2F-1 is essential for normal epidermal wound repair. *J Biol Chem* 27710626-10632).

DeCaprio, J.A. (2009). How the Rb tumor suppressor structure and function was revealed by the study of Adenovirus and SV40. *Virology* 384, 2.(274-284).

del Pozo, J.C., Boniotti, M.B., and Gutierrez, C. (2002). Arabidopsis E2Fc functions in cell division and is degraded by the ubiquitin-SCF(AtSKP2) pathway in response to light. *Plant Cell* 14, 12.(3057-3071).

Deschenes, C., Alvarez, L., Lizotte, M.E., Vezina, A., and Rivard, N. (2004). The nucleocytoplasmic shuttling of E2F4 is involved in the regulation of human intestinal epithelial cell proliferation and differentiation. *J Cell Physiol* 199, 2.(262-273).

Dick, F.A. (2007). Structure-function analysis of the retinoblastoma tumor suppressor protein - is the whole a sum of its parts? *Cell Div* 226).

Dominguez-Brauer, C., Chen, Y.J., Brauer, P.M., Pimkina, J., and Raychaudhuri, P. (2009). ARF stimulates XPC to trigger nucleotide excision repair by regulating the repressor complex of E2F4. *EMBO Rep* 10, 9.(1036-1042).

Dynlacht, B.D., Moberg, K., Lees, J.A., Harlow, E., and Zhu, L. (1997). Specific regulation of E2F family members by cyclin-dependent kinases. *Mol Cell Biol* 17, 7.(3867-3875).

Dyson, N and Harlow, E. (1992). Adenovirus E1A targets key regulators of cell proliferation. *Cancer Surveys* 12.(161-195).

Eckner, R., Ludlow, J.W., Lill, N.L., Oldread, E., Arany, Z., Modjtahedi, N., DeCaprio, J.A., Livingston, D.M., and Morgan, J.A. (1996). Association of p300 and CBP with simian virus 40 large T antigen. *Mol Cell Biol* 16, 7.(3454-3464).

Fagan, R., Flint, K.J., and Jones, N. (1994). Phosphorylation of E2F-1 modulates its interaction with the retinoblastoma gene product and the adenoviral E4 19 kDa protein. *Cell* 78799-811).

Farhana, L., Dawson, M., Rishi, A.K., Zhang, Y., Van Buren, E., Trivedi, C., Reichert, U., Fang, G., Kirschner, M.W., and Fontana, J.A. (2002). Cyclin B and E2F-1 expression in prostate carcinoma cells treated with the novel retinoid CD437 are regulated by the ubiquitin-mediated pathway. *Cancer Res* 623842-3849).

Field, S.J., Tsai, F.Y., Kuo, F., Zubiaga, A.M., Kaelin, W.G., Jr., Livingston, D.M., Orkin, S.H., and Greenberg, M.E. (1996). E2F-1 functions in mice to promote apoptosis and suppress proliferation. *Cell* 85, 4.(549-561).

Flemington, E.K., Speck, S.H., and Kaelin, W.J. (1993). E2F-1 mediated transactivation is inhibited by complex formation with the retinoblastoma susceptibility gene product. *Proc Natl Acad Sci USA*, 90.(

Galbiati, L., Mendoza-Maldonado, R., Gutierrez, M.I., and Giacca, M. (2005). Regulation of E2F1 after DNA damage by p300-mediated acetylation and ubiquitination. *Cell Cycle* 4930-939).

Garcia-Alvarez, G., Ventura, V., Ros, O., Gil, J., and Tauler, A. (2007). Glycogen synthase kinase-3beta binds to E2F1 and regulates its transcriptional activity. *Biochim Biophys Acta* 1773375-382).

Giangrande, P., Hallstrom, T.C., Tunyaplin, C., Calame, K., and Nevins, J.R. (2003). Identification of E-box factor TFE3 as a functional partner for the E2F3 transcription factor. *Mol Cell Biol* 233707-3720).

Giangrande, P., Zhu, W., Rempel, R. E., Laasko, N., and Nevins, J.R. (2004). Combinatorial gene control involving E2F and E box family members. *EMBO J* 231336-1347).

Ginsberg, D., Vairo, G., Chittenden, T., Xiao, Z.X., Xu, G., Wydner, K.L., DeCaprio, J.A., Lawrence, J.B., and Livingston, D.M. (1994). E2F-4, a new member of the E2F transcription factor family, interacts with p107. *Genes Dev* 8, 22.(2665-2679).

Griffiths-Jones, S. (2004). The microRNA Registry. *Nucleic Acids Res* 32, Database issue.(D109-111).

Guo, J., Longshore, S., Nair, R., and Warner, B.W. (2009). Retinoblastoma protein (pRb), but not p107 or p130, is required for maintenance of enterocyte quiescence and differentiation in small intestine. *J Biol Chem* 284, 1.(134-140).

Guo, R., Chen, J., Mitchell, D.L., and Johnson, D.G. (2010a). GCN5 and E2F1 stimulate nucleotide excision repair by promoting H3K9 acetylation at sites of damage. *Nucleic Acids Res*.

Guo, R., Chen, J., Zhu, F., Biswas, A.K., Berton, T.R., Mitchell, D.L., and Johnson, D.G. (2010b). E2F1 localizes to sites of UV-induced DNA damage to enhance nucleotide excision repair. *J Biol Chem* 285, 25.(19308-19315).

Halazonetis, T.D., Gorgoulis, V.G., and Bartek, J. (2008). An oncogene-induced DNA damage model for cancer development. *Science* 319, 5868.(1352-1355).

Hallstrom, T.C., and Nevins, J.R. (2009). Balancing the decision of cell proliferation and cell fate. *Cell Cycle* 8, 4.(532-535).

Hardy, S., Engel, D.A., and Shenk, T. (1989). An adenovirus early region 4 gene product is required for induction of the infection-specific form of cellular E2F activity. *Genes Dev* 3, 7.(1062-1074).

Hateboer, G., Kerkhoven, R.M., Shvarts, A., Bernards, R., and Beijersbergen, R.L. (1996). Degradation of E2F by the ubiquitin-proteasome pathway: regulation by retinoblastoma family proteins and adenovirus transforming proteins. *Genes Dev* 10, 23.(2960-2970).

Helin, K., and Harlow, E. (1994). Heterodimerization of the transcription factors E2F-1 and DP-1 is required for binding to the adenovirus E4 (ORF6/7) protein. *J Virol* 68, 8.(5027-5035).

Heriche, J.K., Ang, D., Bier, E., and O'Farrell, P.H. (2003). Involvement of an SCFSlmb complex in timely elimination of E2F upon initiation of DNA replication in Drosophila. *BMC Genet* 49).

Herrera, R.E., Chen, F., and Weinberg, R.A. (1996). Increased histone H1 phosphorylation and relaxed chromatin structure in Rb-deficient fibroblasts. *Proc Natl Acad Sci U S A* 93, 21.(11510-11515).

Hijmans, E.M., Voorhoeve, P.M., Beijersbergen, R.L., van 't Veer, L.J., and Bernards, R. (1995). E2F-5, a new E2F family member that interacts with p130 in vivo. *Mol Cell Biol* 15, 6.(3082-3089).

Hofmann, F., Martelli, F., Livingston, D.M., and Wang, Z. (1996). The retinoblastoma gene product protects E2F-1 from degradation by the ubiquitin-proteasome pathway. *Genes Dev* 10, 23.(2949-2959).

Hsu, S.I.-H., Yang, C.M., Sim, K.G., Hentschel, D.M., O'Leary, E., and Bonventre, J.V. (2001). Trip-Br: a novel family of PHD zing finger- and bromodomain-interacting proteins that regulate the transcriptional activity of E2F-1/DP-1. *EMBO J* 202273-2285).

Huang, Y.C., Misquitta, S., Blond, S.Y., Adams, E., and Colman, R.F. (2008). Catalytically active monomer of glutathione S-transferase pi and key residues involved in the electrostatic interaction between subunits. *J Biol Chem* 283, 47.(32880-32888).

Ianari, A., Gallo, R., Palma, M., Alesse, E., and Gulino, A. (2004). Specific role for p300/CREB-binding protein-associated factor activity in E2F1 stabilization in response to DNA damage. *J Biol Chem* 27930830-30835).

Ingram, L., Munro, S., Coutts, A.S., and La Thangue, N.B. (2011). E2F-1 regulation by an unusual DNA damage-responsive DP partner subunit. *Cell Death Differ* 18, 1.(122-132).

Ishida, H., Masuhiro, Y., Fukushima, A., Argueta, J.G., Yamaguchi, N., Shiota, S., and Hanazawa, S. (2005). Identification and characterization of novel isoforms of human DP-1: DP-1{alpha} regulates the transcriptional activity of E2F1 as well as cell cycle progression in a dominant-negative manner. *J Biol Chem* 280, 26.(24642-24648).

Ishida, S., Huang, E., Zuzan, H., Spang, R., Leone, G., West, M., and Nevins, J.R. (2001). Role for E2F in control of both DNA replication and mitotic functions as revealed from DNA microarrays. *Mol Cell Biol* 214684-4699).

Ivanova, I.A., D'Souza, S.J.A., and Dagnino, L. (2005). Signalling in the epidermis: The E2F cell cycle regulatory pathway in epidermal morphogenesis, regeneration and transformation. *Int J Biol Sci* 187-95).

Ivanova, I.A., D'Souza, S.J.A., and Dagnino, L. (2006). E2F stability is regulated by a novel-PKC/p38® MAP kinase signalling pathway during keratinocyte differentiation. *Oncogene* 25430-437).

Ivanova, I.A., and Dagnino, L. (2007). Activation of p38- and CRM1-dependent nuclear export promotes E2F1 degradation during keratinocyte differentiation. *Oncogene* 261147-1154).

Ivanova, I.A., Nakrieko, K.A., and Dagnino, L. (2009). Phosphorylation by p38 MAP kinase is required for E2F1 degradation and keratinocyte differentiation. *Oncogene* 2852-63).

Ivanova, I.A., Vespa, A., and Dagnino, L. (2007). A novel mechanism of E2F1 regulation via nucleocytoplasmic shuttling: determinants of nuclear import and export. *Cell Cycle* 62186-2195).

Johnson, P.F. (2005). Molecular stop signs: regulation of cell-cycle arrest by C/EBP transcription factors. *J Cell Sci* 118, Pt 12.(2545-2555).

Judah, D., Chang, W.Y., and Dagnino, L. (2010). EBP1 is a novel E2F target gene regulated by transforming growth factor-beta. *PLoS One* 5, 11.(e13941).

Kachhap, S.K., Rosmus, N., Collis, S.J., Kortenhorst, M.S., Wissing, M.D., Hedayati, M., Shabbeer, S., Mendonca, J., Deangelis, J., Marchionni, L., *et al.* (2010). Downregulation of homologous recombination DNA repair genes by HDAC inhibition in prostate cancer is mediated through the E2F1 transcription factor. *PLoS One* 5, 6.(e11208).

Kontaki, H., and Talianidis, I. (2010). Lysine methylation regulates E2F1-induced cell death. *Molec Cell* 39, 2. (152-160)

Kishore, R., Luedemann, C., Bord, E., Goukassian, D., and Losordo, D.W. (2003). Tumor necrosis factor-mediated E2F1 suppression in endothelial cells: differential requirement of c-Jun N-terminal kinase and p38 mitogen-activated protein kinase signal transduction pathways. *Circ Res* 93, 10.(932-940).

Kong, H.J., Yu, H.J., Hong, S., Park, M.J., Choi, Y.H., An, W.G., Lee, J.W., and Cheong, J. (2003). Interaction and functional cooperation of the cancer-amplified transcriptional coactivator activating signal cointegrator-2 and E2F-1 in cell proliferation. *Mol Cancer Res* 1, 13.(948-958).

Kozaki, K., Imoto, I., Mogi, S., Omura, K., and Inazawa, J. (2008). Exploration of tumor-suppressive microRNAs silenced by DNA hypermethylation in oral cancer. *Cancer Res* 68, 7.(2094-2105).

Krek, W., Xu, G., and Livingston, D.M. (1995). Cyclin A-kinase regulation of E2F-1 DNA binding function underlies suppression of an S phase checkpoint. *Cell* 83, 7.(1149-1158).

Lammens, T., Li, J., Leone, G., and De Veylder, L. (2009). Atypical E2Fs: new players in the E2F transcription factor family. *Trends Cell Biol* 19, 3.(111-118).

Lanigan, F., Geraghty, J.G., and Bracken, A.P. (2011). Transcriptional regulation of cellular senescence. *Oncogene*.

Lario, L.D., Ramirez-Parra, E., Gutierrez, C., Casati, P., and Spampinato, C.P. (2011). Regulation of plant MSH2 and MSH6 genes in the UV-B-induced DNA damage response. *J Exp Bot*.

Lee, B.K., Bhinge, A.A., and Iyer, V.R. (2011). Wide-ranging functions of E2F4 in transcriptional activation and repression revealed by genome-wide analysis. *Nucleic Acids Res*.

Lee, J.O., Russo, A.A., and Pavletich, N.P. (1998). Structure of the retinoblastoma tumour-suppressor pocket domain bound to a peptide from HPV E7. *Nature* 391, 6670.(859-865).

Lee, K.H., Chen, Y.L., Yeh, S.D., Hsiao, M., Lin, J.T., Goan, Y.G., and Lu, P.J. (2009). MicroRNA-330 acts as tumor suppressor and induces apoptosis of prostate cancer cells through E2F1-mediated suppression of Akt phosphorylation. *Oncogene* 28, 38.(3360-3370).

Levitt, N.C., and Hickson, I.D. (2002). Caretaker tumour suppressor genes that defend genome integrity. *Trends Mol Med* 8, 4.(179-186).

Li, W.W., Fan, J., Hochhauser, D., and Bertino, J.R. (1997). Overexpression of p21waf1 leads to increased inhibition of E2F-1 phosphorylation and sensitivity to anticancer drugs in retinoblastoma-negative human sarcoma cells. *Cancer Res* 57, 11.(2193-2199).

Liao, C.C., Tsai, C.Y., Chang, W.C., Lee, W.H., and Wang, J.M. (2010). RB.E2F1 complex mediates DNA damage responses through transcriptional regulation of ZBRK1. *J Biol Chem* 285, 43.(33134-33143).

Lin, P.S., McPherson, L.A., Chen, A.Y., Sage, J., and Ford, J.M. (2009). The role of the retinoblastoma/E2F1 tumor suppressor pathway in the lesion recognition step of nucleotide excision repair. *DNA Repair (Amst)* 8, 7.(795-802).

Lin, W.-C., Lin, F.-T., and Nevins, J.R. (2001). Selective induction of E2F1 in response to DNA damage, mediated by ATM-dependent phosphorylation. *Genes Dev* 151833-1844).

Liu, X. and R. Marmorstein (2007). Structure of the retinoblastoma protein bound to adenovirus E1A reveals the molecular basis for viral oncoprotein inactivation of a tumor suppressor. *Genes Dev* 21,21. (2711-2716).

Lopez, R.G., Garcia-Silva, S., Moore, S.J., Bereshchenko, O., Martinez-Cruz, A.B., Ermakova, O., Kurz, E., Paramio, J.M., and Nerlov, C. (2009). C/EBPalpha and beta couple interfollicular keratinocyte proliferation arrest to commitment and terminal differentiation. *Nat Cell Biol* 11, 10.(1181-1190).

Ludlow, J.W., DeCaprio, J.A., Huang, C.M., Lee, W.H., Paucha, E., and Livingston, D.M. (1989). SV40 large T antigen binds preferentially to an underphosphorylated member of the retinoblastoma susceptibility gene product family. *Cell* 56, 1. (57-65).

Mann, D.J., and Jones, N.C. (1996). E2F-1 but not E2F-4 can overcome p16-induced G1 cell-cycle arrest. *Curr Biol* 6, 4.(474-483).

Marabese, M., Vikhanskaya, F., Rainelli, C., Sakai, T., and Broggini, M. (2003). DNA damage induces transcriptional activation of p73 by removing C-EBPalpha repression on E2F1. *Nucleic Acids Res* 31, 22.(6624-6632).

Marti, A., Wirbelauer, C., Scheffner, M., and Krek, W. (1999). Interaction between ubiquitin-protein ligase SCFSKP2 and E2F-1 underlies the regulation of E2F-1 degradation. *Nature Cell Biol* 114-19).

Martinez-Balbas, M.A., Bauer, U.M., Nielsen, S.J., Brehm, A., and Kouzarides, T. (2000). Regulation of E2F1 activity by acetylation. *Embo J* 19, 4.(662-671).

Marzio, G., Wagener, C., Gutierrez, M.I., Cartwright, P., Helin, K., and Giacca, M. (2000). E2F family members are differentially regulated by reversible acetylation. *J Biol Chem* 275, 15.(10887-10892).

Masuhiro, Y., Kayama, K., Fukushima, A., Baba, K., Soutsu, M., Kamiya, Y., Gotoh, M., Yamaguchi, N., and Hanazawa, S. (2008). SOCS-3 inhibits E2F/DP-1 transcriptional activity and cell cycle progression via interaction with DP-1. *J Biol Chem* 283, 46.(31575-31583).

McClellan, K.A., and Slack, R.S. (2007). Specific in vivo roles for E2Fs in differentiation and development. *Cell Cycle* 6, 23.(2917-2927).

Meng, R.D., Phillips, P., and El-Deiry, W.S. (1999). p53-independent increase in E2F-1 expression enhances the cytotoxic effects of etoposide and of adriamycin. *Int J Oncol* 14, 1.(5-14).

Milton, A., Luoto, K., Ingram, L., Munro, S., Logan, N., Graham, A.L., Brummelkamp, T.R., Hijmans, E.M., Bernards, R., and La Thangue, N.B. (2006). A functionally distinct member of the DP family of E2F subunits. *Oncogene* 25, 22.(3212-3218).

Moberg, K., Starz, M.A., and Lees, J.A. (1996). E2F-4 switches from p130 to p107 and pRB in response to cell cycle reentry. *Mol Cell Biol* 161436-1449).

Moody, C.A., and Laimins, L.A. (2010). Human papillomavirus oncoproteins: pathways to transformation. *Nat Rev Cancer* 10, 8.(550-560).

Morris, L., Allen, K.E., and La Thangue, N.B. (2000). Regulation of E2F transcription by cyclin E-Cdk2 kinase mediated through p300/CBP co-activators. *Nat Cell Biol* 2, 4.(232-239).

Morrison, A.J., Sardet, C., and Herrera, R.E. (2002). Retinoblastoma protein transcriptional repression through histone deacetylation of a single nucleosome. *Mol Cell Biol* 22, 3.(856-865).

Nemethova, M., Smutny, M., and Wintersberger, E. (2004). Transactivation of E2F-regulated genes by polyomavirus large T antigen: evidence for a two-step mechanism. *Mol Cell Biol* 24, 24.(10986-10994).

Nevins, J.R. (2001). The Rb/E2F pathway and cancer. *Human Mole Genet* 10699-703).

Nouspikel, T. (2009). DNA repair in mammalian cells : Nucleotide excision repair: variations on versatility. *Cell Mol Life Sci* 66, 6.(994-1009).

O'Donnell, K.A., Wentzel, E.A., Zeller, K.I., Dang, C.V., and Mendell, J.T. (2005). c-Myc-regulated microRNAs modulate E2F1 expression. *Nature* 435, 7043.(839-843).

Ohta, T., and Xiong, Y. (2001). Phosphorylation- and SKP1-independent in vitro ubiquitination of E2F1 by multiple ROC-Cullin ligases. *Cancer Res* 611347-1353).

Ormondroyd, E., de la Luna, S., and La Thangue, N.B. (1995). A new member of the DP family, DP-3, with distinct protein products suggests a regulatory role for alternative splicing in the cell cycle transcription factor DRTF1/E2F. *Oncogene* 11, 8.(1437-1446).

Ozaki, T., Okoshi, R., Sang, M., Kubo, N., and Nakagawara, A. (2009). Acetylation status of E2F-1 has an important role in the regulation of E2F-1-mediated transactivation of tumor suppressor p73. *Biochem Biophys Res Commun* 386, 1.(207-211).

Peart, M.J., Poyurovsky, M.V., Kass, E.M., Urist, M., Verschuren, E.W., Summers, M.K., Jackson, P.K., and Prives, C. (2010). APC/C(Cdc20) targets E2F1 for degradation in prometaphase. *Cell Cycle* 9, 19.(3956-3964).

Pediconi, N., Ianari, A., Costanzo, A., Belloni, L., Gallo, R., Cimino, L., Porcellini, A., Screpanti, I., Balsano, C., Alesse, E., *et al.* (2003). Differential regulation of E2F1 apoptotic target genes in response to DNA damage. *Nat Cell Biol* 5552-558).

Petrocca, F., Visone, R., Onelli, M.R., Shah, M.H., Nicoloso, M.S., de Martino, I., Iliopoulos, D., Pilozzi, E., Liu, C.G., Negrini, M., *et al.* (2008). E2F1-regulated microRNAs impair TGFbeta-dependent cell-cycle arrest and apoptosis in gastric cancer. *Cancer Cell* 13, 3.(272-286).

Pickering, M.T., Stadler, B.M., and Kowalik, T.F. (2009). miR-17 and miR-20a temper an E2F1-induced G1 checkpoint to regulate cell cycle progression. *Oncogene* 28, 1.(140-145).

Polager, S., and Ginsberg, D. (2008). E2F - at the crossroads of life and death. *Trends Cell Biol* 18, 11.(528-535).

Prost, S., Lu, P., Caldwell, H., and Harrison, D. (2007). E2F regulates DDB2: consequences for DNA repair in Rb-deficient cells. *Oncogene* 26, 24.(3572-3581).

Rabinovich, A., Jin, V.X., Rabinovich, R., Xu, X., and Farnham, P.J. (2008). E2F in vivo binding specificity: comparison of consensus versus nonconsensus binding sites. *Genome Res* 18, 11.(1763-1777).

Raychaudhuri, P., Bagchi, S., Devoto, S.H., Knaus, V. B., Moran, E. and Nevins, J. R. (1991). Domains of the adenovirus E1A protein required for oncogenenic activity are also required for dissociation of E2F transcription factor complexes. *Genes Dev* 5, 7.(1200-1211).

Ramathal, C., Bagchi, I.C., and Bagchi, M.K. (2010). Lack of CCAAT enhancer binding protein beta (C/EBPbeta) in uterine epithelial cells impairs estrogen-induced DNA replication, induces DNA damage response pathways, and promotes apoptosis. *Mol Cell Biol* 30, 7.(1607-1619).

Real, S., Espada, L., Espinet, C., Santidrian, A.F., and Tauler, A. (2010). Study of the in vivo phosphorylation of E2F1 on Ser403. *Biochim Biophys Acta* 1803, 8.(912-918).

Schlisio, S., Halperin, T., Vidal, M., and Nevins, J.R. (2002). Interaction of YY1 with E2Fs, mediated by RYBP, provides a mechanism for specificity of E2F function. *EMBO J* 21, 21.(5775-5786).

Shiloh, Y. (2003). ATM and related protein kinases: safeguarding genome integrity. *Nat Rev Cancer* 3, 3.(155-168).

Shin, E.K., Shin, A., Paulding, C., Schaffhausen, B., and Yee, A.S. (1995). Multiple change in E2F function and regulation occur upon muscle differentiation. *Mol Cell Biol* 15, 4.(2252-2262).

Siafakas, R., and Richardson, D.R. (2009). Growth arrest and DNA damage-45 alpha (GADD45alpha). *Int J Biochem Cell Biol* 41986-989).

Sim, K.G., Zang, Z., Yang, C.M., Bonventre, J.V., and Hsu, S.I. (2004). TRIP-Br links E2F to novel functions in the regulation of cyclin E expression during cell cycle progression and in the maintenance of genomic stability. *Cell Cycle* 3, 10.(1296-1304).

Smith, E.J., Leone, G., DeGregori, J., Jakoi, L., and Nevins, J.R. (1996). The accumulation of an E2F-p130 transcriptional repressor distinguishes a G0 cell state from a G1 cell state. *Mol Cell Biol* 16, 12.(6965-6976).

Sullivan, C.S., Cantalupo, P., and Pipas, J.M. (2000). The molecular chaperone activity of simian virus 40 large T antigen is required to disrupt Rb-E2F family complexes by an ATP-dependent mechanism. *Mol Cell Biol* 20, 17.(6233-6243).

Sylvestre, Y., De Guire, V., Querido, E., Mukhopadhyay, U.K., Bourdeau, V., Major, F., Ferbeyre, G., and Chartrand, P. (2007). An E2F/miR-20a autoregulatory feedback loop. *J Biol Chem* 282, 4.(2135-2143).

Tazawa, H., Tsuchiya, N., Izumiya, M., and Nakagama, H. (2007). Tumor-suppressive miR-34a induces senescence-like growth arrest through modulation of the E2F pathway in human colon cancer cells. *Proc Natl Acad Sci U S A* 104, 39.(15472-15477).

Timchenko, N.A., Wilde, M., Iakova, P., Albrecht, J.H., and Darlington, G.J. (1999). E2F/p107 and E2F/p130 complexes are regulated by C/EBPalpha in 3T3-L1 adipocytes. *Nucleic Acids Res* 27, 17.(3621-3630).

Truscott, M., Harada, R., Vadnais, C., Robert, F., and Nepveu, A. (2008). p110 CUX1 cooperates with E2F transcription factors in the transcriptional activation of cell cycle-regulated genes. *Mol Cell Biol* 28, 10.(3127-3138).

van Ginkel, P.R., Hsiao, K.M., Schjerven, H., and Farnham, P.J. (1997). E2F-mediated growth regulation requires transcription factor cooperation. *J Biol Chem* 272, 29.(18367-18374).

Vandel, L., and Kouzarides, T. (1999). Residues phosphorylated by TFIIH are required fpr E2F-1 degradation during S-phase. *EMBO J* 184280-4291).

Wang, B., Liu, K., Lin, F.T., and Lin, W.C. (2004). A role for 14-3-3 tau in E2F1 stabilization and DNA damage-induced apoptosis. *J Biol Chem* 279, 52.(54140-54152).

Wang, G.L., Salisbury, E., Shi, X., Timchenko, L., Medrano, E.E., and Timchenko, N.A. (2008). HDAC1 cooperates with C/EBPalpha in the inhibition of liver proliferation in old mice. *J Biol Chem* 283, 38.(26169-26178).

Wang, H., Larris, B., Peiris, T.H., Zhang, L., Le Lay, J., Gao, Y., and Greenbaum, L.E. (2007). C/EBPbeta activates E2F-regulated genes in vivo via recruitment of the coactivator CREB-binding protein/P300. *J Biol Chem* 282, 34.(24679-24688).

Wan, Z., Zhi, N., Wong, S., Keyvanfar, K., Liu, D., Raghvachari, N., Munson, P. J., Su, S. Malide, D., Kajigaya, S., and Young, N. S.(2010). Human parvovirus B19 cauyses cell cycle arrest of human erythroid progenitors via deregulation of the E2F family of transcription factors. *J Clin Invest* 120, 10.(3530-3544).

Weber, J.D., Taylor, L.J., Roussel, M.F., Sherr, C.J., and Bar-Sagi, D. (1999). Nucleolar Arf sequesters Mdm2 and activates p53. *Nat Cell Biol* 1, 1.(20-26).

Weintraub, S.J., Chow, K.N., Luo, R.X., Zhang, S.H., He, D., and Dean, D.C. (1995). Mechanism of active transcriptional repression by the retinoblastoma protein. *Nature* 375812-815).

Wikonkal, N.M., Remenyik, E., Knezevic, D., Zhang, W., Liu, M., Zhao, H., Berton, T.R., and Johnson, D.G. (2003). Inactivating E2f1 reverts apoptosis resistance and cancer sensitivity in Trp53-deficient mice. *Nat Cell Biol* 5655-660).

Wong, C.F., Barnes, L.M., Dahler, A.L., Smith, L., Serewko-Auret, M.M., Popa, C., Abdul-Jabbar, I., and Saunders, N.A. (2003). E2F modulates keratinocyte squamous differentiation: Implications for E2F inhibition in squamous cell carcinoma. *J Biol Chem* 27828516-28522).

Woods, K., Thomson, J.M., and Hammond, S.M. (2007). Direct regulation of an oncogenic micro-RNA cluster by E2F transcription factors. *J Biol Chem* 282, 4.(2130-2134).

Xiao, B., Spencer, J., Clements, A., Ali-Khan, N., Mittnacht, S., Broceno, C., Burghammer, M., Perrakis, A., Marmorstein, R., and Gamblin, S.J. (2003). Crystal structure of the retinoblastoma tumor suppressor protein bound to E2F and the molecular basis of its regulation. *Proc Natl Acad Sci U S A* 100, 5.(2363-2368).

Xie, Q., Bai, Y., Wu, J., Sun, Y., Wang, Y., Zhang, Y., Mei. P., and Yuan, Z. (2011) Methylation-mediated regulation of E2F1 in DANN damage-induced cell death. *J Recept Signal Transduct Res* 31, 2. (139-146).

Xu, T., Zhu, Y., Xiong, Y., Ge, Y.Y., Yun, J.P., and Zhuang, S.M. (2009). MicroRNA-195 suppresses tumorigenicity and regulates G1/S transition of human hepatocellular carcinoma cells. *Hepatology* 50, 1.(113-121).

Yang, J., Song, K., Krebs, T. L., Jackson, M. W., and Danielpour, D. (2008). Rb/E2F4 and Smad2/3 link survivin to TGF-beta-induced apoptosis and tumor progression. *Oncogene* 27, 40.(5226-5238).

Zaragoza, K., Begay, V., Schuetz, A., Heinemann, U., and Leutz, A. (2010). Repression of transcriptional activity of C/EBPalpha by E2F-dimerization partner complexes. *Mol Cell Biol* 30, 9.(2293-2304).

Zhang, X., Wan, G., Berger, F.G., He, X., and Lu, X. (2011). The ATM Kinase Induces MicroRNA Biogenesis in the DNA Damage Response. *Mol Cell* 41, 4.(371-383).

Zhu, J.W., DeRyckere, D., Li, F.X., Wan, Y.Y., and DeGregori, J. (1999). A role for E2F1 in the induction of ARF, p53, and apoptosis during thymic negative selection. *Cell Growth Differ* 10, 12.(829-838).

10

Posttranslational Modifications of Rad51 Protein and Its Direct Partners: Role and Effect on Homologous Recombination – Mediated DNA Repair

Milena Popova*, Sébastien Henry* and Fabrice Fleury
Unité U3B, UMR 6204 CNRS, 2, rue de la Houssinière, University of Nantes
France

1. Introduction

Double-strand breaks (DSB) are probably the most deleterious form of DNA alteration in a cell. They may arise from ionizing radiation, free radicals, chemicals, or during replication of single-strand breaks. There are two distinct and complementary mechanisms for DSB repair: non-homologous end-joining (NHEJ) and homologous recombination (HR). Both repair pathways are important for the elimination of DSBs in eukaryotes.

Although the mechanisms of the cellular choice between these two pathways remain unclear, there is evidence that it depends on the cell cycle, as well as on mechanisms such as posttranslational modifications. When an intact DNA copy is available, HR is preferred and it is mainly active during late S and G2 phases of the cell cycle, while NHEJ is predominant during G0 and early S phases. The NHEJ pathway is characterised by a phosphorylation cascade where the first step is the activation of DNA-PKc protein which comprises a catalytic subunit and which is essential to complete the repair process. In contrast to NHEJ, the role of posttranslational modifications of proteins involved in the HR pathway is not clearly defined. Rad51 is a central protein in HR repair and its activity is based on pairing and strand exchange between homologous DNAs. The molecular regulation of Rad51 levels and activity has not been completely established. However, the kinase-induced phosphorylation of this protein modulates its recombinase activity by changing its interface and recognition sites and probably its intracellular distribution. Indeed, Rad51 associates with its paralogues and with other partner proteins, such as Rad52, Rad54, BRCA2 tumour suppressor, BLM helicase (Fig.1). Rad51 forms distinct subnuclear complexes called foci, which represent the functional units in DNA repair by HR. This accumulation of repair proteins to sites of double-strand break repair is closely dependant on protein-protein interactions which can be regulated by posttranslational modification processes including tyrosine, serine and threonine phosphorylations. This underlines the high complexity of HR regulation in mammalian cells.

Regulation of Rad51 recombinase activity and its interactions following DNA damage are poorly understood. In this chapter we have summarized the posttranslational modifications

---

* M.P. and S.H. contributed equally to this work

of Rad51 and of the proteins interacting physically with Rad51 during HR repair. We then attempt to relate the impact of these modifications on HR DNA repair and on the intracellular distribution of DNA repair proteins.

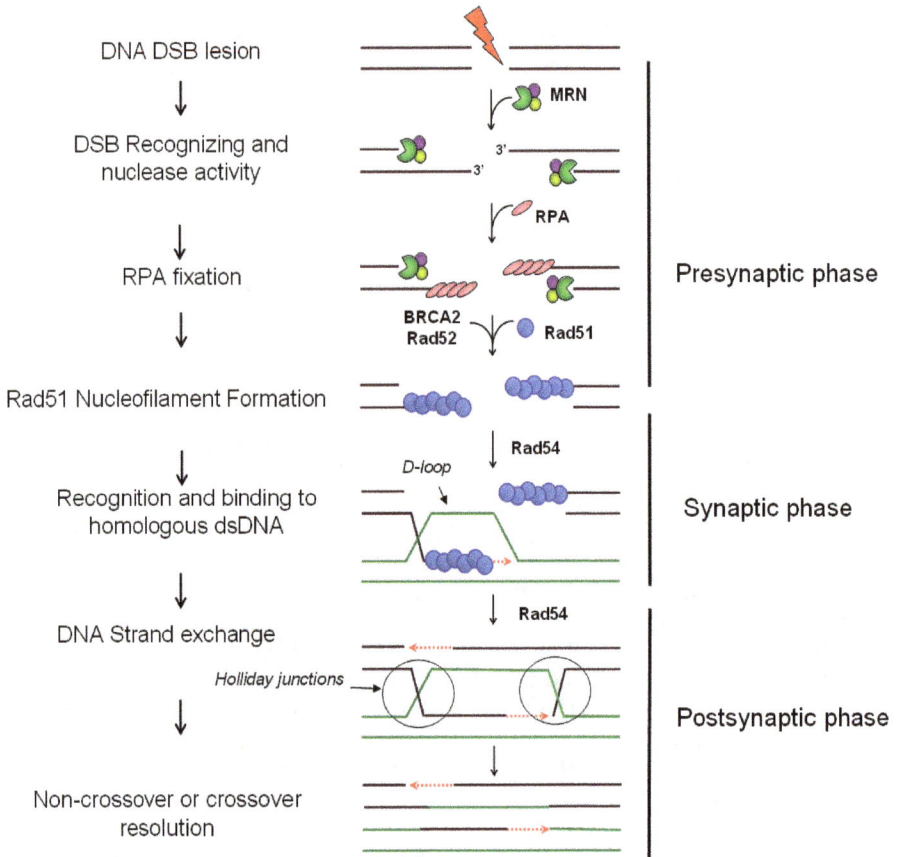

Fig. 1. Schematic representation of the mechanism of DNA DSB repair by homologous recombination.

2. Post-translational modifications of Rad51

2.1 Tyrosine phosphorylation of Rad51 by the c-Abl family of tyrosine kinases

Several studies have shown that Rad51 can be phosphorylated on tyrosine but until recently there were discrepancies on the exact site of phosphorylation. Three studies had shown the phosphorylation of Tyrosine 315 (Y315) and only one the phosphorylation of Tyrosine 54 (Y54). A recent publication demonstrated that both of these tyrosines can be phosphorylated. The kinases which phosphorylate Rad51 belong to the c-Abl family which has two members, c-Abl and Arg. The oncogenic fusion tyrosine kinase BCR/Abl has also been shown to phosphorylate Rad51. However, other tyrosine kinases can also phosphorylate Rad51 at a different site than Tyrosine 315 in MEF cAbl-/- cells (Chen et al., 1999b).

2.1.1 Phosphorylation on Tyrosine 54

The first study showing that Rad51 can be phosphorylated was published in 1998 by Yuan and colleagues. Using co-immunoprecipitation, the authors observed that human Rad51 (hRad51) binds to c-Abl in cells. This association was unaffected by irradiation of the cells and was not dependent on DNA binding. Pull-down assays were performed with a GST-c-Abl fusion protein or a GST-c-Abl SH3 domain fusion peptide. These were incubated with cell lysates or purified hRad51. The results confirmed the association between hRad51 and c-Abl *in vitro* and showed that the binding is direct and is mediated by the SH3 domain of c-Abl.

In vitro phosphorylation assays with purified c-Abl and hRad51 demonstrated that hRad51 is a substrate for this kinase. Immunoprecipitation of Rad51 was performed with lysates from irradiated cells overexpressing hRad51 and c-Abl. The analyses of the immunoprecipitated protein with an anti-phosphoTyrosine antibody confirmed the phosphorylation of Rad51 *in vivo*. The *in vivo* and *in vitro* phosphorylated hRad51 proteins were then purified and analyzed by mass spectroscopy. The detected peaks indicated that the phosphorylation is located on Tyrosine 54 on both *in vivo* and *in vitro* phosphorylated Rad51 (Chen et al., 1999a; Chen et al., 1999b; Chen et al., 1999c; Dong et al., 1999; Yuan et al., 1999; Zhong et al., 1999).

2.1.2 Phosphorylation on Tyrosine 315 by c-Abl

Two years after Yuan and colleagues published their study, another group demonstrated that Rad51 can be phosphorylated. However Chen and colleagues did not observe the phosphorylation of Tyrosine 54 but detected the phosphorylation of another tyrosine residue, in position 315.

The authors used GST pull-down assays and immunoprecipitation to show that Rad51 forms a complex with c-Abl and ATM in cells. The association between the three proteins was independent of irradiation and DNA binding. The level of phosphorylation of Rad51 after irradiation of cells was investigated. The analyses of immunoprecipitated Rad51 with an anti-phosphoTyrosine antibody showed that the level of phosphorylation increases after irradiation. Rad51 was a direct substrate for c-Abl and the phosphorylation was dependent on both c-Abl and ATM. In order to determine which tyrosine residue was phosphorylated, the authors co-expressed c-Abl and wild type or mutated Rad51 in cells. Different tyrosine to phenylalanine Rad51 mutants were performed. Phenylalanine is an amino acid that cannot be phosphorylated. Thus, a signal would no longer be detected by the anti-phosphoTyrosine antibody when the phosphorylated residue is mutated. The mutation of Y315 to phenylalanine abolished Rad51 phosphorylation, indicating that c-Abl phosphorylates Rad51 on this residue (Yuan et al., 1998).

2.1.3 Phosphorylation on Tyrosine 315 by BCR/Abl

Rad51 can also be phosphorylated by the oncogenic fusion tyrosine kinase BCR/Abl. BCR/Abl is expressed in most cases of chronic myeloid leukemia and in some cases of acute myeloid leukemia and possesses constitutive kinase activity.

Slupianek and colleagues suggested that Rad51 and BCR/Abl interact physically since a portion of Rad51 co-localizes with the fusion tyrosine kinase in the cytoplasm of BCR/Abl overexpressing cells. This interaction was confirmed by the co-immunoprecipitation of the two proteins.

Rad51 was immunoprecipitated from cells overexpressing BCR/Abl and its phosphorylation state was examined with an anti-phosphoTyrosine antibody. The interaction between

BCR/Abl and Rad51 resulted in the constitutive phosphorylation of Rad51 on tyrosine. Rad51 was also phosphorylated by c-Abl after treatment of cells with cisplatin and mitomycin C. In order to determine the position of phosphorylation, the authors transiently co-expressed BCR/Abl and wild type or mutated Rad51 in cells. Tyrosine to phenylalanine mutations were performed at Tyrosine 54 or Tyrosine 315. The analysis of the Rad51 immunoprecipitates with an anti-phosphoTyrosine antibody revealed the phosphorylation of the wild type and the Y54F Rad51 protein. A substantial reduction in the phosphorylation level of Rad51 was observed when Y315 was mutated to phenylalanine, indicating that the majority of the phosphorylation of Rad51 occurred on Y315. To further confirm the phosphorylation of the Y315 residue, Slupianek and colleagues prepared an antiserum using a phosphorylated Y315 peptide. Western blots were then performed with lysates from cells overexpressiong Rad51 alone or with BCR/Abl. The antiserum did not recognize Rad51 when the protein was overexpressed in cells alone. In contrast, in cells co-expressing BCR/Abl a strong signal was observed. This confirms that the fusion tyrosine kinase BCR/Abl phosphorylates Rad51 on Tyrosine 315 (Slupianek et al., 2001).

2.1.4 Phosphorylation by Arg
The only other member of the c-Abl family, the kinase Arg, also phosphorylates Rad51. Arg shares considerable structural and sequence homology with c-Abl in the N-terminal SH3 and SH2 domains, as well as in the tyrosine kinase domain (Kruh et al., 1990). Co-immunoprecipitation of Rad51 from cells overexpressing Rad51 and Arg indicated that Arg can interact with Rad51 *in vivo*. An anti-phosphoTyrosine antibody showed that Rad51 is phosphorylated by Arg and this phosphorylation seemed to be more effective than the phosphorylation by c-Abl. However, the position of phosphorylation was not determined (Li et al., 2002).

2.1.5 Phosphorylation of both Tyrosine 54 and Tyrosine 315 by c-Abl
The study conducted by Popova and colleagues has allowed to reconcile the discrepancies on which tyrosine residue is phosphorylated in Rad51. The authors purified specific anti-phosphoTyrosine antibodies for each site of phosphorylation. These antibodies were used to analyze the phosphorylation state of Rad51 by immunoblotting of lysates from cells overexpressing Rad51 and c-Abl. The ability of these specific antibodies to detect distinctively the phosphorylation of the two tyrosine residues has allowed to observe the phosphorylation of both Y54 and Y315 in the same experiment. This confirmed that both Tyrosine 54 and 315 can be phosphorylated (Popova et al., 2009).

In all previous studies the phosphorylation of only one site was observed, either Y54 or Y315. The fact that Yuan and colleagues observed only the phosphorylation of Y54 and did not detect the phosphorylation of Y315 could be due to the technique they used. In their study, the *in vitro* or *in vivo* phosphorylated Rad51 protein, as well as the unphosphorylated protein were digested by trypsin. The obtained fragments were then analyzed by mass spectroscopy and the spectra of the unphosphorylated and the phosphorylated proteins were compared. The lack of a phosphorylation peak in the fragment containing Y315 could be explained by its biophysical characteristics. Following trypsin digestion, the peptide containing Tyrosine 54 is 17 amino acids long and has a pHi of 4,83. On the contrary, the peptide containing Tyrosine 315 is 28 amino acids long and its pHi is 4,03. Thus, the Y315 peptide is longer and more negatively charged compared to the Y54 peptide which could interfere with its detection by mass spectroscopy (Raggiaschi et al., 2005).

Another possible explanation could be the proximity of the digestion and the phosphorylation sites. The presence of phosphorylation near a digestion site may decrease its digestion efficiency (Benore-Parsons et al., 1989; Kjeldsen et al., 2007). Thus the phosphorylated protein would be partially digested resulting in a longer phospho-peptide. A corresponding peptide would not be obtained from the digestion of the unphosphorylated protein. A phosphorylation peak would not be observed in these conditions. In the amino acid sequence of Rad51, only one residue separates the trypsin digestion site from Tyrosine 315. Due to the proximity of the two sites, Rad51 would rather be digested at arginine 310 than on lysine 313. This would result in the generation of a phosphopeptide which would be 3 amino acids longer than the corresponding peptide from the unphosphorylated protein. Consequently, the phosphorylation of Rad51 on Y315 would not be detected by mass spectroscopy.

2.1.6 Model of sequential phosphorylation

Popova and co-authors have established a possible mechanism by which Rad51 is phosphorylated by c-Abl. They co-expressed c-Abl and wild type or mutated hRad51 in cells. In the amino acid sequence of hRad51, Tyrosine 54 or Tyrosine 315 were mutated to phenylalanine, thus rendering the residue at this position nonphosphorylatable. Western blot analysis of the cell lysates, revealed with their specific anti-phosphoTyrosine antibodies, showed a relationship between the phosphorylation of Y54 and Y315. When residue 315 was mutated to phenylalanine and nonphosphorylatable, Tyrosine 54 was no longer phosphorylated. On the contrary, the mutation of residue 54 had no effect on the phosphorylation of Tyrosine 315. The authors hypothesized that the phosphorylation of Tyrosine 315 is needed for the phosphorylation of Tyrosine 54.

The c-Abl kinase possesses a SH3 and a SH2 domain in its N-terminal region. The SH3 domain recognizes and binds preferentially to proline rich regions containing the sequence PXXP. The SH2 domain recognizes pYXXP sequences. hRad51 has two PXXP motifs in its amino acid sequence – between amino acids 283 and 286, and between amino acids 318 and 321. When Tyrosine 315 is phosphorylated, a pYXXP motif is revealed between amino acids 315 and 318. This motif might be recognized by the SH2 domain of c-Abl.

According to this model of sequential phosphorylation, c-Abl recognizes a PXXP motif in the sequence of Rad51 through its SH3 domain and phosphorylates Tyrosine 315. The phosphorylation of this residue reveals the pYXXP binding motif which is recognized by the SH2 domain of c-Abl. This allows the phosphorylation of Tyrosine 54.

To confirm this model, GST pull-down assays were performed. A GST- c-Abl SH2 domain peptide was incubated with lysates from cells overexpressing Rad51 and c-Abl. The results showed that hRad51 binds to the SH2 domain of c-Abl and that this interaction takes place when Rad51 is phosphorylated on Tyrosine 315. Therefore a model of sequential phosphorylation of Rad51, where the phosphorylation of Tyrosine 315 by c-Abl reveals a novel binding site for the kinase thus allowing the phosphorylation of Tyrosine 54, is highly plausible.

2.2 Role of Rad51 phosphorylation

Even though the process of phosphorylation seems to be of considerable importance in the regulation of Rad51 activity, its exact roles and consequences have not been elucidated yet. Moreover, the existing data is contradictory.

In their study, Yuan and colleagues investigated the possible effect of Y54 phosphorylation on Rad51 activity. Strand exchange assays showed that phosphorylation of S. cerevisiae

Rad51 (ScRad51) results in the inhibition of dsDNA conversion to joint molecules and nicked circular dsDNA. An inhibition of the binding of phospho-ScRad51 and phospho-hRad51 to ssDNA was also observed. Because Rad51 exerts its activity by binding to and forming nucleofilaments with ssDNA, the authors concluded that by inhibiting the binding to ssDNA, phosphorylation inhibits Rad51 function (Yuan et al., 1998).

In the search of a possible role for Y315 phosphorylation, Chen and colleagues investigated if the phosphorylation impacts the interaction between Rad51 and Rad52. Rad52 is a protein needed in the presynaptic stage of homologous recombination (Fig. 1). Binding assays with purified *in vitro* phosphorylated Rad51 and Rad52, as well as co-immunoprecipitation of Rad51 and Rad52 from irradiated cells were performed. The results indicated that phosphorylation enhances the interaction between these two proteins *in vitro* and *in vivo*. The authors hypothesized that this irradiation-induced phosphorylation of Rad51 on tyrosine residues and the concomitant increase in association with Rad52 may lead to increased DNA repair efficiency (Chen et al., 1999b). *In vitro* studies with different Y315 mutants suggest that the phosphorylation of this residue is important for the binding of Rad51 to dsDNA and for nucleofilament formation (Takizawa et al., 2004). Moreover, Y315 is located near the polymerisation site of the protein, a region which is essential for the filament formation of Rad51 on DSBs, (Conilleau et al., 2004).

Slupianek and colleagues analyzed the role of Rad51 phosphorylation in the resistance of cells to DNA damaging agents. The resistance of BCR/Abl expressing cells to cisplatin and mitomycin C was decreased upon overexpression of nonphosphorylatable Rad51 Y315F. The mutation of Y54 had no effect on resistance. These results link the phosphorylation of Y315 to the resistance to DNA cross-linking agents and suggest that it has an important impact on DNA repair (Slupianek et al., 2001).

Recently, the same team reported an implication of Y315 phosphorylation in the regulation of BCR/Abl-Rad51 interaction. BCR/Abl-mediated phosphorylation of Y315 appears to be important for the dissociation of Rad51 from BCR/Abl in chronic myeloid leukemia cells (Slupianek et al., 2009). The authors studied the intracellular localization of wild type and mutated Rad51 in response to DSBs induced by genotoxic treatment. The nonphosphorylatable Rad51 Y315F mutant remained mostly in the cytoplasm, while the wild-type protein accumulated in the nucleus in BCR/Abl-positive cells. This indicates that phospho-Y315 stimulates abundant nuclear localization of Rad51 on DSBs.

2.3 Phosphorylation on Threonine 309 by Chk1

Rad51 can also be phosphorylated on threonine. Sorensen and colleagues observed that a Chk1 signal is necessary for efficient homologous recombination. The inhibition of this kinase decreased the level of homologous recombination and of DNA DSB repair. The inhibition of Chk1 also impaired the formation of Rad51 foci which was not due to decreased Rad51 levels. The interaction of Rad51 with chromatin was dependent on Chk1 activity. Using immunoprecipitation, Sorensen and colleagues showed that Chk1 and Rad51 can interact physically in cells. Chk1 phosphorylates Rad51 on Threonine 309 which is located in a Chk1 consensus phosphorylation site. Cells transfected with a nonphosphorylatable Rad51 mutant were more sensitive to hydroxyurea which confirms that Chk1 signaling is required for homologous recombination repair (Sorensen et al., 2005).

2.4 Sumoylation – Ubiquitination of Rad51

Yeast two-hybrid assays have shown that Rad51 can interact with HsUbc9, later named
UBE21. HsUbc9/UBE21 is the human homologue of *S. cerevisiae* UBC9 and *S. pombe* Hus5
ubiquitin conjugating enzymes (Kovalenko et al., 1996; Shen et al., 1996). In mammalian
cells the downregulation of Ubc9 was associated with defects in cytokinesis and an
increased number of apoptotic cells. Furthermore, its gene inactivation is lethal in mouse
embryos (Moschos and Mo, 2006). Nuclear depletion of Ubc9 disrupts the intracellular
trafficking of Rad51 and thus inhibits the formation of Rad51 nuclear foci following DNA
damage (Saitoh et al., 2002).

Rad51 also interacts with UBL1 (ubiquitin like 1), also called PIC1, GMP1, SUMO-1 and
Sentrin (Shen et al., 1996). The yeast homologue of UBL1, SMT3, inhibits a centrosome
protein involved in centrosome segregation (Shen et al., 1996). UBL1 interacts with
HsUBC9/UBE21 (Shen et al., 1996). Studies have shown that HsUbc9/UBE21 is a UBL1-
conjugating enzyme, rather than an ubiquitin-conjugating enzyme. Immunoprecipitation
essays in HeLa cells and GST pull-down essays have shown that the interaction between
Rad51 and Ubl1 is mediated by Rad52 and/or Ubc9. This suggests that Ubc9 can conjugate
UBL1 to Rad51. The overexpression of UBL1 in mammalian cells decreases DSB-induced HR
and resistance to IR (Li et al., 2000).

3. Rad51-interacting proteins involved in the nuclear translocation of Rad51 and in the HR process

The number and size of Rad51 nuclear foci is a hallmark of the cellular response to
genotoxic stress. These nuclear foci characterize the formation of Rad51 filaments. Indeed
Rad51 is recruited to sites of DNA DSBs in response to damage where it promotes DNA
strand invasion and strand exchange. Impaired formation of Rad51 foci in response to DNA
damage has been demonstrated in hamster or chicken cells defective in the Rad51 paralogs
XRCC2, XRCC3, Rad51B, Rad51C, and in mammalian BRCA1 or BRCA2-defective cells
(Chen et al., 1999c; Takata et al., 2001; Yuan et al., 1999).

The foci formation requires the translocation of Rad51 into the nucleus after DSB induction
by genotoxic stress or stalled replication forks (Haaf et al., 1995).) This process is often
accompanied by posttranslational modifications of Rad51 partners which cooperate to
achieve the fidelity of DNA repair. Several works have shown that these modifications can
modulate protein interactions involving Rad51 and can affect Rad51 foci formation.

3.1 Nuclear translocation of Rad51

The first stage of DNA DSB repair by HR requires the delivery of Rad51 at the sites of DNA
damage. Since Rad51 does not have a Nuclear Localisation Signal (NLS) sequence, its
nuclear entry likely requires the interaction with other proteins containing functional NLS
sequences (Gildemeister et al., 2009). BRCA1 and BRCA2 proteins have both been described
as primordial recombination mediators for the nuclear translocation of Rad51.

3.1.1 Involvement of BRCA1/Akt1

Several studies have demonstrated that the overexpression of Rad51 results in its
cytoplasmic accumulation (Mladenov et al., 2006) but genotoxic stress triggers the
translocation of Rad51 from the cytoplasm to the nucleus (Gildemeister et al., 2009). Plo and

colleagues have reported that the nuclear translocation of Rad51 was impaired by AKT1 which repressed HR (Plo et al., 2008). In tumour cells with high levels of active AKT1, BRCA1 and Rad51 are retained in the cytoplasm. However, BRCA1 phosphorylation by AKT1 was not required for this retention. Interestingly, 77% of tumours containing high levels of AKT1 exhibited also cytoplasmic retention of Rad51 (Plo et al., 2008). This shows that AKT1 activation strongly favors the cytoplasmic localization of both BRCA1 and Rad51 proteins.

3.1.2 BRCA2-mediated nuclear translocation of Rad51

Like BRCA1, BRCA2 is a tumour suppressor implicated in familial breast cancer. BRCA2 protein contains six highly conserved BRC repeats which are involved in the interaction between BRCA2 and Rad51 (Marmorstein et al., 1998; Mizuta et al., 1997; Wong et al., 1997). It has been proposed that the BRCA2 protein is directly involved in the regulation of the nucleofilament formation and in the nuclear transport of Rad51 (Davies et al., 2001).

Medova and colleagues have demonstrated that the inhibition of the MET receptor tyrosine kinase by a small inhibitor molecule impairs the formation of the Rad51-BRCA2 complex. By targeting MET, the authors have shown the incapacity of tumour cells to repair DNA DSBs through homologous recombination. This was due to the impaired translocation of Rad51 into the nucleus (Medova et al.).

The pancreatic adenocarcinoma cell line CAPAN-1 is the best characterized BRCA2 defective human cell line (Jasin, 2002). CAPAN-1 cells have indeed lost a wild-type BRCA2 allele and presents a 6174delT mutation on the other allele. This mutation causes the premature C-terminal truncation of the protein. This results in the deletion of the BRCA2 domains for DNA repair and the nuclear localization signals (Holt et al., 2008). Rad51 exhibits impaired nuclear translocation in CAPAN-1 cells. Therefore it has been proposed that Rad51 requires BRCA2 for its nuclear translocation and that C-terminally truncated BRCA2 retains Rad51 in the cytoplasm.

Another group has however observed a DNA damage-induced increase in nuclear Rad51 in the BRCA2-defective cell line CAPAN-1. Moreover, chromatin-associated Rad51 levels were found to be increased (2-fold) following IR exposure (Gildemeister et al., 2009).

To analyze a possible BRCA2-independent mechanism for Rad51 nuclear transport, the authors studied two other Rad51-interacting proteins, Rad51C and Xrcc3. Both of these proteins contain a functional NLS. In contrast to Xrcc3, subcellular distribution of Rad51C was affected by DNA damage since nuclear Rad51C was significantly increased following IR exposure. Furthermore, the depletion of Rad51C in HeLa and CAPAN-1 cells by RNA interference resulted in lower levels of nuclear Rad51. These results provide an important overview of the cellular regulation of Rad51 nuclear entry. This data underlines the potential role for Rad51C in the nuclear translocation of Rad51, which suggests a BRCA2-independent mechanism for Rad51 nuclear entry both before and after DNA damage. Other studies have also demonstrated that an interaction between Rad51 and BRCA2 is not required for nuclear transport of Rad51 but it may prevent the formation of Rad51 filaments in the cytoplasm.

3.2 Recruitment of Rad51 at the damage site – Presynaptic phase of HR

Following damage, DSB are recognized by the MRN complex (MRE11-Rad51-NSB1 complex). MRN binds to and resects the extremities of the DSB through its nuclease activity.

This results in the generation of 3′ single-stranded DNA (ssDNA). RPA (Replication Protein A) binds to the 3′ overhangs and thus protects them from further resection. This protein also removes secondary structures present on the ssDNA which allows efficient Rad51 nucleofilament formation (McIlwraith et al., 2000).

During the presynaptic phase Rad51 is loaded on the ssDNA ends with the help of BRCA2 (Huen et al., 2010). Rad51 recognizes and binds to the BRC repeats and the TR2 domain of BRCA2 (Fig.2). The Oligonucleotide Binding Folds (OB Folds) in the C-terminal region of the protein are also required for the recruitment of Rad51 (O'Donovan and Livingston, 2010; Wong et al., 1997).

The interaction of BRCA2 with two other proteins, BRCA1 and the bridging factor PALB2, is necessary for its role in the presynaptic phase of HR. These proteins along with other factors form a macro-complex named BRCC whose role in DNA repair has been described elsewhere (Dong et al., 2003).

In addition to its linking function between BRCA1 and BRCA2, PALB2 also interacts with a domain in Rad51 which is comprised between amino acids 184 and 257 (Fig.3) (Buisson et al., 2010). Thus, PALB2 cooperates with BRCA2 to stimulate Rad51 filament assembly during HR. The stimulation of the filament assembly by PALB2 is also mediated by its interaction with another co-factor, Rad51AP1 (Dray et al., 2010).

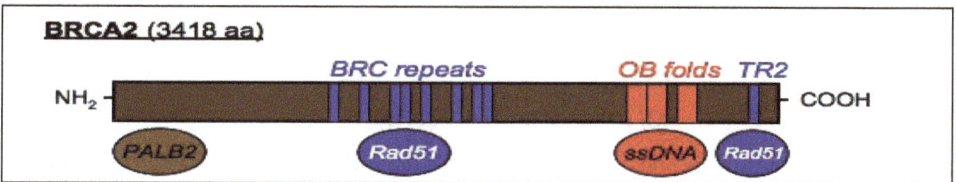

Fig. 2. Domain organization of BRCA2. Schematic drawing indicating the interaction sites with Rad51, PALB2 and DNA.

According to these data, BRCA2 plays an essential role in recruiting and loading Rad51 on sites of DSB and in initiating the HR process.

In order for the Rad51 presynaptic filament to assemble, Rad52 has to displace RPA from the ssDNA (Sugiyama and Kowalczykowski, 2002). RPA is a single-stranded DNA binding protein composed of three subunits, with sizes of respectively 70, 32 and 14 kDa (Wold, 1997). It has previously been shown by co-immunoprecipitation experiments that each of the three subunits of RPA interacts with Rad51, and that the RPA-Rad51 interaction is regulated by the 70kDa subunit (Golub et al., 1998). The co-localization of Rad51 and RPA foci in response to ionizing radiation was observed in a mice fibroblast model and suggests a possible *in vivo* interaction between the two proteins. Furthermore, a recent study has shown that depletion of RPA in mammalian cells leads to the impairment of Rad51 foci formation following DSB induced by hydroxyurea treatment. This confirms the importance of RPA in the presynaptic assembly of Rad51 (Sleeth et al., 2007).

Because RPA binding on ssDNA may prevent Rad51 access to DSB, the presynaptic filament formation needs to be time-regulated by the mediator Rad52. Rad52 is a key member of the RAD52 epistasis group, which includes Rad51, and whose function in HR has been previously described (Symington, 2002). The human Rad52 (hRad52) protein contains 418 amino acids. It has a highly conserved region in its N-terminus, and possesses a ssDNA/dsDNA binding region and a RPA binding site (Kagawa et al., 2002; Park et al., 1996). Shen and colleagues have demonstrated both *in vitro* and *in vivo* that hRad52 physically interacts with hRad51. The Rad51 binding domain on Rad52 has been identified between residues 291 to 330 (Fig.3) located in the C-terminal region of the protein (Shen et al., 1996).

Furthermore, five amino acid residues of hRad51 have been shown to participate in the Rad51-Rad52 interaction. These residues are located in the C-terminal region of hRad51 (Kurumizaka et al., 1999). Interestingly, the Rad52 binding site on Rad51 is not the same in *Homo Sapiens* and *Saccahromyces cerevisiae*, suggesting that this interaction is not conserved among species.

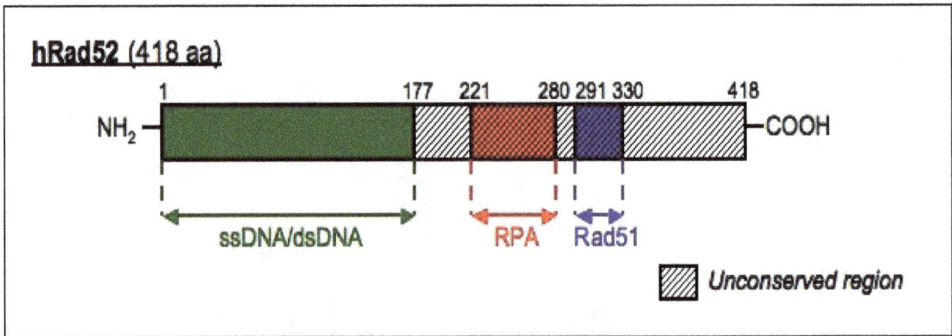

Fig. 3. Human Rad52 (hRad52) domains involved in HR.

The capacity to bind RPA and DNA confers to Rad52 the ability to displace RPA from the ssDNA and thus helps the formation of the Rad51 presynaptic filament (Plate et al., 2008; San Filippo et al., 2008).

The posttranslational modifications of RPA and Rad52 could modulate the formation of the presynaptic filament. Indeed, RPA is phosphorylated on one of its three subunits in a DNA damage-dependent manner and the resulting hyperphosphorylated RPA proteins directly interact with Rad51 (Binz et al., 2004; Wu et al., 2005). More recently, Shi and colleagues demonstrated by mutating the phosphorylation site of RPA that this posttranslational modification is required for Rad51 assembly (Shi et al., 2010). The importance of RPA phosphorylation during the presynaptic phase of HR was confirmed by Deng and colleagues who proposed a model in which RPA phosphorylation promotes Rad52 function and thus prepares DSB to be processed by Rad51 (Deng et al., 2009).

Phosphorylation of the Rad52 mediator in a c-Abl dependant manner has also been described in response to ionizing treatment (Kitao and Yuan, 2002). There is no evidence for the direct effect of Rad52 phosphorylation on Rad51 assembly. However, anterior studies have shown that the phosphorylation of Rad51 by c-Abl has an impact on the interaction between Rad51 and Rad52 (Chen et al., 1999b).

Another important posttranslational modification which plays a role in this stage of the HR process is SUMOylation. SUMOylation is already known to regulate the properties and stability of different proteins (Hay, 2005). It has recently been shown that the 70 kDa subunit of RPA can be SUMOylated and this process may regulate Rad51 presynaptic filament formation (Dou et al., 2010).

3.3 Regulation of Rad51 nucleofilament stability and enhancement of the strand exchange activity - Synaptic phase

Once the Rad51 nucleofilament is assembled, it has to be stabilized before Rad51 strand exchange activity may occur. This is mainly achieved by the Rad54 protein, which interacts both *in vitro* and *in vivo* with Rad51 during the synaptic phase of HR (Golub et al., 1997; Mazin et al., 2010). This protein-protein interaction is mediated by the Rad54 N-terminal region. It can occur either with the free Rad51 protein or with the assembled nucelofilament (Mazin et al., 2003; Raschle et al., 2004). Furthermore, using mouse embryonic stem cells, Tan and colleagues have demonstrated that Rad54 is required for Rad51 IR-induced foci formation (Tan et al., 1999). Rad54 functions in an ATP-independent manner to stabilize the Rad51 nucleofilament (Wolner and Peterson, 2005). However, it can also disrupt the assembled Rad51 complex (Li et al., 2007; Solinger et al., 2002). Thus, Rad54 modulates the stability of the Rad51 filament.

Another important consequence of the Rad51-Rad54 interaction is that Rad54 stimulates the recombinase and strand exchange activities of Rad51 (Mazina and Mazin, 2004; Sigurdsson et al., 2002). An additional protein interacting with Rad51 in the mature synaptic filament has been discovered. First identified as Pir51 (for Protein interacting with Rad51), this cofactor was later renamed Rad51AP1 (Rad51 Associated Protein 1). This protein was first characterized for its DNA crosslink repair activity (Henson et al., 2006; Kovalenko et al., 1997). Modesti and colleagues proposed a model in which Rad51AP1 could stimulate the formation of the D-loop by Rad51, which is the final step of the synaptic phase (Modesti et al., 2007).

To this day, the potential effect of Rad54 posttranslational modifications on Rad51 activity during this late stage of HR has not been demonstrated. Recent results obtained in yeast show that Rad54 phosphorylation leads to a reduction in Rad51-Rad54 complexes (Niu et al., 2009). It is not excluded that a similar mechanism could exist in superior eukaryotes.

3.4 Post-synaptic phase of HR – Resolution of Holliday junction

Following the synaptic phase, D-loops can be eliminated by different subpathways, each requiring different proteins. Here we will present only the pathways involving double Holliday junctions (dHJ) (Bzymek et al., 2010). Double HJ are structural intermediates which are resolved by specific endonucleases and result in either crossover or non-crossover products. The dHJ intermediates can also be resolved by helicases (RecQ helicase family) combined with topoisomerase action. In human cells, this pathway combines BLM helicase and topoisomerase IIIa, both of which catalyze dHJ dissolution (Wu and Hickson, 2003). Interestingly, BLM helicase is phosphorylated by different kinases, such as Chk1, at different stages of the cell cycle or in response to DNA damage. BLM can interact with 53PB1, a signal transducer, and with Topoisomerase IIIa during the presynaptic and the postsynaptic phases of HR respectively. It has been shown that BLM and 53BP1 can interact physically with Rad51 and regulate HR by modulating the assembly of Rad51 filaments. The *in vivo* phosphorylation of both BLM and 53BP1 affects negatively Rad51 foci formation (Tripathi et al., 2007). Concerning Topoisomerase IIIa, Rao and colleagues suggested that the BLM phosphorylation on T99 results in its dissociation from topoisomerase IIIa, thereby modulating the resolution of dHJ (Rao et al., 2005).

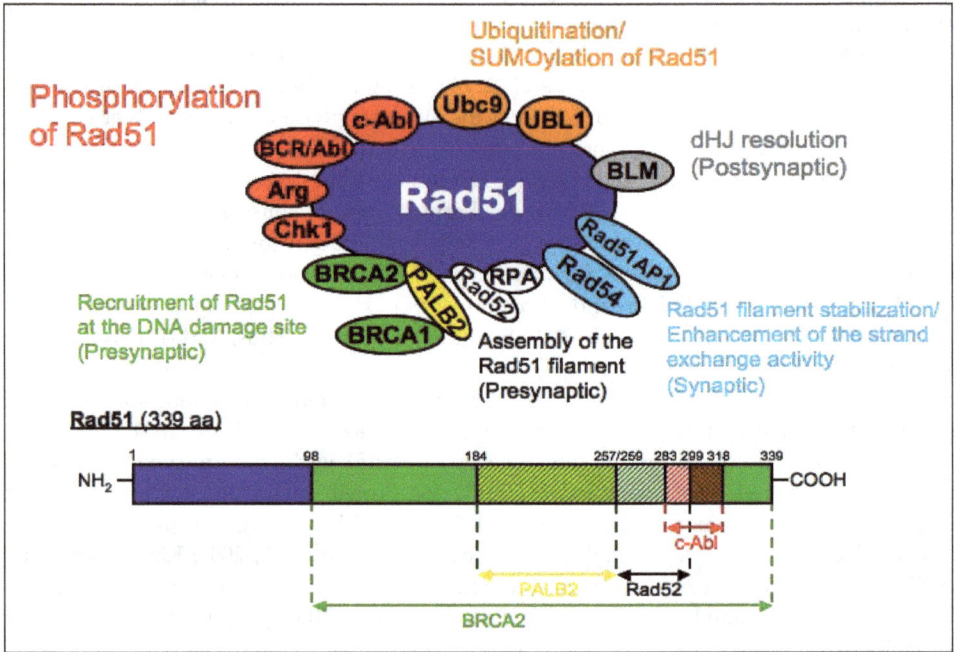

Fig. 4. Schematic representation of Rad51 interactions with its direct partners involved in its posttranslational modification and the steps of HR (top). Localization of binding sites in the hRad51 sequence (bottom).

4. Conclusion

In all living organisms HR is strictly regulated in time and in space to maintain the stability of the genome. Rad51 is the central protein in the HR process. The regulation of HR involves many protein interactions (Fig. 4) which are strongly dependent on posttranslational modifications. Indeed, almost all key mediator proteins of HR are subject to phosphorylation by specific kinases, thereby modulating some stage of this process (e.g. the nucleofilament formation). Hence, these posttranslational reactions underline the complexity of the regulation of HR. Despite of the several studies on the mechanism of Rad51 phosphorylation, its biochemical role in the HR reaction remains unclear.

The impact of phosphorylation on the interactions of Rad51 with its partners still needs to be determined. In order to better understand the regulation of HR, the future challenge will be to identify the complete interaction network of Rad51, the motor protein of HR.

5. Acknowledgment

This work was supported by grants from the Ligue contre le Cancer Comité de Loire Atlantique et du Morbihan. SH is supported by a fellowship from the Region Pays de la Loire (CIMATH2 grant). MP was supported by a fellowship from Conseil Général des Pays de Loire-Atlantique (Atlanthèse grant).

6. References

Benore-Parsons, M., Seidah, N.G., & Wennogle, L.P. (1989). Substrate phosphorylation can inhibit proteolysis by trypsin-like enzymes. Arch Biochem Biophys 272, 274-280.

Binz, S.K., Sheehan, A.M., & Wold, M.S. (2004). Replication protein A phosphorylation and the cellular response to DNA damage. DNA Repair (Amst) 3, 1015-1024.

Buisson, R., Dion-Cote, A.M., Coulombe, Y., Launay, H., Cai, H., Stasiak, A.Z., Stasiak, A., Xia, B., & Masson, J.Y. (2010). Cooperation of breast cancer proteins PALB2 and piccolo BRCA2 in stimulating homologous recombination. Nat Struct Mol Biol 17, 1247-1254.

Bzymek, M., Thayer, N.H., Oh, S.D., Kleckner, N., & Hunter, N. (2010). Double Holliday junctions are intermediates of DNA break repair. Nature 464, 937-941.

Chen, C.F., Chen, P.L., Zhong, Q., Sharp, Z.D., & Lee, W.H. (1999a). Expression of BRC repeats in breast cancer cells disrupts the BRCA2-Rad51 complex and leads to radiation hypersensitivity and loss of G(2)/M checkpoint control. J Biol Chem 274, 32931-32935.

Chen, G., Yuan, S.S., Liu, W., Xu, Y., Trujillo, K., Song, B., Cong, F., Goff, S.P., Wu, Y., Arlinghaus, R., et al. (1999b). Radiation-induced assembly of Rad51 and Rad52 recombination complex requires ATM and c-Abl. J Biol Chem 274, 12748-12752.

Chen, J.J., Silver, D., Cantor, S., Livingston, D.M., & Scully, R. (1999c). BRCA1, BRCA2, & Rad51 operate in a common DNA damage response pathway. Cancer Res 59, 1752s-1756s.

Conilleau, S., Takizawa, Y., Tachiwana, H., Fleury, F., Kurumizaka, H., & Takahashi, M. (2004). Location of tyrosine 315, a target for phosphorylation by cAbl tyrosine kinase, at the edge of the subunit-subunit interface of the human Rad51 filament. J Mol Biol 339, 797-804.

Davies, A.A., Masson, J.Y., McIlwraith, M.J., Stasiak, A.Z., Stasiak, A., Venkitaraman, A.R., & West, S.C. (2001). Role of BRCA2 in control of the RAD51 recombination and DNA repair protein. Mol Cell 7, 273-282.

Deng, X., Prakash, A., Dhar, K., Baia, G.S., Kolar, C., Oakley, G.G., & Borgstahl, G.E. (2009). Human replication protein A-Rad52-single-stranded DNA complex: stoichiometry and evidence for strand transfer regulation by phosphorylation. Biochemistry 48, 6633-6643.

Dong, Y., Hakimi, M.A., Chen, X., Kumaraswamy, E., Cooch, N.S., Godwin, A.K., & Shiekhattar, R. (2003). Regulation of BRCC, a holoenzyme complex containing BRCA1 and BRCA2, by a signalosome-like subunit and its role in DNA repair. Mol Cell 12, 1087-1099.

Dong, Z., Zhong, Q., & Chen, P.L. (1999). The Nijmegen breakage syndrome protein is essential for Mre11 phosphorylation upon DNA damage. J Biol Chem 274, 19513-19516.

Dou, H., Huang, C., Singh, M., Carpenter, P.B., & Yeh, E.T. (2010). Regulation of DNA repair through deSUMOylation and SUMOylation of replication protein A complex. Mol Cell 39, 333-345.

Dray, E., Etchin, J., Wiese, C., Saro, D., Williams, G.J., Hammel, M., Yu, X., Galkin, V.E., Liu, D., Tsai, M.S., *et al.* (2010). Enhancement of RAD51 recombinase activity by the tumor suppressor PALB2. Nat Struct Mol Biol *17*, 1255-1259.

Gildemeister, O.S., Sage, J.M., & Knight, K.L. (2009). Cellular redistribution of Rad51 in response to DNA damage: novel role for Rad51C. J Biol Chem *284*, 31945-31952.

Golub, E.I., Gupta, R.C., Haaf, T., Wold, M.S., & Radding, C.M. (1998). Interaction of human rad51 recombination protein with single-stranded DNA binding protein, RPA. Nucleic Acids Res *26*, 5388-5393.

Golub, E.I., Kovalenko, O.V., Gupta, R.C., Ward, D.C., & Radding, C.M. (1997). Interaction of human recombination proteins Rad51 and Rad54. Nucleic Acids Res *25*, 4106-4110.

Haaf, T., Golub, E.I., Reddy, G., Radding, C.M., and Ward, D.C. (1995). Nuclear foci of mammalian Rad51 recombination protein in somatic cells after DNA damage and its localization in synaptonemal complexes. Proc Natl Acad Sci U S A *92*, 2298-2302.

Hay, R.T. (2005). SUMO: a history of modification. Mol Cell *18*, 1-12.

Henson, S.E., Tsai, S.C., Malone, C.S., Soghomonian, S.V., Ouyang, Y., Wall, R., Marahrens, Y., & Teitell, M.A. (2006). Pir51, a Rad51-interacting protein with high expression in aggressive lymphoma, controls mitomycin C sensitivity and prevents chromosomal breaks. Mutat Res *601*, 113-124.

Holt, J.T., Toole, W.P., Patel, V.R., Hwang, H., & Brown, E.T. (2008). Restoration of CAPAN-1 cells with functional BRCA2 provides insight into the DNA repair activity of individuals who are heterozygous for BRCA2 mutations. Cancer Genet Cytogenet *186*, 85-94.

Huen, M.S., Sy, S.M., & Chen, J. (2010). BRCA1 and its toolbox for the maintenance of genome integrity. Nat Rev Mol Cell Biol *11*, 138-148.

Jasin, M. (2002). Homologous repair of DNA damage and tumorigenesis: the BRCA connection. Oncogene *21*, 8981-8993.

Kagawa, W., Kurumizaka, H., Ishitani, R., Fukai, S., Nureki, O., Shibata, T., & Yokoyama, S. (2002). Crystal structure of the homologous-pairing domain from the human Rad52 recombinase in the undecameric form. Mol Cell *10*, 359-371.

Kitao, H., & Yuan, Z.M. (2002). Regulation of ionizing radiation-induced Rad52 nuclear foci formation by c-Abl-mediated phosphorylation. J Biol Chem *277*, 48944-48948.

Kjeldsen, F., Savitski, M.M., Nielsen, M.L., Shi, L., & Zubarev, R.A. (2007). On studying protein phosphorylation patterns using bottom-up LC-MS/MS: the case of human alpha-casein. Analyst *132*, 768-776.

Kovalenko, O.V., Golub, E.I., Bray-Ward, P., Ward, D.C., & Radding, C.M. (1997). A novel nucleic acid-binding protein that interacts with human rad51 recombinase. Nucleic Acids Res *25*, 4946-4953.

Kovalenko, O.V., Plug, A.W., Haaf, T., Gonda, D.K., Ashley, T., Ward, D.C., Radding, C.M., & Golub, E.I. (1996). Mammalian ubiquitin-conjugating enzyme Ubc9 interacts with Rad51 recombination protein and localizes in synaptonemal complexes. Proc Natl Acad Sci U S A *93*, 2958-2963.

Kruh, G.D., Perego, R., Miki, T., & Aaronson, S.A. (1990). The complete coding sequence of
    arg defines the Abelson subfamily of cytoplasmic tyrosine kinases. Proc Natl Acad
    Sci U S A 87, 5802-5806.
Kurumizaka, H., Aihara, H., Kagawa, W., Shibata, T., & Yokoyama, S. (1999). Human Rad51
    amino acid residues required for Rad52 binding. J Mol Biol 291, 537-548.
Li, W., Hesabi, B., Babbo, A., Pacione, C., Liu, J., Chen, D.J., Nickoloff, J.A., & Shen, Z. (2000).
    Regulation of double-strand break-induced mammalian homologous
    recombination by UBL1, a RAD51-interacting protein. Nucleic Acids Res 28, 1145-
    1153.
Li, X., Zhang, X.P., Solinger, J.A., Kiianitsa, K., Yu, X., Egelman, E.H., & Heyer, W.D. (2007).
    Rad51 and Rad54 ATPase activities are both required to modulate Rad51-dsDNA
    filament dynamics. Nucleic Acids Res 35, 4124-4140.
Li, Y., Shimizu, H., Xiang, S.L., Maru, Y., Takao, N., & Yamamoto, K. (2002). Arg tyrosine
    kinase is involved in homologous recombinational DNA repair. Biochem Biophys
    Res Commun 299, 697-702.
Marmorstein, L.Y., Ouchi, T., & Aaronson, S.A. (1998). The BRCA2 gene product
    functionally interacts with p53 and RAD51. Proc Natl Acad Sci U S A 95, 13869-
    13874.
Mazin, A.V., Alexeev, A.A., & Kowalczykowski, S.C. (2003). A novel function of Rad54
    protein. Stabilization of the Rad51 nucleoprotein filament. J Biol Chem 278, 14029-
    14036.
Mazin, A.V., Mazina, O.M., Bugreev, D.V., & Rossi, M.J. (2010). Rad54, the motor of
    homologous recombination. DNA Repair (Amst) 9, 286-302.
Mazina, O.M., & Mazin, A.V. (2004). Human Rad54 protein stimulates DNA strand
    exchange activity of hRad51 protein in the presence of Ca2+. J Biol Chem 279,
    52042-52051.
McIlwraith, M.J., Van Dyck, E., Masson, J.Y., Stasiak, A.Z., Stasiak, A., & West, S.C. (2000).
    Reconstitution of the strand invasion step of double-strand break repair using
    human Rad51 Rad52 and RPA proteins. J Mol Biol 304, 151-164.
Medova, M., Aebersold, D.M., & Zimmer, Y. MET inhibition in tumor cells by PHA665752
    impairs homologous recombination repair of DNA double strand breaks. Int J
    Cancer.
Mizuta, R., LaSalle, J.M., Cheng, H.L., Shinohara, A., Ogawa, H., Copeland, N., Jenkins,
    N.A., Lalande, M., & Alt, F.W. (1997). RAB22 and RAB163/mouse BRCA2: proteins
    that specifically interact with the RAD51 protein. Proc Natl Acad Sci U S A 94,
    6927-6932.
Mladenov, E., Anachkova, B., & Tsaneva, I. (2006). Sub-nuclear localization of Rad51 in
    response to DNA damage. Genes Cells 11, 513-524.
Modesti, M., Budzowska, M., Baldeyron, C., Demmers, J.A., Ghirlando, R., & Kanaar, R.
    (2007). RAD51AP1 is a structure-specific DNA binding protein that stimulates joint
    molecule formation during RAD51-mediated homologous recombination. Mol Cell
    28, 468-481.
Moschos, S.J., & Mo, Y.Y. (2006). Role of SUMO/Ubc9 in DNA damage repair and
    tumorigenesis. J Mol Histol 37, 309-319.

Niu, H., Wan, L., Busygina, V., Kwon, Y., Allen, J.A., Li, X., Kunz, R.C., Kubota, K., Wang, B., Sung, P., *et al.* (2009). Regulation of meiotic recombination via Mek1-mediated Rad54 phosphorylation. Mol Cell *36*, 393-404.

O'Donovan, P.J., & Livingston, D.M. (2010). BRCA1 and BRCA2: breast/ovarian cancer susceptibility gene products and participants in DNA double-strand break repair. Carcinogenesis *31*, 961-967.

Park, M.S., Ludwig, D.L., Stigger, E., & Lee, S.H. (1996). Physical interaction between human RAD52 and RPA is required for homologous recombination in mammalian cells. J Biol Chem *271*, 18996-19000.

Plate, I., Hallwyl, S.C., Shi, I., Krejci, L., Muller, C., Albertsen, L., Sung, P., & Mortensen, U.H. (2008). Interaction with RPA is necessary for Rad52 repair center formation and for its mediator activity. J Biol Chem *283*, 29077-29085.

Plo, I., Laulier, C., Gauthier, L., Lebrun, F., Calvo, F., & Lopez, B.S. (2008). AKT1 inhibits homologous recombination by inducing cytoplasmic retention of BRCA1 and RAD51. Cancer Res *68*, 9404-9412.

Popova, M., Shimizu, H., Yamamoto, K., Lebechec, M., Takahashi, M., & Fleury, F. (2009). Detection of c-Abl kinase-promoted phosphorylation of Rad51 by specific antibodies reveals that Y54 phosphorylation is dependent on that of Y315. FEBS Lett *583*, 1867-1872.

Raggiaschi, R., Gotta, S., & Terstappen, G.C. (2005). Phosphoproteome analysis. Biosci Rep *25*, 33-44.

Rao, V.A., Fan, A.M., Meng, L., Doe, C.F., North, P.S., Hickson, I.D., & Pommier, Y. (2005). Phosphorylation of BLM, dissociation from topoisomerase IIIalpha, and colocalization with gamma-H2AX after topoisomerase I-induced replication damage. Mol Cell Biol *25*, 8925-8937.

Raschle, M., Van Komen, S., Chi, P., Ellenberger, T., & Sung, P. (2004). Multiple interactions with the Rad51 recombinase govern the homologous recombination function of Rad54. J Biol Chem *279*, 51973-51980.

Saitoh, H., Pizzi, M.D., & Wang, J. (2002). Perturbation of SUMOlation enzyme Ubc9 by distinct domain within nucleoporin RanBP2/Nup358. J Biol Chem *277*, 4755-4763.

San Filippo, J., Sung, P., & Klein, H. (2008). Mechanism of eukaryotic homologous recombination. Annu Rev Biochem *77*, 229-257.

Shen, Z., Cloud, K.G., Chen, D.J., & Park, M.S. (1996). Specific interactions between the human RAD51 and RAD52 proteins. J Biol Chem *271*, 148-152.

Shi, W., Feng, Z., Zhang, J., Gonzalez-Suarez, I., Vanderwaal, R.P., Wu, X., Powell, S.N., Roti Roti, J.L., & Gonzalo, S. (2010). The role of RPA2 phosphorylation in homologous recombination in response to replication arrest. Carcinogenesis *31*, 994-1002.

Sigurdsson, S., Van Komen, S., Petukhova, G., & Sung, P. (2002). Homologous DNA pairing by human recombination factors Rad51 and Rad54. J Biol Chem *277*, 42790-42794.

Sleeth, K.M., Sorensen, C.S., Issaeva, N., Dziegielewski, J., Bartek, J., & Helleday, T. (2007). RPA mediates recombination repair during replication stress and is displaced from DNA by checkpoint signalling in human cells. J Mol Biol *373*, 38-47.

Posttranslational Modifications of Rad51 Protein and Its Direct Partners: Role and Effect
on Homologous Recombination – Mediated DNA Repair

181

Slupianek, A., Dasgupta, Y., Ren, S., Cramer, K., & Skorski, T. (2009). Targeting BCR/ABL-RAD51 Interaction to Prevent Unfaithful Homeologous Recombination Repair In 51st ASH Annual Meeting and Exposition.

Slupianek, A., Schmutte, C., Tombline, G., Nieborowska-Skorska, M., Hoser, G., Nowicki, M.O., Pierce, A.J., Fishel, R., & Skorski, T. (2001). BCR/ABL regulates mammalian RecA homologs, resulting in drug resistance. Mol Cell 8, 795-806.

Solinger, J.A., Kiianitsa, K., & Heyer, W.D. (2002). Rad54, a Swi2/Snf2-like recombinational repair protein, disassembles Rad51:dsDNA filaments. Mol Cell 10, 1175-1188.

Sorensen, C.S., Hansen, L.T., Dziegielewski, J., Syljuasen, R.G., Lundin, C., Bartek, J., & Helleday, T. (2005). The cell-cycle checkpoint kinase Chk1 is required for mammalian homologous recombination repair. Nat Cell Biol 7, 195-201.

Sugiyama, T., & Kowalczykowski, S.C. (2002). Rad52 protein associates with replication protein A (RPA)-single-stranded DNA to accelerate Rad51-mediated displacement of RPA and presynaptic complex formation. J Biol Chem 277, 31663-31672.

Symington, L.S. (2002). Role of RAD52 epistasis group genes in homologous recombination and double-strand break repair. Microbiol Mol Biol Rev 66, 630-670, table of contents.

Takata, M., Sasaki, M.S., Tachiiri, S., Fukushima, T., Sonoda, E., Schild, D., Thompson, L.H., & Takeda, S. (2001). Chromosome instability and defective recombinational repair in knockout mutants of the five Rad51 paralogs. Mol Cell Biol 21, 2858-2866.

Takizawa, Y., Kinebuchi, T., Kagawa, W., Yokoyama, S., Shibata, T., & Kurumizaka, H. (2004). Mutational analyses of the human Rad51-Tyr315 residue, a site for phosphorylation in leukaemia cells. Genes Cells 9, 781-790.

Tan, T.L., Essers, J., Citterio, E., Swagemakers, S.M., de Wit, J., Benson, F.E., Hoeijmakers, J.H., & Kanaar, R. (1999). Mouse Rad54 affects DNA conformation and DNA-damage-induced Rad51 foci formation. Curr Biol 9, 325-328.

Tripathi, V., Nagarjuna, T., & Sengupta, S. (2007). BLM helicase-dependent and -independent roles of 53BP1 during replication stress-mediated homologous recombination. J Cell Biol 178, 9-14.

Wold, M.S. (1997). Replication protein A: a heterotrimeric, single-stranded DNA-binding protein required for eukaryotic DNA metabolism. Annu Rev Biochem 66, 61-92.

Wolner, B., & Peterson, C.L. (2005). ATP-dependent and ATP-independent roles for the Rad54 chromatin remodeling enzyme during recombinational repair of a DNA double strand break. J Biol Chem 280, 10855-10860.

Wong, A.K., Pero, R., Ormonde, P.A., Tavtigian, S.V., & Bartel, P.L. (1997). RAD51 interacts with the evolutionarily conserved BRC motifs in the human breast cancer susceptibility gene brca2. J Biol Chem 272, 31941-31944.

Wu, L., & Hickson, I.D. (2003). The Bloom's syndrome helicase suppresses crossing over during homologous recombination. Nature 426, 870-874.

Wu, X., Yang, Z., Liu, Y., & Zou, Y. (2005). Preferential localization of hyperphosphorylated replication protein A to double-strand break repair and checkpoint complexes upon DNA damage. Biochem J 391, 473-480.

Yuan, S.S., Lee, S.Y., Chen, G., Song, M., Tomlinson, G.E., & Lee, E.Y. (1999). BRCA2 is required for ionizing radiation-induced assembly of Rad51 complex in vivo. Cancer Res 59, 3547-3551.

Yuan, Z.M., Huang, Y., Ishiko, T., Nakada, S., Utsugisawa, T., Kharbanda, S., Wang, R., Sung, P., Shinohara, A., Weichselbaum, R., & Kufe, D. (1998). Regulation of Rad51 function by c-Abl in response to DNA damage. J Biol Chem *273*, 3799-3802.

Zhong, Q., Chen, C.F., Li, S., Chen, Y., Wang, C.C., Xiao, J., Chen, P.L., Sharp, Z.D., & Lee, W.H. (1999). Association of BRCA1 with the hRad50-hMre11-p95 complex and the DNA damage response. Science *285*, 747-750.

11

Eidetic Analysis of the Premature Chromosome Condensation Process

Dorota Rybaczek

Department of Cytophysiology, University of Łódź
Poland

'Why does this written doe bound through these written woods?
(...) Perched on four slim legs borrowed from the truth,
She pricks up her ears beneath my fingertips (...)'
(Szymborska, 1993)

1. Introduction

An exact transfer of genetic information depends on the accuracy of mechanisms duplicating DNA molecules in the S-phase and the precise division sister chromosomes during mitosis. The regulation systems of these processes (checkpoints) not only control the activation course of the factors imposing different metabolic specificity on each of the cell cycle phases, but first of all – supervising the proper chronology of events – they condition the behavior of the structural and functional genome integrity. Checkpoints receive signals of all abnormalities or structural damages to DNA and in response evoke reactions inhibiting successive transitions through the cell cycle to enable the expression of specific genes and activation of DNA repair factors. One of the easily perceptible effects of disorders in this signaling system is the induction of premature chromosome condensation (PCC).

The present chapter is a review of the ways and mode of the induction of PCC. The term 'PCC' is inseparably associated with Johnson & Rao (1970) and their experiments on the premature mitosis induced by fusion of interphase and mitotic HeLa cells (G1/M, S/M and G2/M) which were originally carried out using Sendai virus. PCC process can be also induced by chemical signals. Drug-induced PCC provides the new knowledge that DNA replication is tightly coupled with the premature chromosome condensation and that the genome stability results first of all from the alternation of the S-phase and mitosis. The main objective of this review is to show that the PCC induction is possible from various subperiods of cell cycle. Moreover, it has been shown that there are cause-and-effect relationships between the chromosome structure defining 'PCC phenotype' and subperiods, e.g. of the S-phase, initiating the biosynthesis of 'early' or 'late' replicons. Attempts have been made to find answers to questions such as: How to force cells to break out of the rules being developed by Nature for billions of years? How – despite the interrupted, still unterminated process of genome replication – to force a cell to initiate its division? What mechanisms annihilate the subordination principle verified in the course of evolution: first create (DNA-duplicating S-phase) and then divide (mitosis – a stage of DNA condensation and formation of sister descendant nuclei)? Interference in the regulatory systems of cell cycle is not a simple matter. The gene pool, whose products participate in the creation of these systems, is constantly changing with time to continually form new systems and new interactions. Huge difficulties in the development of effective and selective methods that would arrest the proliferation of cancer cells result from their multiplicity and complication degree, as well as from the possibilities of starting the mechanisms of substitutive and biochemical emergency systems. Studies on the mechanisms inhibiting cell divisions seem to be the shortest way to reach the desired end. This chapter shows the usefulness of attempts to force divisions in cells, simultaneously taking into account the strategy of anticancer therapy.

Therefore the PCC phenomenon constitutes in reality not only a significant fundamental problem in the biology of cell cycle, but it is also an issue of paramount importance in view of practical applications. The radio- and chemotherapy methods used in the treatment of malignant diseases lead to extensive damages to DNA, arresting the replication process of genetic material. Despite this fact, the inhibition of cancer cell proliferation most often is of temporary character or it comprises only part of their population. Drug-induced PCC gives a novel tool to characterize the role of the chromosome instability in cancer development. In this chapter, an attempt is also made to explain the molecular base of PCC induction, for which the starting-point is the biochemical organization of the S-phase checkpoints that block mitosis initiation and the mechanisms which make it possible to suppress their restrictive interactions.

2. Discovery of the premature chromosome condensation (PCC)

The process of separate mitotic chromosome formation from chromatin of an interphase cell nucleus is associated with the construction of giant complexes or macromolecules this consequence being a specific expression of molecular morphogenesis. Simultaneously, it results from action of a complicated regulatory system, causing long, replicated DNA molecules to assume a form adapted to the biomechanical processes of mitosis. Control over these processes is provided on many planes of molecular chromatin organization, e.g. by the association of their components with the nuclear matrix, by specific phosphorylations and dephosphorylations conditioned by changes in the activity of protein kinases and phosphatases or by translocations of some molecules along the length of the fibrils of condensing chromatin or along the arms of existing chromosomes. The degree of packing achieved by chromatin throughout its domain organization before the G2→M transition falls short of the culminant metaphase condensation, the concomitant structural changes always resulting from biochemical modifications that proceed in the protein scaffold of the chromosome under formation.

Attentiveness to the integrity of genome determines the fundamental principles governing the regulation of cell cycle: the replication of each DNA molecule during S-phase can take place only once. The second condition involves the initiation of mitosis: this cannot begin before the complete termination of DNA replication. Control over the course of successive phases of cell cycle is extraordinarily precise and rigorous since initiation of the S-phase is restricted exclusively to unreplicated post-mitotic chromatin. Meanwhile, it is known that competence to initiate mitosis is not always conditioned by the replicated state of chromatin (this does not mean however that control over this process is not precise; simply, it results from the closely specified timing involved in setting-up factors liberating the activity of MPF [i.e. Cdk1 kinase and cyclin B complex or maturation/mitosis promoting factor] and its co-operation with aspects of the activators' and inhibitors' character).

Carefully designed experiments by Johnson & Rao (1970) have resulted in the first correct interpretation of the phenomenon of premature chromosome condensation (PCC). The term PCC appeared in the description of phenomena observed during the fusion of interphase and mitotic malignant HeLa cells. It has been shown that the PCC phenomenon is accompanied by disappearance of the nuclear envelope, chromatin condensation and the formation of mitotic spindle. Subsequent investigations have shown that the induction of PCC is inseparably connected with the activity of MPF complex. The experiments, involving fusion of interphase and mitotic cells have clearly shown that the cells in all the subperiods of interphase are characterized by capability to induce PCC. Thus: (i) if the fusion took place between a mitotic cell and an interphase cell in G1 phase, then PCC resulted in the formation of chromosomes consisting of a single chromatid only (univalent chromosomes); (ii) if the fusion took place between a mitotic cell and an interphase cell in S-phase, strongly fragmented chromosomes were formed in the PCC process, and their morphology constituted a specific reminder of the nucleoplasm organization in the period of the activity of the cell replication apparatus (a 'pulverized' appearance that consisted of univalent and bivalent chromosomes); (iii) if a mitotic cell and interphase cell in G2-phase participated in the combination, PCC resulted in the formation of chromosomes consisting of two chromatids apparently non-differentiated morphologically as compared to normal chromosomes, although probably less condensed (bivalent chromosomes). Despite the fact that the first observation of PCC was reported by Kato & Sandberg (1967) in virus mediated multinucleated fused cells of interphase and mitotic cells, it was Johnson & Rao who in 1970 properly defined the observed phenomenon as 'premature chromosome condensation' (PCC), and the condensed interphase chromatin as 'prematurely condensed chromosomes' (PCCs) (Gotoh & Durante, 2006). Generally, the PCC method facilitates the visualization of interphase chromatin as a condensed form of chromosome structure (Gotoh, 2007). Nowadays, the efficiency and scope of PCC induction have been proved by combining techniques.

Premature chromosome condensation became a method used to: (1) distinguish a cell cycle stage (Cadwell et al., 2011) and the Rabl-orientation of interphase chromosomes, e.g. G1-PCCs and G2-PCCs (Cremer et al., 1982); (2) investigations in chromosome dynamics, also that in interphase chromatin, chromosome replication studies and DNA repair analysis (Gotoh & Durante, 2006, as cited in Cornforth & Bedford, 1983; Hittelman & Pollard, 1982; Hittelman & Rao, 1974; Mullinger & Johnson, 1983; Schor et al., 1975); (3) perform mutagenic assay (Gotoh, 2009, as cited in Cornforth & Bedford, 1983; Durante et al., 1996); (4) chromosome instability analysis (Bezrookove et al., 2003); (5) prenatal diagnosis (Gotoh & Durante, 2006, as cited in Srebniak et al., 2005); (6) karyotyping of chromosomes (Kowalska et al., 2003); and (6) cytogenetic analysis of cancers (Darroudi et al., 2010).

2.1 Induction of PCC in historical and methodological terms

Virus-mediated PCC was first reported more than 40 years ago. In 1983, cell fusion could be achieved by means of polyethylene glycol (PEG-induced or chemically-mediated; Pantelias & Maillie, 1983). This allowed the external MPF to migrate from the inducing mitotic cell to the interphase recipient. A few years later it was possible to obtain chemically-induced PCC (drug-induced PCC): initially from synchronized cells and later still from each phase of the cell cycle (Gotoh & Durante, 2006, as cited in Schlegel & Pardee, 1986, Schlegel et al., 1990; Yamashita et al., 1990). Detailed data concerning PCC induction reported so far in the world literature are presented in Table 1.

The way of induction of PCC	References
A. Fusion-induced PCC	
Exploits the action of external MPF	
A1. Virus-induced fusion	
(a) UV inactivated Sendai virus-induced PCC	Kato & Sandberg, 1967
	The first observation of PCC
	Johnson & Rao, 1970
	The first correct interpretation of PCC
(b) PCC-type remodeling of the donor nucleus	Le Bourhis et al., 2010
after somatic cell nuclear transfer (SCNT)	
A2. Chemically-induced fusion	
(a) (PEG)-induced fusion	Pantelias & Maillie, 1983
	The first successfully applied polyethylene glycol (PEG)-induced fusion just before PCC induction
B. Drug-induced PCC	
Exploits the activation of endogenous MPF	
B1. With synchronization	
Cells had to be synchronized	
in G1- or S- or G2-phase	
before PCC induction	
B1.1. Induced by protein phosphatase inhibitors	
(a) Okadaic acid	Schlegel & Pardee, 1986
	The first one successfully applied in chemically-induced PCC in S phase cells
	Ghosh et al., 1998;
	Schlegel et al., 1990;
	Yamashita et al., 1990
(b) Sodium metavanadate	Ghosh et al., 1998;
	Rybaczek & Kowalewicz-Kulbat, 2011;
B1.2. Induced by protein kinase inhibitors	
(a) Caffeine	Schlegel & Pardee, 1986
	The first one successfully applied in chemically-induced PCC in S phase cells
	Schlegel et al., 1990;
	Yamashita et al., 1990;
	Nghiem et al., 2001
(b) Caffeine	Sen & Ghosh, 1998;
2-aminopurine	Rybaczek et al., 2008;
Staurosporine	Rybaczek & Kowalewicz-Kulbat, 2011;
6-dimethylaminopurine	Steinmann et al., 1991
B2. Without synchronization	
PCC induction occurs without synchronization	
(in any phase of cell cycle)	
B2.1. Induced by protein phosphatase inhibitors	
(a) Okadaic acid	Bialojan & Takai, 1988; Cohen et al., 1990; Gotoh et
Calyculin A	al., 1995; Gotoh & Tanno, 2007;
	Ishihara et al., 1989; Prasanna et al., 2000
B2.2. Adriamycin (Doxorubicin)	Hittelman & Rao, 1975
C. Spontaneous PCC (SPCC)	
(a) In the ontogenesis of generative cells and during the development of endosperm	cited by Tam & Schlegel, 1995
(b) During heat exposure	Mackey et al., 1988; Swanson et al., 1995;
(c) In normal and transformed mammal cells	Kovaleva et al., 2007
(d) After X-irradiation of HeLa cells	Ianzini & Mackey, 1997

Table 1. Data concerning PCC induction reported in the word literature

A specific change occurred when the preliminary cell-free cytoplasmic extracts (obtained by the centrifugation of *Xenopus* egg cells) were used *in vitro* to study the assembly of 'synthetic nuclei' involving the conversion of chromatin sperms into mitotic chromosomes. The resulting expansion of knowledge concerning the action of inhibitors and/or activators used for PCC induction has created new opportunities and perspectives for researchers (Prokhorova et al., 2003). In cells with disturbed DNA structure or those blocked during DNA biosynthesis, PCC-type processes can be induced by various chemical compounds such as: (i) inhibitors of protein kinases, e.g. 2-aminopurine (Herbig et al., 2004), caffeine (Wang et al., 1999, Gabrielli et al., 2007), staurosporine, 7-hydroxystaurosporine (UCN-01), CEP-3891 (Kohn et al., 2002; Syljuåsen et al., 2005), wortmannin (WORT) (Liu et al., 2007) and Gö6976 (Jia et al., 2008) as well as (ii) inhibitors of protein phosphatases, e.g. calyculin A (CalA), okadaic acid (OA) and sodium metavanadate (Van) (Hosseini & Mozdarani, 2004; Rybaczek & Kowalewicz-Kulbat, 2011). In order to induce the PCC phenomenon, an incubation with a strong inhibitor, e.g. of a given kinase or phosphatase or alternating incubation in two different inhibitors, of which the former slows down the course of S phase (e.g. hydroxyurea or aphidicolin) and the latter specifically influences the activity of selected kinases or phosphatases is a frequently used approach. Such substances restore the activity of protein kinases and control the phosphorylation of subordinate proteins in the regulatory pathways of cell cycle simultaneously creating conditions necessary for the initiation of prophase chromosome condensation, thereby fulfilling the role of the inductors of Cdk1-cyclin B complexes and realising mitotic phosphorylations (Sturgeon et al., 2008).

Disturbance of the efficiency of cell cycle checkpoints can result from the action of many factors followed by overriding or breakage of control over genome integrity and the course of various interphase subperiods and finally PCC induction. The premature mitosis induced in meristems of *V. faba* roots by prolonged incubation with a mixture of hydroxyurea and caffeine is characterized by strong differentiation between the morphological forms of chromosomes, which allows one to separate several different cell classes. The degree of chromosome fragmentation (number of sections lost in anaphase) probably determines the level of genetic material disintegration in cells blocked by hydroxyurea in S-phase. Such observations suggest that there is a relationship between the chromosome structure determining 'PCC phenotype' and S-phase subperiod (initiating the biosynthesis of 'early' or 'late' replicons), in which replication block has taken place. This assumes that the period of time elapsing between the beginning of PCC induction and the appearance of first cells with symptoms of premature mitosis depends on the number of replication units in which the biosynthesis processes of complementary DNA strands have been initiated but not terminated. The same mechanism may also explain the considerable differentiation between cells showing morphological features of premature mitosis (Figure 1A).

Modern models assume that phosphorylation of the N-terminal fragments of H3 histones constitutes a preliminary molecular signal which makes initiation of chromosome condensation possible by creating conditions for the assembly of other proteins directly engaged in the structural metamorphoses of chromatin (e.g. condesin complexes). Therefore, the post-translation modifications of H3 molecules are only one of the symptoms reflecting the action of complicated regulatory system which leads to the increase in chromatin packing during prophase chromosome condensation. This view is consistent with observations pointing to considerable asynchrony of the period of intensified H3 histone phosphorylation and the initial stages of mitotic condensation among various organisms, as well as to the phosphorylation of H3 molecules in plant cells taking place in condensed chromosomes. Meanwhile, S10 phosphorylation of H3 histone has turned out to be an excellent marker in phenotyping PCC from various subperiods of e.g. S-phase (Figure 1B).

The G2→M transition is a period of an increased sensitivity of cells also to the action of factors that directly do not lead DNA damage. For example, when cells entering the early stages of prophase are subjected to hypothermia, anoxia, osmotic shock or other stresses, their chromosomes are decondensed followed by a return to late G2 phase. This phenomenon usually is reversible following a period of stress adaptation or regression of mitosis (Mikhailov & Rieder, 2002). So, the question - what is the beginning of mitosis? – is not trivial, especially in the context of its irreversibility. It is known that almost always the very fact of initiation of S phase determines future mitosis. On the other hand, the fact of DNA replication involving a successful transition through the cell cycle checkpoints gives the process of genetic material segregation to two daughter cells the status of authority or even necessity. It helps when seeking the answer to the question 'what is the beginning of mitosis?' to recall the term 'antephase', historically associated with the paper by Furlough & Johnson published in 1951 (Pines & Rieder, 2001, as cited in Furlough & Johnson, 1951).

The term 'antephase' is used to describe the final stage of G2 phase directly preceding the first perceptible symptoms of prophase chromosome condensation. Cells in the middle stages of prophase (cf. neuroblasts of grasshopper, cells of newt or PtK_1 cell line) subjected to ionizing radiation gradually decondense the chromosomes under formation, their proteins are dephosphorylated and the course of cell cycle is arrested. The same cells irradiated during late stages of prophase, despite strong chromosome fragmentation, initiate further mitosis stages. A similar process – gradual chromosome decondensation – is also observed when the disassembly of microtubules by nocodazole takes place before cell transition through the middle stage of prophase. The action of nocodazole in later periods of prophase induces colcemid-mitoses – the division of chromosomes deprived of communication with the microtubular spindle apparatus leading to the formation of polyploid nuclei. Thus there is a specified point, i.e. the final period of 'antephase', after which a cell is unable to return to the interphase condition. In the cells of many animal species (vertebrates), 'the point of the last chance to return to G2 phase' occurs during the final stages of prophase, the border here between interphase and mitosis being set considerably later than usual (Pines & Rieder, 2001).

Thus, there are cells in which advanced stages of chromosome condensation are an easily recognizable indicator of G2 phase termination ('antephase'), whence they become convenient subjects for studying the transient G2→M period at a molecular level. The activity of kinase Cdk1-cyclin A complexes increases during the G2 phase and reaches its maximum at the moment of nuclear envelope decomposition. The microinjection of these complexes into G2 cells induces a violent chromosome condensation but inhibitors of Cdk1 kinase (such as p21[Waf1/Cip1]) block the transition to mitosis and then the early-prophase cells return to the interphase state. This is a period during which inactive Cdk1-cyclin B complexes remain in the area of cytoplasm (Furuno et al., 1999). Thus, it seems that preparation for the entry into mitosis taking place in vertebrate cells during 'antephase' is mainly controlled by Cdk1-cyclin A complex and not by the association of Cdk1 with cyclin B. An important role is also played by Plk1 kinase, which controls organization of the cell centrosomal apparatus and, while phosphorylating Cdc25 phosphatase, indirectly activates Cdk1-cyclin B complexes (Kumagai & Dunphy, 1996). Another protein kinase, Aurora B (animals) or its homologue in yeast cells (Ipl1) induces chromosome condensation via phosphorylation of S10 in H3 histone (Hsu et al., 2000). In many types of cells, changes occur at the level of chromosome condensation causing centrosomes to interact with a growing number of γ-tubulin molecules. Meanwhile, normal cells (and not the transformed ones) are subjected to the control mechanisms of G2 phase which, by the intervention of ATM/ATR kinases, monitor the condition of DNA structure and make it possible to arrest the cell cycle just before the initiation of mitosis.

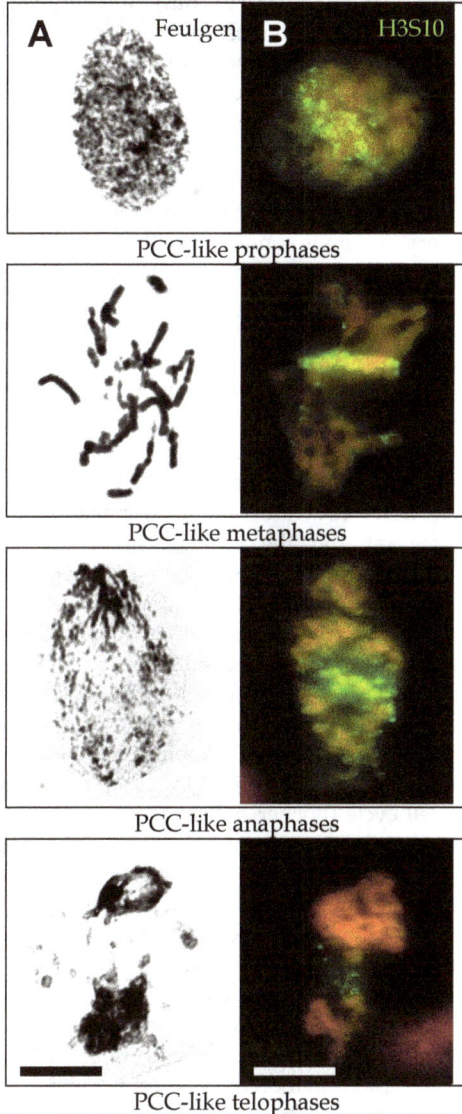

Fig. 1. A. Caffeine-induced PCC in Feulgen-stained root meristem cells of *Vicia faba*. The full array of aberrations included: chromosomal breaks and gaps, lost and lagging chromatids and chromosomes, acentric fragments and micronuclei. B. Immunofluorescence of phospho-H3 (S10; Cell Signaling) after caffeine-induced PCC in root meristem cells of *Vicia faba*. *Bar* 20 μm

3. Molecular origin of PCC induction

Hypotheses describing the mechanism of PCC induction are based on the results of genetic and biochemical analyses of yeast cells and human and animal cells *in vitro*. Among plants, the induction of premature chromosome condensation sometimes spontaneously occurs in the ontogenesis of generative cells and during the development of endosperm (cited by Tam & Schlegel, 1995). The special nature of these phenomena, the role played in them by phytohormonal factors and conditioning connected with the specificity of plant cell mitotic divisions have not yet been recognized. Neither have the organization of checkpoints blocking the initiation of mitosis, nor the mechanism that could overcome their restrictive interactions, been explained.

During the cell cycle, there are mechanisms governing transition 'to' and 'from' proliferation and coordinating of the complexes responsible for the successive stages of cell cycle transition. The first regulatory system is connected with the transition from G0 phase to cell cycle, which results in the transcription of appropriate genes, this being realization of the program responsible for leaving the resting condition. This leads to the commencement of DNA synthesis and consequently the initiation of mitotic division. The action of the second regulatory system involves arresting the course of G1 phase – before the start of DNA replication – or G2 phase – before the initiation of mitosis.

In cell cycles of the G1-S-G2-M type, the function of two main checkpoints is directly associated with the coordination of transient stages determining the maintenance of genetic identity. The G1 phase checkpoint monitors the metabolic conditions of cells in the G1 phase, the integrity of the nuclear DNA structure prior to the beginning of the synthesis of the complementary chromosome set and the level of necessary substrates, enzymes and replication factors. In plant cytology, it is known as the 'Principal Control Point 1' (PCP1), being the counterpart of the START point in yeast and of the Restriction Point in animal cells. The second checkpoint, PCP2, of a more conservative character, plays a similar role in G2 phase. Its functions are connected with the assessment of internal and external conditions of the cell environment during mitosis and cytokinesis. PCP2 controls if the replication of DNA is properly terminated and also monitors the integrity of chromatin structure before its mitotic condensation (Del Campo et al., 2003). The checkpoint pathways functioning in G1 and G2 phases of cell cycle constitute however only part of the complex system formed by checkpoint mechanisms of the whole cell cycle (Bucher & Britten, 2008). For example, the mechanisms determining S-phase initiation and development are functionally connected with the system of three S-phase checkpoints: (i) the intra-S-phase checkpoint or DSB-induced replication-independent intra-S-phase checkpoint which block mitosis initiation in the case of structural DNA damage; (ii) the replication checkpoint or replication-dependent intra-S-phase checkpoint which block mitosis initiation in the case of inhibition of DNA biosynthesis; and (iii) the replication-dependent S-M checkpoint which ensures that the G2 phase and mitosis can begin after complete genome replication (Bartek et al., 2004).

In many respects, the control mechanisms of G2 phase are similar to the regulatory systems that protect a cell against too rapid initiation of DNA replication. Still more analogies can be found by comparing both of them with the complex molecular systems of the S phase checkpoints. Admitting some simplification, it can be assumed that this continual repetition is not unusual: if monitoring of the DNA structure integrity is the basic aim of complicated biochemical mechanisms, the means serving that end do not have to be adapted for the realization of many targets, but just one – the most important. On the other hand, their molecular construction must take into account the specificity of the successive phases of the cell cycle.

The mentioned analogies have probably contributed to the fact that the best known function of checkpoints in G2 phase consists in blocking the initiation of mitosis in case of DNA damage. Detection of structural anomalies liberates the action of two complementary and partly convergent molecular pathways centred on ATM/ATR kinases and a common purpose – maintenance the Cdk1 kinase complexes with cyclin B inactive (Mikhailov & Rieder, 2002; Paulsen & Cimprich, 2007). In one pathway, ATM directly (or indirectly) activates part of the regulatory system, with a key factor p53. This protein induces the synthesis of inhibitors of cyclin-dependent kinases, e.g. p21$^{Waf1/Cip1}$. In the other pathway, independent of p53, ATM activates effector kinases Chk1 and Chk2 that, in turn, prevent the activation of Cdc25C phosphatase (sometimes by Plk1 kinase). Undoubtedly, we have here an excellent example of two coincident regulatory systems that mutually intensify the effectiveness of their individual interactions, while at the same time providing an example of 'economical' use of the same metabolic networks in three different cell cycle phases (Brown & Baltimore, 2007; Matthews et al., 2007; Peddibhotla et al., 2009).

Mitosis is the most dramatic and potentially most 'dangerous' cell cycle phase – involving the condensation of replicated chromosomes, their association with the kinetochores and finally the segregation of sister chromatids to opposite poles. Not only chromosomes replicated in S phase are divided (karyokinesis) but also all organelles (cytokinesis). Thus the checkpoints of mitosis monitor all the transient stages of this complicated process but primarily they control the condition of mitotic spindle by detecting abnormalities in its structural and functional organization (Cortez & Elledge, 2000). Chromatin condensation forming mitotic chromosomes under physiological conditions takes place only during mitosis. Condensins are the key elements in this process. Condensin I and II occur among vertebrates. Condensin I obtains access to chromosomes always after the nuclear envelope breakdown (NEBD) and then, in cooperation with condensin II responsible for the initial stages of condensation in prophase nuclei, promotes the assembly of metaphase chromosomes. Condensins are regulated by phosphorylation dependent on Cdk1, which was demonstrated in the studies utilizing *Xenopus* egg extract. Active MPF initiates the nuclear envelope breakdown to allow condensin I to acquire chromosomes. Hypothetically, it is assumed that Cdk1-cyclin A complex phosphorylates and activates condensin II to initiate the early stage of chromosome condensation inside the prophase nucleus. Next, directly after the nuclear envelope breakdown, Cdk1-cyclin B complex phosphorylates and activates condensin I (Hirano, 2005).

Cell cycle checkpoints act via the principle of establishing a cause-effect relationship between separate biochemical processes (Hartwell & Weinert, 1989) involving feedback loops (Elledge, 1996). The term 'checkpoint' refers to a definite subset of internal and external regulatory mechanisms that link further processes to the realization of earlier ones. At the same time, one has to remember to always take into account the existence of hysteresis in the molecular interaction of the control network in the cell cycle system. Hysteresis means that it takes more to push a system from point A to point B than it does to keep the system at stage B (Sha et al., 2003; Solomon, 2003). Generally, there are two classes of regulatory systems in the cell cycle: (i) intrinsic systems of a constitutive character sorting out the events directly connected with the cell cycle, and (ii) extrinsic systems that are revealed under the influence of inducing factors and are engaged only when DNA damage is detected (Elledge, 1996).

Each of the cell cycle checkpoints comprises three essential parts: (i) capability of sensing that a cell cycle event is aberrant or incomplete, (ii) means by which this signal is transmitted, and finally, (iii) effectors that delay or block the cell cycle transitions until the problem is resolved. The position of arrest within the cell cycle varies depending on the phase during which the damage is sensed. Since the main role of all these checkpoints is to make a decision whether or not the cell division cycle can be continued, their particular elements deserve special attention as promising targets for pharmacological treatment of cancer (Deckert et al., 2009). The action of checkpoints of cell cycle crucially depends on the effectiveness of the system transmitting signals released by the cell sensory apparatus. The activation of mitotic protein kinases (M-Cdk) is then blocked which makes it possible to effect repairs or to terminate DNA replication or apoptosis induction (Khanna & Jackson, 2001; Zhou & Elledge, 2000).

The PCC phenomenon results from the overriding of S-M checkpoint. It blocks the ability of cells to make mitotic divisions after extensive DNA damage or under conditions of unfinished replication. DNA damage activates in the first place sensor kinases from the PIKK family, ATM and/or ATR which subsequently inhibit the formation of active MPFs by the phosphorylation of Chk2 and Chk1 kinases, which blocks the onset of mitosis. The initiated cascade of signals simultaneously activates repair factors including DNA-PK kinase which is essential for the repair of DNA suffering NHEJ-type damage. Blocking of the function of sensor ATM and/or ATR kinases can bring about avoidance of the restrictive interactions of S phase checkpoints causing premature mitosis (Block et al., 2004).

Knowledge of PCC mechanisms and S-M checkpoint action goes hand in hand with studies of malignant diseases and eagerness to maximize the beneficial effects of radio- and chemotherapy (Erenpreisa & Cragg, 2007). Despite evidence that overexpression of Cdk1-cyclin B complexes can promote PCC, it is not clear if these complexes initiate PCC in human cells. Studies of the import of cyclin B1 in human beings and in asteroid oocytes have shown that it is imported within the period of last several minutes of prophase, just after the initiation of chromosome condensation. These observations suggest that normal chromosome condensation is not initiated by Cdk1-cyclin B1 complexes. On the other hand, there is evidence indicating that the chromosome condensation and other phenomena occurring in the early prophase are initiated by Cdk2-cyclin A complexes present in this period within the cell nucleus. The initiation of cytoplasmic phenomena of mitotic character such as reorganization of Golgi apparatus and microtubular changes probably requires no import of Cdk1-cyclin B1 complexes into the nucleus (Takizawa & Morgan, 2000). On the other hand, the initiation of mitotic division occurs once Cdc25 phosphatase has dephosphorylated phosphate groups (both Y15 and T14) from the area of exposed pocket binding ATP within kinase p34^{cdc2}. Thus Cdc25 phosphatase is the activator of mitosis.

Dysfunctions of S phase checkpoints also occur in mutated cells (Krause et al., 2001). Deletions of *wee1* and *mik1* genes cause disappearance of proteins transmitting signals about DNA structure damage or blocked replication. Such cells initiate mitosis, but underreplications result in broken chromosomes being lost in the central spindle zone. Similar results follow from changes in the activity balance of protein kinases and phosphatases caused by overexpression of *cdc25* genes. Disappearance of intra-S-phase checkpoint function, caused e.g. by the lack (or mutation) of *atr*, also brings about the initiation of mitosis by cells that contain partly replicated genetic material (as opposed to normal cells in which replication forks activate ATR if they meet a defect impeding the biosynthesis of DNA which leads to the activation of S phase checkpoints and cell cycle inhibition). In the embryonic evolution of *Drosophila melanogaster*, grapes (*grp*) – whose product is a homolog of Chk1 kinase – is one of the genes of checkpoints functioning in this period of morphogenesis. Grapes mutants show a shortened course of interphase, a defective chromosome condensation and delayed metaphase initiation (Yu et al., 2000). In this case, premature mitosis is caused firstly by overriding of function of S phase checkpoints (then chromosome condensation is not dependent on the termination of S phase) and secondly, by the moment of initiation of chromosome condensation (ICC) occurring with no delay. Thus, only the period between ICC and metaphase is shortened, which seems to be the direct cause of incomplete chromosome condensation. To sum up, in *D. melanogaster* the delay in *grp* embryo entry into metaphase is caused by chromatin condensation defects rather than by partial DNA replication (Royou et al., 2005).

The facts presented above that barely outline extensive problems connected with the cellular control mechanisms show a huge complexity of both stimulating and inhibiting biochemical systems. The associated regulatory network comprises processes that activate the expression of genes at the moment desired for the cell and block the course of chemical reactions when their products could accumulate in excessive amount or prematurely encourage cell cycle transitions. It seems that all eukaryotic cells are equipped with sensory factors, signal transmitting systems and effector factors. The significance of these consists in the fact that they make DNA replication and chromosome division possible without endangering the whole information contained in them which is indispensable for the organism development and maintenance of species continuity.

4. PCC and DNA damages

The successive phases of prematurely initiated mitosis follow an aberrated course. The loss of relatively large chromosome sections suggests that disturbances of post-replication repair processes in the G2 (G2-PCC) are responsible for this. On the other hand, much higher disintegration of genetic material in some chromosomes leads to the conclusion that this is symptomatic of mitosis initiated from cell subpopulations that have not yet finished the DNA replication process (S-PCC; Rybaczek et al., 2008). Some influence on the degree of DNA fragmentation can be exerted by chemical compounds used in studies of PCC induction, e.g. caffeine which additionally intensifies chromosome destruction during their individualization. It is certain that the losses or breaks in the chromosome continuity: (i) either illustrate the unreplicated areas of genome; (ii) or result from physical stresses created during the mitotic condensation and segregation of chromosomes; (iii) or originate from relative fragility of single-stranded DNA sections generated during retardation of replication forks under the conditions of nucleotide triphosphate deficiency (i.e. after the use of replication inhibitors, e.g. hydroxyurea) (Cimprich, 2003; El Achkar et al., 2005).

The appearance of double strand breaks (DSBs) in DNA molecules is connected with the formation of immunofluorescence foci associated with the phosphorylated form of H2AX histones at S139 (DSB marker). The family of lysine-like H2A histones includes three subfamilies of proteins (H2A1-H2A2, H2AZ and H2AX). In mammalian cells, H2AX amounts to about 2 to 25% (probably almost 100% in *S. cerevisiae*) (Rogakou et al., 1998). The C-terminal motif of AQ(D/E)(I/L/Y) is the sequence distinguishing H2AX among H2A histones, while S139, ahead of it, is the site of γ-phosphorylation. Formation of γ-H2AX molecules resulting from the exposure of mammalian cells (or living mice) to the action of ionizing radiation (in sublethal or lethal doses) is an extremely violent process (Kurose et al.,

2006). Half of the γ-H2AX histones appears after a 1-minute period of irradiation, while the maximal level is reached within 3 to 10 minutes of exposure (per 1 Gy of radiation about 1% of H2AX histone molecules undergoes γ-phosphorylation, which corresponds to about 2×10^6 bp DNA in the area of the double-strand break of its structure) (Paull et al., 2000; Rogakou et al., 1998). It is assumed that each group of these molecules indicates a single DSB region, therefore the formation of H2AX foci phosphorylated at S139 constitutes a sensitive test revealing the presence of structural genome damage (Friesner et al., 2005). Literature reports in recent years show the occurrence of fluorescence foci of H2AX molecules within the area of S139 during PCC induction (Huang et al., 2006; Rybaczek et al., 2007; Rybaczek & Kowalewicz-Kulbat, 2011). On the other hand, Stevens and his co-workers (2010) clearly indicate that PCC is γ-H2AX negative, and that γ-H2AX phosphorylation is only a hallmark of a chromosome fragmentation phenomenon. However, simultaneously they define PCC as occurring in interphase cells exposed to active MPF vs that occurring during mitosis chromosome fragmentation. According to other investigations, however, the actual definition of drug-induced PCC involves both the described phenomena, i.e. PCC and chromosome fragmentation (Riesterer et al., 2009; Rybaczek et al., 2008; Terzoudi et al., 2010).

To protect against disturbances during DNA biosynthesis, cells have developed a network of biochemical reactions known as DNA-replication-stress-response. The basic strategy of this response is retardation of processes, whose continuation would result, among other things, in the transfer of affected DNA molecules to a new cell generation. Therefore under the conditions of replication stress, the DNA biosynthesis rate is slowed down and onset of mitosis is - most frequently - completely blocked. This continues until the expression of specific genes and activation of repair factors (Deckert et al., 2009; Mosesso et al., 2010). Each structural disturbance (e.g. DSB) causes the rate of production of replication forks to slow down. Additionally, any limitation of the replication apparatus effectiveness (resulting, e.g. from the deficiency of nucleotide triphosphate or polymerase dysfunction) can facilitate DNA damage. In such a situation, the action of checkpoint sensory factors releases a cascade of signals supplied to various effector proteins through intermediary elements.

The detection of double-strand breaks in DNA molecules activates the biochemical pathway in which ATM kinase is the superior element. The processes proceeding with its contribution are triggered in all the phases of cell cycle, while the factors participating here may also assist the second pathway going in parallel. ATR is also activated by damage resulting in disruption of continuity in both DNA strands, but here the induction process occurs more slowly (Scott & Pandita, 2006). The pathway subordinated to ATR kinase is specialized first of all in reacting to disturbance of the replication fork function. These disturbances can results from endogenous interactions, chemotherapy or experimental procedures leading to the inhibition or disturbance of replication processes caused by hydroxyurea (HU) or ultraviolet radiation (UV). In both biochemical pathways, Cdc25 phosphatase is the target substrate.

Mec1 and Tel1 kinases in *S. cerevisiae* and Rad3 and Tel1 kinases in humans are homologues of both conservative signaling ATM and ATR kinases (Garber et al., 2005). Their substrates, kinases of the CHK family, are subordinate factors activated by ATM/ATR/Mec1/Rad3. Such groups jointly form the central pathway module of response to replication stress, which both records incoming information on the DNA condition and sends signals to replication forks. In human and *Xenopus* cells, ATR kinase is indispensable in the phosphorylation of Chk1 kinase and it occurs in stable combination with ATR-interacting protein (ATRIP), there complexes being concentrated in those areas of cell nucleus showing DNA damage. The activator of ATR-ATRIP complexes in vertebrate cells consists of TopBP1 protein. Investigations utilizing cytoplasmic extracts of *Xenopus* egg cells have also shown that the association of ATR with chromatin takes place during the period of DNA replication, whereas it disappears after the termination of replication (Freire et al., 2006; Harper & Elledge, 2007).

It is uncertain whether disturbances of DNA replication are detected by means of only one, universal, sensory mechanism. Maybe, regardless of the type of the factor blocking S phase, the DNA structure generated by replication forks is identical at the damage site. However, one cannot exclude the possibility that each type of the DNA structure disturbance exerts different, specific influence on replication forks and that the factors associated with particular types of damage are also different.

5. Consequences of PCC induction

Literature shows that proper functioning of all multicellular organisms, including *Homo sapiens,* depends not only on their ability to produce new cells but also on the ability of each cell to annihilate itself when it becomes unwanted or damaged. This takes place also when the control mechanisms of the cell cycle have been overridden during a simultaneous strong and/or long-lasting action of stress stimulus. Thus, among cells induced to enter premature, unauthorized division, there are also those that choose the apoptosis pathway (Sahu et al., 2009). Therefore premature mitoses/PCC are described as a mitotic catastrophe, abortive or suicidal. During the induction of apoptosis – following PCC induction – chromatin undergoes drastic changes: previously usually dispersed, it suddenly begins to condense into one or more aggregates in the vicinity of a nuclear membrane (Figure 2A). Changes connected with the initial phase of apoptosis also involve formation of intranuclear membranous structures (sometimes strongly developed and multi-layered) adhering to the nuclear envelope (Figure 2B-D).

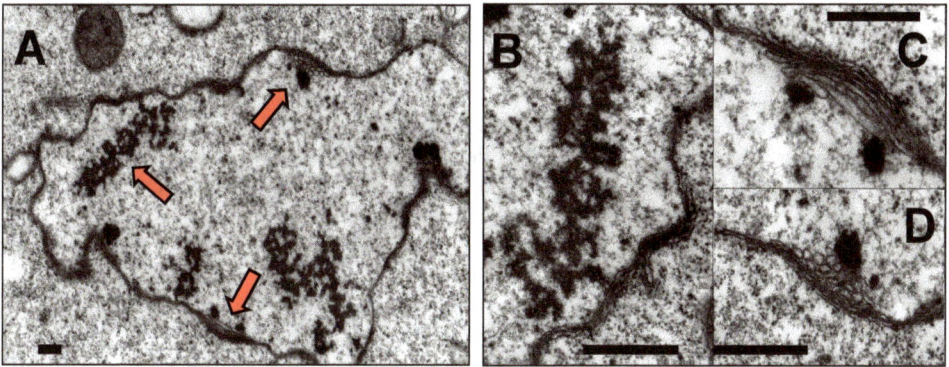

Fig. 2. Ultrastructure of an apoptotic cell of *V. faba* root meristem (after PCC induction caused by 5 mM caffeine). Selected fragments of the above micro-photography (marked with arrows) at magnification. *Bar* 1μm

Among unicellular organisms, irregularities in the organization of cell cycle control systems result in a decreased reproduction potential, while among multicellular mechanisms they cause uncontrolled proliferation, cancer development and genetic disease transfer (Hartwell & Weinert, 1989; Russell, 1998).

6. Conclusion

What are the practical implications and prospects of PCC induction? Could PCC induction serve as a novel anti-cancer approach? Undoubtedly, the phenomenon of premature mitosis is an essential characteristic of cell biology, therefore an important issue in respect of potential medical applications. It is so because, according to many researchers, the chemotherapy commonly used for the treatment of malignancy leads to extensive DNA damage, whereas PCC induction (resulting from the stimulation of biochemical mechanisms overriding the action of S-M checkpoint) can intensify therapeutic effects. Recently, drug-induced PCC was optimized to assist analysis of the behavior of cancer cells with minimal side effects. However, PCC will also contribute to the understanding of normal cellular processes.

7. Acknowledgement

The author expresses her cordial thanks to Professor Joanna Deckert for her support and inspiration, to Anna Pastucha for her assistance with immunocytochemical assays employing anti-H3S10, and to Małgorzata Fronczak M.Sc. for her help in preparing this manuscript in English. This work was funded by the Ministry of Science and Higher Education, grant N N303 355935 (contract no. 3559/B/P01/2008/25).

8. References

Bartek, J., Lukas, C. & Lukas, J. (2004) Checking on DNA damage in S phase. *Nat Rev Mol Cell Biol* 5, 792-804.

Bezrookove, V., Smits, R., Moeslein, G., Fodde, R., Tanke, H.J., Raap, A.K. & Darroudi, F. (2003) Premature chromosome condensation revisited: a novel chemical approach permits efficient cytogenetic analysis of cancers. *Genes Chromosomes Cancer* 38, 177-186.

Bialojan, C. & Takai, A. (1988) Inhibitory effect of a marine-sponge toxin, okadaic acid, on protein phosphatases. Specificity and kinetics. *Biochem J* 256, 283-290.

Block, W.D., Merkle, D., Meek, K. & Lees-Miller, S.P. (2004) Selective inhibition of the DNA-dependent protein kinase (DNA-PK) by the radiosensitizing agent caffeine. *Nucleic Acids Res* 32, 1967-1972.

Brown, E.J. & Baltimore, D. (2007) Essential and dispensable roles of ATR in cell cycle arrest and genome maintenance. *Genes Dev* 17, 615-628.

Bucher, N. & Britten, C.D. (2008) G2 treatment abrogation and checkpoint kinase-I targeting in the treatment of cancer. *Br J Cancer* 98, 532-528.

Cadwell, K.K., Curwen, G.B., Tawn, E.J., Winther, J.F., Boice, J.D. Jr. (2011) G2 checkpoint control and G2 chromosomal radiosensitivity in cancer survivors and their families. *Mutagenesis* 26, 291-294.

Cimprich, K.A. (2003) Fragile sites: Breaking up over a slowdown. *Curr Biol* 13, R231-R233.

Cohen, P., Holmes, C.F. & Tsukitani, Y. (1990) Okadaic acid: A new probe for the study of cellular regulation. *Trends Biochem Sci* 15, 98-102.

Cortez, D. & Elledge, S.J. (2000) Conducting the mitotic symphony. *Nature* 406, 354-356.

Cremer, T., Cremer, C., Baumann, H., Luedtke, E-K., Sperling, K., Teuber, V. & Zorn, C. (1982) Rabl's model of the interphase chromosome arrangement tested in Chinese hamster cells by premature chromosome condensation and laser-UV-microbeam experiments. *Hum Genet* 60, 46-56.

Darroudi, F., Bergs, J.W.J., Bezrookove, V., Buist, M.R., Stalpers, L.J. & Franken, N.A.P. (2010) PCC and COBRA-FISH a new tool to characterize primary cervical carcinomas: To asses hall-marks and stage specificity. *Cancer Lett* 287, 67-74.

Deckert, J., Pawlak, S. & Rybaczek, D. (2009) The nucleus as a 'headquarters' and target in plant cell stress reactions, In: *Compartmentation of Responses to Stresses in Higher Plants, True or False*, Waldemar Maksymiec, pp. 61-90, Transworld Research Network, ISBN: 978-81-7895-422-6, Kerala, India.

Del Campo, A., Samaniego, R., Giménez-Abián, J.F., Giménez-Martín, G., López-Sáez, J.F., Moreno Díaz de la Espina, S. & De la Torre C. (2003) G2 checkpoint targets late replicating DNA. *Biology of the Cell* 95, 521-526.

El Achkar, E., Gerbault-Seureau, M., Muleris, M., Dutrillaux, B. & Debatisse, M. (2005) Premature condensation induces breaks at the interface of early and late replicating chromosome bands bearing common fragile sites. *Proc Natl Acad Sci USA* 102, 18069-18074.

Elledge, S.J. (1996) Cell cycle checkpoint: preventing an identity crisis. *Science* 274, 1664-1672.

Erenpreisa, J. & Cragg, M.S. (2007) Cancer: A matter of life cycle? *Cell Biol Int* 31, 1507-1510.

Freire, R., van Vugt, M.A.T.M., Mamely, I. & Medema R.H. (2006) Claspin. Timing the cell cycle arrest when the genome is damaged. *Cell Cycle* 5, 2831-2834.

Friesner, J.D., Liu, B., Culligan, K. & Britt, A.B. (2005) Ionizing radiation-dependent γ-H2AX focus formation requires ataxia telangiectasia mutated and ataxia telangiectasia mutated and Rad3-related. *Mol Biol Cell* 16, 2566-2576.

Furuno, N., den Elzen, N. & Pines, J. (1999) Human cyclin A is required for mitosis until mid-prophase. *J Cell Biol* 147, 295-306.

Gabrielli, B., Chau, Y.Q., Giles, N., Harding, A., Stevens, F. & Beamish, H. (2007) Caffeine promotes apoptosis in mitotic spindle checkpoint-arrested cells. *J Biol Chem* 10, 6954-6964.

Garber, P.M., Vidanes, G.M. & Toczyski, D.P. (2005) Damage in transition, *Trends Biochem Sci* 30, 63-66.

Ghosh, S., Paweletz, N. & Schroeter, D. (1998) Cdc2-independent induction of premature mitosis by okadaic acid in HeLa cells. *Exp Cell Res* 242, 1-9.

Gotoh, E. & Durante, M. (2006) Chromosome condensation outside of mitosis: mechanisms and new tools. *J Cell Physiol* 209, 297-304.

Gotoh, E. & Tanno, Y. (2007) Simple biodosimetry method for cases of high-dose radiation exposure using the ratio of the longest/shortest length of Giemsa-stained drug-induced prematurely condensed chromosomes (PCC). *Int J Radiat Biol* 81, 379-385.

Gotoh, E. (2007) Visualizing the dynamics of chromosome structure formation coupled with DNA replication. *Chromosoma* 116, 453-462.

Gotoh, E. (2009) Drug-induced premature chromosome condensation (PCC) protocols : cytogenetic approaches in mitotic chromosome and interphase chromatin, In : *Chromatin Protocols: Second Edition, series: Methods in Molecular Biology,* Srikumar P. Chellappan, 523, 83-92, Humana Press, 523, 83-92.

Gotoh, E., Asakawa, Y. & Kosaka, H. (1995) Inhibition of protein serine/threonine phosphatases directly induces premature chromosome condensation in mammalian somatic cells. *Biomed Res* 16, 63-68.

Harper, J.W. & Elledge, S.J. (2007) The DNA damage response: ten years after. *Mol Cell* 28, 739-745

Hartwell, L.H. & Weinert, T.A. (1989) Checkpoints: controls that ensure the order of cell cycle events. *Science* 246, 629-634.

Herbig, U., Jobling, W.A., Chen, B.P.C., Chen, D.J. & Sedivy, J.M. (2004) Telomere shortening triggers senescence of human cells through a pathway involving ATM, p53, and p21[CIP1], but not p16[INK4a]. *Mol Cell* 4, 501-513.

Hirano, T. (2005) Condensins: organizing and segregating the genome. *Curr Biol* 15, 265-275.

Hittelman, W.N. & Rao, P.N. (1975) The nature of adriamycin-induced cytotoxicity in Chinese hamster cells as revealed by premature chromosome condensation. *Cancer Res* 35, 27-35.

Hosseini, S. & Mozdarani, H. (2004) Induction of premature chromosome condensation (PCC) by calyculin A for biodosimetry. *Iran J Radiat Res* 1, 1-6.

Hsu, J-Y., Sun, Z-W., Li, X., Reuben, M., Tatchell, K., Bishop, D.K., Grushcow, J.M., Brame, C.J., Caldwell, J.A. & Hunt, D.F. (2000) Mitotic phosphorylation of histone H3 is governed by Ipl1/aurora kinase and glc7/PP1 phosphatase in budding yeast and nematodes. *Cell* 102, 279-291.

Huang, X., Kurose, A., Tanaka, T., Traganos, F., Dai, W. & Darzynkiewicz, Z. (2006) Sequential phosphorylation of *Ser*-10 on histone H3 and *Ser*-139 on histone H2AX and ATM activation during premature chromosome condensation: relationship to cell-cycle phase and apoptosis. *Cytometry Part A* 69A, 222-229.

Ianzini, F. & Mackey, M.A. (1997) Spontaneous premature chromosome condensation and mitotic catastrophe following irradiation of HeLa S3 cells. *Int J Radiat Biol* 72, 409-421.

Ishihara, H., Martin, B.L., Brautigan, D.L., Karaki, H., Ozaki, H., Kato, Y., Fusetani, N., Watabe, S., Hashimoto, K., Uemura, D., et al. (1989) Calyculin A and okadaic acid: Inhibitors of protein phosphatase activity. *Biochem Biophys Res Commun* 159, 871-877.

Jia, X.Z., Yang, S.Y., Zhou, J., Li, S.L., Ni, J.H., An, G.S. & Jia, H.T. (2008) Inhibition of CHK1 kinase by Gö6976 converts 8-chloro-adenosine-induced G2/M arrest into S arrest in human myelocytic leukemia K562 cells. *Biochem Pharmacol* 5, 770-780.

Johnson, R.T. & Rao, P.N. (1970) Mammalian cell fusion: Induction of premature chromosome condensation in interphase nuclei. *Nature* 226, 717-722.

Kato, H. & Sandberg, A. (1967) Chromosome pulverization in human binucleate cells following colcemid treatment. *J Cell Biol* 34, 35-46.

Khanna, K.K. & Jackson, S.P. (2001) DNA double-strand breaks: signaling, repair and the cancer connection. *Nat Genet* 27, 247-254.

Kohn, E.A., Ruth, N.D., Brown, M.K., Livingstone, M. & Eastman, A. (2002) Abrogation of the S phase damage checkpoint results in S phase progression or premature mitosis depending on the concentration of 7-hydroxystaurosporine and the kinetics of Cdc25C activation. *J Biol Chem* 277, 26553-26564.

Kovaleva, O.A., Glazko, T.T., Kochubey, T.P., Lukash, L.L. & Kudryavets, Y.I. (2007) Spontaneous premature condensation of chromosomes in normal and transformed mammal cells. *Exp Oncol* 29, 18-22.

Kowalska, A., Srebniak, M., Wawrzkiewicz, A. & Kaminski, K. (2003) The influence of calyculin A on lymphocytes in vitro. *J Appl Genet* 44, 413-418.

Krause, S.A., Loupert, M.-L., Vass. S., Schoefelder, S., Harrison, S. & Heck, M.M.S. (2001) Loss of cell cycle checkpoint control in *Drosophila* RFC4 mutants. *Mol Cell Biol* 21, 5156-5168.

Kumagai, A. & Dunphy, W.G. (1996) Purification and molecular cloning of Plx1, a Cdc25-regulatory kinase from *Xenopus* egg extracts. *Science* 273, 1377-1380.

Kurose, A., Tanaka, T., Huang, X., Traganos, F. & Darzynkiewicz, Z. (2006) Synchronization in the cell cycle by inhibitors of DNA replication induces histone H2AX phosphorylation : an induction of DNA damage. *Cell Prolif* 39, 231-240.

Le Bourhis, D., Beaujean, N., Ruffini, S., Vignon, X. & Gall, L. (2010) Nuclear remodeling in bovine somatic cell nuclear transfer embryos using MG132-treated recipient oocytes. *Cell Reprogram* 12, 729-738.

Liu, Y., Jiang, N., Wu, J., Dai, W. & Rosenblum, J.S. (2007) Polo-like kinases inhibited by wortmannin: labeling site and downstream effects. *J Biol Chem* 4, 2505-2511.

Mackey, M.A., Morgan, W.F. & Dewey, W.C. (1988) Nuclear fragmentation and premature chromosome condensation induced by heat shock in S-phase Chinese hamster ovary cells. *Cancer Res* 48, 6478-6483.

Matthews, D.J., Yakes, F.M., Chen, J., Tadano, M., Bornheim, L., Clary, D.O., Tai, A., Wagner, J.M., Miller, N., Kim, Y.D., Robertson, S., Murray, L. & Karnitz, L.M. (2007) Pharmacological abrogation of S-phase checkpoint enhances the anti-tumor activity of gemcitabine in vivo. *Cell Cycle* 6, 104-110.

Mikhailov, A. & Rieder, C.L. (2002) Cell cycle: stressed out of mitosis. *Curr Biol* 12, R331-R333.

Mosesso, P., Palitti, F., Pepe, G., Pinero, J., Bellacima, R., Ahnstrom, G. & Natarajan, A.T. (2010) Relationship between chromatin structure, DNA damage and repair following X-irradiation of human leukocytes. *Mutat Res* 701, 86-91.

Nghiem, P., Park, P.K., Kim, Y-S, Vaziri, C. & Schreiber, S.L. (2001) ATR inhibition selectively sensitizes G_1 checkpoint-deficient cells to lethal premature chromatin condensation. *Proc Natl Acad Sci USA* 98, 9092-9097.

Pantelias, G.E. & Maillie, H.D. (1983) A simple method for premature chromosome condensation induction in primary human and rodent cells using polyethylene glycol. *Somatic Cell Genet* 9, 533-547.

Paull, T.T., Rogakou, E.P., Yamazaki, V., Kirchgessner, C.U., Gellert, M. & Bonner, W.M. (2000) A critical role for histone H2AX in recruitment of repair factors to nuclear foci after DNA damage. *Curr Biol* 10, 886-895.

Paulsen, R.D. & Cimprich, K.A. (2007) The ATR pathway: Fine-tuning the fork. *DNA Repair* 6, 953-966.

Peddibhotla, S., Lam, M.H., Gonzalez-Rimbau , M. & Rosen J.M. (2009) The DNA-damage effector checkpoint kinase 1 is essential for chromosome segregation and cytokinesis. *Proc Natl Acad Sci USA* 106, 5159-5164.

Pines, J. & Rieder, C.L. (2001) Re-staging mitosis: a contemporary view of mitotic progression. *Nat Cell Biol* 3, E3-E6.

Prasanna, P.G.S., Escalada, N.D. & Blakely, W.F. (2000) Induction of premature chromosome condensation by a phosphatase inhibitor and a protein kinase in unstimulated human peripheral blood lymphocytes: a simple and rapid technique to study chromosome aberrations using specific whole-chromosome DNA hybridization probes for biological dosimetry. *Mutation Res* 466, 131-141.

Prokhorova, T.A., Mowrer, K., Gilbert, C.H. & Walter, J.C. (2003) DNA replication of mitotic chromatin in Xenopus egg extracts. *Proc Natl Acad Sci USA* 100, 13241-13246.

Riesterer, O., Matsumoto, F., Wang, L., Pickett, J., Molkentine, D., Giri, U., Milas, L. & Raju, U. (2009) A novel Chk inhibitor, XL-844, increases human cancer cell rediosensitivity through promotion of mitotic catastrophe. *Invest New Drugs* 29, 514-522.

Rogakou, E.P., Pilch, D.R., Orr, A.H., Ivanova, V.S. & Bonner, W.M. (1998) DNA double-stranded breaks induce histone H2AX phosphorylation on serine 139. *J Biol Chem* 273, 5858-5868.

Royou, A., Macias, H. & Sullivan, W. (2005) The Drosophila Grp/Chk1 DNA damage checkpoint controls entry into anaphase. *Curr Biol* 15, 334-339.

Russell, P. (1998) Checkpoint on the road to mitosis. *Trends Biol Sci* 23, 399-402.

Rybaczek, D. & Kowalewicz-Kulbat, M. (2011) Premature chromosome condensation induced by caffeine, 2-aminopurine, staurosporine and sodium metavanadate in S-phase arrested HeLa cells is associated with a decrease in Chk1 phosphorylation, formation of phospho-H2AX and minor cytoskeletal rearrangements. *Histochem Cell Biol* 135, 263-280.

Rybaczek, D., Bodys, A., Maszewski, J. (2007) H2AX foci in late S/G2- and M-phase cells after hydroxyurea- and aphidicolin-induced DNA replication stress in *Vicia*. *Histochem Cell Biol* 128, 227-241.

Rybaczek, D., Żabka, A., Pastucha, A. & Maszewski, J. (2008) Various chemical agents can induce premature chromosome condensation in *Vicia faba*. *Acta Physiol Plant* 30, 663-672.

Sahu, R.P., Batra, S. & Srivastava, S.K. (2009) Activation of ATM/Chk1 by cucrumin causes cell cycle arrest and apoptosis in human pancreatic cancer cells. *Br J Cancer* 100, 1425-1433.

Schlegel, R. & Pardee A.B. (1986) Caffeine-induced uncoupling of mitosis from the completion of DNA replication in mammalian cells. *Science* 232, 1264-1266.

Schlegel, R., Belinsky, G.S., Harris, MO. (1990) Premature mitosis induced by mammalian cells by the protein kinase inhibitors 2-aminopurine and 6-dimentylaminopurine. *Cell Growth Differ* 1, 171-178.

Scott, S.P. & Pandita, T.K. (2006) The cellular control of DNA double-strand breaks. *J Biol Chem* 99, 1463-1475.

Sen, R. & Ghosh, S. (1998) Induction of premature mitosis in S-blocked onion cells. *Cell Biol Int* 22, 867-874.

Sha,W., Moore, J., Chen, K., Lassaletta, A.D., Yi, C-S. & Tyson, J.J., (2003) Hysteresis drives cell-cycle transitions in *Xenopus laevis* egg extracts. *Proc Natl Acad Sci USA* 100, 975-980.

Solomon, M.J. (2003) Hysteresis meets the cell cycle. *Proc Natl Acad Sci USA* 100, 771-772.

Steinmann, K.E., Belinsky, G.S., Lee, D. & Schlegel, R. (1991) Chemically induced premature mitosis: Differential response in rodent and human cells and the relationship to cyclin B synthesis and p34^{cdc2}/cyclin B complex formation. *Cell Biology* 88, 6843-6847.

Stevens, J.B., Abdallah, b.Y., Regan, S.M., Liu, G., Bremer, S.W., Ye, C.J. & Heng, H.H. (2010) Comparison of mitotic cell death by chromosome fragmentation to premature chromosome condensation. *Molecular Cytogenetics* 3, 20-30.

Sturgeon, C.M., Cinel, B., Díaz-Marrero, A.R., McHardy, L.M., Ngo, M., Andersen, R.J. & Roberge, M. (2008) Abrogation and inhibition of nuclear export by *Cryptocarya* pyrones. *Cancer Chemother Pharmacol* 3, 407-413.

Swanson, P.E., Carroll, S.B., Zhang, X.F. & Mackey, M.A. (1995) Spontaneous premature chromosome condensation, micronucleus formation, and non-apoptotic cell death in heated HeLa S3 cells. *Am J Pathol* 146, 963-971.

Syljuåsen, R.G., Sørennsen, C.S., Hansen, L.T., Fugger, K., Lundin, C., Johansson, F., Helleday, T., Sehested, M., Lukas, J. & Bartek, J. (2005) Inhibition of human Chk1 causes increased initiation of DNA replication, phosphorylation of ATR targets, and DNA breakage. *Mol Cell Biol* 9, 3553-3562.

Szymborska, W. (1993). *Poems New and Collected*. English translation by Barańczak S. & Cavanagh C. (1998), Harcourt Inc., ISBN 0-15-601146-8, Orlando, Florida, USA

Takizawa, C.G. & Morgan D.O. (2000) Control of mitosis by changes in the subcellular location of cyclin-B1-Cdk1 and Cdc25. *Curr Op Cell Biol* 12, 658-665.

Tam, S.W. & Schlegel, R. (1995) Quantification of premature and normal mitosis, In: *Cell Cycle - Materials and Methods*, Michele Pagano, pp. 93-99, Springer-Verlag, ISBN 3-540-58066-2, Berlin-Heidelberg-New York.

Terzoudi, G.I., Hatzi, V.I., Donta-Bakoyianni D. & Pantelias G.E. (2010) Chromatin dynamics during cell cycle mediate conversion of DNA damage into chromatin breaks and affect formation of chromosomal aberrations: Biological and clinical significance. *Mutation Res* 711, 174-186.

Waldren, C.A. & Johnson, R.T. (1974) Analysis of interphase chromosome damage by means of premature chromosome condensation after X- and ultraviolet-irradiation. *Proc Nat Acad Sci USA* 71, 1137-1141.

Wang, S.-W., Norbury, C., Harris, A.L. & Toda, T. (1999) Caffeine can override the S-M checkpoint in fission yeast. *J Cell Sci* 112, 927-937.

Yamashita, K., Yasuda, H., Pines, J., Yasumoto, K., Nishitani, H., Ohtsubo, M., Hunter, T., Sugimura, T. & Nishimoto, T. (1990) Okadaic acid, a potent inhibitor of type 1 and type 2A protein phosphatases, activates cdc2/H1 kinase and transciently induces a premature mitosis-like state in BHK21 cells. *EMBO J* 9, 4331-4338.

Yu, K.R., Saint, R.B. & Sullivan, W. (2000) The Grapes checkpoint coordinates nuclear envelope breakdown and chromosome condensation. *Nat Cell Biol* 2, 609-615.

Zhou B.-B.S. & Elledge S.J. (2000) The DNA damage response: putting checkpoints in perspective. *Nature* 408, 433-439.

12

A DNA Repair Protein BRCA1 as a Potentially Molecular Target for the Anticancer Platinum Drug Cisplatin

Adisorn Ratanaphan
Prince of Songkla University
Thailand

1. Introduction

Cancer is a leading cause of death in the world. The incidence of cancers is related to environmental factors, behavioral patterns, and genetic disorders. Cancer therapy usually aims to selectively destroy cancer cells while sparing normal tissue. Most chemotherapeutic agents function by damaging cancer cell DNA. The cellular responses to DNA damage are thus critical factors for determining the effectiveness of most cancer therapies (Ashworth, 2008). When normal cells are exposed to damage, DNA repair mechanism is induced. The DNA repair processes are the cellular responses associated with the restoration of the normal DNA nucleotide sequences. The DNA repair activity of the cell is an important determinant of a cells sensitivity to chemotherapeutic agents. It is known that resistance to DNA-damaging agents can be associated with increased cellular repair activities, while defects in DNA repair pathways result in hypersensitivity to damage (Kelley & Fishel, 2008; Quinn et al., 2003, 2009). Several studies have clearly demonstrated that the impairment or absence of genes or proteins responsible for DNA damage repair, frequently causes genomic instability, cell cycle arrest and apoptosis. The importance of these repair pathways is highlighted by the fact that more than 100 genes have been found in mammalian cells that are involved in some way in DNA damage repair pathways. The breast cancer susceptibility gene 1 (*BRCA1*) is a tumor suppressor gene involved in maintaining genomic integrity through multiple functions in DNA damage repair, transcriptional regulation, a cell cycle checkpoint and protein ubiquitination (Brzovic et al., 2001; Hashizume et al., 2001; Mark et al., 2005; Varma et al., 2005; Williams et al., 2004). In cancer cells, damage to *BRCA1* by the anticancer platinum drug cisplatin may lead to a loss of such functions and ultimately results in cancer cell death. In addition, preclinical and clinical studies have recently revealed that inactivation of the BRCA1 protein in cancer cells leads to chemosensitivity. Therefore, approaching the BRCA1 protein as a potential therapeutic target for cisplatin or other such platinum based drugs might be of interest for molecular-targeted cancer therapy. In this chapter, the biophysical characterization and functional consequences of the human *BRCA1* gene and the BRCA1 RING protein induced by cisplatin are described.

2. Breast cancer susceptibility gene 1 (*BRCA1*) and its encoded protein

In 1990, chromosome 17q21 was identified by linkage analysis as the location of a breast cancer susceptibility gene 1 or *BRCA1* (Hall et al., 1990). The entire gene covers approximately 100 kb of genomic sequence, and was subsequently cloned four years later (Miki et al., 1994). BRCA1 is a tumor suppressor gene composed of 24 exons, with an mRNA that is 7.8 kb in length, and 22 coding exons that translate into a protein of 1863 amino acids (Fig. 1) with a molecular weight of 220 kDa (Brzovic et al., 1998). It has 3 major domains, including (1) the N-terminal RING finger domain (BRCA1 RING domain), (2) the large central segment with the nuclear localization signal (NLS), and (3) the BRCA1 C-terminal domain (BRCT). The BRCA1 protein plays an essential role in maintaining genomic stability associated with a number of cellular processes, including DNA repair, a cell cycle checkpoint, transcriptional regulation, and protein ubiquitination (Huen et al., 2010; O'Donovan & Livingston, 2010).

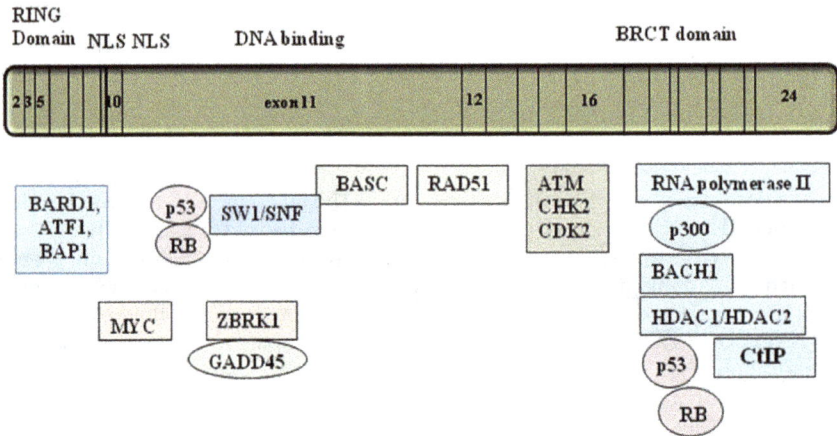

Fig. 1. Scheme of BRCA1 mRNA and sites of protein interaction

2.1 The BRCA1 RING domain

The N-terminal RING finger domain contains the conservative sequences of cysteine and histidine residues (C_3HC_4) necessary for specific coordination with two Zn^{2+} ions. The first 109 amino acids of BRCA1 protein constitute a protease-resistance domain. The solution structure of the BRCA1 RING domain revealed the existence of antiparallel α-helices at both ends, flanking the central RING motif (residues 24-64) and was characterized by a short antiparallel three-stranded β-sheet, and two large Zn^{2+}-binding loops, and a central α-helix (Brzovic et al., 2001)The two Zn^{2+}-binding sites are formed in an interleaved fashion in which the first and third pairs of cysteines (Cys24, Cys27, Cys44, and Cys47) form site I, and the second and fourth pairs of cysteines and a histidine (Cys39, His41, Cys61, and Cys64) form site II. It is an important domain since it might mediate a central role in macromolecular interactions to exert the tumor suppression functions. The solution structure together with yeast-two-hybrid studies revealed that the BRCA1 RING domain preferentially formed a heterodimeric complex with another RING domain BARD1 (BRCA1-associated RING domain 1) through an extensive four-helix-bundle interface (Brzovic et al., 2001; Wu et al., 1996). The binding interface is composed of residues 8-22 and 81-96 of BRCA1, and residues 36-48 and 101-116 of BARD1. The BRCA1-BARD1 complex requires each other for their mutual stabilities, and they are co-localized in nuclear dots during S phase but not the G phase of the cell cycle and in nuclear foci (Hashizume et al., 2001). The progression to S phase by aggregation of nuclear BRCA1 and BARD1 implied the importance of both proteins for a DNA repair function (Jin et al., 1997). The BRCA1-BARD1 complex also exhibits enzymatic activity of an E3 ubiquitin ligase that specifically transfers ubiquitin to protein substrates that are essential for cellular viability (Hashizume et al., 2001; Xia et al., 2003). Cancer-predisposing mutations in the Zn^{2+}-binding sites were demonstrated not only to alter the affinity for Zn^{2+} and the native BRCA1 RING structure but also abolished the interaction with BARD1 and the E3 ligase activity (Morris et al., 2006). The results supported the importance of Zn^{2+} as a structural component, as it obviously played a critical role in the stabilization of the structure and function of the BRCA1 RING domain.

2.2 The large central segment of BRCA1

The central segment of BRCA1 covers exon 11 (approximately 3500 bp) and constitutes approximately 60 percent of the coding region of the gene. Deletion of exon 11 results in removal of the nuclear localization signal of BRCA1. Biophysical characterization revealed that this domain was intrinsically disordered or natively unfolded under physiological conditions. This might potentially allow the BRCA1 central region to act as a long flexible scaffold, to mediate interactions with DNA, and perhaps a number of other proteins involved in the DNA damage response and repair (Mark et al., 2005). The reported binding partners to the central region were c-Myc, RB, p53, FANCA, RAD50, RAD51, JunB, and BRCA2 (Rosen et al., 2003). Recently, the BRCA1 central region has been shown to efficiently interact with p53, and stimulate p53-mediated DNA binding and transcriptional activities (Buck, 2008). This result indicated that the BRCA1 central segment facilitated the induction of cell cycle arrest and apoptosis in response to DNA damage. Furthermore, the association between the central region of BRCA1 and PALB2 (partner and localizer of BRCA2, also known as FANCN) was observed primarily through apolar bonding between their respective coiled-coil domains (Sy et al., 2009). PALB2 binds directly to BRCA1, and serves as the molecular scaffold for the formation of the BRCA1-PALB2-BRCA2 complex. BRCA1 mutations (L1407P and M1411T) identified in cancer patients were shown to disrupt the specific interaction between BRCA1 and PALB2, resulting in a defective homologous recombination (HR) repair and a compromised cell survival after DNA damage (Sy et al., 2009).

2.3 The BRCA1 C-terminal domain

The C-terminal region (residues 1646-1863) of BRCA1 contains two BRCT (BRCA1 C-terminal) domains in tandem (motif 1: amino acids 1653-1736; motif 2: amino acids 1760-1855). Each BRCT domain is characterized by a central, parallel four-stranded β-sheet with a pair of α-helices (α1 and α3) packed against one face, and a single α-helix (α2) packed against the opposite face of the sheet (Williams et al., 2001). The two BRCA1-BRCT repeats interact in a head-to-tail fashion. This domain serves as a multipurpose protein-protein interaction module that binds to other BRCT repeats or other protein domains with apparently unrelated structures (Watts & Brissett, 2010). Based on its physical interactions with other proteins, BRCA1 has been implicated in a wide array of cellular functions, including cell cycle regulation, DNA damage response, transcriptional regulation, replication and recombination, and higher chromatin hierarchical control (Starita & Parvin, 2003). The BRCA1-BRCT domain has been identified as a phosphopeptide recognition module, and is demonstrated to bind to the phosphorylated protein partners (BACH1 and

CtIP, containing the consensus sequence pSer-X-X-Phe) that is involved mainly in the control of the G2/M phase checkpoint and DNA damage repair (Varma et al., 2005; Williams et al., 2004). Several cancer-predisposing mutations in the BRCA1-BRCT domain resulted in destabilization of the structural integrity at the BRCT active sites, and abolished their affinities to synthetic BACH1 and CtIP phosphopeptides (Rowling et al., 2010). These findings provide a better understanding of the pathogenic BRCA1 mutations on functional mechanisms and tumorigenesis.

3. BRCA1 and DNA damage repair

A substantial amount of evidence that has implicated BRCA1 in the DNA damage repair pathways has been documented. BRCA1 co-localizes with RAD51 and BARD1 to nuclear foci (sites associated with repair of DNA caused by the damaging agents or γ-irradiation) (Hashizume et al., 2001; Scully et al., 1997). The nuclear foci is marked by the histone variant H2AX that was phosphorylated on Ser139 (known as γH2AX) (Rogakou et al., 1998). γH2AX is one of the initial recruiting factors for various checkpoints and DNA repair proteins, including Abraxas, RAP80, and BRCA1, at sites of DNA breaks (Foulkes, 2010). The H2AX signaling cascade begins to emerge with the finding that MDC1 (mediator of DNA damage checkpoint 1) is the main downstream factor in the pathway, and is required for the damage-induced focal accumulation of a number of DNA damage repair factors at the DNA breaks (Stucki et al., 2005).

BRCA1 plays a role in maintaining genome integrity through its role in DNA damage repair. Several observations have implicated BRCA1 in homologous recombination (HR), non-homologous end-joining (NHEJ and nucleotide excision repair (NER). A role for BRCA1 in HR-mediated repair is involved through its stable complex formation with BRCA2, which has a well-defined role in HR through its direct interaction with RAD51 (Bhattacharyya et al., 2000). RAD51 (the mammalian homolog of the Escherichia coli RecA protein) is a DNA recombinase that catalyzes strand exchange in an early step of HR (Baumann et al., 1996). PALB2 (the partner and localizer of BRCA2) has recently been identified as the bridging factor required for the BRCA1-BRCA2 association (Rahman et al., 2007). The BRCA1-PALB2 interaction was mediated by their respective coiled-coil domains, and was found to promote HR-mediated repair (Rahman et al., 2007). Importantly, missense mutations identified in the PALB2-binding region on BRCA1 disrupted the specific interaction of BRCA1 with PALB2, and compromised DNA repair in a gene conversion assay (Sy et al., 2009). Although these studies have revealed a molecular link between BRCA1 function and HR-mediated repair, the mechanism by which BRCA1 promotes HR through the PALB2-BRCA2-RAD51 axis remains unclear.

As an alternative to HR, there is a growing body of evidences, to indicate that a component of NHEJ is regulated by BRCA1. The exact role of BRCA1 in NHEJ, however, has not been well defined (Zhang & Powell, 2005). In the NHEJ pathway, the DNA-dependent protein kinase catalytic subunit (DNA-PKcs) and a Ku heterodimer of Ku80 and Ku70 are recruited to the sites of DNA DSBs for preparing the DNA ends before ligation by the XRCC4 ligase IV. The most possible explanation for BRCA1 being involved in NHEJ is its association with a NHEJ factor Ku80 (Chiba & Parvin, 2001; Wei et al., 2008). Many studies have provided strong evidences that the NHEJ pathway was impaired, both in vivo and in vitro, in BRCA1-deficient mouse embryonic fibroblasts and in the human breast cancer cell line HCC1937 which carries a homozygous mutation in the BRCA1 gene (Bau et al., 2004; Zhong et al., 2002).

4. BRCA1 and transcriptional regulation

As described earlier, BRCA1 contains a C-terminal transactivation domain as was first defined using the yeast two-hybrid system (Chapman & Verma, 1996; Monteiro et al., 1996). The transactivation domain was mapped to the region of the protein encoded by exons 21-24 using deletion constructs of BRCA1 fused to the GAL4 DNA binding domain. The BRCA1-BRCT domain has been implicated in the regulation of transcription of several genes responsible for DNA damage. The ability of BRCA1 to act as either a co-activator or a co-repressor of transcription may involve its ability to recruit the basal transcriptional machinery and other proteins that have been implicated in chromatin remodeling (Mullan et al., 2006). BRCA1 was capable of activating the p21 promoter (Somasundaram et al., 1997). One report claimed that BRCA1 participated in the stabilization of p53 in response to DNA damage, and served as a co-activator for p53 (Zhang et al., 1998). The interaction of BRCA1 and p53 potentially resulted in the redirection of p53-mediated transactivation from a pro-apoptotic target to genes involved in DNA repair and cell cycle arrest (Zhang et al., 1998). In addition, BRCA1 has been shown to interact with the RNA polymerase II holoenzyme (Scully et al., 1997). However, BRCA1 could repress the transcription of an estrogen receptor α (ERα) and its downstream estrogen responsive genes (Fan et al., 1999). The transcriptional repression activity of BRCA1 for ERα occurs by the association of the N-terminus of BRCA1 (residues 1-300) with the C-terminal activation function (AF-2) of ERα. Breast cancer-associated mutations of BRCA1 were found to abolish its ability to inhibit ERα activity (Fan et al., 2001). The repression activity exerted by BRCA1 involved the ability of BRCA1 to down-regulate levels of the transcriptional coactivator p300, which has also been shown to interact with the AF-2 domain of ERα (Fan et al., 2002). Further investigations revealed that overexpression of BRCA1 could inhibit the recruitment of the co-activators [steroid receptor co-activator 1 (SRC1), and amplified breast cancer 1 (AIB1)], and enhanced the recruitment of a co-repressor [histone deacetylase 1 (HDAC1)] to the progesterone response elements (PRE) of c-Myc.

5. BRCA1 and protein ubiquitination

The BRCA1 protein displays an E3 ubiquitin ligase activity through its RING domain, and this activity is enhanced when it exists as a heterodimer with the BARD1 RING domain (Xia et al., 2003). In vitro and in vivo studies have indicated that the BRCA1-BARD1 complex was capable of autoubiquitination that paradoxically stabilized the protein complex, and that also activated its in vitro E3 ligase activity with other proteins (Chen et al., 2002; Wu-Baer et al., 2010). However, the substrate specificity of the BRCA1 E3 ligase activity and its biological relevance to tumor suppression function are still unknown. Putative substrates for ubiquitination by the BRCA1-BARD1 RING complexes have recently emerged from in vitro and in vivo studies such as the nucleosomal histones H2A and its variant H2AX, RNA polymerase II, γ-tubulin, nucleophosmin/B23, and estrogen receptor α (ERα) (Eakin et al., 2007; Horwitz et al., 2007; Parvin, 2009; Sato et al., 2004; Starita et al., 2005; Thakar et al., 2010). BRCA1 can form a RING heterodimer E3 ligase activity with BARD1, and this is required for the recruitment of BRCA2 and RAD51 to damaged sites for HR repair (Ransburgh et al., 2010). Many cancer-predisposing mutations in the BRCA1 RING domain, that inhibited the E3 ligase activity and its ability to accumulate at damaged sites, were defective in homologous recombination that is critical for tumor suppression (Morris et al.,

2006, 2009; Ransburgh et al., 2010). Moreover, BRCA1 accumulation at the sites of DSBs occurred rapidly (within 20 s), and the RING structure was required (residue 1-200 of BRCA1) for the rapid recruitment with Ku80 at damaged sites in response to non-homologous end joining (Wei et al., 2008). Missense mutations in the BRCA1 RING domain significantly reduced their accumulations at DSBs, and abolished the association with Ku80. Therefore, the loss of the BRCA1 E3 ligase activity rendered cancerous cells hypersensitive to DNA-damaging agents,and clearly demonstrated a significant role for ubiquitnation in the DNA damage response and DNA repair activity (Ransburgh et al., 2010; Ruffner et al., 2001). Thus ubiquitination is involved in key steps that properly conduct the DNA repair process after DSBs.

Several reports have shown that the BRCA1 E3 ligase was capable of in vitro monoubiquitination of histones H2A and its variant H2AX (Thakar et al., 2010). This implied a BRCA1 function in regulating chromatin structure in the context of transcriptional regulation and DNA repair. Hyperphosphorylated RNA polymerase II (RNAPII) at its carboxyl terminal domain (CTD), consists of multiple repeats of the heptapeptide (YSPTSPS), involved in a generalized response to UV irradiation. It also served as a substrate for the BRCA1-dependent ubiquitination that was proposed to facilitate BRCA1 function in DNA repair by inhibiting DNA transcription, and then recruiting other DNA repair proteins at a lesion (Starita et al., 2005). Recently, It was found that the BRCA1-mediated ubiquitination of RNAPII prevented a stable association of some transcription factors (TFIIE and TFIIH) in the transcriptional preinitiation complex, and thus blocked the initiation of mRNA synthesis (Horwitz et al., 2007). Ubiquitination of the preinitiation complex was not targeting proteins for degradation by proteasome but rather the ubiquitin moiety itself interfered with the assembly of basal transcription factors at the promoter (Horwitz et al., 2007). Nucleoplasmin B23 and γ-tubulin were found to be the candidate substrates of the BRCA1 E3 ligase activity in vivo (Parvin, 2009; Sato et al., 2004). Both proteins were present in centrosomes, and apparently were not targeted for degradation by BRCA1-mediated modifications. The results indicated that ubiquitination of nucleoplasmin B23 and γ-tubulin played a vital role in regulating the centrosome number and maintenance of genomic stability by unknown mechanisms. Recently, the BRCA1 protein has been shown to inhibit ERα transcriptional activity, and to induce repression of estrogen response genes and cell proliferation (Xu et al., 2005). A potential explanation for the regulation of estrogen signaling by BRCA1 was the ERα ubiquitination and degradation mediated by the BRCA1 E3 ligase activity (Dizin & Irminger-Finger, 2010; Eakin et al., 2007). Conversely, the BRCA1-associated protein 1 (BAP1) is a deubiquitinating enzyme that can interact with the BRCA1 RING domain (Jensen et al., 1998). It was shown that BAP1 inhibited the BRCA1 autoubiquitination, and the nucleophosmin/B23 ubiquitination mediated by the BRCA1 E3 ligase activity (Nishikawa et al., 2009). Down-regulation of BAP1 in cells also resulted in the retardation of the S phase and ionizing irradiation hypersensitivity, a phenotype similar to BRCA1 deficiency. This again indicated that the BRCA1-BARD1 complex and the BAP1 protein coordinately regulated ubiquitination during a DNA damage response and the cell cycle.

6. Cisplatin

Cisplatin [cis-diamminedichloroplatinum(II)] is the platinum-based anticancer drug and is most effective in the treatment of metastatic testiscular cancers, ovarian, head, neck, bladder, cervical and lung cancers (Kelland, 2007). Although widely used as a well established anticancer drug in cancer chemotherapy, cisplatin displays major toxic side effects, such as nephrotoxicity, nausea and vomiting and neurotoxicity. In addition to its toxic side effects, a major limitation of cisplatin chemotherapy is the development of genetic mechanisms of resistance. The effectiveness of cisplatin depends on the drug uptake, and the actual amount that reacts with cellular targets.

6.1 Cisplatin-DNA adducts

It is generally accepted that DNA is the most important intracellular target of cisplatin. When cisplatin is dissolved in aqueous solution, chloride ions are displaced to allow the formation of aquated species, which are the reactive forms of the compound (Pinto and Lippard, 1985). The concentration of chloride ions influences the reactivity of cisplatin. After intravenous administration it is relatively less reactive in the extracellular space where the physiological chloride concentration is about 100 mM, but on crossing the plasma membrane, it is activated in the intracellular space where the chloride concentration drops to 2-3 mM. Chlorine groups of cisplatin are easily replaced by water molecules to allow the formation of aqauted species in a stepwise manner. Activated cisplatin is a potent electrophile that will react with any nucleophile, including the sulfhydryl groups on proteins and nucleophilic groups on nucleic acids. DNA is attacked by activated cisplatin at guanine residues in position N7, in double stranded DNA from the side of the major groove. The attack is apparently preceeded by an electrostatic attraction between the positively charged platinum (II) complex and the negatively charged phosphodeoxyribose DNA backbone and facilitated by bidirectional diffusion along the backbone. The initial attack of DNA by activated cisplatin is followed by the replacement of the remaining chloro ligand before the adduct forms an intramolecular attack on a second purine residue (either guanine or adenine). The hydration rate constant of the monoaqua form was faster than that of diaqua form (2.38×10^{-5} s^{-1} compared to 1.4×10^{-5} s^{-1}) (Cubo et al., 2009).

The anticancer activity of cisplatin potentially results from the modification of DNA through a covalent cross-link or platinum (Pt)-DNA adduct (Fig. 2). The DNA adducts interfere with DNA replication and transcription, and ultimately lead to cell death by cancer (Ahmad, 2010; Wang & Lippard, 2005). The predominant adducts formed by cisplatin in vitro are 1,2-intrastrand crosslinks. Quantitative studies show that the 1,2-intrastrand d(GpG), and d(ApG) crosslinks account for 65% and 25%, respectively (Fichtinger-Schepman et al., 1985; Eastman, 1986). They alter the DNA structure, block replication and transcription and activate a programmed cell death (apoptosis). X-ray diffraction of the crosslinked dinucleotide *cis*-Pt(NH$_3$)$_2$[d(pGpG)] reveals that the intrastrand cisplatin crosslink produces a severe local distortion in the DNA double helix, leading to unwinding and kinking. These crosslinks bend and unwind the duplex. The altered structure is recognized by high-mobility-group (HMG) proteins and other proteins. The binding of HMG proteins to cisplatin-modified DNA has been postulated to potentiate the anticancer activity of the drug.

6.2 Cisplatin-protein adducts

The interaction of cisplatin with proteins is of particular significance, and is believed to play an important role in distribution of the drug and the inactivation responsible for determining its efficacy and toxicity (Casini et al., 2008; Sun et al., 2009; Timerbaev et al., 2006). It is intriguing, that protein adducts affect some crucial aspects of protein structure

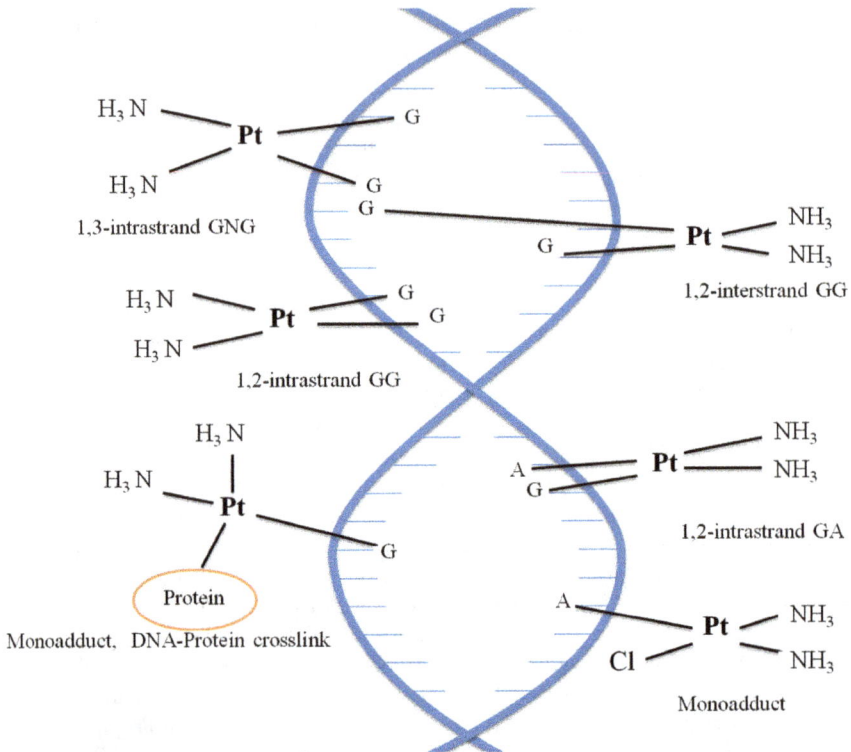

Fig. 2. Common cisplatin-DNA adducts

and functions. For instance, the platination of human serum albumin caused partial unfolding of the protein structure at a high drug concentration, and induced intermolecular crosslinks possibly at Cys34 and/or Met298 via bifunctional adducts or via NH_3 release (Ivanov et al., 1998; Neault & Tajmir-Riahi, 1998). Myoglobin, a small protein, containing a heme group required for the transport of oxygen in skeletal muscles and myocardial cells, formed intramolecular mono- and bi-functional adducts with cisplatin. Its putative platinum-binding sites were His116 and His119 (Zhao & King, 2010). A number of intramolecular crosslinks also occurred with ubiquitin adducts (Casini et al., 2009). The loss of activity of the C-terminal heat shock protein 90 after protein aggregation was reported to be a consequence of cisplatin binding but it did not exhibit any conformational change (Ishidaa et al., 2008). It is intriguing, that cisplatin can cause a structural perturbation of a synthetic peptide containing a Zn^{2+} finger domain. The platinum coordinates to Zn^{2+}-binding sites to induce Zn^{2+} ejection and subsequently the loss of the protein tertiary structure. This implies that cisplatin can inhibit critical biological functions regulated by Zn^{2+} finger proteins. Such a mechanism has been discussed in the apoptosis process mediated by the interaction of cisplatin and platinum-based compounds with Zn^{2+} finger transcriptional factors (Bose et al., 2005). Likewise, the nucleocapsid Zn^{2+} finger NCp7 protein, a protein required for the recognition and packaging of viral RNA, became attached to some platinum compounds, when its ability to bind nucleic acid was changed and prevented viral infectivity (de Paula et al., 2009; Musah, 2004).

7. *BRCA1* and its encoded product as potentially molecular targets for cisplatin for cancer therapy

In recent years, there has been significant progress made in evaluating what happens when BRCA1 is inactivated so it cannot respond to DNA damage in cancer cells, in other words, taking advantage of the inherent weakness of the BRCA1 dysfunction in cancer cells. These cells have increased sensitivity to DNA-damaging agents that eventually result in major genomic instability and cell death (Amir et al., 2010; Ashworth, 2008; Helleday et al., 2008; Lieberman, 2008; Powell & Bindra, 2009; Quinn et al., 2009; Tassone et al., 2009; Zhu et al., 2009). Cancerous cells with inactivated BRCA1 had defects in DNA repair of double strand breaks (DSBs) (Farmer et al., 2005; Kennedy et al., 2004; Litman et al., 2008). Moreover, extensive investigations have revealed the relevance of the BRCA1-mediated ubiquitination to DNA repair functions. Mutations in the BRCA1 RING domain resulted in the loss of the E3 ubiquitin ligase activity, and conferred hypersensitivity of the cancerous cells to DNA-damaged chemotherapy and γ-irradiation (Ransburgh et al., 2010; Ruffer et al., 2001; Wei et al., 2008).

It was initially reported that overexpression of BRCA1 in the human breast cancer MCF7 cell line resulted in an increased resistance to cisplatin (Husain et al., 1998). Furthermore, antisense or siRNA-based inhibition of endogenous BRCA1 expression promoted the increased sensitivity to cisplatin that was associated with the decreased DNA repair by NER and an increased apoptosis (Lafarge et al., 2001; Quinn et al., 2003). This indicates that the reduced BRCA1 expression observed in sporadic cancers may also be exploited for DNA damage-based chemotherapy (James et al., 2007; Quinn et al., 2009). In a similar situation, BRCA1-deficient mouse embryonic stem cells displayed defective DNA repair and a 100-fold increased sensitivity to the alkylating agent mitomycin C and cisplatin than those containing wild-type BRCA1 (Bhattacharyya et al., 2000; Moynahan et al., 2001). This sensitivity was reversed upon correction of the BRCA1 mutation in mouse embryonic fibroblast cells with a disrupted BRCA1 (Fedier et al., 2003). Reconstitution of BRCA1 in the cells via transfection meant that BRCA1 functions were regained, and resulted in a reduced level of cancer cell death, following treatment with cisplatin or other DNA damaging agents (Quinn et al., 2003). Moreover, more recent evidence has revealed the implication of BRCA1 in cisplatin-resistant breast and ovarian cancer cell lines. These cells that acquired resistance to DNA-damaging agents was mediated by a secondary mutation in BRCA1. This mutation restored the BRCA1 protein expression and function for DNA repair, causing the cancer cells to become more tolerant to cisplatin (Swisher et al., 2008; Tassone et al., 2003; Wang & Figg, 2008). Recently, a number of clinical studies have examined the utilization of this BRCA1 dysfunction in response to the DNA-damaging drug cisplatin. A pathological complete response (pCR) with excellent compliance was observed in cancer patients with BRCA1 mutations (Byrski et al., 2009; Font et al., 2010; Quinn et al., 2007; Silver et al., 2010; Taron et al., 2004). This indicates that patients with BRCA1 dysfunction gain more benefit from treatments that exert their effects by causing DNA damage.

Therefore, it is important to continue elucidating *BRCA1*/BRCA1-dependent pathways to design molecular–targeted therapy for the platinum treatment of cancer cells by taking advantage of their impairment of the *BRCA1*/BRCA1 repair capacity and BRCA1-dependent ubiquitination inactivated by cisplatin.

7.1 Cellular repair of cisplatin-damaged *BRCA1*

Preliminary results from our laboratory have indicated that the cisplatin-modified *BRCA1* gene sequence was resistant to restriction endonuclease cleavage, and indicated that cisplatin preferentially formed 1,2-intrastrand d(GpG) cross-links (Ratanaphan et al., 2009). The drug inhibited *BRCA1* amplification in a dose-dependent manner. It has been found that cisplatin-treated, *BRCA1* exon 11, of adenocarcinoma MCF-7 cells exhibited a time dependent recovery after drug exposure to the cells at 37°C for 6 h, with an initial low level of lesion removal during the first 4 h (Fig. 3). A more complete lesion removal was observed with over 90% of 50 μM cisplatin after 18 h of repair time. However, only 30% of the lesion repair was observed at a higher cisplatin concentration of 200 μM (Ratanaphan et al., 2009).

From a host cell reactivation assay, the result indicated that a reduction in cellular reactivation of the drug-damaged reporter gene encoding plasmid was a consequence of an increase in platination levels within the transcribed reporter gene. This indicated that the cellular response to cisplatin reflected its intrinsically low capacity for removal of cisplatin-*BRCA1* adducts. Following cisplatin-induced *BRCA1* adducts, a number of cellular repair proteins, excluding BRCA1, are responsible for recognizing and processing the removal of DNA damage. NER is a major process for removing platinum-damaged DNA. This process requires an ATP-dependent multiple protein complex that recognizes the bending induced on DNA by cisplatin. The NER complex has a dual role that can unwind the DNA strands (helicase), and excise the damage strand (endonuclease) of about 24-32 nucleotides in length, containing a platinum lesion. DNA resynthesis factors are recruited at the site of the incised DNA, and employ the opposite strand as template to fill in the gap in concert with DNA ligases. Two distinct sub-pathways of NER that may be involved, are transcription-coupled repair (TCR) and/or global genomic repair (GGR). TCR preferentially repairs transcribed strands of the RNA polymerase II-transcribed active gene, while GGR repairs throughout the genome (Shuck et al., 2008). Recently, the suppression of ERCC1 expression in a HeLa S3 cell line by small interfering RNA (siRNA) led to a decrease in the repair activity of cisplatin-induced DNA damage along with a decrease in cell viability against platinum-

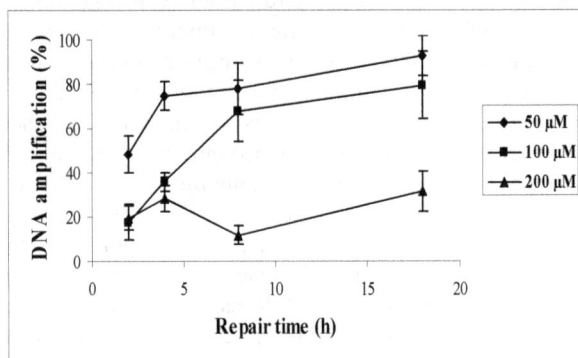

Fig. 3. Cellular repair of cisplatin-damaged 3,426-bp *BRCA1* exon 11. MCF-7 cells were incubated with medium plus cisplatin at various concentrations (50–200 μM) for 6 h. The cells were washed twice with PBS and fresh medium was added. The genomic DNA was then extracted at 2, 4, 8 and 18 h and used as the template for the QPCR assay (Ratanaphan et al., 2005, 2009).

based drugs (Chang et al., 2005). Recombination pathways can also be involved as repair systems responsible for DNA damage induced by the anticancer drug cisplatin. Recombination-deficient *E. coli* mutants were sensitive to cisplatin and exhibited a decreased survival by four orders of magnitude in comparison with the parental strain at a cisplatin concentration of 75-80 µM (Zdraveski et al., 2000). Many recombination-deficient strains showed a sensitivity to the drug equal to that of the NER-deficient strains. Double mutations in recombination and NER proteins were approximately 4-fold more sensitive to cisplatin than the corresponding single mutants. This indicates that recombination and NER pathways play roles that are independent of each other in protecting cells from cisplatin-induced damage. Impaired recombination DNA repair in yeast and prostate cancer cell lines also showed an increased sensitivity to cisplatin (Wang et al., 2005).

7.2 Inhibition of *BRCA1* transcriptional transactivation

The one hybrid GAL4 transcription assay is used to study the effect of cisplatin on transcriptional transactivation. The level of transcriptional transactivation is inversely proportional to the amount of platinum-*BRCA1* adducts. The results are most likely due to inhibition of transcription of the reporter plasmid that resulted from interstrand crosslinks (Ratanaphan et al., 2009). The transcriptional transactivation activity of *BRCA1* has previously been reported by fusing the C-terminal domain of *BRCA1* to a heterogenous DNA-binding domain (Chapman and Verma, 1996). The BRCT domain (amino acids 1380-1863) of human *BRCA1* scores positively in transcriptional activation trap experiments using various forms of so-called "one hybrid assay".The *BRCA1*-fused DNA-binding domain activates transcription in a cell-free system to a similar extent as a dose of the powerful activator, VP16 (Scully et al., 1997). A *GAL4:BRCA1* has also been introduced in yeast- and mammalian-based transcription assays to characterize the deleterious mutations in the 3/ - terminal region of the *BRCA1* (Vallon-Christersson et al., 2001). The transcriptional activity reflects a tumor-suppressing function of the BRCA1 protein.

In order to investigate whether the drug-damaged *BRCA1* is able to transactivate the expression of a firefly luciferase gene, DNA repair-proficient MCF-7 cells were transiently transfected with the cisplatin-damaged pBIND-BRCT along with the reporter plasmid pG5Luc. The firefly luciferase activity was significantly decreased at a cisplatin concentration of 12.5 µM (Fig. 4).

It has been hypothesized that the BRCT domain could transactivate the expression of another reporter gene. The reporter gene pSV-β-galactosidase was used for this purpose. It was of interest, that the level of transactivation was significantly higher when co-transfected with the pBIND-BRCT than with the parental pBIND (Fig. 5). This indicated that the GAL4-BRCT domain may stimulate the pSV-β-galactosidase. However, the expression of β-galactosidase was decreased to the level of β-galactosidase alone when co-transfected with the platinated pBIND-BRCT. It was again of interest that, β-galactosidase expression was dramatically diminished when both the pSV-β-galactosidase and the pBIND-BRCT were platinated (Fig. 6). Expression of β-galactosidase from the pSV-β-galactosidase can be transactivated both by the GAL4 domain of the pBIND and pBIND-BRCT. Acting upon the GAL4 DNA sequence similarity, the GAL4 protein alone can stimulate the expression of β-galactosidase. However, the degree of transactivation was slightly higher by the pBIND-BRCT. This indicates that the BRCT domain on the fusion protein is able to transactivate the β-galactosidase gene-bearing pSV-β-galactosidase. When platinated pSV-β-galactosidase is co-transfected with the pBIND or the pBIND-BRCT, a relatively lower expression of β-galactosidase was observed. The transcription level of β-galactosidase expression was reduced from 2-2.5 fold to 1.3 fold in both plasmids. Considering the data from the proficiency in repairing cisplatin-*BRCA1* adducts, it demonstrated that over 80% of the DNA lesion was repaired 8 h after cisplatin removal. Thus, it is possible that, during the repair time, RNA polymerase II or other transcriptional machineries may be blocked at any lesion on DNA (Jung & Lippard, 2003, 2006; Tornaletti et al., 2003).

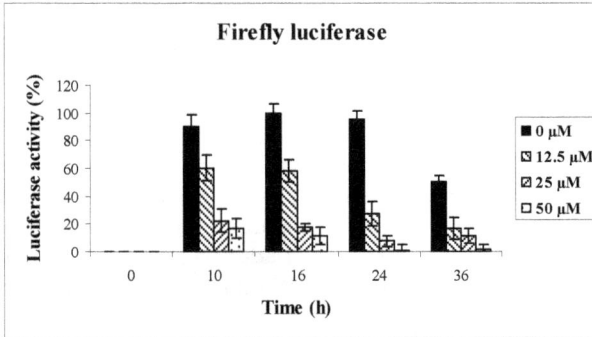

Fig. 4. Time course of firefly luciferase expression. The pBIND-BRCT was incubated with cisplatin at concentrations of 0, 12.5, 25 and 50 μM and then co-transfected with the pG5Luc plasmid into MCF-7 cells. A cell lysate was prepared at 10, 16, 24 and 36 h after transfection. Firefly luciferase expression is detected by the Dual-Luciferase® Reporter Assay System. The data were derived from four independent experiments ± standard deviations (SD) (Ratanaphan et al., 2009).

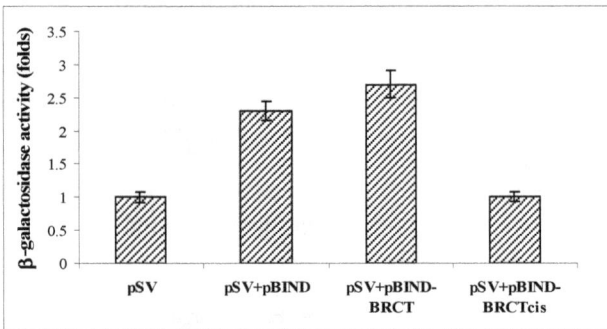

Fig. 5. Transcriptional transactivation. The pBIND or pBIND-BRCT was co-transfected with pSV-β-galactosidase. Cell lysates were prepared at 16 h after transfection. β-galactosidase activity was detected using the β-galactosidase assay. The data were derived from four independent experiments ± standard deviations (SD) (Ratanaphan et al., 2009).

Several investigations have revealed transcriptional inhibition on DNA templates, containing the site-specific Pt-DNA adducts. The mammalian RNA polymerase II and *E. coli* RNA polymerase did not catalyze the transcriptional reactions when the DNA template strands carried the 1,2-intrastrand d(GpG) and d(ApG) adducts, whereas those polymerases

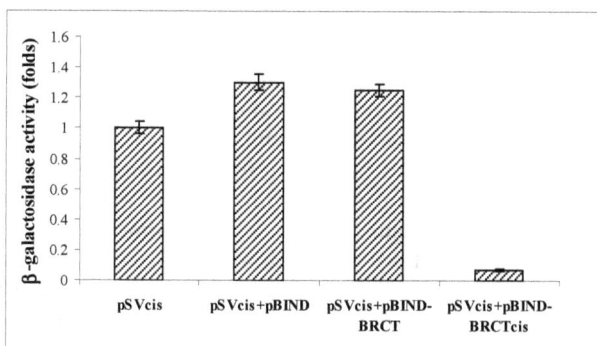

Fig. 6. Transcriptional transactivation of platinated pBIND-BRCT on platinated pSV-β-galactosidase. The platinated pSV-β-galactosidase (with cisplatin at a concentration of 12.5 µM) was co-transfected with non-platinated pBIND and pBIND-BRCT or platinated pBIND-BRCT. Cell lysates were prepared at 16 h after transfection. β-galactosidase expression was detected using the β-galactosidase assay. The data were derived from four independent experiments ± standard deviations (SD) (Ratanaphan et al., 2009).

could transcribe the complementary templates which had no DNA lesions on the template strands (Corda et al., 1991). Transcription of globally platinated DNA templates by SP6 and T7 RNA polymerases were also blocked primarily at 1,2-d(GpG) and d(ApG) Pt adducts, and to a lesser extent at the interstrand crosslink (Tornaletti, 2005). Bifunctional Pt-DNA adducts were much more effective at impeding transcription progression than monofunctional DNA adducts (Tornaletti, 2005). Moreover, cisplatin caused a dose-dependent inhibition of mRNA synthesis. Treatment of human fibroblast cells with 50 µM cisplatin for 24 h resulted in a 55% decrease in mRNA level and a reduced expression of p21[WAF1] protein. This indicated that cisplatin inhibited the transcription of the *p21[WAF1]* gene (Ljungman et al., 1999). Recently, the processing of site-specific Pt-DNA crosslinks in mammalian cells was investigated (Ang et al., 2010). Site-specific platinated oligonucleotides, containing 1,2-d(GpG) and 1,3-d(GpTpG) adducts, were inserted into an expression vector between its promoter and a luciferase reporter gene. Transcription inhibitions that occurred by blocking passage of the RNA polymerase complex through the 1,2-d(GpG) and 1,3-d(GpTpG) adducts were 50% and 37.7% of the unplatinated controls for vectors, repectively. An X-ray crystal structure of RNA polymerase II showed stalling at the 1,2-intrastrand d(GpG) crosslink to explain the physical block of transcription by the cisplatin-DNA adduct (Damsma et al., 2007). Disruption of chromatin remodeling was another mechanism by which a cisplatin adduct could interfere with transcription. Nucleosomal DNA, containing the 1,2-d(GpG) or 1,3-d(GpTpG) intrastrand crosslinks, enforced a characteristic rotational positioning of the DNA around the histone octamer such that the Pt adduct faced inward towards the histone core (Ober and Lippard, 2008). Increased solvent accessibility of the platinated DNA strand was observed, and this indicated it might be caused by a structural perturbation in proximity of the DNA lesion. In addition, the nucleosomes treated with cisplatin exhibited a significant decrease in heat-induced mobility (Wu et al., 2008). These effects also indicated that a cisplatin assault could inhibit transcription by altering the native nucleosomal organization, and limiting the nucleosomal sliding that protected access of the RNA polymerase to the DNA template.

It has been suggested that inhibition of transcription by cisplatin was a critical determinant of cell-cycle arrest in the G2 phase because cells could not synthesize the mRNA necessary to pass into mitosis, and this eventually led to apoptosis. Possible mechanisms to explain this inhibitory process can be divided into three categories; (1) hijack of transcription factors (2) physical block of RNA polymerase, and (3) inhibition of chromatin remodeling (Todd & Lippard, 2009). A number of proteins have been identified that specifically recognize the distorted Pt-DNA adducts, including transcription factors. The upstream binding factor (UBF), a member of the HMG-domain proteins, is a ribosomal RNA transcription factor. hUBF can bind the 1,2-intrastrand adducts with a high K_d of 60 pM (Jordan & Carmo-Fonseca, 1998). Treatment of DNA with cisplatin inhibited ribosomal RNA synthesis by competing with hUBF for its natural binding site in an in vitro transcription assay (Zhai et al., 1998). The TATA-binding protein (TBP) is a critical transcription factor for all three mammalian RNA polymerases (pol I, II, and III). TBP binding to the DNA duplex, containing the 1,2-intrastrand d(GpG) crosslinks of cisplatin, was similar to that of the TATA-promoter binding in terms of structural and affinity aspects with a K_d of 0.3 nM (Jung et al., 2001). It was shown that TBP interacted directly with cisplatin-damaged DNA, and the introduction of exogenous cisplatin-modified DNA into the HeLa whole cell extract could sequester TBP and inhibit transcription 3-to 4-fold more than undamaged DNA (Vichi et al., 1997). Collectively, the failure of RNA synthesis resulted from the hijack of transcription factors by Pt-DNA adducts, that prevented the assembly of transcriptional elongation complexes at their normal promoter sequence and inhibited the transcriptional process. Significant reduction in transcriptional transactivation of cisplatin-modified *BRCA1* in the presence of a second expression vector containing multiple cisplatin-damaged sites could address the lack of or the unavailability of cellular transcription factors at cisplatin-*BRCA1* lesions. Damage of *BRCA1*, if not properly repaired, may lead to its functional impairment in cancerous cells which ultimately induce programed cell death.

7.3 Cisplatin binding to the BRCA1 RING domain

The types of adduct formed with cisplatin are distinctive and dependent on the accessibility of the platinum center and protein side-chains (Ivanov et al., 1998; Peleg-Shulman et al., 2002). The BRCA1 RING domain has been found to form favourable intramolecular and intermolecular cross-links caused by cisplatin (Atipairin et al., 2010). Although cisplatin has been demonstrated to induce protein dimerization and has caused perturbations in some protein structures, the secondary structure of the BRCA1 RING domain in the apo-form was maintained and underwent more folded structural rearrangement after increasing cisplatin concentrations as judged by an increase in the negative CD spectra at 208 and 220 nm. It was possible that cisplatin might bind to the unoccupied Zn^{2+}-binding sites and caused the structural changes. The binding constant of the in vitro platination was $3.00 \pm 0.11 \times 10^6$ M^{-1}, and the free energy of binding (ΔG) was -8.68 kcal Mol^{-1}. In addition, the CD spectra of BRCA1 pre-incubated with Zn^{2+} gave identical profiles to indicate that cisplatin could interact with other residues rather than the Zn^{2+}-binding sites and barely affected the overall conformation of the Zn^{2+}-bound BRCA1. In order to locate the binding site of cisplatin on the BRCA1 (1-139) protein, in-gel tryptic digestion of the free BRCA1 and the cisplatin-BRCA1 adducts (molar ratio 1:1) were subjected to analysis by LC-MS. A unique fragment ion of 656.29^{2+} was obtained from the cisplatin-BRCA1 adduct digests. Tandem mass spectrometric analyses of this fragment ion indicated that the ion arose from

$[Pt(NH_3)_2(OH)]^+$ that was attached to a BRCA1 peptide [111]ENNSPEHLK[119] (Fig. 7) (Berners-Price et al., 1992).

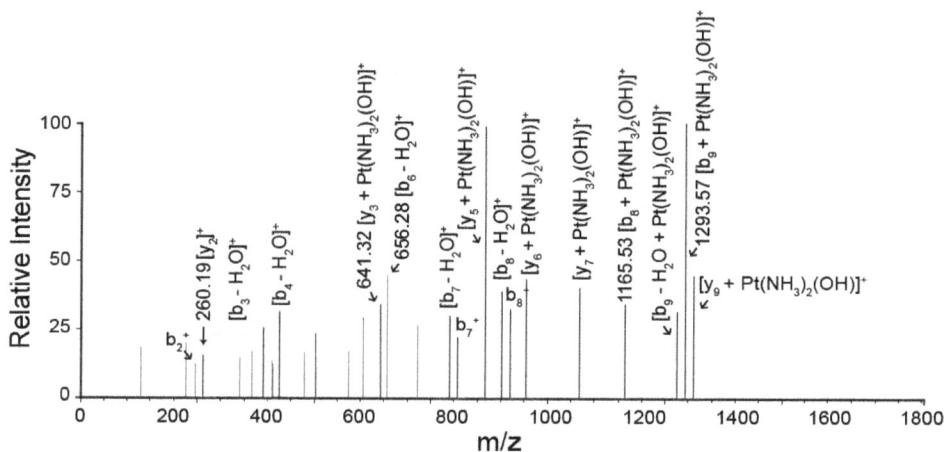

Fig. 7. The product-ion spectrum of the MS/MS analysis for the 656.29^{2+} ion. It indicated that $[Pt(NH_3)_2(OH)]^+$ is attached to a peptide [111]ENNSPEHLK[119] of BRCA1 (Atipairin et al., 2010).

7.4 Thermal stability of the cisplatin-BRCA1 adducts

Thermal denaturation was monitored by circular dichroism (CD) to follow heat-induced unfolding which determined the effect of cisplatin binding on the stability of the BRCA1 RING domain. The BRCA1(1-139) protein pre-incubated with or without Zn^{2+} was incubated with cisplatin, and the CD spectra showed identical changes with an increase in ellipticity when the temperature was raised from 15°C to 95°C (Fig. 8). It indicated that the folded proteins gradually lost their ordered structures. When cooling to 20°C after being heated at 95°C, the CD spectrum partially recovered. This indicated that the reversibility of the unfolding/refolding process was incomplete. The melting temperatures of the BRCA1(1-139) proteins were about 74°C and 83°C in the absence and presence of Zn^{2+}, respectively (Fig. 9). This indicated that the BRCA1 RING domain was more thermostable by about 9°C upon Zn^{2+}-binding. Thus, it supported the important role of Zn^{2+} in the determination and stabilization of the local secondary structure in the RING domain. It was notable that cisplatin at a concentration of 10 µM had similar melting temperatures to those observed for Zn^{2+} binding to the BRCA1 RING domain. However, higher melting temperatures were observed at a 10-fold concentration of cisplatin (100 µM). These data indicated that cisplatin binding to the BRCA1 RING domain conferred an enhanced thermostability by 13°C. Resistance to thermal denaturation of the cisplatin-modifed BRCA1 RING domain might result from the favourably intramolecular and intermolecular crosslinks driven by the free energy (Atipairin et al., 2010).

7.5 Inactivation of BRCA1 E3 ligase activity by cisplatin

To gain further insights into the functional consequences of cisplatin-induced BRCA1, the BRCA1 RING protein was platinated in vitro by cisplatin at various concentrations. The results showed that the relative E3 ligase activity was inversely proportional to the

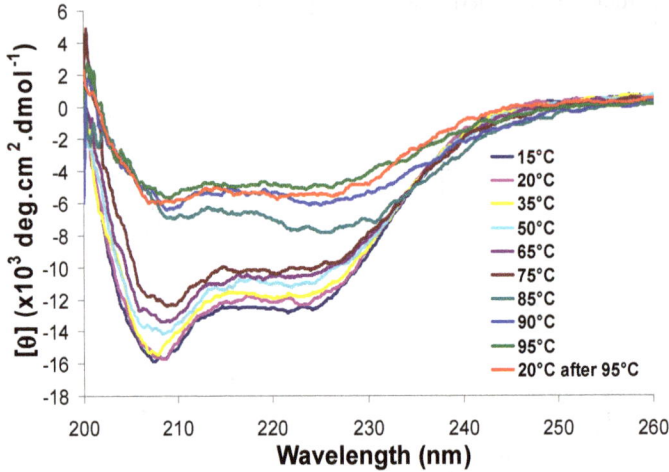

Fig. 8. Thermal transition of the cisplatin-BRCA1 adducts in the presence of Zn^{2+}. The BRCA1(1-139) proteins (10 µM) after pre-incubation with a 3 molar equivalent ratio of Zn^{2+} to protein were mixed with cisplatin concentrations of 10 µM. Samples were incubated in the dark at ambient temperature for 24 h. The measurements were performed from 15°C to 95°C with a heating rate of 1°C/min. After heating to 95°C, the measurement at 20°C was also performed. The CD spectra were plotted between the mean residue ellipticity and wavelength (Atipairin et al., 2010).

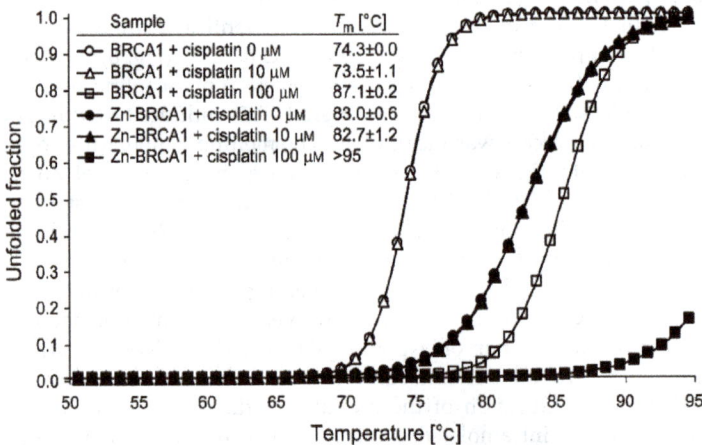

Fig. 9. Thermal denaturation curves of the cisplatin-BRCA1 adducts. The BRCA1(1-139) protein (10 µM) without Zn^{2+} and after pre-incubation with a 3 molar equivalent ratio of Zn^{2+} to protein were mixed with various concentrations of cisplatin (0, 10, and 100 µM). Samples were incubated in the dark at ambient temperature for 24 h before CD measurements. The CD signals at 208 nm were measured, and the unfolded fraction as a function of temperature was plotted (Atipairin et al., 2010).

concentration of the drug (Fig. 10). An increase in platinum concentration was accompanied by a high amount of BRCA1 adducts and a low amount of native BRCA1 protein. To address whether the inhibition of the E3 ligase activity resulted from the formation of BRCA1 adducts or a reduced amount of the BRCA1 subunit, a ten-fold excess amount of the platinated BRCA1 was assayed for the E3 ligase activity. The result demonstrated that platination of BRCA1 was indeed involved in the inhibition of the E3 ligase activity (Atipairin et al., 2011a).

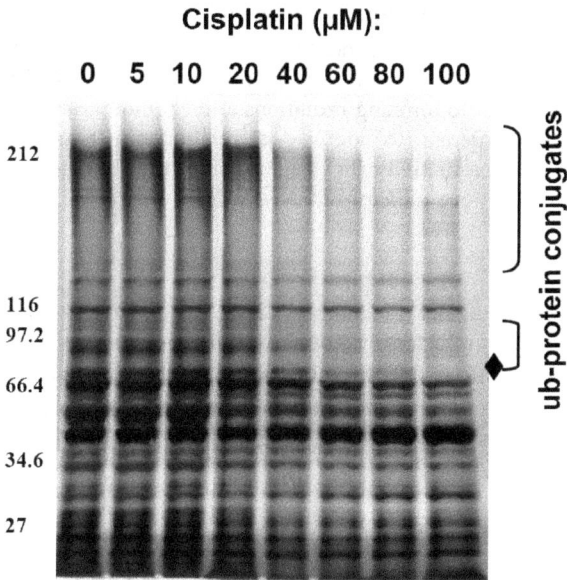

Fig. 10. *In vitro* ubiquitin ligase activity of cisplatin-BRCA1 complexes. Two µg of the drug-BRCA1 adducts with a number of defined concentrations of cisplatin was assayed for the ubiquitin ligase activitiy. An apparent ubiquitinated product (as indicated by the filled diamond) was markedly reduced as the concentration of platinum increased (Atipairin et al., 2011a).

8. Conclusion

We have demonstrated an in vitro inactivation of *BRCA1*/BRCA1 by the anticancer platinum drug cisplatin. The transcriptional activation of cisplatin-modified *BRCA1*, when tested in a "one-hybrid GAL4 transcriptional assay", was inversely proportional to cisplatin doses and was dramatically diminished in the presence of a second expression vector containing multiple cisplatin-damaged sites. This indicates a repair-mediated transcriptional transactivation of cisplatin-damaged *BRCA1* as well as the lack or unavailability of cellular transcription factors at cisplatin-*BRCA1* lesions. The BRCA1 protein contained a preformed structure in the apo-form with structural changes and more resistance to limited proteolysis after Zn^{2+} binding. Cisplatin-bound protein exhibited an enhanced thermostability, resulting from the favourable intermolecular crosslinks driven by the free energy. Only the apo-form, not the holo-form, of BRCA1 underwent a more folded structural rearrangement with the retention of protein structure upon cisplatin binding with the preferential His117 site of the BRCA1 peptide [111]Glu-Asn-Asn-Ser-Pro-Glu-His-Leu-Lys[119]. BRCA1 E3 ubiquitin ligase activity was also inactivated by the drug. These data could raise the possibility of selectively targeting the BRCA1 DNA repair for cisplatin in cancer chemotherapy.

As mentioned earlier, the BRCA1-BARD1 RING complex has an E3 ubiquitin ligase function that plays essential roles in response to DNA damage and DNA repair. Evidence from several preclinical and clinical studies have provided data showing that many cancer-predisposing mutations within the BRCA1 RING domain demonstrated a loss of ubiquitin ligase and repair of DNA double-strand break activities (Atipairin, et al., 2011a, 2011b; Morris et al., 2006, 2009; Ransburgh et al., 2010). Furthermore, the BRCA1-associated cancers conferred a hypersensitivity to ionizing radiations and chemotherapeutic agents. Therefore, it would be of great interest to identify a relationship between BRCA1-mediated ubiquitination and chemosensitivity by approaching the BRCA1 RING domain as a potentially molecular target or predictor with cisplatin.

9. Acknowledgments

The author would like to thank all the members of BRCA1 lab for discussion and comments. This work was financial supported by grants of the Synchrotron Light Research Institute (Public organization) (1-2548/LS01), the National Research Council of Thailand (02011420-0003, PHA530097S), Prince of Songkla University (PHA530188S), and Dr. Brian Hodgson for the proof reading and assistance with the English.

10. References

Ahmad, S. (2010). Platinum-DNA interactions and subsequent cellular processes controlling sensitivity to anticancer platinum complexes. *Chem Biodivers.* 7 (3): 543-566.

Amir, E., Seruga, B., Serrano, R. & Ocana A. (2010). Targeting DNA repair in breast cancer: A clinical and translational update. *Cancer Treat Rev.* 36 (7): 557-565.

Ang, W.H., Myint, M. & Lippard, S.J. (2010). Transcription inhibition by platinum-DNA cross-links in live mammalian cells. *J Am Chem Soc.* 132 (21): 7429-7435.

Ashworth, A. (2008). A synthetic lethal therapeutic approach: poly(ADP) ribose polymerase inhibitors for the treatment of cancers deficient in DNA double-strand break repair. *J Clin Oncol.* 26 (22): 3785-3790.

Atipairin, A., Canyuk, B. & Ratanaphan, A. (2010). Cisplatin affects the conformation of apo-form, not holo-form, of BRCA1 RING finger domain and confers thermal stability. *Chem Biodivers.* 7 (8): 1949-1967.

Atipairin, A., Canyuk, B. & Ratanaphan, A. (2011a). The RING heterodimer BRCA1-BARD1 is a ubiquitin ligase inactivated by the platinum-based anticancer drugs. *Breast Cancer Res Treat.* 126 (1): 203-209.

Atipairin, A., Canyuk, B. & Ratanaphan, A. (2011b). Substitution of aspartic acid with glutamic acid at position 67 of BRCA1 RING domain retains ubiquitin ligase activity and zinc(II)-binding with a reduced transition temperature. *J Biol Inorg Chem.* 16 (2): 217-226.

Bau, D.T., Fu, Y.P., Chen, S.T., Cheng, T.C., Yu, J.C., Wu, P.E., et al. (2004). Breast cancer risk and the DNA double-strand break end-joining capacity of nonhomologous end-joining genes are affected by BRCA1. *Cancer Res.* 64 (14): 5013-5019.

Baumann, P., Benson, F.E. & West, S.C. (1996). Human Rad51 protein promotes ATP-dependent homologous pairing and strand transfer reactions *in vitro*. *Cell*. 87 (4): 757-766.

Berners-Price, S.J., Frenkiel, T.A., Frey, U., Ranford, J.D. & Sadler, P.J. (1992). Hydrolysis products of cisplatin: pKa determinations via [^{1}H,^{15}N] NMR spectroscopy. *J Chem Soc., Chem Commun.* 789-791.

Bhattacharyya, A., Ear, U.S., Koller, B.H., Weichselbaum, R.R. & Bishop, D.K. (2000). The breast cancer susceptibility gene *BRCA1* is required for subnuclear assembly of Rad51 and survival following treatment with the DNA cross-linking agent cisplatin. *J Biol Chem.* 275 (31): 23899-23903.

Bose, R.N., Yang, W.W. & Evanics, F. (2005). Structural perturbation of a C4 zinc-finger module by *cis*-diamminedichloroplatinum(II): insights into the inhibition of transcription processes by the antitumor drug. *Inorg Chim Acta.* 358 (10): 2844-2854.

Brzovic, P.S., Meza, J., King, M.C. & Klevit, R.E. (1998). The cancer-predisposing mutation C61G disrupts homodimer formation in the NH2-terminal BRCA1 RING finger domain. *J Biol Chem.* 273(14): 7795-7799.

Brzovic, P.S., Rajagopal, P., Hoyt, D.W., King, M.C. & Klevit, R.E. (2001). Structure of a BRCA1-BARD1 heterodimeric RING-RING complex. *Nat Struct Biol.* 8(10): 833-837.

Byrski, T., Huzarski, T., Dent, R., Gronwald, J., Zuziak, D., Cybulski, C., et al. (2009) Response to neoadjuvant therapy with cisplatin in BRCA1-positive breast cancer patients. *Breast Cancer Res Treat.* 115(2): 359-363.

Casini, A., Gabbiani, C., Michelucci, E., Pieraccini, G., Moneti, G., Dyson, P.J., et al. (2009). Exploring metallodrug-protein interactions by mass spectrometry: comparisons between platinum coordination complexes and an organometallic ruthenium compound. *J Biol Inorg Chem.* 14(5): 761-770.

Casini, A., Guerri, A., Gabbiani, C. & Messori, L. (2008). Biophysical characterisation of adducts formed between anticancer metallodrugs and selected proteins: New insights from X-ray diffraction and mass spectrometry studies. *J Inorg Biochem.* 102(5-6): 995-1006.

Chang, I.Y., Kim, M.H., Kim, H.B., Lee, D.Y., Kim, S.H., Kim, H.Y,, et al., (2005). Small interfering RNA-induced suppression of ERCC1 enhances sensitivity of human cancer cells to cisplatin. *Biochem Biophys Res Commun.* 327(1): 225-233.

Chapman, M.S. & Verma, I.M. (1996). Transcriptional activation by BRCA1. *Nature.* 382 (6593): 678-679.

Chen, A., Kleiman, F.E., Manley, J.L., Ouchi, T. & Pan, Z.Q. (2002). Autoubiquitination of the BRCA1-BARD1 RING ubiquitin ligase. *J Biol Chem.* 277(24): 22085-22092.

Chiba, N. & Parvin, J.D. (2001). Redistribution of BRCA1 among four different protein complexes following replication blockage. *J Biol Chem.* 276(42): 38549-38554.

Corda, Y., Job, C., Anin, M-F, Leng, M. & Job, D. (1991). Transcription by eukaryotic and prokaryotic RNA polymerases of DNA modified at a d(GG) or a d(AG) site by the antitumor drug cis-diamminedichloroplatinum(II). *Biochemistry.* 30(1): 222-230.

Cubo, L., Thomas, D.S., Zhang, J., Quiroga, A.G., Navarro-Ranninger, C. & Berners-Price, S.J. (2009). [^{1}H,^{15}N] NMR studies of the aquation of cis-diamine platinum(II) complexes. *Inorg Chim Acta.* 362(3): 1022-1026.

Damsma, G.E., Alt, A., Brueckner, F., Carell, T. & Cramer, P. (2007). Mechanism of transcriptional stalling at cisplatin-damaged DNA. *Nat Struc Mol Biol.* 14(12): 1127-1133.

de Paula, Q.A., Mangrum, J.B. & Farrell, N.P. (2009). Zinc finger proteins as templates for metal ion exchange: Substitution effects on the C-finger of HIV nucleocapsid NCp7 using M(chelate) species (M=Pt, Pd, Au). *J Inorg Biochem.* 103(10): 1347-1354.

Dizin, E. & Irminger-Finger, I. (2010). Negative feedback loop of BRCA1-BARD1 ubiquitin ligase on estrogen receptor α stability and activity antagonized by cancer-associated isoform of BARD1. *Int J Biochem Cell Biol.* 42(5): 693-700.

Eakin, C.M., Maccoss, M.J., Finney, G.L. & Klevit, R.E. (2007). Estrogen receptor α is a putative substrate for the BRCA1 ubiquitin ligase. *Proc Natl Acad Sci USA.* 104(14): 5794-5799.

Eastman, A. (1986). Reevaluation of interaction of cis-dichloro(ethylenediamine)platinum(II) with DNA. *Biochemistry.* 25 (13): 3912-3915.

Fan, S., Ma, Y.X., Wang, C., Yuan, R.Q., Meng, Q., Wang, J.A., et al. (2002). p300 Modulates the BRCA1 inhibition of estrogen receptor activity. *Cancer Res.* 62(1): 141-151.

Fan, S., Ma, Y.X., Wang, C., Yuan, R.Q., Meng, Q., Wang, J-N., et al. (2001). Role of direct interaction in BRCA1 inhibition of estrogen receptor activity. *Oncogene.* 20(1): 77-87.

Fan, S., Wang, J., Yuan, R., Ma, Y., Meng, Q., Erdos, M.R., et al. (1999). BRCA1 inhibition of estrogen receptor signaling in transfected cells. *Science.* 284(5418): 1354-1356.

Farmer, H., McCabe, N., Lord, C.J., Tutt, A.N., Johnson, D.A., Richardson, T.B., et al. (2005). Targeting the DNA repair defect in BRCA mutant cells as a therapeutic strategy. *Nature.* 434(7035): 917-921.

Fedier, A., Steiner, R.A., Schwarz, V.A., Lenherr, L., Haller, U. & Fink, D. (2003). The effect of loss of Brca1 on the sensitivity to anticancer agents in p53-deficient cells. *Int J Oncol.* 22(5): 1169-1173.

Fichtinger-Schepman, A.M., Van der Veer, J.L., den Hartog, J.H., Lohman, P.H. & Reedijk, J. (1985). Adducts of the antitumor drug cis-diamminedichloroplatinum(II) with DNA: formation, identification, and quantitation. *Biochemistry.* 24 (3): 707-713.

Font, A., Taron, M., Gago, J.L., Costa, C., Sánchez, J.J., Carrato, C., et al. (2010). BRCA1 mRNA expression and outcome to neoadjuvant cisplatin-based chemotherapy in bladder cancer. Ann Oncol. doi:10.1093/annonc/mdq333

Hall, J.M., Lee, M.K., Newman, B., Morrow, J.E., Anderson, L.A., Huey, B., et al. (1990). Linkage of early onset familial breast cancer to chromosome 17q21. *Science.* 250(4988): 1684-1689.

Hashizume, R., Fukuda, M., Maeda, I., Nishikawa, H., Oyake, D., Yabuki, Y., et al. (2001). The RING heterodimer BRCA1-BARD1 is a ubiquitin ligase inactivated by a breast cancer-derived mutation. *J Biol Chem.* 276(18): 14537-14540.

Helleday, T., Petermann, E., Lundin, C., Hodgson, B. & Sharma, R.A. (2008). DNA repair pathways as targets for cancer therapy. *Nat Rev Cancer.* 8(3): 193-204.

Horwitz, A.A., Affar, E.B., Heine, G.F., Shi, Y. & Parvin, J.D. (2007). A mechanism for transcriptional repression dependent on the BRCA1 E3 ubiquitin ligase. *Proc Natl Acad Sci USA.* 104(16): 6614-6619.

Huen, M.S.Y., Sy, S.M.H. & Chen, J. (2010). BRCA1 and its toolbox for the maintenance of genome integrity. *Nat Rev Mol Cell Biol.* 11(2): 138-148.

Husain, A., He, G., Venkatraman, E.S. & Spriggs, D.R. (1998). BRCA1 up-regulation is associated with repair-mediated resistance to cis-diamminedichloroplatinum(II). *Cancer Res.* 58(6): 1120-1123.

Ishidaa, R., Takaokaa, Y., Yamamotoa, S., Miyazakia, T., Otakab, M., Watanabeb, S., et al. (2008). Cisplatin differently affects amino terminal and carboxyl terminal domains of HSP90. *FEBS Lett.* 582(28): 3879-3883.

Ivanov, A.I., Christodoulou, J., Parkinson, J.A., Barnham, K.J., Tucker, A., Woodrow, J., et al. (1998). Cisplatin binding sites on human albumin. *J Biol Chem.* 273(24): 14721-14730.

James, C.R., Quinn, J.E.,Mullan, P.B., Johnston, P.G. & Harkin, D.P. (2007). BRCA1, a potential predictive biomarker in the treatment of breast cancer. *Oncologist.* 12(2): 142-150.

Jensen, D.E., Proctor, M., Marquis, S.T., Gardner, H.P., Ha, S.I., Chodosh, L.A., et al. (1998). BAP1: a novel ubiquitin hydrolase which binds to the BRCA1 RING finger and enhances BRCA1-mediated cell growth suppression. *Oncogene.* 16(9): 1097-1112.

Jin, Y., Xu, X.L., Yang, M.C., Wei, F., Ayi, T.C., Bowcock, A.M. et al., (1997). Cell cycle-dependent colocalization of BARD1 and BRCA1 proteins in discrete nuclear domains. *Proc Natl Acad Sci USA* 94 (22): 12075-12080.

Jordan, P. & Carmo-Fonseca, M. (1998). Cisplatin inhibits synthesis of ribosomal RNA *in vivo.* *Nucleic Acids Res.* 26(12): 2831-2836.

Jung, Y. & Lippard, S.J. (2003). Multiple states of stalled T7 RNA polymerase at DNA lesions generated by platinum anticancer agents. *J Biol Chem.* 278(52): 52084-52092.

Jung, Y., Mikata, Y. & Lippard, S.J. (2001). Kinetic studies of the TATA-binding protein interaction with cisplatin-modified DNA. *J Biol Chem.* 276(47): 43589-43596.

Kelland, L. (2007). The resurgence of platinum-based cancer chemotherapy. *Nat Rev Cancer.* 7(8): 573-584.

Kelley, M.R., & Fishel, M.L. (2008). DNA repair proteins as molecular targets for cancer therapeutics. *Anticancer Agents Med Chem.* 8(4): 417-425.

Kennedy, R.D., Quinn, J.E., Mullan, P.B., Johnston, P.G. & Harkin, D.P. (2004). The role of BRCA1 in the cellular response to chemotherapy. *J Natl Cancer Inst.* 96(22): 1659-1668.

Lafarge, S., Sylvain, V., Ferrara, M. & Bignon, Y.J. (2001). Inhibition of BRCA1 leads to increased chemoresistance to microtubule-interfering agents, an effect that involves the JNK pathway. *Oncogene.* 20(45): 6597-6606.

Lieberman, H.B. (2008). DNA damage repair and response proteins as targets for cancer therapy. *Curr Med Chem.* 15(4): 360-367.

Litman, R., Gupta, R., Brosh, Jr R.M. & Cantor, S.B. (2008). BRCA-FA pathway as a target for anti-tumor drugs. *Anticancer Agents Med Chem.* 8(4): 426-430.

Ljungman, M., Zhang, F., Chen, F., Rainbow, A.J. & McKay, B.C. (1999). Inhibition of RNA polymerase II as a trigger for the p53 response. *Oncogene.* 18(3): 583-592.

Mark, W.Y., Liao, J.C., Lu, Y., Ayed, A., Laister, R., Szymczyna, B., et al. (2005). Characterization of segments from the central region of BRCA1: an intrinsically disordered scaffold for multiple protein-protein and protein-DNA interactions? *J Mol Biol.* 345(2): 275-287.

Miki, Y., Swensen, J., Shattuck-Eidens, D., Futreal, P.A., Harshman, K., Tavtigian, S., et al. (1994). A strong candidate for the breast and ovarian cancer susceptibility gene BRCA1. *Science.* 266(5182): 66-71.

Monteiro, A.N., August, A. & Hanafusa, H. (1996). Evidence for a transcriptional activation function of BRCA1 C-terminal region. *Proc Natl Acad Sci USA* 93(24): 13595-13599.

Morris, J.R., Boutell, C., Keppler, M., Densham, R., Weekes, D., Alamshah, A., et al. (2009). The SUMO modification pathway is involved in the BRCA1 response to genotoxic stress. *Nature.* 462(7275): 886-890.

Morris, J.R., Pangon, L., Boutell, C., Katagiri, T., Keep, N.H. & Solomon, E. (2006). Genetic analysis of BRCA1 ubiquitin ligase activity and its relationship to breast cancer susceptibility. *Hum Mol Genet.* 15(4): 599-606.

Moynahan, M.E., Cui, T.Y. & Jasin, M. (2001). Homology-directed dna repair, mitomycin c resistance, and chromosome stability is restored with correction of a Brca1 mutation. *Cancer Res.* 61(12): 4842-4850.

Mullan, P.B., Quinn, J.E. & Harkin, D.P. (2006). The role of BRCA1 in transcriptional regulation and cell cycle control. *Oncogene.* 25(43): 5854-5863.

Musah, R.A. (2004). The HIV-1 nucleocapsid zinc finger protein as a target of antiretroviral therapy. *Curr Top Med Chem.* 4(15): 1605-1622.

Neault, J.F. & Tajmir-Riahi, H.A. (1998). Interaction of cisplatin with human serum albumin. Drug binding mode and protein secondary structure. *Biochim Biophys Acta.* 1384(1): 153-159.

Nishikawa, H., Wu, W., Koike, A., Kojima, R., Gomi, H., Fukuda, M., et al. (2009). BRCA1-associated protein 1 interferes with BRCA1/BARD1 RING heterodimer activity. *Cancer Res.* 69(1): 111-119.

Ober, M. & Lippard, S.J. (2008). A 1,2-d(GpG) cisplatin intrastrand cross-link influences the rotational and translational setting of DNA in nucleosomes. *J Am Chem Soc.* 130 (9): 2851-2861.

O'Donovan, P.J. & Livingston, D.M. (2010). BRCA1 and BRCA2: breast/ovarian cancer susceptibility gene products and participants in DNA double-strand break repair. *Carcinogenesis.* 31(6): 961-967.

Parvin, J.D. (2009). The BRCA1-dependent ubiquitin ligase, gamma-tubulin, and centrosomes. *Environ Mol Mutagen.* 50(8): 649-653.

Peleg-Shulman, T., Najajreh, Y. & Gibson, D. (2002). Interactions of cisplatin and transplatin with proteins: Comparison of binding kinetics, binding sites and reactivity of the Pt-protein adducts of cisplatin and transplatin towards biological nucleophiles. *J Inorg Biochem.* 91(1): 306-311.

Pinto, A.L. & Lippard, S.J. (1985). Sequence-dependent termination of in vitro DNA synthesis by cis- and trans-diamminedichloroplatinum(II). *Proc Natl Acad Sci USA* 82(14): 4616-4619.

Powell, S.N. & Bindra, R.S. (2009). Targeting the DNA damage response for cancer therapy. *DNA Repair.* 8(9): 1153-1165.

Quinn, J.E., Carser, J.E., James, C.R., Kennedy, R.D. & Harkin, D.P. (2009). BRCA1 and implications for response to chemotherapy in ovarian cancer. *Gynecol Oncol.* 113(1): 134-142.

Quinn, J.E., James, C,R., Stewart, G.E., Mulligan, J.M., White, P., Chang, G.K., et al. (2007). BRCA1 mRNA expression levels predict for overall survival in ovarian cancer after chemotherapy. *Clin Cancer Res.* 13(24): 7413-7420.

Quinn, J.E., Kennedy, R.E., Mullan, P.B., Gilmore, P.M., Carty, M., Johnston, P.G., et al. (2003). BRCA1 functions as a differential modulator of chemotherapy-induced apoptosis. *Cancer Res.* 63(19): 6221-6228.

Rahman, N., Seal, S., Thompson, D., Kelly, P., Renwick, A., Elliott, A., et al. (2007). PALB2, which encodes a BRCA2-interacting protein, is a breast cancer susceptibility gene. *Nature Genet.* 39(2): 165-167.

Ransburgh, D.J., Chiba, N., Ishioka, C., Toland, A.E. & Parvin, J.D. (2010). Identification of breast tumor mutations in BRCA1 that abolish its function in homologous DNA recombination. *Cancer Res.* 70(3): 988-995.

Ratanaphan, A., Canyuk, B., Wasiksiri, S. & Mahasawat, P. (2005). In vitro platination of human breast cancer suppressor gene 1 (BRCA1) by the anticancer drug carboplatin. Biochim Biophys Acta. 1725(2): 145-151.

Ratanaphan, A., Wasiksiri, S., Canyuk, B. & Prasertsan, P. (2009). Cisplatin-damaged BRCA1 exhibits altered thermostability and transcriptional transctivation. *Cancer Biol Ther.* 8(10): 890-898.

Rogakou, E.P., Pilch, D.R., Orr, A.H., Ivanova, V.S. & Bonner, W.M. (1998). DNA double-stranded breaks induce histone H2AX phosphorylation on serine 139. *J Biol Chem.* 273(10): 5858-5868.

Rosen, E.M., Fan, S., Pestell, R.G. & Goldberg, I.D. (2003). BRCA1 gene in breast cancer. *J Cell Physiol.* 196(1): 19-41.

Rowling, P.J.E., Cook, R. & Itzhaki, L.S. (2010). Toward classification of BRCA1 missense variants using a biophysical approach. *J Biol Chem.* 285(26): 20080-20087.

Ruffner, H., Joazeiro, C.A., Hemmati, D., Hunter, T. & Verma, I.M. (2001). Cancer-predisposing mutations within the RING domain of BRCA1: loss of ubiquitin protein ligase activity and protection from radiation hypersensitivity. *Proc Natl Acad Sci USA.* 98(9): 5134-5139.

Sato, K., Hayami, R., Wu, W., Nishikawa, T., Nishikawa, H., Okuda, Y., et al. (2004). Nucleophosmin/B23 is a candidate substrate for the BRCA1-BARD1 ubiquitin ligase. *J Biol Chem.* 279(30): 30919-30922.

Scully, R., Chen, J., Plug, A., Xiao, Y., Weaver, D., Feunteun, J., et al. (1997). Association of BRCA1 with Rad51 in mitotic and meiotic cells. *Cell.* 88(2): 265-275.

Shuck, S.C., Short, E.A. & Turchi, J.J. (2008). Eukaryotic nucleotide excision repair: from understanding mechanisms to influencing biology. *Cell Res.* 18(1): 64-72.

Silver, D.P., Richardson, A.L., Eklund, A.C., Wang, Z.C., Szallasi, Z., Li, O., et al. (2010). Efficacy of neoadjuvant cisplatin in triple-negative breast cancer. *J Clin Oncol.* 28(7): 1145-1153.

Somasundaram, K., Zhang, H., Zeng, Y.X., Houvras, Y., Peng, Y., Zhang, H. et al., (1997). Arrest of the cell cycle by the tumour-suppressor BRCA1 requires the CDK-inhibitor p21WAF1/CiP1. *Nature.* 389 (6647): 187-190.

Starita, L.M., Horwitz, A.A., Keogh, M.C., Ishioka, C., Parvin, J.D. & Chiba, N. (2005). BRCA1/BARD1 ubiquitinate phosphorylated RNA polymerase II. *J Biol Chem.* 280(26): 24498-24505.

Starita, L.M. & Parvin, J.D. (2003). The multiple nuclear functions of BRCA1: transcription, ubiquitination and DNA repair. *Curr Opin Cell Biol.* 15(3): 345-350.

Stucki, M., Clapperton, J.A., Mohammad, D., Yaffe, M.B., Smerdon, S.J. & Jackson, S.P. (2005). MDC1 directly binds phosphorylated histone H2AX to regulate cellular responses to DNA double-strand breaks. *Cell.* 123(7): 1213-1226.

Sun, X., Tsang, C-N. & Sun, H. (2009). Identification and characterization of metallodrug binding proteins by (metallo)proteomics. *Metallomics.* 1(1): 25-31.

Swisher, E.M., Sakai, W., Karlan, B.Y., Wurz, K., Urban, N. & Taniguchi, T. (2008). Secondary BRCA1 mutations in BRCA1-mutated ovarian carcinomas with platinum resistance. *Cancer Res.* 68(8): 2581-2586.

Sy, S.M.H., Huen, M.S.Y. & Chen, J. (2009). PALB2 is an integral component of the BRCA complex required for homologous recombination repair. *Proc Natl Acad Sci USA.* 106(17): 7155-7160.

Taron, M., Rosell, R., Felip, E., Mendez, P., Souglakos, J., Ronco, M.S., et al. (2004). BRCA1 mRNA expression as an indicator of chemoresistance in lung cancer. *Hum Mol Genet.* 13(20): 2443-2449.

Tassone, P., Martino, M.T.D., Ventura, M., Pietragalla, A., Cucinotto, I., Calimeri, T., et al. (2009). Loss of BRCA1 function increases the antitumor activity of cisplatin against human breast cancer xenografts in vivo. *Cancer Biol Ther.* 8(7): 648-653.

Tassone, P., Tagliaferri, P., Perricelli, A., Blotta, S., Quaresima, B., Martelli, M.L., et al. (2003). BRCA1 expression modulates chemosensitivity of BRCA1-defective HCC1937 human breast cancer cells. *Br J Cancer.* 88(8): 1285-1291.

Thakar, A., Parvin, J.D. & Zlatanova, J. (2010). BRCA1/BARD1 E3 ubiquitin ligase can modify histones H2A and H2B in the nucleosome particle. *J Biomol Struct Dyn.* 27(4): 399-406.

Timerbaev, A.R., Hartinger, C.G., Aleksenko, S.S. & Keppler, B.K. (2006). Interactions of antitumor metallodrugs with serum proteins: advances in characterization using modern analytical methodology. *Chem Rev.* 106(6): 2224-2248.

Todd, R.C. & Lippard, S.J. (2009). Inhibition of transcription by platinum antitumor compounds. *Metallomics.* 1(4): 280-291.

Tornaletti, S. (2005). Transcription arrest at DNA damage sites. *Mutat Res.* 577(1-2): 131-145.

Vallon-Christersson, J., Cayanan, C., Haraldsson, K., Loman, N., Bergthorsson, J.T., Brondum-Nielsen, K. et al. (2001). Functional analysis of BRCA1 C-terminal missense mutations identified in breast and ovarian cancer families. *Hum Mol Genet* 10(4): 353-360.

Varma, A.K., Brown, R.S., Birrane, G. & Ladias, J.A. (2005). Structural basis for cell cycle checkpoint control by the BRCA1-CtIP complex. *Biochemistry.* 44(33): 10941-10946.

Vichi, P., Coin, F., Renaud, J-P., Vermeulen, W., Hoeijmakers, J.H.J., Moras, D., et al. (1997). Cisplatin- and UV-damaged DNA lure the basal transcription factor TFIID/TBP. *EMBO J.* 16(24): 7444-7456.

Wang, D., & Lippard, S. (2005). Cellular processing of platinum anticancer drugs. *Nat Rev Drug Discov.* 4(4): 307-320.

Wang, W. & Figg, W.D. (2008). Secondary BRCA1 and BRCA2 alterations and acquired chemoresistance. *Cancer Biol Ther.* 7(7): 1004-1005.

Wei, L., Lan, L., Hong, Z., Yasui, A., Ishioka, C. & Chiba, N. (2008). Rapid recruitment of BRCA1 to DNA double-strand breaks is dependent on its association with Ku80. *Mol Cell Biol.* 28(24): 7380-7393.

Williams, R.S., Green, R. & Glover, J.N. (2001). Crystal structure of the BRCT repeat region from the breast cancer-associated protein BRCA1. *Nat Struct Biol.* 8(10): 838-42.

Williams, R.S., Lee, M.S., Hau, D.D. & Glover, J.N. (2004). Structural basis of phosphopeptide recognition by the BRCT domain of BRCA1. *Nat Struct Mol Biol.* 11(6): 519-525.

Wu, B., Dröge, P. & Davey, C.A. (2008). Site selectivity of platinum anticancer therapeutics. *Nat Chem Biol.* 4(2): 110-112.

Wu, L.C., Wang, Z.W., Tsan, J.T., Spillman, M.A., Phung, A., Xu, X.L., et al. (1996). Identification of a RING protein that can interact in vivo with the *BRCA1* gene product. *Nat Genet.* 14(4): 430-440.

Wu-Baer, F., Ludwig, T. & Baer, R. (2010). The UBXN1 protein associates with autoubiquitinated forms of the BRCA1 tumor suppressor and inhibits its enzymatic function. *Mol Cell Biol.* 30(11): 2787-2798.

Xia, Y., Pao, G.M., Chen, H-W., Verma, I.M. & Hunter, T. (2003). Enhancement of BRCA1 E3 ubiquitin ligase activity through direct interaction with the BARD1 protein. *J Biol Chem.* 278(7): 5255-5263.

Xu, J., Fan, S. & Rosen, E.M. (2005). Regulation of the estrogen-inducible gene expression profile by the breast cancer susceptibility gene BRCA1. *Endocrinology.* 146(4): 2031-2047.

Zdraveski, Z.Z., Mello, J.A., Marinus, M.G. & Essigmann, J.M. (2000). Multiple pathways of recombination define cellular responses to cisplatin. *Chem Biol.* 7(1): 39-50.

Zhai, X., Beckmann, H., Jantzen, H-M. & Essigmann, J.M. (1998). Cisplatin-DNA adducts inhibit ribosomal RNA synthesis by hijacking the transcription factor human upstream binding factor. *Biochemistry.* 37(46): 16307-16315.

Zhang, H., Somasundaram, K., Peng, Y., Tian, H., Bi, D., Weber, B.L., et al. (1998) BRCA1 physically associates with p53 and stimulates its transcriptional activity. *Oncogene.* 16(13): 1713-1721.

Zhang, J. & Powell, S.N. (2005). The role of the BRCA1 tumor suppressor in DNA double-strand break repair. *Mol Cancer Res.* 3(10): 531-539.

Zhao, T. & King, F.L. (2010). A mass spectrometric comparison of the interactions of cisplatin and transplatin with myoglobin. *J Inorg Biochem.* 104(2): 186-192.

Zhong, Q., Boyer, T.G., Chen, P.L. & Lee, W.H. (2002). Deficient nonhomologous end-joining activity in cell-free extracts from Brca1-null fibroblasts. *Cancer Res.* 62(14): 3966-3970.

Zhu, Y., Hub, J., Hu, Y. & Liu, W. (2009). Targeting DNA repair pathways: a novel approach to reduce cancer therapeutic resistance. *Cancer Treat Rev.* 35(7): 590-596.

13

Cell Cycle and DNA Damage Response in Postmitotic Neurons

Inna I. Kruman
Texas Tech University Health Sciences Center
USA

1. Introduction

Cellular DNA copes with constant exposure to different hazards, environmental and intrinsic. This leads to DNA lesions which interfere with transcription and replication and if not repaired or repaired incorrectly, can produce mutations or large-scale genome aberrations that may lead to cell malfunction or cell death and contribute to different pathologies (Jackson, 2009; Sancar et al., 2004). For this reason, virtually every organism is equipped with highly conserved genome surveillance network known as the DNA damage response (DDR) whose function is to sense genome damage and activate several downstream pathways, including cell cycle checkpoints, DNA repair and apoptotic signaling (Rouse & Jackson, 2002; Zhou & Elledge., 2000). The DDR has been investigated mainly in mitotic cells, in which the cell cycle checkpoints are a major contributor to the DDR, required for DNA repair (Stracker et al., 2008). Not much is known about the DDR in postmitotic neurons. It is known, however, that all eukaryotic DNA repair systems operating in proliferating cells also operate in neurons (Fishel et al., 2007; Lee & McKinnon, 2007; Sharma, 2007; Weissman et al., 2007; Wilson, & McNeill, 2007) and that dysfunctional DDR plays an important role in neurodegeneration and is associated with syndromes (e.g. ataxia telangiectasia) characterized by neurological abnormalities (Barzilai, 2010; Rass et al., 2007; Shiloh, 2003, 2006). This suggests the importance of DDR for postmitotic neurons. While the cell cycle checkpoints are part of DDR involved in DNA repair, apoptotic signaling, and cell fate decisions in mitotic cells, their contribution to the DDR of postmitotic neurons remains unclear. Nonetheless, evidence accumulates that DNA damage-initiated apoptosis of postmitotic neurons is associated with cell cycle signaling. Recently, we have demonstrated the importance of the cell cycle activation for DNA repair in postmitotic neurons (Tomashevski et al., 2010). This suggests that the expression of cell-cycle markers (Schmetsdorf et al., 2007, 2009) and DNA repair activity (Sharma, 2007) observed in the brain under physiological conditions may contribute to DNA repair. The involvement of the cell cycle machinery to both DNA repair and DNA damage-initiated apoptosis in postmitotic neurons suggests a potential function of cell cycle checkpoints in the DDR of these postmitotic cells.

This review focuses on the DDR of postmitotic neurons in the context of what is known about the DDR of mitotic cells.

2. DNA damage response in mitotic cells

The genome of eukaryotic cells is continuously exposed to chemicals, ultraviolet (UV) or ionizing radiation (IR), as well as to by-products of intracellular metabolism (e.g.

oxyradicals). The resulting DNA lesions can block genome replication and transcription and result in loss or incorrect transmission of genetic information. If left unrepaired or are repaired incorrectly, DNA lesions lead to mutations or cell death resulting in different abnormalities, including tumorigenesis and neurodegeneration. To maintain genomic integrity during cell division, cells are equipped with highly efficient defense mechanism, the DDR (Reinhardt & Yaffe, 2009) which functions to recognize and remove DNA lesions by DNA repair and eliminate the irreparably damaged cells by apoptosis (Ciccia & Elledge, 2010; Jackson, 2009; Jackson & Bartek 2009). The DDR cascade senses genome damage and activates several pathways, including cell cycle checkpoints, DNA repair and apoptotic programs. Defects in DDR or DNA repair contribute to aging and various disorders, including neurodegenerative diseases and cancer (Jackson & Bartek 2009). This underlines the critical importance of DDR as a regulator of both cell death and survival processes.

2.1 Formation of DDR foci

The earliest events of the DDR involve alterations in chromatin structure (Berkovich et al., 2007; Downs et al., 2007; Smerdon et al., 1978) and the formation of DDR foci. The biochemical details of these processes are poorly understood. Since DDR foci are the sites where DDR signaling originates, the understanding of their formation and functioning is crucial to understanding how DDR activities are exerted. Among the first events of the DDR is recruitment of a mediator complex MRN consisting of Mre11, Rad50, and Nbs1, and phosphorylation of a variant H2A histone - H2AX - near the break, extending for distances up to several megabases (Fernandez-Capetillo et al., 2004). Working together, MRN and phosphorylated H2AX (γH2AX) act as a signal amplifier that recruits additional signaling molecules to the DSB lesion. The MRN complex serves as an initial DSB sensor, at least one component of which (Nbs1) localizes to the break in an H2AX-independent manner (Celeste et al., 2002, 2003) and facilitates the recruitment and activation of the apical DDR phosphoinositide-3-kinase related kinase (PIKK) ataxia telangiectasia mutated (ATM) (Falck et al., 2005; Lee & Paull 2005; Uziel et al., 2003). This is an important step in the DDR. ATM phosphorylates a number of proteins essential in the control of cell-cycle checkpoints, DNA repair and, in the case of excessive DNA damage, cell death (Khanna et al., 2001; Shiloh, 2003). The widely accepted model of ATM activation is its autophosphorylation at Ser 1981 which releases it from the inhibitory homodimer structure, leading to its recruitment to sites of DNA double-strand breaks (DSBs) (Dupre et al., 2006; Lavin & Kozlov, 2007). Among the first proteins recruited to DNA breaks are direct sensors of DNA breaks such as PARP-1 and PARP-2 whose catalytic activity is triggered by their binding to single-strand breaks (SSBs) and DSBs (D'Amours et al. 1999; de Murcia & Ménissier de Murcia, 1994). The Ku70–Ku80 heterodimer and the MRN complex, DSB sensors, directly bind to DSBs (de Jager et al., 2001; Kim et al., 2005; Lisby et al., 2004; Mimori & Hardin, 1986). Ku heterodimer possibly competes with MRN and PARP-1 for binding to DSBs (Clerici et al., 2008; Wang et al., 2006; Zhang et al., 2007). The direct binding of DNA breaks by factors such as Ku and MRN is crucial for the DDR. The recruitment and activation of the apical DDR kinases ATM, ATM rad3-related (ATR), and DNA-dependent protein kinase (DNA-PK) have also well-known significance at sites of DNA breaks and in DDR foci formation (Polo & Jackson, 2011). The functional importance of downstream DDR factors is not well understood which can be explained by complexity and diversity of downstream DDR events, and the fact that multiple systems appear to cooperate to control the formation of DDR foci. However, it is clear the DDR foci formation is critical for the maintenance of genome integrity.

Downstream from direct sensors of DNA breaks, mediator of DNA damage checkpoint protein1 (MDC1) is recruited. This DDR component serves as a binding platform for DNA damage checkpoint and repair proteins (Jungmichel & Stucki, 2010). For example, ATM-dependent phosphorylation of MDC1 creates binding sites for the FHA domain of the ubiquitin E3 ligase RNF8, which in turn promotes the focal accumulation of another mediator of the DNA damage checkpoint, 53BP1 and breast cancer 1 (BRCA1) at DSB sites (Huen et al., 2008; Kolas et al., 2007; Mailand et al. 2006). Constitutive phosphorylation of MDC1 by casein kinase 2 (CK2) mediates DSB focus formation by MRN (Chapman & Jackson, 2008; Melander et al. 2008; Spycher et al., 2008).

The building of multiprotein DDR foci at DNA breaks is tightly controlled by posttranslational protein modifications, including phosphorylation, ubiquitylation, sumoylation, methylation, acetylation, and PARylation (Polo & Jackson, 2011).

ATM, ATR, and DNA-PK phosphorylate H2AX (Burma et al., 2001; Downs et al., 2000; Rogakou et al., 1998; Stiff et al. 2004; Ward & Chen 2001) which is followed by the recruitment of downstream DDR components, including checkpoint mediators such as MDC1 (Hammet et al., 2007; Nakamura et al., 2004; Sanders et al., 2010; Sofueva et al., 2010; Stucki et al., 2005). Phosphorylated H2AX also promotes the recruitment of chromatin modifying complexes, such as p400 (Downs et al., 2004; Kusch et al., 2004; van Attikum et al., 2004, 2007; Xu et al., 2010). In some cases, phosphorylation promotes the dissociation of proteins from sites of DNA breaks. For example, autophosphorylation of DNA-PK catalytic subunit (DNA-PKcs) induces its dissociation from Ku (Chan & Lees-Miller, 1996; Merkle et al., 2002). Recent studies have revealed the critical importance of ubiquitylation, the process whereby ubiquitin (monoubiquitylation) or polyubiquitin (polyubiquitylation) is covalently attached to proteins in the assembly of DDR proteins at DSB sites (Al-Hakim et al., 2010; Messick & Greenberg, 2009; Pickart, 2001). Another critical modification involved in control of DDR foci is histone acetylation near DSBs. Acetylation of histones H3 and H4 is essential for DNA repair (Averbeck & Durante, 2011). The importance of this modification is underlined by the recruitment of several histone acetyltransferases including Hat1, and NuA4 and deacetylases such as Sir2 and Hst1 in budding yeast (Downs et al., 2004; Qin & Parthun, 2006; Tamburini & Tyler 2005) and the Tip60 acetyltransferase and deacetylases (HDAC1, HDAC2, HDAC4, SIRT1 and SIRT6) in mammalian cells (Kaidi et al., 2010; Miller et al., 2010; Murr et al., 2006; O'Hagan et al., 2008; Oberdoerffer et al., 2008). SIRT1 binding in the DSB area has been found to promote the recruitment of NBS1 and RAD51 (Oberdoerffer et al., 2008). Histone H3K56 deacetylation by HDAC1 and HDAC2 regulates recruiting DNA repair factors of nonhomologous end-joining pathway to DSB regions (Miller et al., 2010). Additionally, MOF (males absent on the first)-dependent acetylation of histone H4K16 is important for IR-induced focus formation of MDC1, 53BP1, and BRCA1 in mammalian cells (Li et al., 2010; Sharma et al., 2010). H2AX acetylation by Tip60 promotes H2AX eviction from damaged chromatin, as shown in both Drosophila and mammalian cells (Ikura et al., 2007; Kusch et al., 2004). The acetylation of histone proteins in the DNA break area can regulate the assembly of DDR factors indirectly by modulating chromatin compaction (Lee et al., 2010).

The covalent protein modification process of binding with ADP-ribose polymers, known as PARylation, is one of the earliest events in the DDR. The PARylation is catalyzed by PARP enzymes (Hakme et al. 2008) comprising a large family of proteins, several members of which have clearly identified DDR functions (Citarelli et al., 2010). PARylation is quickly suppressed by PARG (PARG) (Gagne et al., 2006; Hakme et al. 2008; Krishnakumar & Kraus, 2010). It is involved in buildup of the chromatin remodeling factors ALC1 and CHD4 (Ahel et al., 2009; Chou et al., 2010; Gottschalk et al., 2009; Polo et al., 2010), the Polycomb histone-modifying complex (Chou et al., 2010), and the histone variant macro H2A (Timinszky et al., 2009). A contribution of PARylation to the early recruitment of MRN has also been reported (Haince et al., 2008). PARylation can also promote protein dissociation from DNA damage, as shown for the histone chaperone FACT which facilitates chromatin transcription (Heo et al., 2008; Huang et al., 2006).

The mobilization of DDR factors to SSBs or DSBs is very rapid and transient (Gagne et al. 2006; Hakme et al., 2008; Lieber, 2010; Mahaney et al., 2009; Mortusewicz et al., 2007). Responses to DSBs can be markedly influenced by cell cycle status. While accumulation of DDR factors such as γH2AX, MRN, and MDC1 occurs regardless of the cell cycle phase, others - including BRCA1 and RAD51 accumulate effectively only in S/G2 cells (Bekker-Jensen et al., 2006; Jazayeri et al., 2006; Lisby et al., 2004; Sartori et al., 2007). Studies in yeast and mammalian systems have demonstrated that colocalization of DDR proteins rather than DNA damage per se is critical for DNA damage signaling (Bonilla et al., 2008; Soutoglou & Misteli, 2008). One of important regulatory functions of DDR foci is to contribute to the proper coordination of DNA damage signaling and repair with other DNA metabolic activities by inhibiting replication and transcription. In this regard, DNA and histone modifications at sites of DNA breaks have been proposed to contribute to silencing of damaged chromatin (O'Hagan et al., 2008; Shanbhag et al., 2010).

It is now clear that chromatin modifications are an important component DDR network (van Attikum & Gasser, 2009). Recent electron microscopy studies revealed that generation of DSB leads to a rapid, ATP-dependent, local decondensation of chromatin that occurs in the absence of ATM activation. ATM activation itself leads to chromatin relaxation at DSB sites (Ziv et al., 2006). The local and global changes in chromatin organization facilitate recruitment of damage-response proteins and remodeling factors, which further modify chromatin in the vicinity of the DSB and propagate the DNA damage response, thereby providing functional crosstalk between chromatin modification and proteins involved in DDR (Peterson & Cote, 2004; van Attikum & Gasser, 2005).

2.2 DNA repair

DNA repair is essential for maintaining the integrity of the genome. The complicated network of DNA repair mechanisms functions to remove DNA damage by DNA repair pathways. This network include base excision repair (BER), mismatch repair (MMR) and nucleotide excision repair (NER) (Hakem, 2008). One of the most powerful activators of the DDR are DSBs, the most cytotoxic DNA lesions which potentially induce gross chromosomal aberrations, often linked to cell death or cancer (Hopfner, 2009). It has been estimated that a single unrepaired DSB is sufficient for cell lethality (Khanna & Jackson, 2001). DSBs in eukaryotic cells are repaired by two major mechanisms: nonhomologous end-joining (NHEJ), an error-prone ligation mechanism, and a high-fidelity process based on homologous recombination (HR) between sister chromatids that operate in the S and G2 phases of the cell cycle (Pardo et al., 2009; van Gent & der Burg, 2007). DNA damage-induced recruitment of the protein MDC1 dramatically enhances activation of ATM which in turn recruits 53BP1 and BRCA (Bekker-Jensen et al., 2005; Stewart et al., 2003; Stucki et al., 2005). 53BP1 facilitates DNA repair by NHEJ pathway, predominant in mammalian cells

(Moynahan et al., 1999). Several proteins are required for efficient repair of DSB by NHEJ. The core complex consists of the DNA-PK and the ligase IV/XRCC4/XLF complexes. NHEJ initiates upon the binding of two ring-shaped Ku70/Ku80 heterodimers to both DNA broken ends within seconds of the creation of the DNA damage (Lieber, 2010; Mahaney et al., 2009). DNA-PKcs is also recruited to this DNA-Ku scaffold and probably enables the formation of a synaptic complex. In the synaptic complex, the DNA broken ends are positioned next to each other. Depending on the properties of the lesion, some DNA ends must be processed before the final ligation step. For example, a damaged DNA end can contain an aberrant 5' hydroxyl group, aberrant 3' phosphate, damaged base and/or damaged backbone sugar residue. Several enzymes can process such lesions (Chappell et al., 2002; Koch et al., 2004). Werner helicase, associated with Ku70 and Artemis, a structure-specific nuclease, which can cleave DNA hairpin structures and remove 3' overhang DNA may prepare DNA ends (Perry et al., 2006). When the end processing has been accomplished, ligase IV/XRCC4 can catalyze the final ligation reaction. For NHEJ, the Ku70–Ku80 heterodimer plays a central role in recruiting other NHEJ components. In particular, Ku recruits the protein kinase DNA-PKcs (Dvir et al., 1992; Gottlieb & Jackson, 1993) via a specific interaction between DNA-PKcs and the Ku80 C terminus (Gell & Jackson, 1999; Singleton et al., 1999), as well as the downstream NHEJ complex XLF–XRCC4–LigaseIV and the nuclease Artemis (Calsou et al., 2003; Yano et al., 2008).

2.3 Cell cycle checkpoints

Checkpoints are complex kinase signaling pathways that prevent further progression through the cell cycle and coordinate DNA repair with chromosome metabolism and cell-cycle transitions (Houtgraaf et al., 2006; Poehlmann & Roessner, 2010). In response to DNA damage, the checkpoints delay or stop the cell cycle at critical points before or during DNA replication (G1/S and intra-S checkpoints) and before cell division (G2/M checkpoint), thereby preventing replication and segregation of damaged DNA. The critical importance of the cell cycle checkpoint pathways in maintaining genomic integrity is highlighted by the observation that loss or mutation of checkpoint genes is frequently observed in cancer (Kastan & Bartek, 2004). Recent evidence suggests mutually integrated roles of the checkpoint machinery in the activation of DNA repair, chromatin remodelling, modulation of transcriptional programmes and the optional triggering of permanent cell cycle withdrawal by cellular senescence or apoptosis (Bartek & Lukas, 2001; Shiloh, 2003; Zhou & Elledge, 2000). The canonical DDR network has traditionally been divided into two major kinase signaling branches utilizing the upstream kineses ATM and ATR. These kinases control the G1/S, intra-S, and G2/M checkpoints through activating their downstream effector kinases Chk2 and Chk1, respectively (Reinhardt & Yaffe, 2009). The ATM/Chk2 module is activated after DNA DSBs and the ATR/Chk1 pathway responds primarily to DNA SSBs or bulky lesions. Both pathways converge on cell division cycle 25 homolog **A** (Cdc25A), a positive regulator of cell cycle progression, which is inhibited by Chk1- or Chk2-mediated phosphorylation (Poehlmann & Roessner, 2010). Post-translational modifications, such as checkpoint- and cyclin-dependent kinase (CDK)-dependent phosphorylation, ubiquitylation and sumoylation were shown to be crucial for regulation of stability and activity of important components of the checkpoint machinery, thereby regulating important cell cycle events. These post-translational modifications may affect the recruitment of repair proteins to damaged DNA or tune the efficiency or the specificity of the repair machinery towards a certain type of DNA damage and facilitate repair in a specific cell-cycle phase (Branzei & Foiani, 2008). Chromatin structure and compaction is also regulated throughout the cell cycle, and can be influenced by checkpoints and post-translational modifications (Groth et al., 2007; Karagiannis & El-Osta, 2007; Kouzarides, 2007). Thus, cell cycle checkpoints induce G1, S, and G2 cell cycle arrest, recruit repair machinery to the sites of damage, and target irreversibly damaged cells for apoptosis (Kastan & Bartek, 2004; Reinhardt & Yaffe, 2009). ATM and DNA-PK respond mainly to DSBs, whereas ATR is activated in response to incomplete DNA replication due to stalled replication forks (Bartek, & Lukas, 2007; Reinhardt & Yaffe, 2009). During replication, single-stranded DNA becomes opsonized by the replication protein A, which recruits ATR via the ATR-interacting protein to the DNA lesions exposed by stalled forks and orchestrates DNA-topoisomerase II beta-binding protein (TopBP1)-dependent ATR- (Kumagai & Dunphy, 2006) and checkpoint activation (Elledge, 1996). Following activation, the checkpoint transducers transmit and amplify the checkpoint signal to downstream targets such as the DNA-repair apparatus and the cell-cycle machinery (Branzei & Foiani, 2008). DNA synthesis is frequently associated with nucleotide misincorporation, accumulation of nicks and gaps, slippage at repetitive sequences, fork collapse at DNA breaks and aberrant transitions at collapsed forks that cause reversed and/or resected forks. Replication-fork collapse during S phase can often induce DSBs (Branzei & Foiani, 2008). ATR activation by DSBs requires ATM and MRN (Jazayeri et al., 2006; Sartori et al., 2007). It is possible that activation of the tumor suppressor protein, p53, following this replication fork collapse could be detrimental per se, taking into account its implication in apoptosis (Brady & Attardi, 2010). However, there are mechanisms that operate in the S phase to prevent p53 from a death-related activation of p53 transcription programme. It has been speculated that induction of such program within S phase, when the E2F-1 transcription factor (known to cooperate with p53 to induce apoptosis) is highly active, could promote unwanted cell death (Gottifredi et al., 2001).

A major target of ATM in checkpoint pathways is the effector kinase Chk2 that functions to arrest the cell cycle after DSBs by inactivating phosphatases of the Cdc25 family through catalytic inactivation, nuclear exclusion, and/or proteasomal degradation (Aressy & Ducommun, 2008; Busino et al., 2004). This, in turn, prevents Cdc25 family members from dephosphorylating and activating cyclin-CDK complexes, thereby initiating G1/S and G2/M cell cycle checkpoints. In contrast to G1/S or G2/M arrest, cells that experience genotoxic stress during DNA replication only delay their progression through S phase in a transient manner, and if damage is not repaired during this delay they exit S phase and are arrested later when reaching the G2 checkpoint (Bartek et al., 2004).

Following DNA repair, cells must extinguish the DNA damage signal to allow the cells to reenter the cell cycle, but the mechanisms through which this occurs, particularly with respect to the ATM-Chk2 pathway, are poorly understood. Since DNA damage checkpoints respond to as little as a single DNA DSB (Lobrich &, Jeggo, 2007), it has long been assumed that human cells also maintain the G2/M checkpoint until all of the breaks are repaired. Recent evidence, however, shows that the G2 checkpoint in immortalized human cells in culture displays a defined threshold of approximately 10–20 DSBs (Deckbar et al., 2007). Limited checkpoint control was not only apparent in response to IR doses that cause very few DNA DSBs, but also in response to more extensive amounts of DNA damage where checkpoint release occurred at fewer than 10–20 unrepaired DSBs (Deckbar et al., 2007).

Although the fate of cells that continue proliferating in the presence of unrepaired DNA breaks is unclear, and the identity of the rate-limiting DNA damage checkpoint components has yet to be revealed, accumulating evidence suggests that the DNA damage checkpoint machinery can be overridden. G2 checkpoint escape in the presence of unrepaired DNA damage may be particularly common during the evolution of cancer cells (Bartek & Lukas, 2007; Bartkova et al., 2005; Gorgoulis et al., 2005; Kastan & Bartek 2004; Shiloh, 2003).

In mammalian cells, p53 is an important player of the cell cycle checkpoint machinery (Polager & Ginsberg, 2009). During checkpoint control following DNA damage, p53 can either be phosphorylated directly by ATM or ATR (Banin et al., 1998; Hammond et al., 2002; Tibbetts et al., 1999), or indirectly via Chk1 and Chk2 (Hirao et al., 2000, Shieh et al., 2000). Certain cancer-related mutations in the *Chk2* gene can prevent phosphorylation of p53 (Falck et al., 2001; Jazayeri et al., 2006). The effects of Chk1 and Chk2 in the regulation of p53 also depend on the site where p53 is phosphorylated (Polo & Jackson, 2011). A target of p53 in cell cycle checkpoints is the CDK inhibitor p21 (Deng et al., 1995; el-Deiry et al., 1993; Gu et al., 1993; Xiong et al., 1993). P21 functions by inhibiting several CDKs, including CDK4/6, and CDK2 (Harper et al., 1993; Xiong et al., 1993). The silencing of cyclin E - CDK2 activity in late G1 occurs even in cells lacking p53 or p21 (Bartek & Lukas, 2001). These facts argue for a two-wave model of the G1 checkpoint response in mammalian cells, in which the initial, rapid, transient and p53-independent response (Chk2 - Cdc25A – CDK2 axis) is followed by the delayed but more sustained G1 arrest imposed by the Chk1/Chk2–p53–p21–CDK pathway centered on p53 (Bartek & Lukas, 2001; Polager & Ginsberg, 2009). G2 arrest following DNA damage is dependent on the actions of several proteins such as 14-3-3δ which is strongly induced by DNA damage (Chan et al., 1999; Laronga et al., 2000). It acts by sequestering CDK1 -cyclin B complex to prevent entry into mitosis and by modulating the p53-Mdm2 axis (Chan et al., 1999; Yang et al., 2008). *14-3-3δ* is a valid tumor suppressor gene that is frequently inactivated in a number of human malignancies (Ferguson et al., 2000; Henrique et al., 2005; Kuroda et al., 2007). P21 and 14-3-3δ cooperate to maintain G2 arrest following DNA damage. CDK1-cyclin B is subsequently inactivated by p21 in the nucleus (Chan et al., 2000).

The G1/S checkpoint generated through the Chk1/Chk2 - Cdc25A - CDK2 pathway is executed by the active unphosphorylated Cdc25A phosphatase through dephosphorylation of the CDK2–cyclin E complex (Poehlmann & Roessner, 2010). As a consequence, the CDK2 - cyclin E complex is kept in its active form, which causes G1-S transition. Following DNA damage, Chk1 and Chk2 phosphorylate Cdc25A, inducing its degradation. Due to the degradation of the Cdc25A phosphatase, the CDK2-Cyclin E complex remains in its hyperphosphorylated inactive form, culminating in G1/S arrest. P21 potentially participates in the G1/S checkpoint by blocking directly DNA synthesis due to its ability to bind the central region of proliferating cell nuclear antigen (PCNA), a protein that acts as a processivity factor for DNA synthesis in eukaryotic cells (Oku et al., 1998). *In vitro* studies showed that the C-terminal domain of p21 is sufficient to displace DNA replication enzymes from PCNA, thereby blocking DNA synthesis (Chen et al., 1996; Warbrick et al., 1995). The main role of p21 in the G1 checkpoint lies in its ability to inhibit the activity of cyclin E- and cyclin A-CDK2 compexes required for the G1-S transition (Brugarolas et al., 1999). Consequently, pRb remains hypophosphorylated thereby sequestering the transcription factor E2F, whose activity is required for S-phase entry (Ewen et al., 1993).

The G2/M checkpoint generated through the Chk1/Chk2 - Cdc25C - CDK1 pathway is executed by Cdc25C through. dephosphorylation of CDK1-Cyclin B1 complex (Reagan-Shaw et al., 2005; Roshak et al., 2000). Since activating dephosphorylation of only a small amount of CDK1- Cyclin B1 complex is the initiating step for mitotic entry (Hoffmann et al., 1993), and the maintenance of the Cdk1–Cyclin B1 complex in its inactive state blocks entry into mitosis (Poehlmann and Roessner, 2010), CDK1 is the ultimate target of the G2 checkpoint regulation. CDK1 is phosphorylated at two positions by protein kinases Wee1 and Myt1, and is dephosphorylated by Cdc25C phosphatase. G2/M DNA damage checkpoint arrest may be induced by increased phosphorylation of CDK1 by Wee1/Myt1 or by preventing CDK1 dephosphorylation by Cdc25C phosphatase triggered by activated Chk1.

In response to DNA damage, p53 can be phosphorylated at multiple sites by several different protein kinases such as ATM, ATR, DNA-PK, and Chk1/Chk2 (Meek et al., 1994; Milczarek et al., 1997). Phosphorylation impairs the ability of Mdm2 to bind p53, promoting p53 accumulation and activation (Shieh et al., 1997, Tibbetts et al., 1999). Activated p53 upregulates a number of target genes, such as Gadd45 and p21. The accumulation of p21 inhibits CDK2–cyclin E kinase activity, which results in G1 arrest (Bartek et al., 2007). Thus, G1 arrest is a consequence of preventing pRb phosphorylation via inhibition of CDK2. P53 also has functions in the G2/M checkpoint via activating by Chk1/Chk2 which may trigger induction of p21 and by blocking the activity of the mitotic CDK1-Cyclin B1 complex (Stark & Taylor, 2006; Stewart et al., 1995; Taylor & Stark, 2001). In general, one key function of Chk1 and Chk2 activated by ATR and ATM, respectively, manifests in the inactivation of different members of the Cdc25 family by phosphorylation, resulting in a stop of cell cycle progression after DNA damage in the G1/S - or G2/M phases of the cell cycle.

In order for cells to survive following DNA damage, it is important that cell cycle arrest is not only initiated but also maintained for the duration of time necessary for DNA repair (Van Vugt et al., 2010). Mechanisms governing checkpoint initiation versus maintenance appear to be molecularly distinct. This was initially demonstrated by the observation that interference with specific checkpoint components can leave checkpoint initiation intact but disrupt checkpoint maintenance, leading to premature cell cycle reentry accompanied by death by mitotic catastrophe (Bekker-Jensen et al., 2005; Castedo et al., 2004; Deckbar et al., 2007; Lal et al., 2006; Lobrich & Jeggo, 2007). Although the process of checkpoint termination and cell cycle reentry has not been studied extensively, the existing data suggest that inactivation of a checkpoint response is an active process that requires dedicated signaling pathways, such as the the the polo-like kinase 1 (Plk1) pathway (Bartek & Lukas, 2007; van Vugt & Medema, 2004). Interestingly, a number of proteins involved in terminating the maintenance phase of a DNA damage checkpoint also play critical roles in later mitotic events, suggesting the existence of a positive feedback in which the earliest events of mitosis involve the DNA damage checkpoint through unclear mechanism(s). Resumption of cell cycle progression following DNA repair involves switching off the DDR, including disassembly of DDR foci (Bartek & Lukas, 2007. This occurs mainly by reversing the posttranslational modifications associated with focal DDR protein assembly such as PARG-induced erasing PARylation (Gagne et al., 2006) or γH2AX dephosphorylation which plays an important role in terminating checkpoint signaling (Bazzi et al. 2010; Cha et al. 2010; Chowdhury et al., 2008; Macurek et al. 2010; Nakada et al. 2008). Deubiquitylating enzymes have also been implicated in terminating DDR processes (Nicassio et al., 2007; Shao et al. 2009). Deubiquitylation of histone H2A was shown to relieve the inhibition of RNA polymerase II transcription at DSBs (Shanbhag et al., 2010). Automodification is coupled to its dissociation from DNA damage sites, such as DNA-PKcs autophosphorylation and its dissociation from Ku (Chan & Lees-Miller 1996; Hammel et al. 2010; Merkle et al., 2002) and auto-PARylation of PARP-1 and its dissociation from DNA damage sites (Mortusewicz et al. 2007). Checkpoint silencing has been best studied in the budding yeast S. cerevisiae (Leroy et al., 2003; Toczyski et al., 1997; Vaze et al., 2002). The Plk Cdc5 is required for silencing checkpoint signaling, and this requirement appears to be widely conserved, since S. cerevisiae, and human cells all depend on Plks for silencing of the S- or G2 checkpoints, respectively (Syljuasen et al., 2006; Toczyski et al., 1997; van Vugt et al., 2004; Yoo et al., 2004). The activity of Plks has been shown to be required for inactivation of the ATR-Chk1 pathway and the Wee1 axis of checkpoint signaling (Mailand et al., 2006; Mamely et al., 2006; van Vugt et al., 2004; Yoo et al., 2004). DSBs primarily trigger a checkpoint arrest through the ATM-Chk2 signaling pathway. The CDK- and Plk1-dependent phosphorylation of 53BP1 and Chk2 are critical checkpoint-inactivating events in the sensor and effector arms of the G2/M checkpoint pathway, important for checkpoint termination and cell cycle reentry (Van Vugt et al., 2010). This inactivation can take place on chromatin, as reported in human cells (Chowdhury et al., 2008; Nakada et al., 2008). The reversal of H2AX phosphorylation also involves Tip60-dependent histone acetylation and subsequent histone eviction from damaged chromatin in Drosophila and human cells (Jha et al., 2008; Kusch et al. 2004). This is particularly relevant if one considers that DNA damage checkpoints are to respond to very small numbers of DSBs, with some experimental data indicating that 10 -20 DSBs are enough to elicit G2 arrest in human cells (Deckbar et al., 2007), while very few or even a single unrepaired DSB can be sufficient to trigger p53-dependent G1 arrest in human cells (Huang et al., 1996) or cell death in yeast (Bennett et al., 1993).

2.4 DNA damage-induced apoptosis

Programmed cell death, or apoptosis, is a natural process of removing unnecessary or damaged cells, and is required for the proper execution of the organism's life cycle (Chowdhury et al., 2006; Zimmermann et al., 2001). Apoptosis was shown to be involved in numerous processes including embryonic development, response to cellular damage, aging and as a mechanism of tumor suppression (Blank &, Shiloh, 2007; Cohen et al., 2004; Lee et al., 2007; Mazumder et al., 2007; Rich et al., 2000; Subramanian et al., 2005). Two pathways were shown to induce apoptosis: an extrinsic and an intrinsic pathways. The difference between these two pathways is the mechanism by which the death signal is transduced (Chowdhury et al., 2006). Whereas the extrinsic pathway is activated by binding of ligands to a death receptor, the intrinsic pathway is activated by cellular stress, for example DNA damage. The intrinsic pathway involves the release of cytochrome c from the intermembrane space of the mitochondria. Together with apoptotic protease activating factor 1 (APAF1), cytochrome c activates caspase 9, leading to activation of downstream caspases and the induction of the death response (Bitomsky & Hofmann, 2009). Key players in the regulation of the intrinsic pathway include the Bcl2 protein family, which can influence the permeability of the outer mitochondrial membrane (Reed, 2006). Members of the Bcl2 protein family are divided into proapoptotic proteins such as Bax, Bak and Bok, and antiapoptotic ones including Bcl2, Bcl-X, Bcl-w and Mcl-1. Proteins of a third subfamily, known as the BH3-only proteins, are thought to be initiators of apoptosis, and probably function by regulating Bcl2-like proteins from the other two subfamilies. In healthy cells,

Bax exists as a monomer, either in the cytosol or weakly bound to the outer mitochondrial membrane. Upon stimulation of apoptosis, Bax translocates to the mitochondria, where it becomes anchored into the mitochondrial membrane. Following its translocation, Bax oligomerizes into large complexes, which are essential for the permeabilization of the mitochondrial membrane (Antignani & Youle, 2006; Bitomsky & Hofmann, 2009; Reed, 2006). Given its central role in mediating apoptosis, several mechanisms have been proposed for Bax regulation and retention in the cytosol, both by binding to other proteins and through posttranslational modifications. One of the first proteins that were shown to sequester Bax away from the mitochondria was Ku70 (Cohen et al., 2004; Lee et al., 2007; Mazumder et al., 2007; Subramanian et al., 2005). Thus, in addition to its role in regulating NHEJ DNA-repair, Ku70 functions in regulating Bax-mediated apoptosis. Overexpression of Ku70 lowered levels of cell death after apoptotic stimuli, while reducing Ku70 levels increased sensitivity to Bax-mediated apoptosis (Amsel et al., 2008). Taken together, these results suggest that Ku70 has anti-apoptotic activity. Such activity is associated with its ability to be acetylated (Cohen et al., 2004). Apoptotic stimuli lead to dissociation of the Ku70-Bax complex, resulting in cell death following Bax translocation to the mitochondria. It was suggested that under normal conditions, Bax undergoes ubiquitylation, which negatively regulates its proapoptotic function by labeling it for proteasomal degradation. The association with Ku70 mediates and promotes Bax deubiquitylation. Upon apoptotic stimulus, Ku70 is acetylated and releases Bax which translocates to the mitochondria where induces apoptosis. These findings suggest a complex role for Ku70 with both pro-apoptotic (maintaining an active pool of Bax) and anti-apoptotic (sequestering Bax away from the mitochondria) elements.

In response to DNA damage, deacetylase SIRT1 binds to and deacetylates specific lysine residue of substrate proteins, the modification of which leads to the repression of their transcriptional activities (Luo et al., 2001; Picard et al., 2004; Vaziri et al., 2001). SIRT1 has been suggested to suppress apoptotic responses (Luo et al., 2001; Vaziri et al., 2001). It has been demonstrated that, when exposed to IR, SIRT1 enhances DNA repair activity by binding to Ku70 and subsequently deacetylating this protein. This could facilitate one possible mechanism of cell survival (Jeong et al., 2007).

Another mechanism of cell fate regulation involves p21 (Abbas &. Dutta, 2009; Garner & Raj, 2008; Liu et al., 2003). Under some circumstances (i.e., enforced overexpression), p21 may promote apoptotic signaling that ultimately leads to cell death (Liu et al., 2003). However, DNA-damaged cells can undergo cell cycle arrest followed by apoptosis in the absence of p21 (Waldman et al., 1996, 1997). The mechanism by which p21 negatively regulates DNA damage-induced death machinery relies on its binding to key apoptotic regulatory proteins (Liu et al., 2003). P21 physically interacts, through its first N-terminal 33 aminoacids, with procaspase-3, i.e. the inactive precursor of the apoptotic executioner caspase-3 (Suzuki et al., 1998, 1999). When bound to p21, the inactive pro-caspase cannot be converted into the active protease, and apoptosis is impeded (Suzuki et al.,1999). Caspase 2, which acts upstream of caspase 3, is also kept in a repressed status by p21 (Baptiste-Okoh et al., 2008). The strict interaction between p21 and caspases is supported also by the observation that p21 itself is cleaved by caspases early during DNA damage- induced apoptosis (Jin et al., 2000; Levkau et al., 1998). The anti- or pro-apoptotic role of p21 could depend on the nature of the apoptotic stimulus. For example, apoptosis was enhanced or inhibited by p21, according to whether cells were treated with cisplatin, or methotrexate (Kraljevic Pavelic et al., 2008).

Functions of p21 in response to DNA damage could be also modulated by the extent of genotoxic lesions, through either stabilization or degradation of the protein. Low levels of DNA lesions will allow p21 stabilization and induce cell cycle arrest (thus having anti-apoptotic activity). In contrast, after extensive DNA damage, p21 down-modulation will allow cells to go to apoptosis (Lee et al., 2009; Martinez et al., 2002).

It is well established that p53 is capable of inducing apoptosis by transcription-dependent and transcription-independent mechanisms (Caelles et al., 1994). It has been demonstrated that recombinant p53 is capable of triggering mitochondrial membrane permeabilization in cell-free systems (Ding et al., 1998; Schuler et al., 2000). Later on, p53 has been reported to translocate to the cytoplasm in response to numerous stress signals, including DNA damage, where it drives mitochondrial outer membrane permeabilization and caspase activation (Marchenko et al., 2000; Mihara et al., 2003). Modifications of p53 may affect its transcriptional activity. For example, acetylation at p53 carboxyl-terminal lysine residues enhances its transcriptional activity associated with cell cycle arrest and apoptosis (Yamaguchi et al., 2009). The interaction between p53 and Ku70 is independent of p53 acetylation. However, p53 acetylation at its carboxyl terminus is required for p53 to prevent and/or displace Bax from its inhibitory interaction with Ku70, thus allowing this key proapoptotic protein to target mitochondria and initiate apoptosis (Yamaguchi et al., 2009). P53 has powerful apoptotic effects, and consequently is a subject to tight regulatory control. Normally, p53 protein is maintained at a low level through the Mdm2-mediated ubiquitination and degradation pathway. However, when cells are exposed to stress including genotoxic one, p53 protein is rapidly accumulated and activated for downstream biological functions. The regulatory events that affect the amount, stability and activity of p53 are in part associated with a variety of post-translational modifications, including phosphorylation, ubiquitination and acetylation. In fact, p53 is the first functional non-histone substrate identified for the histone acetyltransferases (HATs) (Yi & Luo, 2010).

Another key molecule critically involved in DNA damage-induced cell death signaling is the p53-related tumour suppressor and transcription factor p73 (Melino et al., 2003). In unstressed cells, p73 forms a complex with the E3 ubiquitin ligase Itch, which marks it for degradation by the ubiquitin-proteasome system. Upon DNA damage, the levels of Itch become reduced and allow the accumulation of p73 (Rossi et al., 2005). Many of p73 pro-apoptotic target genes such as Puma, caspase-6 or CD95, overlap with those of p53 (Dobbelstein et al., 2005). Post-translational modifications of p73 by acetylation through p300 and by phosphorylation by the DNA damage-activated, nonreceptor tyrosine kinase c-Abl were found to be crucial for transactivation of its pro-apoptotic target genes (Costanzo et al., 2002).

The E2F1 transcription factor, which was originally identified as a cell-cycle initiator, mediates apoptosis in response to DNA damage (Iaquinta & Lees, 2007; Polager & Ginsberg, 2008; Yamasaki et al., 1996). Under certain conditions, deregulated E2F1 triggers apoptosis via both p53-dependent and p53-independent mechanisms. To induce p53-dependent apoptosis, E2F1 activates the expression of p14/p19ARF tumor suppressor gene to stabilize p53 (Phillips & Vousden, 2001). Alternatively, E2F1 directly activates various proapoptotic genes or inactivates several antiapoptotic genes (Iaquinta & Lees, 2007; Polager & Ginsberg, 2008). In support of the importance of E2F1 for apoptotic signaling, germline deletion of E2F1 in mice leads to the formation of various tumors, presumably resulting from the lack of E2F1-induced apoptosis (Field et al., 1996; Yamasaki et al., 1996).

2.5 Cell fate decision

Depending on the amount of damage, the DDR activates one of two alternatives: a prosurvival network that includes the damage-induced cell cycle checkpoints and DNA repair or programmed cell death (Barzilai et al., 2008). The mechanistic aspects of this critical choice remain unclear. Activation of p53 in response to DNA damage results in either cell cycle arrest or apoptosis. Although genes that regulate these cellular processes are essentially p53 targets, activation of p53 always results in specific and selective transcription of p53-regulated genes (Riley et al, 2008). Thus, it is likely that unique sets of p53-regulated genes operate in tandem to bring about a desired outcome in response to specific stimuli. How p53 executes these two distinct functions remains largely unclear. Recent reports suggest that activation of specific promoters by p53 is achieved through its interaction with heterologous transcription factors such as Hzf and ASPP family proteins (Das et al, 2007; Tanaka et al, 2007). P53 modifications following stress such as phosphorylation and acetylation stabilize p53, enhancing its sequence-specific DNA binding and transcriptional activity (Sakaguchi et al, 1998). The phosphorylation at amino-terminus is required for p53 stability, while acetylation at carboxyl-terminus is indispensable for p53 transcriptional activation (Tang et al., 2008). The p53 target gene SMAR1 modulates the cellular response to genotoxic stress by a dual mechanism. First, SMAR1 interacts with p53 and facilitates p53 deacetylation through recruitment of deacetylase HDAC1. Then SMAR1 represses the transcription of Bax and Puma by binding to an identical 25 bp MAR element in their promoters (Sinha et al., 2010). A mild DNA damage induces SMAR1-generated anti-apoptotic response by promoting p53 deacetylation and specifically repressing Bax and Puma expression. Reducing the expression of SMAR1 by shRNA leads to significant increase in p53-dependent apoptosis (Sinha et al., 2010). Severe DNA damage results in sequestration of SMAR1, p53 acetylation and transactivation of Bax and Puma leading to apoptosis. Thus, sequestration of SMAR1 into the PML-NBs acts as a molecular switch to p53-dependent cell arrest and apoptosis in response to DNA damage (Sinha et al., 2010). The mechanisms by which moderate damage resulting from mild stress leads to repair, while severe damage results in the 'decision' to kill a cell, remains unclear. Every single cell is therefore continuously confronted with the choice: repair and live or die. Irreparable damage triggers p53's killer functions to eliminate genetically-altered cells. The killer functions of p53 are tightly regulated and balanced against protector functions that promote damage repair and support survival in response to mild damage (Schlereth et al., 2010). In molecular terms, these p53-based cell fate decisions involve protein interactions with factors, which modulate the activation of distinct sets of p53 target genes. The induction of a transient cell cycle arrest that allows for damage repair depends critically on the genes *p21, 14-3-3σ* and *GADD45A*, with *p21* being crucial for cell cycle arrest in the G1 phase, while *14-3-3σ* and *GADD45A* - for arrest in G2 (Levine & Oren, 2009). In the case of prolonged damage, p53-mediated transactivation of the sestrins (*SESN1* and *SESN2*) causes inhibition of the mammalian target of rapamycin (mTOR) signaling and helps to maintain the arrest reversible, while activation of mTOR under these conditions triggers a shift to irreversible cell cycle exit (senescence) (Demidenko et al., 2010; Korotchkina et al., 2010; Steelman & McCubrey, 2009). Another way for p53 to permanently stop cell proliferation without compromising cell viability is induction of differentiation (Schlereth, 2010). Only when cells have irreparable DNA damage that is incompatible with further survival, p53 shifts to the most extreme and irrevocable antiproliferative response - apoptosis (Aylon & Oren, 2007). p53-induced apoptosis does not only require activation of proapoptotic target genes such as *Bax* and *Noxa* but may also involve transcription-independent functions of p53 in the cytoplasm (Green & Kroemer, 2009; Morselli et al., 2009; Vaseva et al., 2009). Discriminatory effects on target can also be exerted by interacting proteins that modulate p53's DNA binding properties via covalent post-translational modifications including phosphorylation, acetylation, methylation, ubiquitylation, and sumoylation. Among the phosphorylation sites, serine 46 (S46) has clear discriminatory function for p53 as a transcriptional activator (Okoshi et al., 2008; Rinaldo et al., 2007). P53 is phosphorylated at this residue by homeodomain interacting protein kinase 2 (HIPK2), dual-specificity tyrosine-phosphorylation-regulated kinase 2 (DYRK2), AMPK, protein kinase C delta or p38 mitogen activated protein kinase in response to severe cellular damage (Okoshi et al., 2008; Rinaldo et al., 2007). While numerous studies have implicated acetylation of lysine residues in the C-terminus of p53 as being important for p53's transcriptional activity in general, acetylation of lysine 120 (K120) in the DNA binding domain by the MYST family histone acetyl transferases, hMOF and Tip60 specifically results in increased binding to proapoptotic targets like *Bax* and *Puma,* while the nonapoptotic targets *p21* and *Mdm2* remain unaffected (Sykes et al., 2006; Tang et al., 2006). On the other hand, acetylation of lysine 320 (K320) by the transcriptional coactivator p300/CBP-associated factor (PCAF) predisposes p53 to activate *p21* and decreases its ability to induce proapoptotic genes. Cells ectopically expressing a mutant p53 where K320 is mutated to glutamine (K320Q) to mimic acetylation, display reduced apoptosis after some forms of DNA damage (Knights et al., 2006). In contrast, K317R knockin mice, where K317 acetylation is missing, consistently display increased apoptosis and higher expression of relevant target genes in several cell types (Chao et al., 2006). However, K320 is not only a target for acetylation but it is also ubiquitylated by the zincfinger protein E4F1 (Le Cam et al., 2006). This modification facilitates p53-dependent activation of *p21* and *Cyclin G1* expression without affecting the expression of the proapoptotic gene *Noxa,* overall resulting in reduced p53-mediated cell death in response to UV. P53-mediated cell cycle arrest is also favored following methylation of at least two arginine residues (R333 and R335) by the arginine methyltransferase PRMT5. Consistently, depletion of PRMT5 by siRNA leads to increased apoptosis following p53 activation (Durant et al., 2009; Jansson et al., 2008).

Another factor which can impact cell fate decision is Chk2. Following DNA damage, Chk2 functions by suppressing apoptosis. In cells that express cell cycle inhibitors such as p21 and 14-3-3δ, cell cycle arrest appears to prevent or slow the onset of cell death. Without these proteins, Chk2-regulated apoptosis is much more apparent. Thus, it seems that the balance between cell cycle inhibitors and Chk2 dictates the outcome following DNA damage (Antoni et al., 2007). The finding that loss of both p21 and 14-3-3δ but not each alone is required to unmask the effect of Chk2 can be understood in the context of how each functions to effect cell cycle arrest. 14-3-3δ is a cytoplasmic protein which in response to DNA damage accumulates and acts by sequestering CDK1 and CDK2 in the cytoplasm and preventing cytokinesis (Chan et al., 1999; Laronga et al., 2000; Wilker et al., 2007). P21 is a nuclear cyclin-dependent kinase inhibitor that directly binds and inactivates cyclin-CDK complexes (el-Deiry et al., 1993; Harper et al., 1993; Xiong et al., 1993). Cooperative effects between these two factors have been shown to dictate the biological response to apoptotic stimuli (Jazayeri et al., 2006; Meng et al., 2009). This implies that the ultimate outcome of

Chk2 activation may depend on the particular cellular context and on molecular determinants of Chk2 function, 14-3-3δ and p21.

3. DNA damage response in postmitotic neurons

Neurons are extremely active cells (Barzilai, 2010; Fishel et al., 2007) and generally exhibit high mitochondrial respiration and production of reactive oxygen species (ROS) that can damage mitochondrial and nuclear DNA (Weissman et al., 2007). For this reason, neurons are particularly susceptible to genotoxic effects generated by ROS (Barzilai et al., 2008). ROS induce the formation of various DNA lesions including oxidative DNA base modifications, SSBs and DSBs (Martin, 2008). DNA damage plays an important role in brain damage (Nagayama et al., 2000). This damage is a common feature of neurodegenerative diseases (Kraemer et al., 2007; Trushina, & McMurray, 2007). The importance of DNA damage in pathogenesis of neurodegenerative diseases is illustrated by the observation that defective DNA repair in various human syndromes such as ataxia telangiectasia is accompanied by neurological abnormalities (Rolig, & McKinnon, 2000). There is a growing interest in the role of DNA damage in neurological dysfunctions associated with cancer treatments (Wefel et al., 2004). Significant evidence points to the critical role of cumulative DNA damage in the aging process of neurons in the central nervous system (CNS) (Coppede` & Migliore, 2010; Fishel et al., 2007; Weissman et al., 2007).

3.1 Cell cycle and neuronal apoptosis

Although accumulating evidence suggests the importance of proper DDR for the nervous system, most of the work to elucidate DDR components has been carried out in proliferating cells. The signal transduction mechanisms in neurons that link DNA damage to apoptosis are not well characterized, and the sensors of DNA damage in neurons are largely unknown (Martin et al., 2009). However, some observations suggest that DDR in postmitotic neurons may have survival checkpoint that serves to eliminate neurons with excessive DNA damage. A loss of function of DDR proteins such as ATM leads to genomic instability and human hereditary diseases, characterized by neurodegeneration (Rass et al, 2007). ATM has a pro-apoptotic function in the developing mouse CNS (Herzog et al., 1998; Lee et al., 2001) and operates similarly to how it operates in proliferating cells (Biton et al., 2006, 2007; Gorodetsky et al., 2007). In addition, neurons in ATM-/- mice are resistant to DNA damage-induced apoptosis (Herzog et al., 1998; Kruman et al., 2004; Lee & McKinnon, 2000; McKinnon, 2001). However, ascribing to ATM and cell cycle checkpoints in neurons the same functions they have in proliferating cells poses certain conceptual difficulties, given the postmitotic nature of these cells.

Another indication of possible cell cycle checkpoint functioning in neurons is extensively documented cell cycle reentry of these postmitotic cells following genotoxic stress. The neurons undergo full or partial DNA replication, showing that they reenter the S phase (Kruman et al., 2004; Yang et al., 2001). This attempt to enter the cell cycle is abortive and does not result in actual division (Athanasiou et al., 1998; Becker & Bonni, 2004; Feddersen et al., 1992) but culminates in apoptotic cell death (Becker & Bonni, 2004; Kruman, 2004; Yang & Herrup, 2001). Cell cycle activation is a common feature of neuronal apoptosis during development and in neurodegenerative disorders (Becker & Bonni, 2004; Herrup et al., 2004; Kruman, 2004; Kruman et al., 2004; Park et al., 1997, 1998). On the other hand, forced cell cycle entry mediated by targeted disruption of the pRb or ectopic E2F1 expression also results in apoptosis of postmitotic neurons (Becker & Bonni, 2004; Feddersen et al., 1995; Johnson et al., 1993; Smith et al., 2000), while preventing cell cycle entry is protective against neurotoxic insults, such as ischemia and kainate-induced excitotoxicity (Kim & Tsai, 2009; Kruman et al., 2004; Zhang et al., 2006). Exposure of mice or mesencephalic neuronal cultures to the dopaminergic cell neurotoxins 1-methyl-4-phenyl-1,2,3,6-tertahydropyridine (MPTP) results in cell cycle activation in post-mitotic neurons prior to their subsequent death, while E2F1 deficiency leads to a significant resistance to MPTP-induced dopaminergic cell death (Hoglinger et al., 2007).

Our recent findings demonstrate the particular role of S phase entry and DNA replication in DNA damage-induced neuronal apoptosis (Kruman et al., 2004; Tomashevski et al., 2010). Expression of S-phase markers was reported in post-mitotic neurons following hypoxia–ischemia (Kuan et al., 2004), in neurons in Alzheimer's disease (Yang et al., 2001) and in neurons ectopically expressing E2F1 (Smith et al., 2000). The special role of S phase might be linked to DNA replication errors which are usually accompanied by DNA damage and activation of cell cycle checkpoints (Elledge, 1996; Kumagai & Dunphy, 2006). Activation of Chk2 following DSB formation was observed in primary neurons exposed to DSB inducer producing repairable DSBs (Sordet et al., 2009). This is consistent with previous finding demonstrating that Chk2, in contrast to Chk1, is expressed and activatable in quiescent cells. This may suggest the survival mechanism by which S phase entry is prevented in postmitotic cells. Since differentiated neurons which enter S phase prior apoptosis predominantly express a highly error prone DNA polymerase β (Copani et al., 2002), the DNA replication might produce additional DNA damage. This may amplify DNA damage and generate apoptotic signaling. The functional link between neuronal cell cycle reentry, DDR, cell cycle checkpoints and apoptosis is supported by data demonstrating that both cell cycle activation and apoptosis in postmitotic neurons exposed to DSB-inducing agents are ATM-dependent (Alvira et al., 2007; Kruman et al., 2004; Otsuka et al., 2004). There is no evidence of entry of neurons under conditions of DNA damage-induced apoptosis into mitosis, although they may progress through DNA synthesis and G2 (Athanasiou et al., 1998; Becker & Bonni, 2004; Feddersen et al., 1992; Yang et al., 2001). This may be explained by activation of G2/M checkpoint induced by replication stress which prevents entry into mitosis. Indeed, expression of G2/M checkpoint markers has been reported in vascular dementia (McShea et al., 1999), and several other neurodegenerative diseases (Husseman et al., 2000).

3.2 Cell cycle and DNA repair in neurons

Terminally differentiated neurons are highly susceptible to oxidative DNA damage (Fishel et al., 2007), and DNA repair is very important for these cells (Biton et al., 2008; Fishel et al., 2007; Lavin & Kozlov, 2007). All eukaryotic DNA repair systems operating in proliferating cells also operate in neurons (Fishel et al., 2007; Lee, & McKinnon, 2007; Sharma, 2007; Weissman et al., 2007; Wilson, & McNeill, 2007). It is believed that most of the lesions inflicted in neuronal genomic and mitochondrial DNA are produced by ROS. These lesions are repaired mainly via the BER pathway, although other types of DNA repair are involved (Fishel et al., 2007; Weissman et al., 2007; Wilson & McNeill, 2007). Although DNA repair activity exists in neurons, it was found that this repair is not as effective as in dividing cells, suggesting that lesions are likely to accumulate (Gobbel et al., 1998; McMurray, 2005; Nouspikel, & Hanawalt, 2000, 2002). Indeed, following cellular differentiation, the levels of many repair factors are reduced (Bill et al., 1992; Nouspikel, & Hanawalt, 2000, 2002).

However, in contrast to global genomic repair (GGR), the repair of transcribed genes is more vigorous (Nouspikel, & Hanawalt, 2000). Thus, DNA repair in the nonessential bulk of the genome of postmitotic neurons is dispensable, and they repair only DNA needed for neuronal functioning (Nouspikel, 2007; Nouspikel, & Hanawalt, 2002). Since neurons are very active and the repair process carries a high energy cost, it is reasonable that these cells preferentially repair transcribed genes. This is important to avoid harming the fidelity of information transcribed to proteins (Fishel et al., 2007; Lu et al., 2004).

It is commonly believed that neurons remain in G0 phase of the cell cycle indefinitely. Cell-cycle reentry, however, is coupled with DNA damage-induced apoptosis of postmitotic neurons (Becker & Bonni, 2004; Herrup et al., 2004; Kruman, 2004; Kruman et al., 2004; Park et al., 1997, 1998). Moreover, recent evidence demonstrates the expression of cell-cycle proteins in differentiated neurons at physiological conditions (Schmetsdorf et al., 2007, 2009). The functional roles of such expression remain unclear. Since DNA repair is generally attenuated by differentiation in most cell types (McMurray, 2005; Narciso et al., 2007), the cell-cycle-associated events in postmitotic cells may reflect the need to reenter the cell cycle to activate DNA repair. Recently, we have demonstrated that the NHEJ activation in postmitotic neurons is associated with G0-G1 transition, driven by cyclin-C-associated pRb-kinase activity, while preventing cell cycle entry attenuated DNA repair (Tomashevski et al., 2010). This suggests the importance of cell cycle entry for DNA repair in postmitotic cells. Previously, quiescent cells, including differentiated cells, were shown to be able to reenter the cell cycle simply by removing appropriate cell cycle inhibitors such as p21. Interference with p21 was sufficient to reactivate the cell cycle and DNA synthesis in terminally differentiated skeletal muscle cells, quiescent fibroblasts and primary cortical neurons (Pajalunga et al., 2007; Tomashevski et al., 2010). Reactivation of cell cycle and DNA replication has also been documented in quiescent cells overexpressing E2F1 and Cdc25A (Pajalunga et al., 2007; Rogoff & Kowalik, 2004; Smith et al., 2000; Zhang et al., 2006). Such reactivation of cell cycle and DNA replication were sufficient to promote neuronal death even in the absence of DNA damage (O'Hare et al., 2000). However, preventing S phase entry, attenuated apoptotic signaling (Tomashevski et al., 2010), suggesting a decisive role of G1-S transition for activation of the apoptotic machinery. Thus, cell cycle activation occurs in response to DNA damage and is involved in both DNA repair and apoptosis in postmitotic neurons. These findings may imply that cell cycle checkpoints may orchestrate both DNA repair and apoptosis of postmitotic neurons, as it occurs in proliferating cells (Bartek & Lukas, 2001; Shiloh, 2003; Zhou & Elledge, 2000).

4. Conclusion and future perspectives

The way that cells react to DNA damage constantly produced by exogenous and endogenous factors is to trigger a complex and coordinated set of events termed the DDR (Reinhardt & Yaffe, 2009). The function of such response is to sense genome damage and activate several downstream pathways, including cell cycle checkpoints, DNA repair and apoptotic programs (Jackson, 2009; Zhou & Elledge, 2000). The earliest events of the DDR are associated with alterations in chromatin structure and the formation of DDR foci facilitating recruitment of proteins involved in DDR propagation (Berkovich et al., 2007; Downs et al., 2007; Smerdon et al., 1978). The biochemical details of these processes are poorly understood. However, studies in yeast and mammalian systems have demonstrated that colocalization of DDR proteins rather than DNA damage per se is critical for DNA damage signaling (Bonilla et al., 2008; Soutoglou & Misteli, 2008). Another important component of DDR network is the cell cycle checkpoint pathway which plays roles in the activation of DNA repair, modulation of transcriptional programmes and the optional triggering apoptosis (Bartek & Lukas, 2001; Shiloh, 2003; Zhou & Elledge, 2000). In response to DNA damage, the checkpoints delay or stop the cell cycle at critical points before or during DNA replication (G1/S and intra-S checkpoints) and before cell division (G2/M checkpoint). This prevents replication and segregation of damaged DNA (Houtgraaf et al., 2006; Poehlmann & Roessner, 2010). DDR is involved in two alternatives: activation of a prosurvival network associated with DNA repair or initiation of programmed cell death removing cells with irrepairable DNA (Barzilai et al., 2008; Kruman, 2004). The checkpoints play important roles in both processes (Bartek & Lukas, 2001; Shiloh, 2003; Zhou & Elledge, 2000). The importance of DDR is illustrated by various pathologies associated with defects in DDR proteins. Mutations in key DDR regulators such as ATM, ATR, MRE11, NBS1 are associated with severe genome instability disorders (Ciccia & Elledge, 2010; Jackson & Bartek 2009).

Due to a high rate of oxygen metabolism and the low levels of antioxidant enzymes compared to other cells, the DNA of postmitotic neurons is under increased risk of damage from free radicals. (Barzilai, 2010; Kruman, 2004). For this reason, DNA repair is critical for the nervous system. While all eukaryotic DNA repair systems operating in proliferating cells also operate in neurons (Fishel et al., 2007; Lee, & McKinnon, 2007; Sharma, 2007; Weissman et al., 2007; Wilson, & McNeill, 2007), differentiation is associated with a decrease in levels of many repair enzymes (Bill et al., 1992; Nouspikel, & Hanawalt, 2000, 2002; Tofilon & Meyn, 1988), and DNA repair in neurons, is not as effective as in dividing cells (Gobbel et al., 1998; McMurray, 2005; Nouspikel, & Hanawalt, 2000, 2002). It raises the question whether DDR in postmitotic neurons is similar to the DDR of mitotic cells. Some evidence such as a contribution of ATM to apoptosis of postmitotic neurons (Herzog et al., 1998; Kruman et al., 2004; Lee & McKinnon, 2000; McKinnon, 2001) points to such similarity. Although postmitotic neurons are quiescent cells, they are capable to reenter the cell cycle before apoptosis induced by genotoxic stress, as was extensively documented (Barzilai, 2010; Kim & Tsai, 2009; Kruman et al., 2004; Yang et al., 2001). Moreover, we recently demonstrated that DNA repair is also depends on cell cycle activation, driven by cyclin-C-associated pRb-kinase activity (Tomashevski et al., 2010). These findings together with observation that Chk2 is expressed and activated in postmitotic neurons and other postmitotic cells following genotoxic stress (Lukas et al., 2001; Sordet et al., 2009), are indications of cell cycle checkpoint functioning in neurons.

Compelling evidence points to similarities in the DDR of proliferating cells and postmitotic neurons. However, neurons are quiescent cells which requires adaptation of the DDR. The major future challenge is to understand the mechanisms by which cell cycle checkpoint machinery operates in postmitotic neurons and involves in DNA repair, apoptosis and cell fate decisions. Further investigation of the DDR in human genomic instability syndromes, neurodegenerative pathologies, and animal models of these conditions, will help to disclose these mechanisms. Clarification of the mechanisms at work will help guide the search for novel treatment modalities for a variety of neurodegenerative conditions.

5. Acknowledgments

This work was supported by Texas Tech University Health Sciences Center grant.

6. References

Abbas, T. & Dutta, A. (2009). p21 in cancer: intricate networks and multiple activities. *Nat Rev Cancer*, Vol.9, No.6, (June 2009), pp.400-414, ISSN 1944-0234

Ahel, D.; Horejsí, Z.; Wiechens, N.; Polo, S.E.; Garcia-Wilson, E.; Ahel, I.; Flynn, H.; Skehel, M.; West, S.C.; Jackson S.P.; Owen-Hughes, T. & Boulton, S.J. (2009). Poly(ADP-ribose)-dependent regulation of DNA repair by the chromatin remodeling enzyme ALC1. *Science*, Vol.325, No.5945, (August 2010), pp.1240-1243, ISSN 1966-1379

Al-Hakim, A.; Escribano-Diaz, C.; Landry, M.C.; O'Donnell, L.; Panier, S.; Szilard, R.K. & Durocher, D. (2010). The ubiquitous role of ubiquitin in the DNA damage response. *DNA Repair (Amst)*, Vol.9, No.12, (November 2010), pp.1229-1240, ISSN 2105-6014

Alvira, D.; Yeste-Velasco, M.; Folch, J.; Casadesús, G.; Smith, M.A.; Pallàs, M. & Camins, A. (2007). Neuroprotective effects of caffeine against complex I inhibition-induced apoptosis are mediated by inhibition of the Atm/p53/E2F-1 path in cerebellar granule neurons. *J Neurosci Res*, Vol.85, No.14, (November 2007), pp.3079-3088, ISSN 1763-8302

Amsel, A.D.; Rathaus, M.; Kronman, N. & Cohen, H.Y. (2008). Regulation of the proapoptotic factor Bax by Ku70-dependent deubiquitylation. *Proc Natl Acad Sci U S A*, Vol.105, No.13, (March 2008), pp. 5117-5122, ISSN 1836-2350

Antignani, A. & Youle, R.J. (2006). How do Bax and Bak lead to permeabilization of the outer mitochondrial membrane? *Curr Opin Cell Biol*, Vol.18, No.6, (October 2006), pp. 685-689, ISSN 1704-6225

Antoni, L.; Sodha, N.; Collins, I. & Garrett, M.D. (2007). CHK2 kinase: cancer susceptibility and cancer therapy-two sides of the same coin? *Nat Rev Cancer*, Vol.7, No.12, (December 2007), pp.925-936, ISSN 1800-4398

Aressy, B. & Ducommun, B. (2008) Cell cycle control by the CDC25 phosphatases. *Anticancer Agents Med Chem*, Vol.8, No.8, (December 2008), pp.818-824, ISSN 1907-5563

Athanasiou, M.C.; Yunis, W.; Coleman, N.; Ehlenfeldt, R.; Clark, H.B.; Orr, H.T. & Feddersen, R.M. (1998). The transcription factor E2F-1 in SV40 T antigen-induced cerebellar Purkinje cell degeneration. *Mol Cell Neurosci*, Vol.12, No.1-2, (September 1998), pp.16-28, ISSN 977-0337

Averbeck, N.B. & Durante, M. (2011). Protein acetylation within the cellular response to radiation. *J Cell Physiol*, Vol.226, No.4, (April 2011), pp.962-967, ISSN 2094-5393

Aylon, Y. & Oren, M. (2007). Living with p53, dying of p53. *Cell*, Vol.130, No.4, (August 2007), pp. 597-600, ISSN 1771-9538

Banin, S.; Moyal, L.; Shieh, S.; Taya, Y.; Anderson, C.W.; Chessa, L.; Smorodinsky, N.I.; Prives, C.; Reiss, Y.; Shiloh, Y.& Ziv, Y. (1998). Enhanced phosphorylation of p53 by ATM in response to DNA damage. *Science*, Vol.281, No.5383, (September 1998), pp. 1674–1677, ISSN 973-3514

Baptiste-Okoh, N.; Barsotti, A.M. & Prives, C. (2008). Caspase 2 is both required for p53-mediated apoptosis and downregulated by p53 in a p21-dependent manner. *Cell Cycle*, Vol.7, No.9, (February 2008), pp.1133-1138, ISSN 1841-8048

Bartek, J. & Lukas, J. (2001). Mammalian G1- and S-phase checkpoints in response to DNA damage. *Curr Opin Cell Biol*, Vol.13, No.6, (December 2001), pp.738-747, ISSN 1169-8191

Bartek, J.; Lukas, C. & Lukas, J. (2004). Checking on DNA damage in S phase. *Nat Rev Mol Cell Biol*, Vol.5, No.10, (October 2007), pp. 792-804, ISSN1545-9660

Bartek, J. & Lukas, J. (2007). DNA damage checkpoints: from initiation to recovery or adaptation. *Curr Opin Cell Biol*, Vol.19, No.2, (February 2007), pp.238-245. ISSN 1730-3408

Bartek, J.; Lukas, J. & Bartkova, J. (2007). DNA damage response as an anti-cancer barrier: damage threshold and the concept of 'conditional haploinsufficiency'. *Cell Cycle*, Vol.6, No.19, (July 2007), pp. 2344-2347, ISSN 1770-0066

Bartkova, J.; Horejsí, Z.; Koed, K.; Krämer, A.; Tort, F.; Zieger, K.; Guldberg, P.; Sehested, M.; Nesland, J.M.; Lukas, C.; Ørntoft, T.; Lukas, J. & Bartek, J. (2005). DNA damage response as a candidate anti-cancer barrier in early human tumorigenesis. *Nature*, Vol.434, No.7035, (April 2005), pp. 864-870. ISSN 1582-9956

Barzilai, A.; Biton, S. & Shiloh, Y. (2008). The role of the DNA damage response in neuronal development, organization and maintenance. *DNA Repair (Amst)*, Vol.7, No.7, (May 2008), pp. 1010-1027. ISSN 1845-8000

Barzilai A. (2010). DNA damage, neuronal and glial cell death and neurodegeneration. *Apoptosis*, Vol.15, No.11, (November 2010), pp.1371-1381. ISSN 2043-7103

Bazzi, M.; Mantiero, D.; Trovesi, C.; Lucchini, G. & Longhese, M.P. (2010). Dephosphorylation of gamma H2A by Glc7/protein phosphatase 1 promotes recovery from inhibition of DNA replication. *Mol Cell Biol*, Vol.30, No.1, (January 2010), pp.131-145, ISSN 1988-4341

Becker, E.B. & Bonni, A. (2004). Cell cycle regulation of neuronal apoptosis in development and disease. *Prog Neurobiol*, Vol.72, No.1, (January 2004), pp.1-25, ISSN 1501-9174

Bekker-Jensen, S.; Lukas, C.; Melander, F.; Bartek, J. & Lukas, J. (2005). Dynamic assembly and sustained retention of 53BP1 at the sites of DNA damage are controlled by Mdc1/NFBD1. *J Cell Biol*, Vol.170, No.2, (July 2005), pp.201–211, ISSN 1600-9723

Bekker-Jensen, S.; Lukas, C.; Kitagawa, R.; Melander, F.; Kastan, M.B.; Bartek, J. & Lukas, J. (2006). Spatial organization of the mammalian genome surveillance machinery in response to DNA strand breaks. *J Cell Biol*, Vol.173, No.2, (April 2006), pp.195-206, ISSN 1661-8811

Bennett, C.B.; Lewis, A.L.; Baldwin, K.K. & Resnick, M.A. (1993). Lethality induced by a single site-specific double-strand break in a dispensable yeast plasmid. *Proc Natl Acad Sci U S A*, Vol.90, No.12, (June 1993), pp. 5613-5617, ISSN 851- 6308

Berkovich, E.; Monnat, R.J. Jr. & Kastan, M.B. (2007). Roles of ATM and NBS1 in chromatin structure modulation and DNA double-strand break repair. *Nat Cell Biol*, Vol.9, No.6, (May 2007), pp.683–690, ISSN 1748-6112

Bill, C.A.; Grochan, B.M.; Vrdoljak, E.; Mendoza, E.A. & Tofilon, P.J. (1992). Decreased repair of radiation-induced DNA double-strand breaks with cellular differentiation. *Radiat Res*, Vol.132, No.2, (November 1992), pp.254–258, ISSN 143- 8708

Bitomsky, N. & Hofmann, T.G. (2009). Apoptosis and autophagy: Regulation of apoptosis by DNA damage signalling - roles of p53, p73 and HIPK2. *FEBS J*, Vol.276, No.21, (September 2007), pp. 6074-6083, ISSN 1978-8416

Biton, S.; Dar, I.; Mittelman, L;. Pereg, Y.; Barzilai, A. & Shiloh, Y. (2006). Nuclear ataxia-telangiectasia mutated (ATM) mediates the cellular response to DNA double strand breaks in human neuron-like cells, *J Biol Chem*, Vol.281, No.25, (April 2006), pp.17482-17491, ISSN 1662-7474

Biton, S.; Gropp, M.; Itsykson, P.; Pereg, Y.; Mittelman, L.; Johe, K.; Reubinoff, B. & Shiloh, Y. (2007). ATM-mediated response to DNA double strand breaks in human neurons derived from stem cells. *DNA Repair (Amst)*, Vol.6, No.1, (December 2006), pp.128-134, ISSN 1717-8256

Biton, S.; Barzilai, A. & Shiloh, Y. (2008). The neurological phenotype of ataxia–telangiectasia: solving a persistent puzzle. *DNA Repair (Amst)*, Vol.7, No.7, (May 2008), pp.1028-1038, ISSN 1845-6574

Blank, M. & Shiloh, Y. (2007). Programs for cell death: apoptosis is only one way to go. *Cell Cycle*, Vol.6, No.6, (March 2007), pp. 686-95, ISSN 1736-1099

Bonilla, C.Y.; Melo, J.A. & Toczyski, D.P. (2008). Colocalization of sensors is sufficient to activate the DNA damage checkpoint in the absence of damage. *Mol Cell*, Vol.30, No.3, (May 2008), pp.267–276, ISSN 1847-1973

Brady, C.A. & Attardi, L.D. (2010). p53 at a glance. *J Cell Sci*, Vol.123, Pt.15, (August 2010), pp.2527-2532, ISSN 2094-0128

Branzei, D. & Foiani, M. (2008). Regulation of DNA repair throughout the cell cycle. *Nat Rev Mol Cell Biol*, Vol.9, No.4, (February 2008), pp.297–308, ISSN 1828-5803

Brugarolas, J.; Moberg, K.; Boyd, S.D.; Taya, Y.; Jacks, T. & Lees, J.A. (1999). Inhibition of cyclindependent kinase 2 by p21 is necessary for retinoblastoma protein-mediated G1arrest after gamma-irradiation. *Proc Natl Acad Sci USA*, Vol.96, No.3, (February 1999), pp.1002–1007, ISSN 992-7683

Burma, S.; Chen, B.P.; Murphy, M.; Kurimasa, A. & Chen, D.J. (2001). ATM phosphorylates histone H2AX in response to DNA double-strand breaks. *J Biol Chem*, Vol.276, No.45, (September 2001), pp.42462-42467, ISSN 1157-1274

Busino, L.; Chiesa, M.; Draetta, G.F. & Donzelli, M. (2004) Cdc25A phosphatase:combinatorial phosphorylation, ubiquitylation and proteolysis. *Oncogene*, Vol.23, No.11, (March 2004), pp. 2050-2056, ISSN 1502-1892

Caelles, C.; Helmberg, A. & Karin, M. (1994) p53-dependentapoptosis in the absence of transcriptional activation of p53- target genes. *Nature*, Vol.370, No.6486, (July 1994), pp. 220-223, ISSN 802-8670

Calsou, P.; Delteil, C.; Frit, P.; Drouet, J. & Salles, B. (2003). Coordinated assembly of Ku and p460 subunits of the DNA- dependent protein kinase on DNA ends is necessary for XRCC4-ligase IV recruitment. *J Mol Biol*, Vol.326, No.1, (February 2003), pp.93-103, ISSN 1254-7193

Castedo, M.; Perfettini, J.L.; Roumier, T.; Andreau, K.; Medema, R. & Kroemer, G. (2004). Cell death by mitotic catastrophe: a molecular definition. *Oncogene*, Vol.23, No.16, (April 2004), pp. 2825-2837, ISSN 1507-7146

Celeste, A.; Petersen, S.; Romanienko, P.J.; Fernandez-Capetillo, O.; Chen, H.T.; Sedelnikova, O.A.; Reina-San-Martin, B.; Coppola, V.; Meffre, E.; Difilippantonio, M.J.; Redon, C.; Pilch, D.R.; Olaru, A.; Eckhaus, M.; Camerini-Otero, R.D.; Tessarollo, L.; Livak, F.; Manova, K.; Bonner ,W.M.; Nussenzweig, M.C. & Nussenzweig, A. (2002). Genomic instability in mice lacking histone H2AX. *Science*, Vol.296, No.5569, (May 2002) pp.922–927, ISSN 1193-4988

Celeste, A.; Fernandez-Capetillo, O.; Kruhlak, M.J.; Pilch, D.R.; Staudt, D.W.; Lee, A.; Bonner, R.F.; Bonner,W.M. & Nussenzweig, A. (2003). Histone H2AX phosphorylation is dispensable for the initial recognition of DNA breaks. *Nat Cell Biol*, Vol.5, No7, (July 2003), pp.675–679, ISSN 1279-2649

Cha, H.; Lowe, J.M.; Li, H.; Lee, .;, Belova, G.;, Bulavin, D.V. & Fornace, A.J. Jr. (2010). Wip1 directly dephosphorylates gamma-H2AX and attenuates the DNA damage response. *Cancer Res*, Vol.70, No.10, (May 2010), pp.4112-2412, ISSN 204-60517

Chan, D.W. & Lees-Miller, S.P. (1996). The DNA-dependent protein kinase is inactivated by autophosphorylation of the catalytic subunit. *J Biol Chem*, Vol.271, No.15, (April 1996), pp.8936-8941, ISSN 86-21537

Chan, T.A.; Hermeking, H.; Lengauer, C.; Kinzler, K.W. & Vogelstein, B. (1999). 14-3-3Sigma is required to prevent mitotic catastrophe after DNA damage. *Nature*, Vol.401, No. 6753, (October 1999), pp. 616-620, ISSN 1052-4633

Chan, T.A.; Hwang, P.M.; Hermeking, H.; Kinzler, K.W. & Vogelstein, B. (2000). Cooperative effects of genes controlling the G(2)/M checkpoint. *Genes Dev*, Vol.14, No.13, (July 2000), pp.1584-1588, ISSN 1088-7152

Chao, C.; Wu, Z.; Mazur, S.J.; Borges, H.; Rossi, M.; Lin, T.; Wang, J.Y.; Anderson, C.W.; Appella, E. & Xu, Y. (2006). Acetylation of mouse p53 at lysine 317 negatively regulates p53 apoptotic activities after DNA damage. *Mol Cell Biol*, Vol.26, No.18, (September 2006), pp. 6859-6869, ISSN 1694-3427

Chapman, J.R. & Jackson, S.P. (2008). Phospho-dependent interactions between NBS1 and MDC1 mediate chromatin retention of the MRN complex at sites of DNA damage. *EMBO Rep*, Vol.9, No.8, (June 2008), pp.795-801, ISSN 1858- 3988

Chappell, C.; Hanakahi, L.A.; Karimi-Busheri, F.; Weinfeld, M. & West, S.C. (2002). Involvement of human polynucleotide kinase in double-strand break repair by non-homologous end joining. *EMBO J*, Vol.21, No.11, (June 2002), pp. 2827- 2832, ISSN 1203-2095

Chen, J.; Peters, R.; Saha, P.; Lee, P.; Theodoras, A.; Pagano, M.; Wagner, G. & Dutta, A. (1996). A 39 amino acid fragment of the cell cycle regulator p21 is sufficient to bind PCNA and partially inhibit DNA replication in vivo. *Nucleic Acids Res*, Vol.24, No.9, (May 1996), pp.1727–1733, ISSN 864-9992

Chou, D.M., Adamson, B, Dephoure, N.E., Tan, X., Nottke, A.C., Hurov, K.E., Gygi, S.P., Colaiácovo, M.P., Elledge, S.J. (2010). A chromatin localization screen reveals poly (ADP ribose)-regulated recruitment of the repressive polycomb and NuRD complexes to sites of DNA damage. *Proc Natl Acad Sci U S A*, Vol.107, No.43, (October 2010), pp. 18475- 18480, ISSN 2093-7877

Chowdhury, I.; Tharakan, B. & Bhat, G.K. (2006). Current concepts in apoptosis: The physiological suicide program revisited. *Cell Mol Biol Lett*, Vol.11, No.4, (September 2006), pp. 506-525, ISSN 1697-7376

Chowdhury, D.; Xu, X.; Zhong, X.; Ahmed, F.; Zhong, J.; Liao, J.; Dykxhoorn, D.M.; Weinstock, D.M.; Pfeifer, G.P. & Lieberman, J. (2008). A PP4-phosphatase complex dephosphorylates gamma-H2AX generated during DNA replication. *Mol Cell*, Vol.31, No.1, (July 2008), pp. 33-46, ISSN 1861-4045

Ciccia, A. & Elledge, S.J. (2010). The DNA damage response: making it safe to play with knives. *Mol Cell*, Vol.40, No.2, (October 2010), pp.179-204, ISSN 2096-5415

Citarelli, M.; Teotia, S. & Lamb, R.S. (2010). Evolutionary history of the poly(ADP-ribose) polymerase gene family in eukaryotes. *BMC Evol Biol*, Vol.10, (October 2010), pp.308-334, ISSN 2094-2953

Clerici, M.; Mantiero, D.; Guerini, I.; Lucchini, G. & Longhese, M.P. (2008). The Yku70-Yku80 complex contributes to regulate double-strand break processing and checkpoint activation during the cell cycle. *EMBO Rep*, Vol.9, No.8, (July 2008), pp.810-818, ISSN 1860-0234

Cohen, H.Y.; Lavu, S.; Bitterman, K.J.; Hekking, B.; Imahiyerobo, T.A.; Miller, C.; Frye, R.; Ploegh, H.; Kessler, B.M. & Sinclair, D.A. (2004). Acetylation of the C terminus of Ku70 by CBP and PCAF controls Bax-mediated apoptosis. *Mol Cell*, Vol.13, No.5, (March 2004), pp. 627-638, ISSN 1502-3334

Copani, A.; Sortino, M.A.; Caricasole, A.; Chiechio, S.; Chisari, M.; Battaglia, G.; Giuffrida-Stella, A.M.; Vancheri, C. & Nicoletti, F. (2002). Erratic expression of DNA polymerases by beta-amyloid causes neuronal death. *FASEB J*, Vol.16, No.14, (October 2002), pp. 2006-2008, ISSN 1239-7084

Coppede, F & Migliore, L. (2010). DNA repair in premature aging disorders and neurodegeneration. *Curr Aging Sci*, Vol.3, No.1, (February 2010), pp.3-19, ISSN 2029-8165

Costanzo, A.; Merlo, P.; Pediconi, N.; Fulco, M.; Sartorelli, V.; Cole, P.A.; Fontemaggi, G.; Fanciulli, M.; Schiltz, L.;

Blandino, G.; Balsano, C. & Levrero, M. (2002). DNA damage-dependent acetylation of p73 dictates the selective activation of apoptotic target genes. *Mol Cell*, Vol.9, No.1, (January 2002), pp.175–186, ISSN 1180-4596

Elledge, S. J. (1996). Cell cycle checkpoints: preventing an identity crisis. *Science*, Vol.274, No.5293, (December 1996), pp. 1664-1672, ISSN 893-9848

D'Amours, D.; Desnoyers, S.; D'Silva, I. & Poirier, G.G. (1999). Poly(ADP-ribosyl)ation reactions in the regulation of nuclear functions. *Biochem J*, Vol.342, Pt. 2, (September 1999), pp.249-268, ISSN 1045-5009

Das, S.; Boswell, S.A.; Aaronson, S.A. & Lee, S.W. (2008). P53 promoter selection: choosing between life and death. *Cell Cycle*, Vol.7, No.2, (October 2007), pp.154-157, ISSN 1821-2532

Deckbar, D.; Birraux, J.; Krempler, A.; Tchouandong, L.; Beucher, A.; Walker, S.; Stiff, T.; Jeggo, P. & Löbrich, M. (2007). Chromosome breakage after G2 checkpoint release. *J Cell Biol*, Vol.176, No.6, (March 2007), pp.749-755, ISSN 1735- 3355

de Jager, M.; van Noort, J.; van Gent, D.C.; Dekker, C.; Kanaar, R. & Wyman, C. (2001). *Mol Cell*, 2001 Nov;Vol.8, No.5, (November 2001), pp.1129-1135, ISSN 1174-1547

de Murcia, G. & Ménissier de Murcia, J. (1994). Poly(ADP-ribose) polymerase: a molecular nick-sensor. *Trends Biochem Sci*, Vol.19, No.4 (April 1994), pp172-176, ISSN 8016-868

Demidenko, Z.N.; Korotchkina, L.G.; Gudkov, A.V. & Blagosklonny, M.V. (2010). Paradoxical suppression of cellular senescence by p53. *Proc Natl Acad Sci USA*, Vol.107, No.21 (May 2010), pp.9660-9664, ISSN 2045-7898

Deng, C.; Zhang, P.; Harper, J.; Elledge, S.J. & Leder, P. (1995). Mice lacking p21CIP1/WAF1 undergo normal development, but are defective in G1 checkpoint control. *Cell*, Vol.82, No.4 (August 1995), pp. 675-684, ISSN 766-4346

Ding, H.F.; McGill, G.; Rowan, S.; Schmaltz, C.; Shimamura, A. & Fisher, D.E. (1998). Oncogene-dependent regulation

of caspase activation by p53 protein in a cell-free system. *J Biol Chem*, Vol.273, No.43 (October 1998), pp.28378-28383, ISSN 977-4464

Dobbelstein, M.; Strano, S.; Roth, J. & Blandino, G. (2005). p73-induced apoptosis: a question of compartments and cooperation. *Biochem Biophys Res Commun*, Vol.331, No.3 (June 2005), pp. 688-693, ISSN 1586-5923

Downs, J.A.; Lowndes, N.F. & Jackson, S.P. (2000). A role for Saccharomyces cerevisiae histone H2A in DNA repair. *Nature*, Vol.408, No.6815, (Decemeber 2000), pp.1001-1004, ISSN 1114-0636

Downs, J.A.; Allard, S.; Jobin-Robitaille, O.; Javaheri, A.; Auger, A.; Bouchard, N.; Kron, S.J.; Jackson, S.P. & Côté, J. (2004). *Mol Cell*, Vol.16, No.6, (December 2004), pp.979-990,ISSN 1561-0740

Downs, J.A.; Allard, S.; Jobin-Robitaille, O.; Javaheri, A.; Auger, A.; Bouchard, N.; Kron, S.J.; Jackson, S.P. & Côté J. (2004). Binding of chromatin-modifying activities to phosphorylated histone H2A at DNA damage sites. *Mol Cell*, 2004 Dec 22;Vol.16, No.6, (December 2004), pp.979-990, ISSN 15610740

Downs, J.A.; Nussenzweig, M.C. & Nussenzweig, A. (2007). Chromatin dynamics and the preservation of genetic information. *Nature*, Vol.447, No.7147, (June 2007), pp.951-958, ISSN 1758-1578

Dupre, A.; Boyer-Chatenet, L. & Gautier, J. (2006). Two-step activation of ATM by DNA and the Mre11-Rad50-Nbs1 complex. *Nature Struct Mol Biol*, Vol. 13, No.5, (May 2006), pp.451–457, ISSN 1662-2404

Durant S.T.; Cho, E.C. & La Thangue, N.B. (2009).p53 methylation—the argument is clear. *Cell Cycle*, Vol.8, No.6, (March 2009), pp. 801-802, ISSN 1922-1494

Dvir, A.; Peterson, S.R.; Knuth, M.W.; Lu, H. & Dynan, W.S. (1992). Ku autoantigen is the regulatory component of a template-associated protein kinase that phosphorylates RNA polymerase II. *Proc Natl Acad Sci U S A*, Vol.89, No.24, (December 1992), pp.11920-11924, ISSN 146-5419

el-Deiry, W.S.; Tokino, T.; Velculescu, V.E.; Levy, D.B.; Parsons, R.; Trent, J.M.; Lin, D.; Mercer, W.E.; Kinzler, K.W. & Vogelstein, B. (1993). WAF1, a potential mediator of p53 tumor suppression. *Cell*, Vol.75, No.4, (September 1993), pp.817-825, ISSN 824-2752

Ewen, M.E.; Sluss, H.K.; Sherr, C.J.; Matsushime, H.; Kato, J. & Livingston, D.M. (1993). Functional interactions of the retinoblastoma protein with mammalian D-type cyclins. *Cell*, Vol.73, No.3, (May 1993), pp. 487–497, ISSN 834-3202

Falck, J.; Lukas, C.; Protopopova, M.; Lukas, J.; Selivanova, G. & Bartek, J. (2001). Functional impact of concomitant versus alternative defects in the Chk2–p53 tumour suppressor pathway. *Oncogene*, Vol.20, No.39, (September 2001), pp.5503- 5510, ISSN 1157-1648.

Falck, J.; Coates, J. & Jackson, S.P. (2005). Conserved modes of recruitment of ATM, ATR and DNA-PKcs to sites of DNA damage. *Nature*, Vol.434, No.7033, (March 2005), pp.605–611, ISSN 1575-8953

Feddersen, R.M.; Ehlenfeldt, R.; Yunis, W.S.; Clark, H.B. & Orr, H.T. (1992). Disrupted cerebellar cortical development and progressive degeneration of Purkinje cells in SV40 T antigen transgenic mice. *Neuron*, Vol.9, No.5, (November 1992),

Feddersen, R.M.; Clark, H.B.; Yunis, W.S. & Orr, H.T. (1995). In vivo viability of postmitotic Purkinje neurons requires pRb family member function. *Mol Cell Neurosci*. Vol.6, No.2, (April 1995), pp.153-167, ISSN 755-1567

Ferguson, A.T.; Evron, E.; Umbricht, C.B.; Pandita, T.; Chan, T.A.; Hermeking, H.; Marks, J.R.; Lambers, A.R.; Futreal, P.A.; Stampfer, M.R. & Sukumar, S. (2000). High frequency of hypermethylation at the 14-3-3sigma locus leads to gene silencing in breast cancer. *Proc Natl Acad Sci USA*, Vol.97, No.22, (May 2000), pp. 6049-6054, ISSN 1081-1911

Fernandez-Capetillo, O.; Lee, A.; Nussenzweig, M.; & Nussenzweig, A. (2004). H2AX: the histone guardian of the genome. *DNA Repair (Amst)*, Vol.3, No.8-9, (Aug-Sep 2004), pp.959–967, ISSN 1527-9782

Field, S.J.; Tsai, F.; Kuo, F.; Zubiaga, A.M.; Kaelin, Jr. W.G.; Livingston, D.M.; Orkin, S.H. & Greenberg, M.E. (1996). E2F-1 functions in mice to promote apoptosis and suppress proliferation. *Cell*, Vol.85, No.4, (May 1996), pp. 549–561, ISSN 865-3790

Fishel, M.L.; Vasko, M.R. & Kelley, M.R. (2007). DNA repair in neurons: so if they don't divide what's to repair? *Mutat Res*, Vol.614, No.1-2, (August 2006), pp.24–36, ISSN 1687-9837

Gagné, J.P.; Hendzel, M.J.; Droit, A. & Poirier, G.G. (2006). The expanding role of poly(ADP-ribose) metabolism: current challenges and new perspectives. *Curr Opin Cell Biol*, Vol.18, No.2, (April 2006), pp.145-151, ISSN 1651-6457

Garner, E. & Raj, K. (2008). Protective mechanisms of p53–p21–pRb proteins against DNA damage-induced cell death. *Cell Cycle*, Vol.7, No.3, (November 2008), pp.277–282, ISSN 1823-5223

Gell, D. & Jackson SP. (1999). Mapping of protein-protein interactions within the DNA-dependent protein kinase complex. *Nucleic Acids Res*, Vol.27, No.17, (September 1999), pp. 3494-3502, ISSN 1044-6239

Gobbel, G.T.; Bellinzona, M.; Vogt, A.R.; Gupta, N.; Fike, J.R. & Chan, P.H. (1998). Response of postmitotic neurons to X- irradiation: implications for the role of DNA damage in neuronal apoptosis. *J Neurosci*, Vol.18, No.1, (January 1998), pp.147–155, ISSN, 941-2495

Gorgoulis, V.; Vassiliou, L.V.; Karakaidos, P.; Zacharatos, P.; Kotsinas, A.; Liloglou, T.; Venere, M.; Ditullio, R.A. Jr.; Kastrinakis, N.G.; Levy, B.; Kletsas, D.; Yoneta, A.; Herlyn, M.; Kittas, C. & Halazonetis, T.D. (2005). Activation of the DNA damage checkpoint and genomic instability in human precancerous lesions. *Nature*, Vol.434, No.7035, (April 2005), pp. 907-913, ISSN 1582-9965

Gorodetsky, E.; Calkins, S.; Ahn, J. & Brooks, P.J. (2007). ATM, the Mre11/Rad50/Nbs1 complex, and topoisomerase I are concentrated in the nucleus of Purkinje neurons in the juvenile human brain. *DNA Repair (Amst)*, Vol.6, No.11, (August 2007), pp.1698-1707, ISSN 1770-6468

Gottlieb, T.M. & Jackson, S.P. (1993). The DNA-dependent protein kinase: requirement for DNA ends and association with Ku antigen. *Cell*, Vol.72, No.1, (January 1993), pp. 131-142, ISSN 842-2676

Gottschalk, A.J.; Timinszky, G.; Kong, S.E.; Jin, J.; Cai, Y.; Swanson, S.K.; Washburn, M.P.; Florens, L.; Ladurner, A.G.; Conaway, J.W. & Conaway, R.C. (2009). Poly(ADP-ribosyl)ation directs recruitment and activation of an ATP- dependent chromatin remodeler. *Proc Natl Acad Sci U S A*, Vol.106, No.33, (August 2009), pp. 13770-13774, ISSN 1966- 6485

Gottifredi, V.; Shieh, S.Y.; Taya, Y. & Prives, C. (2001). p53 accumulates but is functionally impaired when DNA synthesis is blocked. *Proc Natl Acad Sci USA*, Vol.98, No.3, (January 2001), pp.1036-1041, ISSN 1115-8590

Green, D.R. & Kroemer, G. (2009). Cytoplasmic functions of the tumour suppressor p53. *Nature*, Vol.458, No.7242, (April 2009), pp.1127-1130, ISSN 1940-7794

Groth, A.; Rocha, W.; Verreault, A. & Almouzni, G. (2007). Chromatin challenges during DNA replication and repair. *Cell*, Vol.128, (February 2007), pp. 721-733, ISSN 173-20509

Gu, Y.; Turck, C.W. & Morgan, D.O. (1993). Inhibition of CDK2 activity in vivo by an associated 20K regulatory subunit. *Nature*, Vol.366, No.6456, (December 1993), pp.707–710, ISSN 825-9216

Haince, J.F.; McDonald, D.; Rodrigue, A.; Déry, U.; Masson, J.Y.; Hendzel, M.J. & Poirier, G.G. (2008). PARP1-dependent kinetics of recruitment of MRE11 and NBS1 proteins to multiple DNA damage sites. *J Biol Chem*, Vol.283, No.21, (November 2008), pp.1197-1208, ISSN 1802-5084

Hakem, R. (2008). DNA-damage repair; the good, the bad, and the ugly. *EMBO J*, Vol.27, No.4, (February 2008), pp.589-605, ISSN 1828-5820

Hakmé, A.; Wong, H.K.; Dantzer, F. & Schreiber, V. (2008). The expanding field of poly(ADP-ribosyl)ation reactions. 'Protein Modifications: Beyond the Usual Suspects' Review Series. *EMBO Rep*, Vol.9, No.11, (October 2008), pp.1094-100, ISSN 1892-7583

Hammel, M.; Yu, Y.; Mahaney, B.L.; Cai, B.; Ye, R.; Phipps, B.M.; Rambo, R.P.; Hura, G.L.; Pelikan, M.; So, S.; Abolfath, R.M.; Chen, D.J.; Lees-Miller, S.P. & Tainer, J.A. (2009). Ku and DNA-dependent protein kinase dynamic conformations and assembly regulate DNA binding and the initial non-homologous end joining complex. *J Biol Chem*, Vol.285, No.2, (November 2009), pp.1414-1423, ISSN 1989-3054

Hammet, A.; Magill, C.; Heierhorst, J. & Jackson, S.P. (2007). Rad9 BRCT domain interaction with phosphorylated H2AX regulates the G1 checkpoint in budding yeast. *EMBO Rep*, Vol.8, No.9, (August 2007), pp.851-857, ISSN 1772-1446

Hammond, E.M.; Denko, N.C.; Dorie, M.J.; Abraham, R.T. & Giaccia, A.J. (2002). Hypoxia links ATR and p53 through replication arrest. *Mol Cell Biol*, Vol.22, No.6, (March 2002), pp.1834-1843, ISSN 1186-5061

Harper, J.W.; Adami, G.R.; Wei, N.; Keyomarsi, K. & Elledge, S.J. (1993). The p21 Cdk-interacting protein Cip1 is a potent inhibitor of G1 cyclin-dependent kinases. *Cell*, Vol.75, No.4, (November 1993), pp.805-816, ISSN 824-2751

Henrique, R.; Jeronimo, C.; Hoque, M.O.; Carvalho, A.L.; Oliveira, J.; Teixeira, M.R.; Lopes, C. & Sidransky, D. (2005). Frequent 14-3-3sigma promoter methylation in benign and malignant prostate lesions. *DNA Cell Biol*, Vol.24, No.4, (April 2005), pp.264-269, ISSN 1581-2243

Heo, K.; Kim, H.; Choi, S.H.; Choi, J.; Kim, K.; Gu, J.; Lieber, M.R.; Yang, A.S. & An, W. (2008). FACT-mediated exchange of histone variant H2AX regulated by phosphorylation of H2AX and ADP-ribosylation of Spt16. *Mol Cell*, Vol.30, No.1, (April 2008), pp.86-97, ISSN 1840-6329

Herrup, K.; Neve, R.; Ackerman, S.L. & Copani, A. (2004). Divide and die: cell cycle events as triggers of nerve cell death. *J Neurosci*, Vol.24, No.42, (October 2004), pp.9232-9239, ISSN 1549-6657

Herzog, K.H.; Chong, M.J.; Kapsetaki, M.; Morgan, J.I. & McKinnon, P.J. (1998). Requirement for Atm in ionizing radiation- induced cell death in the developing central nervous system. *Science*, Vol.280, No.55366, (May 1998), pp.1089–1091, ISSN 9582124

Hirao, A.; Kong, Y.Y.; Matsuoka, S.; Wakeham, A.; Ruland, J.; Yoshida, H.; Liu, D.; Elledge, S.J. & Mak, T.W. (2000). DNA damage-induced activation of p53 by the checkpoint kinase Chk2. *Science*, Vol.287, No.5459, (March 2000), pp.1824- 1827, ISSN 1071-0310

Hoffmann, I.; Clarke, P.R.; Marcote, M.J.; Karsenti, E. & Draetta, G. (1993). Phosphorylation and activation of human cdc25-C by cdc2-cyclin B and its involvement in the self-amplification ofMPFatmitosis. *EMBOJ*, Vol.12, No.1, (January 1993), pp.53-63, ISSN 842-8594

Höglinger, G.U.; Breunig, J.J.; Depboylu, C.; Rouaux, C.; Michel, P.P.; Alvarez-Fischer, D.; Boutillier, A.L.; Degregori, J.; Oertel, W.H.; Rakic, P.; Hirsch, E.C. & Hunot, S. (2007). The pRb/E2F cell-cycle pathway mediates cell death in Parkinson's disease. *Proc Natl Acad Sci USA*, Vol.104, No.9, (February 2007), pp.25-27, ISSN 1736-0686

Hopfner, K.P. (2009). DNA double-strand breaks come into focus. *Cell*, Vol.139, No.1, (October 2009), pp.25-27, ISSN 1980- 4750

Houtgraaf, J.H.; Versmissen, J. & van der Giessen, W.J. (2006). A concise review of DNA damage checkpoints and repair in mammalian cells. *Cardiovasc Revasc Med*, Vol.7, No.3, (July-September 2006), pp.165-172, ISSN 1694-5824

Huang, J.Y.; Chen, W.H.; Chang, Y.L.; Wang, H.T.; Chuang, W.T & Lee SC. (2006). Modulation of nucleosome-binding activity of FACT by poly(ADP-ribosyl)ation. *Nucleic Acids Res*, Vol.34, No.8, (May 2006), pp.2398-2407, ISSN 1668- 2447

Huen, M.S. & Chen, J. (2008). The DNA damage response pathways: at the crossroad of protein modifications. *Cell Res*, Vol.18, No.1, (January 2008), pp.8-16, ISSN 1808-7291

Husseman, J.W.; D. Nochlin, D. & Vincent, I. (2000). Mitotic activation: a convergent mechanism for a cohort of neurodegenerative diseases. *Neurobiol Aging*, Vol.21, No.6, (November 2000), pp.815-828, ISSN 1112-4425

Iaquinta, P.J. & Lees, J.A. (2007). Life and death decisions by the E2F transcription factors. *Curr Opin Cell Biol*, Vol.19, No.6, (November 2007), pp. 649–657, ISSN 1803-2011

Ikura, T.; Tashiro, S.; Kakino, A.; Shima, H.; Jacob, N.; Amunugama, R.; Yoder, K.; Izumi, S.; Kuraoka, I.; Tanaka, K.; Kimura, H.; Ikura, M.; Nishikubo, S.; Ito, T.; Muto, A.; Miyagawa, K.; Takeda, S.; Fishel, R.; Igarashi, K. & Kamiya, K. (2007). DNA damage-dependent acetylation and ubiquitination of H2AX enhances chromatin dynamics. *Mol Cell Biol*, Vol.27, No.20, August 2007), pp.7028-7040, ISSN 1770-9392

Jackson, S.P. (2009). The DNA-damage response: new molecular insights and new approaches to cancer therapy. *Biochem Soc Trans*, Vol.37, Pt.3, (June 2009), pp. 483–494, ISSN 1984-7258

Jackson, S.P. & Bartek, J. (2009). The DNA-damage response in human biology and disease. *Nature*, Vol.461, No.7267, (October 2009), pp.1071-1078, ISSN 1984-7258

Jansson, M.; Durant, S.T.; Cho, E.C.; Sheahan, S.; Edelmann, M.; Kessler, B. & La Thangue, N.B. (2008). Arginine methylation regulates the p53 response. *Nat Cell Biol*, Vol.10, No.12, (November 2008), pp. 1431-1439, ISSN 1901-1621

Jazayeri, A.; Falck, J.; Lukas, C.; Bartek, J.; Smith, G.C.; Lukas, J. & Jackson, S.P. (2006). ATM- and cell cycle-dependent regulation of ATR in response to DNA double-strand breaks *Nat Cell Biol*, Vol.8, No.1, (December 2005), pp.37-45, ISSN 1632-7781

Jeong, J.; Juhn, K.; Lee, H.; Kim, S.H.; Min, B.H.; Lee, K.M.; Cho, M.H.; Park, G.H. & Lee, K.H. (2007). SIRT1 promotes DNA repair activity and deacetylation of Ku70. *Exp Mol Med*, Vol.39, No.1, (February 2007), pp.8-13, ISSN 1733-4224

Jha, S.; Shibata, E. & Dutta, A. (2008). Human Rvb1/Tip49 is required for the histone acetyltransferase activity of Tip60/NuA4 and for the downregulation of phosphorylation on H2AX after DNA damage. *Mol Cell Biol*, Vol.28, No.8, (February 2008), pp. 2690-2700, ISSN 1828-5460

Jin, Y.H.; Yoo, K.J.; Lee, Y.H. & Lee SK. (2000). Caspase 3-mediated cleavage of p21WAF1/CIP1 associated with the cyclin A- cyclin-dependent kinase 2 complex is a prerequisite for apoptosis in SK-HEP-1 cells. *J Biol Chem*, Vol.275, No.39, (September 2000), pp.30256–30263, ISSN 1088-4382

Johnson, D.G.; Schwarz, J.K.; Cress, W.D. & Nevins, J.R. (1993). Expression of transcription factor E2F1 induces quiescent cells to enter S phase. *Nature*, Vol.365, No.6444, (September, 1993), pp.349-352. ISSN 837-7827

Jungmichel, S. & Stucki, M. (2010). MDC1: The art of keeping things in focus. *Chromosoma*, Vol.119, No.4, (March, 2010), pp.337-349. ISSN 2022-4865

Kaidi, A.; Weinert, B.T.; Choudhary, C. & Jackson, S.P. (2010). Human SIRT6 promotes DNA end resection through CtIP deacetylation. *Science*, Vol.329, No.5997, (September 2010), pp.1348-1353, ISSN 2082-9486

Karagiannis, T. C. & El-Osta, A. (2007). Chromatin modifications and DNA double-strand breaks: the current state of play. *Leukemia*, Vol.21, No.2, (December 2006), pp.195–200, ISSN 1715-1702

Kastan, M.B. & Bartek, J. (2004). Cell-cycle checkpoints and cancer. *Nature*, Vol. 432, No. 7015, (November 2004), pp.316– 323, ISSN 1554-9093

Khanna, K.K. & Jackson, S.P. (2001). DNA double-strand breaks: signaling, repair and the cancer connection. *Nat Genet*, Vol. 27, No.3, (March 2001), pp.247–254, ISSN 1124-2102

Khanna, K. K.; Lavin, M. F., Jackson, S. P. & Mulhern, T. D. (2001). ATM, a central controller of cellular responses to DNA damage. *Cell Death Differ*, Vol. 8, No. 11, (November 2001), pp.1052–1065, ISSN 1168-7884

Kim, D. & Tsai, L.H. (2009). Linking cell cycle reentry and DNA damage in neurodegeneration. *Ann N Y Acad Sci*, Vol. 1170, (July 2009), pp. 674-679, ISSN 1968-6210

Kim, M.Y.; Zhang, T. & Kraus, W.L. (2005). Poly(ADP-ribosyl)ation by PARP-1: 'PAR-laying' NAD+ into a nuclear signal. *Genes Dev*, Vol.19, No.17, (September 2005), pp.1951-1967, ISSN 1614-0981

Knights, C.D.; Catania, J.; Di Giovanni, S.; Muratoglu, S.; Perez, R.; Swartzbeck, A.; Quong, A.A.; Zhang, X.; Beerman, T.; Pestell, R.G. & Avantaggiati, M.L. (2006). Distinct p53 acetylation cassettes differentially influence gene-expression patterns and cell fate. *J Cell Biol*, Vol.173, No.4, (May 2006), pp.533-544, ISSN 1671-7128

Koch, C.A.; Agyei, R.; Galicia, S.; Metalnikov, P.; O'Donnell, P.; Starostine, A.; Weinfeld, M. & Durocher, D. (2004).Xrcc4 physically links DNA end processing by polynucleotide kinase to DNA ligation by DNA ligase IV. *EMBO J*, Vol. 23, No.19, (September 2004), pp. 3874–3885, ISSN 1538-5968

Kolas, N.K.; Chapman, J.R.; Nakada, S.; Ylanko, J.; Chahwan, R.; Sweeney, F.D.; Panier, S.; Mendez, M.; Wildenhain, J.; Thomson, T.M.; Pelletier, L.; Jackson, S.P. & Durocher, D. (2007). Orchestration of the DNA-damage response by the RNF8 ubiquitin ligase. *Science*, Vol.318, No.5856, (November 2007), pp.1637-1640, ISSN 1800-6705

Korotchkina, L.G.; Leontieva, O.V.; Bukreeva, E.I.; Demidenko, Z.N.; Gudkov, A.V. & Blagosklonny, M.V. (2010).The choice between p53-induced senescence and quiescence is determined in part by the mTOR pathway. *Aging (Albany NY)*, Vol.2, No.6, (June 2010), pp. 344-352, ISSN 2060-6252

Kraemer, K.H.; Sander, M. & Bohr, V.A. (2007). New areas of focus at workshop on human diseases involving DNA repair deficiency and premature aging. *Mech Ageing Dev*, Vol.128, No.2, (February 2007), pp.229-235, ISSN 1736-1460

Kraljevic Pavelic, S.; Cacev, T. & Kralj, M. (2008). A dual role of p21waf1/cip1 gene in apoptosis of HEp-2 treated with cisplatin or methotrexate. *Cancer Gene Ther*, Vol.15, No.9, (May 2008), pp. 576-590, ISSN 1848-3502

Krishnakumar, R. & Kraus, W.L. (2010). The PARP side of the nucleus: molecular actions, physiological outcomes, and clinical targets. *Mol Cell*, Vol.39, No.1, (July 2010), pp.8-24, ISSN 2060-3072

Kouzarides, T. (2007).Chromatin modifications and their function. *Cell*, Vol.128, No.4, (February 2007), pp.693-705, ISSN 1732-0507

Kruman, I.I. (2004). Why do neurons enter the cell cycle? *Cell Cycle*, Vol.3, No.6, (June 2004), pp. 769-773, ISSN 1513-6759

Kruman, I.I., Wersto, R.P.; Cardozo-Pelaez, F.; Smilenov, L.; Chan, S.; Chrest, F.; Emokpae, R. Jr.; Gorospe, M. & Mattson, M.P. (2004). Cell cycle activation linked to neuronal cell death initiated by DNA damage. *Neuron*, Vol.41, No.4, (February 2004), pp.549–561, ISSN 1498-0204

Kuan, C.Y.; Schloemer, A.J.; Lu, A.; Burns, K.A.; Weng, W.L.; Williams, M.T.; Strauss, K.I.; Vorhees, C.V.; Flavell, R.A.; Davis, R.J.; Sharp, F.R. & Rakic, P. (2004). Hypoxia-ischemia induces DNA synthesis without cell proliferation in dying neurons in adult rodent brain. *J Neurosci*, Vol.24, No.47, (November 2004), pp. 10763-10772, ISSN 1556-4594

Kumagai, A. & Dunphy, W.G. (2006) How cells activate ATR. *Cell Cycle*, Vol.5, No.12, (June 2006), pp. 1265-1268, ISSN 1676- 0665

Kuroda, Y.; Aishima, S.; Taketomi, A.; Nishihara, Y.; Iguchi, T.; Taguchi, K.; Maehara, Y. & Tsuneyoshi, M. (2007). 14-3- 3sigma negatively regulates the cell cycle, and its downregulation is associated with poor outcome in intrahepatic cholangiocarcinoma. *Hum Pathol*, Vol.38, No.7, (March 2007), pp. 1014-1022, ISSN 1739-1729

Kusch, T.; Florens, L.; Macdonald, W.H.; Swanson, S.K.; Glaser, R.L.; Yates, J.R. 3rd.; Abmayr, S.M.; Washburn, M.P. & Workman, J.L. (2004). Acetylation by Tip60 is required for selective histone variant exchange at DNA lesions. *Science*, Vol.306, No.5704, (November 2004), pp.2084-2087, ISSN 1552-8408

Lal, A.; Abdelmohsen, K.; Pullmann, R.; Kawai, T.; Galban, S.; Yang, X.; Brewer, G. & Gorospe, M. (2006). Posttranscriptional derepression of GADD45alpha by genotoxic stress. *Mol Cell*, Vol.22, No.1, (April 2006), pp.117-128, ISSN 1660-0875

Laronga, C.; Yang, H.Y.; Neal, C. & Lee, M.H. (2000). Association of the cyclin-dependent kinases and 14-3-3sigma negatively regulates cell cycle progression. *J Biol Chem*, Vol. 275, No.30, (July 2000), pp. 23106-23112, ISSN 1076-7298

Lavin, M.F. & Kozlov, S. (2007). DNA damage-induced signalling in ataxia–telangiectasia and related syndromes. *Radiother Oncol*, Vol. 83, No.3, (May 2007), pp.231–237, ISSN 1751-2070

Le Cam, L.; Linares, L.K.; Paul, C.; Julien, E.; Lacroix, M.; Hatchi, E.; Triboulet, R.; Bossis, G.; Shmueli, A.; Rodriguez, M.S.; Coux, O. & Sardet, C. (2006). E4F1 is an atypical ubiquitin ligase that modulates p53 effector functions independently of degradation. *Cell*, Vol.127, No.4, (November 2006), pp. 775-788, ISSN 1711-0336

Lee, E.W.; Lee, M.S.; Camus, S.; Ghim, J.; Yang, M.R.; Oh, W.; Ha, N.C.; Lane, D.P. & Song, J. (2009).Differential regulation of p53 and p21 by MKRN1 E3 ligase controls cell cyclearrest and apoptosis, *EMBO J,*. Vol.28, No.14, (June 2009), pp.2100–2113, ISSN 1953-6131

Lee, J.H. & Paull, T.T. (2005). ATM activation by DNA double-strand breaks through the Mre11-Rad50-Nbs1 complex. *Science*, Vol.308, No. 5721, (March 2005), pp.551-554, ISSN 1579-0808

Lee, J.H., Choy, M.L., Ngo, L., Foster, S.S., Marks, P.A. (2010). Histone deacetylase inhibitor induces DNA damage, which normal but not transformed cells can repair. *Proc Natl Acad Sci U S A*, Vol.107, No.33, (August 2010), pp.14639-14644, ISSN 2067-9231

Lee, S.M.; Bae, J.H.; Kim, M.J.; Lee, H.S.; Lee, M.K.;Chung, B.S.; Kim, D.W.; Kang, C.D. & Kim, S.H. (2007). Bcr-Abl- independent imatinib-resistant K562 cells show aberrant protein acetylation and increased sensitivity to histone deacetylase inhibitors. *J Pharmacol Exp Ther*, Vol.322, No.3, (June 2007), pp.1084-1092, ISSN 1756-9822

Lee, Y. & McKinnon, P.J. (2000). ATM dependent apoptosis in the nervous system. *Apoptosis* Vol.5, No.6, (December 2000), pp.523–529, ISSN 1130-3911

Lee, Y.; Chong, M.J. & McKinnon PJ. (2001). Ataxia telangiectasia mutated–dependent apoptosis after genotoxic stress in the developing nervous system is determined by cellular differentiation status. *J Neurosci*, Vol.21, No.17, (September 2001), pp.6687–6693, ISSN 1151-7258

Lee, Y. & McKinnon, P.J. (2007). Responding to DNA double strand breaks in the nervous system. *Neuroscience*, Vol.145, No.4, (April 2007), pp.1365-1374, ISSN 16934412

Leroy, C.; Lee, S.E.; Vaze, M.B.; Ochsenbien, F.; Guerois, R.; Haber, J.E. & Marsolier-Kergoat, M.C. (2003). PP2C phosphatases Ptc2 and Ptc3 are required for DNA checkpoint inactivation after a double-strand break. *Mol Cell*, Vol.285, No.2, (November 2003), pp.1414-1423, ISSN 1989-3054

Levine, A.J. & Oren, M. (2009). The first 30 years of p53: growing ever more complex. *Nat Rev Cancer*, Vol.9, No.10, (October 2009), pp.749-758, ISSN 1977-6744

Levkau, B.; Koyama, H.; Raines, E.W.; Clurman, B.; Herren, B.; Orth, K.; Roberts, J.M. & Ross, R. (1998). Cleavage of p21Cip1/Waf1 and p27Kip1 mediates apoptosis in endothelial cells through activation of Cdk2: role of a caspase cascade. *Mol Cell*, Vol.1, No.4, (March 1998), pp.553-563, ISSN 966-0939

Li, X.; Corsa, C.A.; Pan, P.W.; Wu, L.; Ferguson, D.; Yu, X.; Min, J. & Dou, Y. (2010). MOF and H4 K16 acetylation play important roles in DNA damage repair by modulating recruitment of DNA damage repair protein Mdc1. *Mol Cell Biol*, Vol.30, No.22, (September 2010), pp.5335-5347, ISSN 2083-7706

Lieber MR. (2010). The mechanism of double-strand DNA break repair by the nonhomologous DNA end-joining pathway. *Annu Rev Biochem*, Vol.79, pp. 181-211, ISSN 2019-2759

Lisby, M; Barlow, J.H.; Burgess, R.C. & Rothstein, R. (2004). *Cell*, Vol.118, No.6, (September 2004), pp.699-713, ISSN 1536- 9670

Liu, Z.M.; Chen, G.G.; Ng, E.K.; Leung, W.K.; Sung, J.J. & Chung, S.C. (2004). Upregulation of heme oxygenase-1 and p21 confers resistance to apoptosis in human gastric cancer cells. *Oncogene*, Vol.23, No.2, (January 2004), pp.503-13, ISSN 1464-7439

Lobrich, M. & Jeggo, P.A. (2007). The impact of a negligent G2/M checkpoint on genomic instability and cancer induction. *Nat Rev Cancer*, Vol. 7, No.11, (November 2007), pp.861–869, ISSN 1794-3134

Lu, T.; Pan, Y.; Kao, S.Y.; Li, C.; Kohane, I.; Chan, J. & Yankner, B.A. (2004). Gene regulation and DNA damage in the ageing human brain. *Nature*, Vol.429, No.6994, (June 2004), pp.883-891, ISSN 1519-0254

Lukas, C.; Bartkova, J.; Latella, L.; Falck, J.; Mailand, N.; Schroeder, T.; Sehested, M.; Lukas, J. & Bartek, J. (2001). DNA damage-activated kinase Chk2 is independent of proliferation or differentiation yet correlates with tissue biology. *Cancer Res*, Vol.61, No.13, (July 2001), pp.4990-4903, ISSN 1143-1331

Luo, J.; Nikolaev, A.Y.; Imai, S.; Chen, D.; Su, F.; Shiloh, A.; Guarente, L. & Gu, W. (2001). Negative control of p53 by Sir2alpha promotes cell survival under stress. *Cell*, Vol.107, No.2, (October 2001), pp. 137-148, ISSN 1167-2522

Macůrek, L.; Lindqvist, A.; Voets, O.; Kool, J.; Vos, H.R. & Medema, R.H. (2010). Wip1 phosphatase is associated with chromatin and dephosphorylates gammaH2AX to promote checkpoint inhibition. *Oncogene*, Vol.29, No.15, (January 2010), pp.2281-2291, ISSN 2010-1220

Mahaney, B.L.; Meek, K. & Lees-Miller, S.P. (2009). Repair of ionizing radiation-induced DNA double-strand breaks by non- homologous end-joining. *Biochem*, Vol.417, No.3, (February 2009), pp.639–650, ISSN 1913-3841

Mailand, N.; Bekker-Jensen,S.; Bartek, J. & Lukas, J. (2006). Destruction of Claspin by SCFbetaTrCP restrains Chk1 activation and facilitates recovery from genotoxic stress. *Mol Cell*, Vol.23, No.3, (August 2006), pp.307–318, ISSN 1688-5021

Mamely, I., van Vugt, M.A., Smits, V.A., Semple, J.I., Lemmens, B., Perrakis, A., Medema, R.H, & Freire, R. (2006). Pololike kinase-1 controls proteasome-dependent degradation of claspin during checkpoint recovery. *Curr Biol*, Vol.16, No.19, (August 2006), pp. 1950-1955, ISSN 1693-4469

Marchenko. N.D.; Zaika, A. & Moll, U.M. (2000) Death signal-induced localization of p53 protein to mitochondria. A potential role in apoptotic signaling. *J Biol Chem*, Vol.275, No.21, (May 2000), pp.16202–16212, ISSN 1082-1866

Martin, L.J. (2008). DNA damage and repair: relevance to mechanisms of neurodegeneration. *J Neuropathol Exp Neurol*, Vol.67, No.5, (May 2008), pp.377-387, ISSN 1843-1258

Martin, L.J.; Liu, Z.; Pipino, J.; Chestnut, B. & Landek, M.A. (2009). Molecular regulation of DNA damage-induced apoptosis of neurons in cerebral cortex. *Cerebral Cortex*, Vol.19, No.6, (September 2008), pp. 1273-1293, ISSN 1882-0287

Martinez, L.A.; Yang, J.; Vazquez, E.S.; Rodriguez-Vargas Mdel, C.; Olive, M.; Hsieh, J.T.; Logothetis, C.J. & Navone, N.M. (2002).p21 modulates threshold of apoptosis induced by DNA-damage and growth factor withdrawal in prostate cancer cells. *Carcinogenesis*, Vol.23, No.8, (August 2002), pp.1289–1296, ISSN 1215-1346

Mazumder, S.; Plesca, D.; Kinter, M. & Almasan, A. (2007). Interaction of a cyclin e fragment with Ku70 regulates bax- mediated apoptosis. *Mol Cell Biol*, Vol.27, No.9, (February 2007), pp.3511-3520, ISSN 1732-5036

McKinnon, P.J. (2001). Ataxia telangiectasia: new neurons and ATM. *Trends Mol Med*, Vol.7, No.6, (June 2001), pp.233-234, ISSN 1137-8498

McMurray, C.T. (2005). To die or not to die: DNA repair in neurons. *Mutat Res*, Vol.577, No.1-2, (September 2005), pp.260– 274, ISSN 1592-1706

McShea, A.; Wahl, A.F. & Smith, M.A. (1999). Re-entry into the cell cycle: a mechanism for neurodegeneration in Alzheimer disease. *Med Hypotheses*, Vol.52, No.6, (June 1999), pp. 525–527, ISSN 1045-9833

Meek, D.W. (1994). Post-translational modification of p53. *Semin Cancer Biol*, Vol.5, No.3, (June 1994), pp.203-210, ISSN 794- 8948

Melander, F.; Bekker-Jensen, S.; Falck, J.; Bartek, J.; Mailand, N. & Lukas, J. (2008). Phosphorylation of SDT repeats in the MDC1 N terminus triggers retention of NBS1 at the DNA damage-modified chromatin. *J Cell Biol*, Vol.181, No.2, (April 2008), pp.213-226, ISSN 1841-1307

Melino, G. (2003) p73, the "assistant" guardian of the genome? *Ann N Y Acad Sci*, Vol.1010, pp.9-15, ISSN 1503-3688

Meng, S.; Arbit, T.; Veeriah, S.; Mellinghoff, I.K.; Fang, F.; Vivanco, I.; Rohle, D. & Chan, T.A. (2009). 14-3-3sigma and p21 synergize to determine DNA damage response following Chk2 inhibition. *Cell Cycle*, Vol.8, No.14, (July 2009), pp. 2238- 2246, ISSN 1950-2805

Merkle, D.; Douglas, P.; Moorhead, G.B.; Leonenko, Z.; Yu, Y.; Cramb, D.; Bazett-Jones, D.P. & Lees-Miller, S.P. (2002). The DNA-dependent protein kinase interacts with DNA to form a protein-DNA complex that is disrupted by phosphorylation. *Biochemistry*, Vol.41, No.42, (October 2002), pp.12706-12714, ISSN 1237-9113

Milczarek, G.J.; Martinez, J. & Bowden, G.T. (1997). p53 Phosphorylation: biochemical and functional consequences. *Life Sci*, Vol.60, No.1, pp.1-11, ISSN 899-5526

Messick TE, Greenberg RA. (2009). The ubiquitin landscape at DNA double-strand breaks. *J Cell Biol*, Vol.187, No.3, (November 2009), pp.319-326. ISSN 1994-8475

Mihara, M.; Erster, S.; Zaika, A.; Petrenko, O.; Chittenden, T.; Pancoska, P. & Moll, U.M. (2003) p53 has a direct apoptogenic role at the mitochondria. *Mol Cell*, Vol.11, No.3, (March 2003), pp. 577-590. ISSN 1266-7443

Miller, S.A.; Mohn, S.E. & Weinmann, A.S. (2010). Jmjd3 and UTX play a demethylase-independent role in chromatin remodeling to regulate T-box family member-dependent gene expression. *Mol Cell*, Vol.40, No.4, (November 2010), pp.594-605, ISSN 2109-5589

Mimori, T. & Hardin, J.A.(1986). Mechanism of interaction between Ku protein and DNA. *J Biol Chem*, Vol.261, No,22, (August 1986), pp.10375-10379, ISSN 301-5926

Morselli, E.; Galluzzi, L.; Kepp, O. & Kroemer, G. (2009). Nutlin kills cancer cells via mitochondrial p53. *Cell Cycle*, Vol.8, No.11, (June 2009), pp. 1647-1648, ISSN 1944-8434

Mortusewicz, O.; Amé, J.C.; Schreiber, V. & Leonhardt, H. (2007). Feedback-regulated poly(ADP-ribosyl)ation by PARP-1 is required for rapid response to DNA damage in living cells. *Nucleic Acids Res*, Vol.35, No.22, (November 2007), pp.7665- 7675, ISSN 1798-2172

Moynahan, M.E.; Chiu, J.W.; Koller, B.H. & Jasin, M. (1999). Brca1 controls homology-directed DNA repair. *Mol Cell*, Vol. 4, No.4, (October 1999), pp.511-518, ISSN 1054-9283

Murr, R.; Loizou, J.I.; Yang, Y.G.; Cuenin, C.; Li, H.; Wang, Z.Q. & Herceg, Z. (2006). Histone acetylation by Trrap-Tip60 modulates loading of repair proteins and repair of DNA double-strand breaks. *Nat Cell Biol*, Vol.8, No.1, (Dec 2006), pp.91–99, ISSN 1634-1205

Nagayama, T.; Lan, J.; Henshall, D.C.; Chen, D.; O'Horo, C.; Simon, R.P. & Chen, J. (2000). Induction of oxidative DNA damage in the peri-infarct region after permanent focal cerebral ischemia. *J Neurochem*, Vol.75, No.4, (October 2000), pp.1716-1728, ISSN 1098-7855

Nakada, S.; Chen, G.I.; Gingras, A.C. & Durocher, D. (2008). PP4 is a gamma H2AX phosphatase required for recovery from the DNA damage checkpoint. *EMBO Rep*, Vol.9, No.10, (August 2008), pp.1019-1026, ISSN 1875-8438

Nakamura, T.M.; Du, L.L.; Redon, C. & Russell, P. (2004). Histone H2A phosphorylation controls Crb2 recruitment at DNA breaks, maintains checkpoint arrest, and influences DNA repair in fission yeast. *Mol Cell Biol*, Vol.24, No.1,14, (July 2004), pp.6215-6230, ISSN 1522-6425

Narciso, L.; Fortini, P.; Pajalunga, D.; Franchitto, A.; Liu, P.; Degan, P.; Frechet, M.; Demple, B.; Crescenzi, M. & Dogliotti, E. (2007). Terminally differentiated muscle cells are defective in base excision DNA repair and hypersensitive to oxygen injury. *Proc Natl Acad Sci U S A*, Vol.104, No.43, (October 2007), pp.17010-17015, ISSN 1794-0040

Nicassio, F.; Corrado, N.; Vissers, J.H.; Areces, L.B.; Bergink, S.; Marteijn, J.A.; Geverts, B.; Houtsmuller, A.B.; Vermeulen, W.; Di Fiore, P.P. & Citterio, E. (2007). Human USP3 is a chromatin modifier required for S phase progression and genome stability. *Curr Biol*, Vol.17, No.22, (November 2007), pp.1972-1977, ISSN 1798-0597

Nouspikel, T. & Hanawalt, P.C. (2000). Terminally differentiatedhuman neurons repair transcribed genes but display attenuated global DNA repair and modulation of repair gene expression.*Mol Cell Biol*, Vol.20, No.5, (March 2000), pp.1562–1570, ISSN 1066-9734

Nouspikel, T. & Hanawalt, P.C. (2002). DNA repair in terminally differentiated cells. *DNA Repair (Amst)*, Vol.1, No.1, (January 2002), pp.59-75, ISSN 1250-9297

Nouspikel, T. (2007). DNA repair in differentiated cells: some new answers to old questions. *Neuroscience*, Vol.145, No.4, (August 2006), pp.1213-1221, ISSN 1692-0273

Oberdoerffer, P.; Michan, S.; McVay, M.; Mostoslavsky, R.; Vann, J.; Park, S.K.; Hartlerode, A.; Stegmuller, J.; Hafner, A.; Loerch, P.; Wright, S.M.; Mills, K.D.; Bonni, A.; Yankner, B.A.; Scully, R.; Prolla, T.A.; Alt, F.W. & Sinclair, D.A. (2008). SIRT1 redistribution on chromatin promotes genomic stability but alters gene expression during aging. *Cell*, Vol.135, No.5, (November 2008), pp.907-918, ISSN 1904-1753

O'Hagan, H.M.; Mohammad, H.P. & Baylin, S.B. (2008). Double strand breaks can initiate gene silencing and SIRT1- dependent onset of DNA methylation in an exogenous promoter CpG island. *PLoS Genet*, Vol.4, No.8, (August 2008), pp.e1000155, ISSN 1870-4159

O'Hare, M.J.; Hou, S.T.; Morris, E.J.; Cregan, S.P.; Xu, Q.; Slack, R.S. & Park, D.S. (2000). Induction and modulation of cerebellar granule neuron death by E2F-1. *J Biol Chem*, Vol.275, No.33, (August 2000), pp. 25358–25364, ISSN 1085- 1232

Okoshi, R.; Ozaki, T.; Yamamoto, H.; Ando, K.; Koida, N.; Ono, S.; Koda, T.; Kamijo, T.; Nakagawara, A. & Kizaki, H. (2008). Activation of AMP-activated protein kinase induces p53-dependent apoptotic cell death in response to energetic stress. *J Biol Chem*, Vol.283, No.7, (December 2007), pp. 3979-3987, ISSN 1805-6705

Oku, T.; Ikeda, S.; Sasaki, H.; Fukuda, K.; Morioka, H.; Ohtsuka, E.; Yoshikawa, H. & Tsurimoto, T. (1998). Functional sites of human PCNA which interact with p21 (Cip1/Waf1), DNA polymerase delta and replication factor C. *Gene Cells*, Vol.3, No.6, (June 1998), pp. 357-369, ISSN 973-4782

Otsuka, Y.; Tanaka, T.; Uchida, D.; Noguchi, Y.; Saeki, N.; Saito, Y. & Tatsuno, I. (2004). Roles of cyclin-dependent kinase 4 and p53 in neuronal cell death induced by doxorubicin on cerebellar granule neurons in mouse. *Neurosci Lett*, Vol.365, No.3, (July 2004), pp. 180-185, ISSN 1524-6544

Pajalunga, D.; Mazzola, A.; Salzano, A.M.; Biferi, M.G.; De Luca, G. & Crescenzi, M. (2007). Critical requirement for cell cycle inhibitors in sustaining nonproliferative states. *J Cell Biol*, Vol.176, No.6, (March 2007), pp.1039-1056, ISSN 1735-3358

Pardo, B.; Gómez-González, B. & Aguilera, A. (2009). DNA repair in mammalian cells: DNA double-strand break repair: how to fix a broken relationship. *Cell Mol Life Sci*, Vol.66, No.6, (March 2009), pp.1039-1056, ISSN 1915-3654

Park, D.S.; Morris, E.J.; Greene, L.A. & Geller, H.M. (1997). G1/S cell cycle blockers and inhibitors of cyclin-dependent kinases suppress camptothecin-induced neuronal apoptosis. *J Neurosci*, Vol.17, No.4, (February 1997), pp. 1256–1270, ISSN 900- 6970

Park, D.S.; Morris, E.J.; Padmanabhan, J.; Shelanski, M.L.; Geller, H.M. & Greene, L.A. (1998). Cyclin-dependent kinases participate in death of neurons evoked by DNA-damaging agents. *J Cell Biol*, Vol.143, No.2, (October 1998), pp.457–467, ISSN 978-6955

Perry, J.J.; Yannone, S.M.; Holden, L.G.; Hitomi, C.; Asaithamby, A.; Han, S.; Cooper, P.K.; Chen, D.J. & Tainer, J.A. (2006). WRN exonuclease structure and molecular mechanism imply an editing role in DNA end processing. *Nat Struct Mol Biol*, Vol.13, No.5, (April 2006), pp.414-422. ISSN 1662-2405

Peterson, C.L. & Cote, J. (2004) Cellular machineries for chromosomal DNA repair. *Genes Dev*, Vol.18, No.6, (March 2004), pp.602-616. ISSN 1507-5289

Phillips, A.C. & Vousden, K.H. (2001). E2F-1 induced apoptosis. *Apoptosis*, Vol.6, No.3, (June 2001), pp.173–182. ISSN 1138- 8666

Pickart, C.M. (2001). Ubiquitin enters the new millennium. *Mol Cell*, Vol.8, No.3, (September 2001), pp.499-504. ISSN 1158- 3613

Picard, F.; Kurtev, M.; Chung, N.; Topark-Ngarm, A.; Senawong, T.; Machado De Oliveira, R.; Leid, M.; McBurney, M.W. & Guarente, L. (2004). Sirt1 promotes fat mobilization in white adipocytes by repressing PPAR-gamma. *Nature,* Vol.429, No.6993, (June 2004), pp. 771-776, ISSN 1517-5761

Poehlmann, A. & Roessner, A. (2010). Importance of DNA damage checkpoints in the pathogenesis of human cancers. *Pathol Res Pract*, Vol.206, No.9, (August 2010), pp. 591-601, ISSN 2067-4189

Polager, S. & Ginsberg, D. (2008). E2F – at the crossroads of life and death. *Trends Cell Biol*, Vol.18, No.11, (September 2008), pp. 528-535, ISSN 1880-5009

Polager, S. & Ginsberg, D. (2009). p53 and E2f: partners in life and death. *Nat Rev Cancer*, Vol.9, No.10, (October 2009), pp.738-748, ISSN 1977-6743

Polo, S.E.; Kaidi, A.; Baskcomb, L.; Galanty, Y.& Jackson, S.P. (2010). Regulation of DNA-damage responses and cell-cycle progression by the chromatin remodelling factor CHD4. *EMBO J*, Vol.29, No.18, (August 2010), pp.3130-3139, ISSN 2069-3977

Polo, S.E. & Jackson, S.P. (2011). Dynamics of DNA damage response proteins at DNA breaks: a focus on protein modifications. *Genes Dev*, Vol.25, No.5, (March, 2011), pp.409-33, ISSN 2136-3960

Qin, S. & Parthun, M.R. (2006). Recruitment of the type B histone acetyltransferase Hat1p to chromatin is linked to DNA double-strand breaks. *Mol Cell Biol*, Vol.26, No.9, (May 2006), pp.3649-3658, ISSN 1661-2003

Rass, U.; Ahel, I. & West, S.C. (2007). Defective DNA repair and neurodegenerative disease. *Cell*, Vol.130, No.6, (September 2007), pp.991-1004, ISSN 1788-9645

Reagan-Shaw, S. & Ahmad, N. (2005). Silencing of polo-like kinase (Plk) 1 via siRNA causes induction of apoptosis and impairment of mitosis machinery in human prostate cancer cells: implications for the treatment of prostate cancer. *FASEB J*, Vol.19, No.6, (January 2005), pp.611-613, ISSN 1566-1849

Reed, J.C. (2006). Proapoptotic multidomain Bcl-2/Bax-family proteins: mechanisms, physiological roles and therapeutic opportunities. *Cell Death Differ*, Vol.13, No.8, (June 2006), pp.1378-1386, ISSN 1672-9025

Reinhardt, H.C. & Yaffe, M.B. (2009). Kinases that control the cell cycle in response to DNA damage: Chk1, Chk2, and MK2. *Curr Opin Cell Biol*, Vol.21, No.2, (February 2009), pp.245-255, ISSN 1923-0643

Rich, T.; Allen, R.L. & Wyllie, A.H. (2000). Defying death after DNA damage. *Nature*, Vol.407, No.6805, (October 2000), pp.777-783, ISSN 1104-8728

Riley, T.; Sontag, E.; Chen, P. & Levine, A. (2008). Transcriptional control of human p53-regulated genes. *Nat Rev Mol Cell Biol*, Vol.9, No.5, (May 2008), pp.402-412, ISSN 1843-1400

Rinaldo, C.; Prodosmo, A.; Mancini, F.; Iacovelli, S.; Sacchi, A.; Moretti, F. & Soddu, S. (2007). MDM2-regulated degradation of HIPK2 prevents p53Ser46 phosphorylation and DNA damage-induced apoptosis. *Mol Cell*, Vol.25, No.5, (March 2007), pp.739-750, ISSN 1734-9959

Rogakou, E.P.; Pilch, D.R.; Orr, A.H.; Ivanova, V.S. & Bonner, W.M. (1998). DNA double-stranded breaks induce histone H2AX phosphorylation on serine 139. *J Biol Chem*, 1998 Mar 6;Vol.273, No.10, (March 1998), pp.5858-5868, ISSN 9488723

Rogoff, H.A. & Kowalik, T.F. (2004). Life, death and E2F: linking proliferation control and DNA damage signaling via E2F1. *Cell Cycle*, Vol.3, No.7, (July 2004), pp. 845-846, ISSN 1519-0206

Rolig, R.L. & McKinnon, P.J. (2000). Linking DNA damage and neurodegeneration. *Trends Neuroscie*, Vol.23, No.9, (September 2000), pp. 417-424, ISSN 1094-1191

Roshak, A.K.; Capper, E.A.; Imburgia, C.; Fornwald, J.; Scott, G. & Marshall, L.A. (2000). The human polo-like kinase PLK, regulates cdc2/cyclin B through phosphorylation and activation of the cdc25C phosphatase, *Cell Signal*, Vol.12, No.6, (June 2000), pp. 405-411, ISSN 1120-2906

Rossi, M.; De Laurenzi, V.; Munarriz, E.; Green, D.R.; Liu, Y.C.; Vousden, K.H.; Cesareni, G. & Melino, G. (2005) The ubiquitin-protein ligase Itch regulates p73 stability. *EMBO J*, Vol.24, No.4, (February 2005), pp. 836-848, ISSN 1567- 8106

Rouse, J. & Jackson SP. (2002). Interfaces between the detection, signaling, and repair of DNA damage. *Science*, Vol.297, No.5581, (July, 2002), pp. 547-551, ISSN 1214-2523]

Sakaguchi, K.; Herrera, J.E.; Saito, S.; Miki, T.; Bustin, M.; Vassilev, A.; Anderson, C.W. & Appella, E. (1998). DNA damage activates p53 through a phosphorylation-acetylation cascade. *Genes Dev*, Vol.12, N0.18, (September 1998), pp.2831-2841, ISSN 9744-860

Sancar, A.; Lindsey-Boltz, L.A.; Unsal-Kacmaz, K. & Linn, S. (July 2004). Molecular mechanisms of mammalian DNA repair and the DNA damage checkpoints. *Annu Rev Biochem, Vol.* 73, No. , (January, 2004), pp.39–85, ISSN 1518-9136

Sanders, S.L.; Arida, A.R. & Phan, F.P. (2010). Requirement for the phospho-H2AX binding module of Crb2 in double-strand break targeting and checkpoint activation. *Mol Cell Biol*, Vol.30, N0.19, (August 2010), pp.4722-4731, ISSN 2067-9488

Sartori, A.A.; Lukas, C.; Coates, J.; Mistrik, M.; Fu, S.; Bartek, J.; Baer, R.; Lukas, J. & Jackson, S.P. (2007). Human CtIP promotes DNA end resection. Nature, Vol.450, No.7169, (October 2007), pp.509–514, ISSN 1796-5729

Schlereth, K.; Beinoraviciute-Kellner, R.; Zeitlinger, M.K.; Bretz, A.C.; Sauer, M.; Charles, J.P.; Vogiatzi, F.; Leich, E.; Samans, B.; Eilers, M.; Kisker, C.; Rosenwald, A. & Stiewe, T. (2010). DNA binding cooperativity of p53 modulates the decision between cell cycle arrest and apoptosis. *Mol Cell*, Vol.38, No.3, (May 2010), pp.356-368, ISSN 2047-1942

Schmetsdorf, S.; Gartner, U. & Arendt, T. (2007). Constitutive expression of functionally active cyclin dependent kinases and their binding partners suggests noncanonical functions of cell cycle regulators in differentiated neurons. *Cereb Cortex*, Vol.17, No.8, (October 2006), pp.1821–1829, ISSN 1705-0646

Schmetsdorf, S.; Arnold, E.; Holzer, M.; Arendt, T.& Gartner, U. (2009). A putative role for cell cycle-related proteins in microtubule-based neuroplasticity. *Eur J Neurosci* , Vol.29, No.6, (March 2009), pp.1096–10107, ISSN 1930-2146

Schuler, M.; Bossy-Wetzel, E.; Goldstein, J.C.; Fitzgerald, P. & Green, D,R. (2000) p53 induces apoptosis by caspase activation through mitochondrial cytochrome c release. *J Biol Chem*, Vol.275, No.10, (March 2000), pp. 7337-7342, ISSN 1070-2305

Shanbhag, N.M.; Rafalska-Metcalf, I.U.; Balane-Bolivar, C.; Janicki, S.M. & Greenberg, R.A. (2010). ATM-dependent chromatin changes silence transcription in cis to DNA double-strand breaks. *Cell*, Vol.1413, No.6, (June 2010), pp.970– 981. ISSN 2055-0933

Shao, G.; Lilli, D.R.; Patterson-Fortin, J.; Coleman, K.A.; Morrissey, D.E. & Greenberg, R.A. (2009). The Rap80-BRCC36 de- ubiquitinating enzyme complex antagonizes RNF8-Ubc13-dependent ubiquitination events at DNA double strand breaks. *Proc Natl Acad Sci U S A*, Vol.106, No.9, (February 2009), pp. 3166-3171, ISSN 1920-2061

Sharma, S. (2007). Age-related nonhomologous end joining activity in rat neurons. *Brain Res Bull*, Vol.73, No1-3, (June 2007), pp.48–54. ISSN 1749-9636

Sharma, G.; Mirza, S.; Parshad, R.; Srivastava, A.; Gupta, S.D.; Pandya, P. & Ralhan, R. (2010). Clinical significance of promoter hypermethylation of DNA repair genes in tumor and serum DNA in invasive ductal breast carcinoma patients. *Life Sci*, Vol.87, No.3-4, (May 2010), pp.83-91, ISSN 2047-0789

Shieh, S-Y.; Ikeda, M.; Taya, Y. & Prives, C. (1997). DNA damage-induced phosphorylation of p53 alleviates inhibition by MDM2. *Cell*, Vol.91, No.3, (October, 2003), pp. 325-334, ISSN 936-3941

Shieh, S.Y.; Ahn, J.; Tamai, K.; Taya, Y. & Prives, C. (2000). The human homologs of checkpoint kinases Chk1 and Cds1 (Chk2) phosphorylate p53 at multiple DNA damage-inducible sites. *Genes Dev*, Vol.14, No.3, (February, 2000), pp. 289- 300, ISSN 1067-3501

Shiloh, Y. (2003). ATM and related protein kinases: safeguarding genome integrity. *Nat Rev Cancer*, 2003; Vol.3, No.3, (March, 2003), pp.155-168, ISSN 1261-2651

Shiloh, Y. (2006). The ATM-mediated DNA-damage response: taking shape. *Trends Biochem Sci*, Vol.31, No.7 (June 2006), pp.402-10. ISSN 1677-4833

Singleton, B.K.; Torres-Arzayus, M.I.; Rottinghaus, S.T.; Taccioli, G.E. & Jeggo, P.A. (1999). The C terminus of Ku80 activates the DNA-dependent protein kinase catalytic subunit. *Mol Cell Biol*, Vol.19, No.15, (May 1999), pp. 3267-3277, ISSN1020- 7052

Sinha, S.; Malonia, S.K.; Mittal, S.P.; Singh, K.; Kadreppa, S.; Kamat, R.; Mukhopadhyaya, R.; Pal, J.K. & Chattopadhyay, S. (2010). Coordinated regulation of p53 apoptotic targets Bax and Puma by SMAR1 through an identical MAR element. *EMBO J*, Vol.29, No.4, (January 2010), pp.830-842, ISSN 2007-5864

Smerdon MJ, Tlsty TD, Lieberman MW. (1978). Distribution of ultraviolet-induced DNA repair synthesis in nuclease sensitive and resistant regions of human chromatin. *Biochemistry*, Vol.17, No.12, (June 1978), pp.2377–2386, ISSN 678515

Smith, D.S.; Leone, G.; DeGregori, J.; Ahmed, M.N.; Qumsiyeh, M.B. & Nevins, J.R. (2000). Induction of DNA replication in adult rat neurons by deregulation of the retinoblastoma/E2F G1 cell cycle pathway. *Cell Growth Differ*, Vol.11, No.12, (December 2000), pp. 625-633, ISSN 1114-9597

Sofueva S, Du LL, Limbo O, Williams JS, Russell P. (2010). BRCT domain interactions with phospho-histone H2A target Crb2 to chromatin at double-strand breaks and maintain the DNA damage checkpoint. *Mol Cell Biol*, Vol.30, No.19, (August 2010), pp.4732-4743, ISSN 2067-9485

Sordet, O.; Redon, C.E.; Guirouilh-Barbat, J.; Smith, S.; Solier, S.; Douarre, C.; Conti, C.; Nakamura, A.J.; Das, B.B.; Nicolas, E.; Kohn, K.; Bonner, W.M. & Pommier, Y. (2009). Ataxia telangiectasia mutated activation by transcription- and topoisomerase I-induced DNA double-strand breaks. *EMBO Rep*, Vol.10, No.8, (June 2009), pp.887-893, ISSN 1955-7000

Soutoglou, E. & Misteli, T. (2008). Activation of the cellular DNA damage response in the absence of DNA lesions. *Science*, Vol.320, No.5882, (May 2008), pp.1507-1510, ISSN 1848-3401

Spycher, C.; Miller, E.S.; Townsend, K.; Pavic, L.; Morrice, N.A.; Janscak, P.; Stewart, G.S. & Stucki, M. (2008). Constitutive phosphorylation of MDC1 physically links the MRE11-RAD50-NBS1 complex to damaged chromatin. *J Cell Biol*, Vol.181, No.2, (April 2008), pp.227-240, ISSN 1841-1308

Stark, G.R. & Taylor, W.R. (2006). Control of the G2/M transition. *Mol. Biotechnol*, Vol.32, No.3, (March 2006), pp.227-248, ISSN 1663-2889

Steelman, L.S. & McCubrey, J.A. (2010). Intriguing novel abilities of Nutlin-3A: induction of cellular quiescence as opposed to cellular senescence - implications for chemotherapy. *Cell Cycle*, Vol.8, No.22, (November 2010), pp. 3634-3635, ISSN 1987-5921

Stewart, G.S.; Wang, B.; Bignell, C.R.; Taylor, A.M. & Elledge, S.J. (2003). MDC1 is a mediator of the mammalian DNA damage checkpoint. *Nature*, Vol.421, No.6926, (February 2003), pp. 961–966, ISSN1260-7005

Stewart, N.; Hicks, G.G.; Paraskevas, F. & Mowat, M. (1995). Evidence for a second cell cycle block at G2/M by p53. *Oncogene*, Vol. 10, No.1, (January 1995), pp. 109-115, ISSN 7529-916

Stiff, T.; O'Driscoll, M.; Rief, N.; Iwabuchi, K.; Löbrich, M. & Jeggo, P.A. (2004). ATM and DNA-PK function redundantly to phosphorylate H2AX after exposure to ionizing radiation. *Cancer Res*, Vol.64, No.7, (April 2004), pp.2390-2396, ISSN 1505-9890

Stracker, T.H.; Couto, S.S.; Cordon-Cardo, C.; Matos, T. & Petrini, J.H. (2008). Chk2 suppresses the oncogenic potential of DNA replication-associated DNA damage. *Mol Cell* Vol.31, No.1, (July 2008), pp.21-32, ISSN 1861-4044

Stucki, M.; Clapperton, J.A.; Mohammad, D.; Yaffe, M.B.; Smerdon, S.J. & Jackson SP. (2005). MDC1 directly binds phosphorylated histone H2AX to regulate cellular responses to DNA double-strand breaks. *Cell*, Vol.123, No.7, (December 2005), pp.1213–1226, ISSN 1637-7563

Subramanian, C.; Opipari, A.W. Jr.; Bian, X.; Castle, V.P. & Kwok, R.P. (2005). Ku70 acetylation mediates neuroblastoma cell death induced by histone deacetylase inhibitors. *Proc Natl Acad SciUSA*, Vol.102, No.13, (March 2005), pp. 4842-4847, ISSN 1577-8293

Suzuki, A.; Tsutomi, Y.; Akahane, K.; Araki, T. & Miura, M. (1998). Resistance to Fas-mediated apoptosis: activation of caspase 3 is regulated by cell cycle regulator p21WAF1 and IAP gene family ILP. *Oncogene*, Vol.17, No.8, (August 1998), pp. 931-939, ISSN 974-7872

Suzuki, A.; Tsutomi, Y.; Miura, M. & Akahane, K. (1999). Caspase 3 inactivation to suppress Fas-mediated apoptosis: identification of binding domain with p21 and ILP and inactivation machinery by p21. *Oncogene*, Vol.18, No.5, (February 1999), pp. 1239-1244, ISSN 1002-2130

Sykes, S.M.; Mellert, H.S.; Holbert, M.A.; Li, K.; Marmorstein, R.; Lane, W.S. & McMahon, S.B. (2006). Acetylation of the p53 DNA-binding domain regulates apoptosis induction. *Mol Cell*, Vol.24, No.6, (December 2006), pp.841-851, ISSN 1718- 9187

Syljuasen, R.G.; Jensen, S.; Bartek, J. & Lukas, J. (2006) Adaptation to the ionizing radiation-induced G2 checkpoint occurs in human cells and depends on checkpoint kinase 1 and Polo-like kinase 1 kinases. *Cancer Res*, Vol.66, No.21, (November 2006), pp.10253–10257, ISSN 1707-9442

Tanaka, T.; Ohkubo, S.; Tatsuno, I. & Prives, C. (2007). hCAS/CSE1L associates with chromatin and regulates expression of select p53 target genes. *Cell*, Vol.130, No.4, (August 2007), pp.638-650, ISSN 1771-9542

Tang, Y.; Luo, J.; Zhang, W. & Gu, W. (2006). Tip60-dependent acetylation of p53 modulates the decision between cell cycle arrest and apoptosis. *Mol Cell*, Vol.24, No.6, (December 2006), pp.827-839, ISSN 1718-9186

Tamburini, B.A. & Tyler, J.K. (2005). Localized histone acetylation and deacetylation triggered by the homologous recombination pathway of double-strand DNA repair. *Mol Cell Biol*, 2005 Jun;Vol.25, No.12, (June 2005), pp.4903-4913, ISSN 1592-3609

Taylor, W.R. & Stark, G.R. (2001). Regulation of the G2/M transition by p53. *Oncogene*, Vol.20, No.15, (April 2001), pp.1803– 1815, ISSN1131-3928

Tibbetts, R.S.; Brumbaugh, K.M.; Williams, J.M.; Sarkaria, J.N.; Cliby, W.A.; Shieh, S.Y.; Taya, Y.; Prives, C. & Abraham, R.T. (1999). A role for ATR in the DNA damageinduced phosphorylation of p53. *Genes Dev*, Vol.13, No.2, (January 1999), pp. 152-157, ISSN 992-5639

Timinszky, G.; Till, S.; Hassa, P.O.; Hothorn, M.; Kustatscher, G.; Nijmeijer, B.; Colombelli, J.; Altmeyer, M.; Stelzer, E.H.; Scheffzek, K.; Hottiger, M.O. & Ladurner, A.G.

(2009). A macrodomain-containing histone rearranges chromatin upon sensing PARP1 activation. *Nat Struct Mol Biol*, Vol.16, No.9, (August 2009), pp.923-929, ISSN 1968-0243

Toczyski, D.P.; Galgoczy, D.J. & Hartwell, L.H. (1997) CDC5 and CKII control adaptation to the yeast DNA damage checkpoint. *Cell*, Vol.90, No.6, (September 1997), pp.1097-10106, ISSN 932-3137

Tofilon, P.J. & Meyn, E. (1988). Reduction in DNA repair capacity following differentiation of murine proadipocytes. *Exp. Cell Res*, Vol.174, No.2, (February 1988), pp.502–510, ISSN 333-8499

Tomashevski, A.; Webster, D.R.; Grammas, P.; Gorospe, M. & Kruman, I.I. (2010). Cyclin-C-dependent cell-cycle entry is required for activation of non-homologous end joining DNA repair in postmitotic neurons. *Cell Death Differ*. Vol.17, No.7 (January 2010), pp.1189-1198, ISSN 2011-1042

Trushina, E. &C McMurray, C.T. (2007). Oxidative stress and mitochondrial dysfunction in neurodegenerative diseases. *Neuroscience*, Vol.145, No.4 (February 2007), pp.1233–1248, ISSN 1730-3344

Uziel T, Lerenthal Y, Moyal L, Andegeko Y, Mittelman L, Shiloh Y. (2003). Requirement of the MRN complex for ATM activation by DNA damage. *EMBO J*, Vol.22, No20, (October 2003), pp.5612-5621, ISSN 1453-2133

van Attikum, H.; Fritsch, O.; Hohn, B. & Gasser, S.M.(2004). Recruitment of the INO80 complex by H2A phosphorylation links ATP-dependent chromatin remodeling with DNA double-strand break repair. *Cell*, Vol.119, No.6,(December 2004), pp.777-788, ISSN 1560-7975

van Attikum, H.; Fritsch, O. & Gasser, S.M. (2007). Distinct roles for SWR1 and INO80 chromatin remodeling complexes at chromosomal double-strand breaks. *EMBO J*, Vol.26, No.18, (August 2007), pp.4113-4125. ISSN 1776-2868

Van Attikum, H. & Gasser, S.M. (2005) The histone code at DNA breaks: a guide to repair? *Nat Rev Mol Cell Biol*, Vol. 6, No.10, (October 2005), pp.757-765, ISSN 1616-7054

van Attikum, H. & Gasser, S.M. (2009). Crosstalk between histone modifications during the DNA damage response. *Trends Cell Biol*, Vol.19, No.5, (April 2009), pp.207-217, ISSN 1934-2239

van Gent, D.C. & van der Burg, M. (2007). Non-homologous end-joining, a sticky affair. *Oncogene*, Vol. 26, No.56, (December 2007), pp. 7731–7740, ISSN 1806-6085

van Vugt, M.A. & Medema, R.H. (2004). Checkpoint adaptation and recovery: back with Polo after the break. *Cell Cycle*, Vol.3, No.11, (November 2004), pp. 1383-1386, ISSN 1549-2511

van Vugt, M.A.; Bras, A. & Medema, R.H. (2004). Polo-like kinase-1 controls recovery from a G2 DNA damage-induced arrest in mammalian cells. *Mol Cell*, Vol. 15, No.5, (September 2004), pp.799-811, ISSN 1535-0223

van Vugt, M.A.; Gardino, A.K.; Linding, R.; Ostheimer, G.J.; Reinhardt, H.C.; Ong, S.E.; Tan, C.S.; Miao, H.; Keezer, S.M.; Li, J.; Pawson, T.; Lewis, T.A.; Carr, S.A.; Smerdon, S.J.; Brummelkamp, T.R. & Yaffe, M.B. (2010). A mitotic phosphorylation feedback network connects Cdk1, Plk1, 53BP1, and Chk2 to inactivate the G(2)/M DNA damage checkpoint. *PLoS Biol*, Vol. 8, No.1, (January 2010), pp.e1000287, ISSN 2012-6263

Vaseva, A.V.; Marchenko, N.D. & Moll, U.M. (2009). The transcription-independent mitochondrial p53 program is a major contributor to nutlin-induced apoptosis in tumor cells. *Cell Cycle*, Vol.8, No.11, (June 2009), pp.1711-1719, ISSN 1941- 1846

Vaze, M.B.; Pellicioli, A.; Lee, S.E.; Ira, G.; Liberi, G.; Arbel-Eden, A.; Foiani, M. & Haber JE. (2002) Recovery from checkpoint-mediated arrest after repair of a double-strand break requires Srs2 helicase. *Mol Cell*, Vol. 10, No.2, (August 2002), pp. 373–385, ISSN 1219-1482

Vaziri, H.; Dessain, S.K.; Ng Eaton, E.; Imai, S.;, Frye, R.A.; Pandita, T.K.; Guarente, L. & Weinberg, R.A. (2001). hSIR2(SIRT1) functions as an NAD-dependent p53 deacetylase. *Cell*, Vol.107, No.2, (October 2001), pp. 149-159, ISSN 1167-2523

Waldman, T.; Lengauer, C.; Kinzler, K.W. & Vogelstein, B. (1996). Uncoupling of S phase and mitosis induced by anticancer agents in cells lacking p21. *Nature*, Vol.381, No.6584, (June 1996), pp.713-716, ISSN 864-9519

Waldman, T.; Zhang, Y.; Dillehay, L.; Yu, J.; Kinzler, K.; Vogelstein, B. & Williams, J. (1997). *Cell cycle*, arrest versus cell death in cancer therapy. *Nat Med*, Vol.3, No.9, (September 1997), pp. 1034-1036, ISSN 928-8734

Wang, M.; Wu, W.; Wu, W.; Rosidi,B.; Zhang,L.; Wang, H. & Iliakis, G. (2006). PARP-1 and Ku compete for repair of DNA double strand breaks by distinct NHEJ pathways. *Nucleic Acids Res*, 2006;Vol.34, No.21, (November 2006), pp.6170- 6182, ISSN 1708-8286

Ward, I.M. & Chen, J. (2001). Histone H2AX is phosphorylated in an ATR-dependent manner in response to replicational stress. *J Biol Chem*, Vol.276, No.51, (October 2001), pp.47759-47762, ISSN 1167-3449

Warbrick, E.; Lane, D.P.; Glover, D.M. & Cox, L.S. (1995). A small peptide inhibitor of DNA replication defines the site of interaction between the cyclin-dependent kinase inhibitor p21WAF1 and proliferating cell nuclear antigen. *Curr Biol*, Vol.5, No.3, (March 1995), pp.275-282, ISSN 778-0738

Wefel, J.S.; Kayl, A.E. & Meyers, C.A. (2004). Neuropsychological dysfunction associated with cancer and cancer therapies: a conceptual review of an emerging target. *Br J Cancer*, Vol. 90, No.9 (May, 2004), pp.1691-1696, ISSN 1515-0608

Weissman, L.; de Souza-Pinto, N.C.; Stevnsner, T. & Bohr, V.A. (2007). DNA repair, mitochondria, and neurodegeneration. *Neuroscience*, Vol. 145, No.4 (November, 2006), pp.1318–1329, ISSN 1709-2652

Wilker, E.W.; van Vugt, M.A.; Artim, S.A.; Huang, P.H.; Petersen, C.P.; Reinhardt, H.C.; Feng, Y.; Sharp, P.A.; Sonenberg, N.; White, F.M. & Yaffe, M.B. (2007). 14-3-3sigma controls mitotic translation to facilitate cytokinesis. *Nature*, Vol.446, No.7133 (March 2007), pp.329-332, ISSN 1736-1185

Wilson, D. M. 3rd. & McNeill, D. R. (2007). Base excision repair and the central nervous system. *Neuroscience,* Vol.145, No.4 (August 2006), pp.1187–1200, ISSN 1693-4943

Xiong, Y.; Hannon, G.J.; Zhang, H.; Casso, D.; Kobayashi, R. & Beach, D. (1993). p21 is a universal inhibitor of cyclin kinases. *Nature*, Vol.366, No.6456, (December 1993), pp.701-704, ISSN 825-9214

Xu Y, Sun Y, Jiang X, Ayrapetov MK, Moskwa P, Yang S, Weinstock DM, Price BD. (2010). The p400 ATPase regulates nucleosome stability and chromatin ubiquitination during DNA repair. *J Cell Biol*, Vol.191, No.1, (September 2010), pp.31-43, ISSN 2087-6283

Yamaguchi, H.; Woods, N.T.; Piluso, L.G.; Lee, H.H.; Chen, J.; Bhalla, K.N.; Monteiro, A.; Liu, X.; Hung, M.C. & Wang, H.G. (2009). p53 acetylation is crucial for Its transcription-independentproapoptotic functions. *J Biol Chem*, Vol.284, No.17, (March 2009), pp. 11171-11183, ISSN 1926-5193

Yamasaki, L.; Jacks, T.; Bronson, R.; Goillot, E.; Harlow, E. & Dyson, N.J. (1996). Tumor induction and tissue atrophy in mice lacking E2F-1. *Cell*, Vol.85, No.4, (May 1996), pp.537–548, ISSN 865-3789

Yang, W.; Dicker, D.T.; Chen, J. & El-Deiry, W.S. (2008). CARPs enhance p53 turnover by degrading 14-3-3sigma and stabilizing MDM2. *Cell Cycle*, Vol.7, No.5, (January 2008), pp. 670-682, ISSN 1838-2127

Yang, Y. & Herrup, K. (2001). Loss of neuronal cell cycle control in ataxia-telangiectasia: a unified disease mechanism, *J Neurosci*, Vol.21, No.8, (April 2001), pp. 2661–2668 , ISSN 1130-6619

Yang, Y.; Geldmacher, D.S. & Herrup, K. (2001) DNA replication precedes neuronal cell death in Alzheimer's disease. *J Neurosci*, Vol.21, No.7, (April 2001), pp. 2661–2668 , ISSN1130-6619

Yano, K.; Morotomi-Yano, K.; Wang, S.Y.; Uematsu, N.; Lee, K.J.; Asaithamby, A.; Weterings, E. & Chen, D.J. (2008). Ku recruits XLF to DNA double-strand breaks. *EMBO Rep*, Vol.9, No.1, (January 2008), pp.91-96, ISSN 1806-4046

Yi, J. & Luo, J. (2010). SIRT1 and p53, effect on cancer, senescence and beyond. *Biochim Biophys Acta*, Vol.1804, No.8, (May 2008), pp. 1684-1689, ISSN 2047-1503

Yoo, H.Y.; Kumagai, A.; Shevchenko, A.; Shevchenko, A. & Dunphy, W.G. (2004). Adaptation of a DNA replication checkpoint response depends upon inactivation of Claspin by the Polo-like kinase. *Cell*, Vol.117, No.5, (May 2004), pp. 575-588, ISSN 1516-3406

Zhang, Y.; Qu, D.; Morris, E.J.; O'Hare, M.J.; Callaghan, S.M.; Slack, R.S.; Geller, H.M. & Park, D.S. (2006). The Chk1/Cdc25A pathway as activators of the cell cycle in neuronal death induced by camptothecin. *J Neurosci*, Vol.26, No.34, (August 2006), pp. 8819–8828, ISSN 1692-8871

Zhang, Y.; Hefferin, M.L.; Chen, L.; Shim, E.Y.; Tseng, H.M.; Kwon, Y.; Sung, P.; Lee, S.E. & Tomkinson, A.E. (2007). Role of Dnl4-Lif1 in nonhomologous end-joining repair complex assembly and suppression of homologous recombination. *Nat Struct Mol Biol*, Vol.14, No.7, (June 2007), pp.639-46, ISSN 1758-9524

Zhou, B.B. & Elledge, S.J. (2000). The DNA damage response: putting checkpoints in perspective. *Nature*, Vol.408, No.6811, (November 2000), pp. 433–439, ISSN 1110-0718

Zimmermann, K.C.; Bonzon, C. & Green, D.R. (2001). The machinery of programmed cell death. *Pharmacol Ther*, Vol.92, No.1, (October 2001), pp.57-70, ISSN 1175-0036

Ziv, Y.; Bielopolski, D.; Galanty, Y.; Lukas, C.; Taya, Y.; Schultz, D.C.; Lukas, J.; Bekker-Jensen, S.; Bartek, J. & Shiloh, Y. (2006). Chromatin relaxation in response to DNA double-strand breaks is modulated by a novel ATM- and KAP-1 dependent pathway. *Nat Cell Biol*, Vol. 8, No.8, (July 2006), pp. 870–876, ISSN 1686-2143

14

Saccharomyces cerevisiae as a Model System to Study the Role of Human DDB2 in Chromatin Repair

Kristi L. Jones, Ling Zhang and Feng Gong
Department of Biochemistry and Molecular Biology, University of Miami
USA

1. Introduction

Genetic and biochemical studies in *Saccharomyces cerevisiae* have made major contributions in elucidating the mechanism of several DNA repair pathways, including the nucleotide excision repair (NER) pathway that remove bulky DNA damage from the genome. Although NER is conserved from yeast to humans, there are differences in NER between yeast and humans. For example, no homolog of the human NER factor DNA damage-binding protein 2 (DDB2) has been identified in the budding yeast *S. cerevisiae*. Here, we present evidence suggesting that *S. cerevisiae* can be used to dissect the roles of DDB2 in initiating NER in chromatin.

Ultraviolet light (UV) is a well studied genotoxic stress that induces bulky DNA damage. These UV lesions are repaired by the NER pathway (Hanawalt, 2002; Sancar & Reardon, 2004). The particular lesions induced by UV irradiation have been characterized, namely, cyclobutane pyrimidine dimers (CPDs) and 6-4 photoproducts (6-4PPs). Both lesions result in the distortion of the DNA double helix, but 6-4PPs result in a greater distortion. Additionally, there are other minor differences between the two types of lesions. CPDs have been consistently shown to have higher incidence than 6-4PPs (Douki & Cadet, 2001). CPDs are induced both in nucleosome core and linker DNA, whereas 6-4PPs are formed with 6-fold greater frequency in linker DNA. In addition, 6-4PPs are repaired much faster than CPDs, as reviewed by Smerdon (Smerdon, 1991).

In humans, a defect in NER results in xeroderma pigmentosum (XP) and several other rare diseases (Kraemer et al., 2007). XP patients are extremely sensitive to UV light and have about 2000-fold higher incidence of sunlight induced skin cancers than the general population. NER lesion recognition is via protein interaction with the structural DNA changes that are induced. Other bulky DNA lesions repaired by NER include those induced by cigarette smoke, cisplatin treatment and a newly identified form of bulky oxidative DNA damage (Zamble et al., 1996; Setlow, 2001; Wang, 2008).

NER has been extensively studied and the basic mechanism is understood. It consists of three main steps: 1) lesion detection, 2) dual incision to remove an oligonucleotide containing the lesion and 3) repair synthesis to fill the gap. There are two sub-pathways of NER, termed transcription coupled repair (TC-NER) and global genome repair (GG-NER) (Hanawalt, 2002). TC-NER is responsible for repair of damage on the actively transcribed strand; while GG-NER is responsible for repair in the remainder of the genome, including lesions on the non-transcribed strand of actively transcribed genes, as well as those in repressed or silent chromatin regions. Both TC-NER and GG-NER consist of all three steps, but, they differ in the lesion recognition step. In TC-NER the lesion is thought to be detected by pausing of RNA polymerase I or II (Conconi et al., 2002; Hanawalt, 2002; Fousteri & Mullenders, 2008). GG-NER, on the other hand, requires a specific lesion recognition hetero-dimeric protein complex, XPC-hRad23 (Xeroderma Pigmentosum complementation group C-human Rad23) in humans and Rad4-Rad23 (RADiation sensitive) in budding yeast (Wood 2010; Guzder et al., 1998; Jansen et al., 1998; Sugasawa, 2009). However, under certain *in vivo* circumstances, DDB2 is the pioneering damage recognition factor during GG-NER (Hwang et al., 1999; Nichols et al., 2000; Sugasawa, 2009). So far, no DDB2 homolog has been identified in the budding yeast (Fig. 1). Of note, the Rad16-Rad7 heterodimer, without a known human homolog, is required for GG-NER in the budding yeast.

	Human	S. cerevisiae
Core Pathway	XPA	Rad14
	XPB	Rad25
	XPD	Rad3
	XPF	Rad1
	XPG	Rad2
	ERCC1	Rad10
GG-NER	DDB1	Mms1
	XPE (DDB2)	no
	no	Rad16
	no	Rad7
	XPC	Rad4
	hRad23	Rad23
TC-NER	CSA	Rad28
	CSB	Rad26

Fig. 1. Conservation of NER pathway between humans and the budding yeast *S. cerevisiae*. Of note, no DDB2 counterpart has been identified in *S. cerevisiae*. Likewise, humans don't have a homolog of the Rad16-Rad7 heterodimer that is essential for GG-NER in *S. cerevisiae*.

Several lines of evidence suggest that DDB2 plays a key role in **chromatin repair** of UV damage. It has been shown that DDB2 is responsible for the lesion detection by directly interacting with the damaged DNA (Tang, et al., 2000; Scrima et al., 2008). Additionally, DDB2 binds the lesion independent of XPC (Wakasugi et al., 2002). DDB2 can co-localize with both CPDs and 6-4 PPs *in vivo*, while XPC seems to bind 6-4 PPs efficiently, but not CPDs. This suggests the necessity of DDB2 in GG-NER is specific for CPD repair (Fitch et al., 2003). Importantly, it has been suggested that the observed high affinity of DDB2 for 6-4PPs aids in the targeting of XPC to 6-4PPs when low levels of damage are present (Nishi et al., 2009).

Additionally, DDB2 is in complex with the E3 ubiquitin ligase complex consisting of DDB1, Cul4 (CULlin 4) and ROC (Ring Of Cullins) (Jackson & Xiong, 2009). E3 ubiquitin ligases transfer ubiquitin to the target protein. DDB2 is thought to be the substrate receptor targeting the E3 ubiquitin ligase complex to DNA lesion sites to facilitate GG-NER. Of note, DDB1 and Cul4 have been shown to be in complex with other proteins, including CSA, a TC-NER specific protein (Jackson & Xiong, 2009). Consistent with its classification as an E3 ubiquitin ligase, XPC, histone H2A, H3, H4, and DDB2 itself have been identified as UV-dependent ubiquitination targets of the DDB1-DDB2 E3 ligase complex (Chen et al., 2001; Nag et al., 2001; Matsuda et al., 2005; Sugasawa et al., 2005; Kapetanaki et al., 2006; Wang et al., 2006). The UV-dependent mono-ubiquitination of histone H2A has been suggested to be involved in both chromatin relaxation and restoration (Kapetanaki et al., 2006; Zhu et al., 2009). Clearly, understanding the role of DDB2 in NER will yield important insights into the mechanisms of NER operation in the context of chromatin.

Chromatin is a hierarchal structure composed of DNA and protein. The core component is the nucleosome. It is a complex of 147 base pairs of DNA wrapped around the core histone octamer. The core histone octamer consists of four subunits, H2A, H2B, H3 and H4 in a 2:2:2:2 ratio (Luger et al., 1997; Kornberg & Lorch, 1999). The innate structure of chromatin restricts DNA protein interactions. ATP-dependent chromatin reconfiguration is an important mechanism to alleviate this tight association. Several groups have demonstrated a requirement for the ATP-dependent chromatin remodeling in chromatin repair (Jiang et al., 2010; Gong et al. 2006; Zhang et al. 2009a; Zhang et al. 2009b; Zhao et al. 2009; Lans et al. 2010; Sarkar et al. 2010). How DNA repair occurs in chromatin is an emerging question and has been discussed in several recent review articles (Osley et al., 2007; Nag & Smerdon, 2009; Waters et al., 2009; Zhang et al., 2009a; Jones et al., 2010).

2. *S. cerevisiae* as a model system to study DDB2-mediated GG-NER in chromatin

It has been demonstrated that DDB2 is the initial lesion detection factor in GG-NER (Tang et al., 2000; Wakasugi et al., 2002; Fitch et al., 2003b; Pines et al., 2009). Although it has been implicated in the recruitment of XPC to CPD sites (Fitch et al., 2003b); how DDB2 transfers these identified lesions to XPC remains controversial. It is believed that ubiquitination of DDB2 leads to its degradation at damage sites and this degradation is required for CPD repair. However, there are several lines of evidence disputing this model, including: 1) inhibition of ubiquitination-mediated DDB2 degradation in mouse via Cul4a ablation enhances CPD repair (Liu et al., 2009), 2) DDB2 degradation is not stimulated by either DNA binding or XPC association (Luijsterburg et al., 2007), and 3) crystal structures suggest that DDB2 and XPC can co-localize on the lesion (Min & Pavletich, 2007; Scrima et al., 2008). Therefore, we try to explore the budding yeast as a simplified, alternative model system to begin to dissect the role(s) of ubiquitination in DDB2-mediated GG-NER.

2.1 Galactose induced expression of DDB2 in *S. cerevisiae*

As discussed in the introduction, DDB2 has no homolog in budding yeast. However, conservation of the GG-NER pathway and interacting partners such as DDB1 are known (Zaidi et al., 2008). Therefore, we hypothesized that DDB2 would act in a physiological relevant manner in budding yeast GG-NER. We first cloned the DDB2 gene into a low copy number, galactose inducible yeast expression vector. The cloning results in a fusion protein;

DDB2 fused with V5His6 tag (Fig. 2A). Both the empty plasmid vector and the DDB2 containing plasmid were transformed into *S. cerevisiae*. As expected, when cells were grown in the presence of galactose, DDB2 protein was produced as identified by Western blot using both V5 and DDB2 antibodies (Fig. 2B and data not shown). No protein was detectable at the calculated molecular weight of DDB2 in the empty vector control using the same Western blot technique (Fig. 2B).

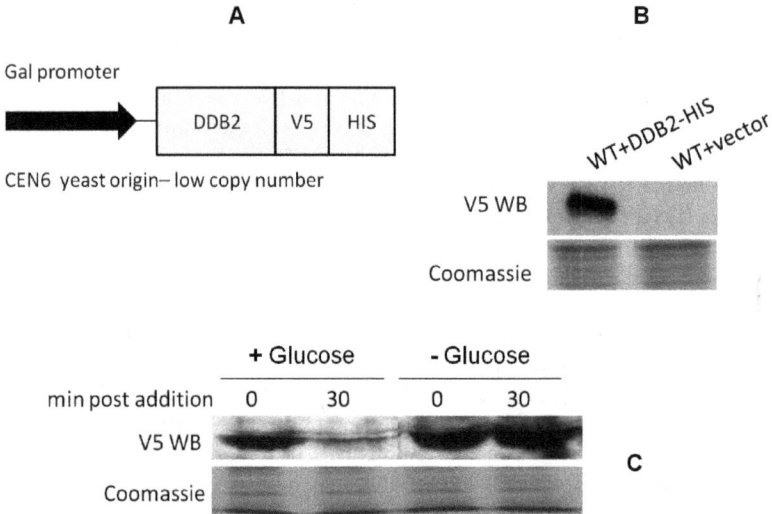

Fig. 2. Expression of DDB2-HIS in S. cerevisiae. (A) Schematic of DDB2 fusion cloned into pYCT/C2 expression vector. (B). Western blot (WB) using V5 antibody to detect expression of DDB2 containing or empty vector. (C) Glucose addition (4%) stops production of DDB2 detected by Western blot using V5 antibody, equal amount of total protein was verified using coomassie blue staining. BY4741 is the wild type (WT) strain used in these experiments.

To access the efficacy of the galactose induction 4% glucose was added to the media. Rapid shut down of the galactose inducible promoter is presumed due to the significant decrease in DDB2 protein levels 30 min post addition of glucose (Fig. 2C). This observed decrease in DDB2 protein levels is likely due to normal protein turnover in the absence of nascent DDB2 transcription and subsequent translation. These data confirm that DDB2 is expressed in *S. cerevisiae* cells under the control of the galactose promoter.

2.2 DDB2 suppresses UV sensitivity of Δ*rad26* cells

Next we identified genetic background in which a DDB2-dependent phenotype could be observed. We screened several yeast strains in which various NER proteins were deleted. The strains tested were Δ*rad7* and Δ*rad16* in which only TC-NER is active, Δ*rad26* in which only GG-NER is active, and Δ*rad1* in which the core pathway is defective and therefore there is no active NER. The spotting assay was used to determine DDB2 dependent suppression of UV sensitivity. Clearly, DDB2 expression suppresses the UV sensitive phenotype of Δ*rad26* cells (Fig. 3A). Survival curve experiments verified these findings (Data not shown).

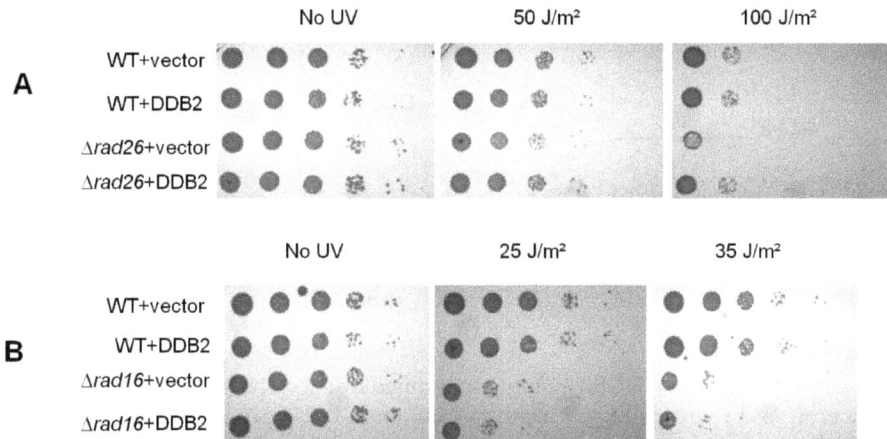

Fig. 3. DDB2 expression suppresses UV sensitivity of Δ*rad26* mutant, but not Δ*rad16* mutant. BY4741 (WT) cells expressing DDB2 or empty vector were diluted 1/10 and plated on galactose media. Cells were exposed to UV irradiation at dose indicated and grown in dark at 30 °C for 48 hours. Δ*rad26* (A). Δ*rad16* (B).

As discussed in the introduction, both DDB2 and Rad16 are necessary for lesion identification *in vivo* and are part of E3 ubiquitin ligase complexes (Verhage et al., 1994; Mueller & Smerdon, 1995; Shiyanov et al., 1999; Tang et al., 2000; Wakasugi et al., 2002; Fitch et al., 2003b; Groisman et al., 2003; Ramsey et al., 2004; Pines et al., 2009). Therefore, it was surprising that DDB2 was unable to suppress the Δ*rad16* UV sensitive (Fig. 3B). Our data suggest that despite similarities in their biochemical properties, on a gross functional level DDB2 and Rad16 are not analogs. It should be noted that Rad16 has also been implicated in post-incision processes (Reed et al., 1998) while DDB2 has not. It is therefore plausible that DDB2 and Rad16 have analogus functions in the lesion identification step of GG-NER, but this post-incision function of Rad16 is unable to be rescued by DDB2 expression.

In addition, we found that DDB2 was not able to significantly suppress UV sensitivity of any other knockout strains, including Δ*rad7* cells (data not shown). These data are consistent with no known DDB2 homolog in budding yeast. The observed DDB2-dependent suppression of TC-NER deficient UV sensitivity is consistent with reported DDB2 stimulation of GG-NER (Wakasugi et al., 2001; Wakasugi et al., 2002).

2.3 DDB2 mutations abrogate its ability to suppress Δ*rad26* UV sensitivity

To assess if DDB2 is functioning in a physiologically relevant manner, we first examined the phenotypic effects of mutant DDB2 on DDB2-dependnet suppression of Δ*rad26* UV sensitive phenotype. Several DDB2 mutations identified in XPE patients are known to interfere with its ability to function properly in GG-NER. It has been reported that a point mutation changing lysine 244 to glutamic acid (DDB2 K244E) results in inability of DDB2 to make contact with DNA lesions (Scrima et al., 2008) (Fig. 4A). However, this mutation does not alter the ability of DDB2 to interact with DDB1 in the Cul4a E3 ubiquitin ligase complex, therefore its role in ubiquitination is not altered. When this damage recognition deficient mutant DDB2 was introduced into Δ*rad26* cells, it was unable to suppress Δ*rad26* UV sensitivity (Fig. 5). This suggests that the observed DDB2-conferred UV resistance is linked to its function in DNA damage detection.

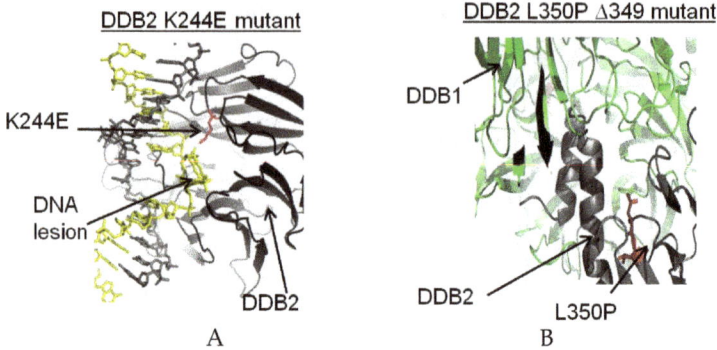

Fig. 4. Crystal structure of DDB2 mutations modified from crystal structure solved by Scrima et al. (A) Lysine to glutamic acid substitution at aa 244 predicted to effect DDB2 DNA interaction. Red residue indicates site of mutation. Yellow indicates damaged DNA strand. (B) Deletion of aa 349 and substitution of proline for leucine at aa 350. This mutation is predicted to effect the DDB2 DDB1 interaction. Red indicates site of mutation. Mutant DDB2 was constructed by site directed mutagenesis.

Another mutation that affects DDB2's function prevents the interaction with its in vivo partner DDB1 (Nichols et al. 2000). This mutation was also constructed and is a complex mutation, consisting of both a deletion of amino acid 349 and a proline substitution for leucine at amino acid 350 (DDB2 L350P) (Fig. 4B). Like DDB2 K244E, this mutation also abrogated DDB2's ability to suppress UV sensitivity in Δ*rad26* cells (Fig. 5). These data suggest DDB2-conferred UV resistance is dependent on a conserved interacting partner.

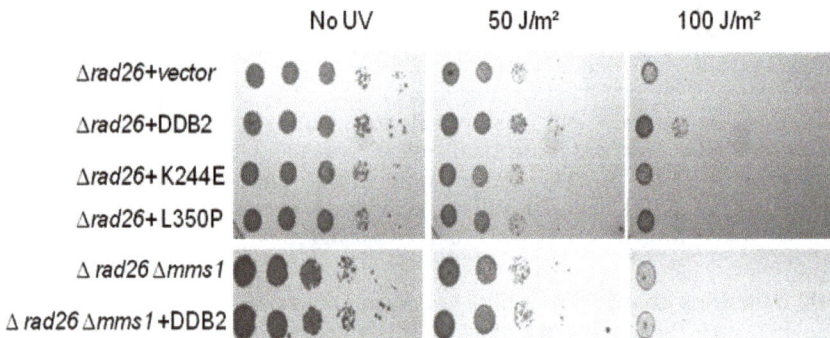

Fig. 5. DDB2 mutations and deletion of Mms1 (DDB1 homolog) abrogate suppression of UV sensitivity in Δ*rad26* cells.

Although Mms1 has been identified as the budding yeast DDB1 homolog (Zaidi et al., 2008), there are no reports of it being involved in NER. However, our previous observation suggesting DDB2 function requires a conserved interacting partner prompted us to test DDB2 function in the absence of Mms1. To test this, wild type DDB2 was expressed in the Δ*rad26*Δ*mms1* double mutant and UV sensitivity was accessed by spotting assays. Indeed, this reciprocal experiment verified that Mms1 is necessary for DDB2-dependent suppression of UV sensitivity (Fig. 5).

Taken together, these data suggest that exogenously expressed DDB2 is acting in a physiologically relevant manner. Additionally, our findings indicate that the DNA damage recognition function of DDB2 is essential for the observed suppression of UV sensitivity. We also found that DDB2 function is dependent on interaction with Mms1, a subunit of an E3 ubiquitin ligase. These observations are consistent with what is reported for DDB2 function in human cells.

3. Conclusion

Studies in *Saccharomyces cerevisiae* have made major contributions to our understanding of NER. Here, we present evidence suggesting that *S. cerevisiae* can be used to dissect the roles of human DDB2 in initiating NER in chromatin. Since DDB2 functions are regulated by the ubiquitin pathway and DDB2 itself is a component of an E3 ligase, it will be interesting to explore the regulation of DDB2 functions by ubiquitination, using yeast mutants with defects in various steps of the ubiquitin pathway.

Ubiquitination is a well studied post-translational modification and recent data suggest multiple fates of ubiquitin modified proteins (Sadowski & Sarcevic, 2010). It will be important to determine if ubiquitination of DDB2 promotes its degradation or controls DDB2 association with chromatin. The budding yeast system described here will also provide an alternative system to screen the effect(s) of various DDB2 lysine mutations to determine which amino acid residue(s) is modified. Additionally, as reviewed by Kirkin et al., ubiquitin signaling is altered in many cancers (Kirkin and Dikic 2010), suggesting a potential role of ubiquitination in regulating DNA binding proteins such as transcription factors and repair proteins. Therefore, it will be interesting to determine what, if any, role ubiquitination plays in the chromatin association of other DNA binding proteins, specifically transcription factors and repair proteins. The utilization of the budding yeast model system will facilitate deciphering such questions.

4. Acknowledgment

The project described in this article was supported by Award R01ES017784 from the National Institute of Environmental Health Sciences and by a young investigator award from the Concern Foundation.

5. References

Chen, X., Zhang, Y., Douglas, L., and Zhou, P. 2001. UV-damaged DNA-binding proteins are targets of CUL-4A-mediated ubiquitination and degradation. J Biol Chem 276(51): 48175-48182.

Conconi, A., Bespalov, V.A., and Smerdon, M.J. 2002. Transcription-coupled repair in RNA polymerase I-transcribed genes of yeast. Proc Natl Acad Sci U S A 99(2): 649-654.

Douki, T. and Cadet, J. 2001. Individual determination of the yield of the main UV-induced dimeric pyrimidine photoproducts in DNA suggests a high mutagenicity of CC photolesions. Biochemistry 40(8): 2495-2501.

Fitch, M.E., Cross, I.V., Turner, S.J., Adimoolam, S., Lin, C.X., Williams, K.G., and Ford, J.M. 2003a. The DDB2 nucleotide excision repair gene product p48 enhances global genomic repair in p53 deficient human fibroblasts. DNA Repair (Amst) 2(7): 819-826.

Fitch, M.E., Nakajima, S., Yasui, A., and Ford, J.M. 2003b. In vivo recruitment of XPC to UV-induced cyclobutane pyrimidine dimers by the DDB2 gene product. J Biol Chem 278(47): 46906-46910.

Fousteri, M. and Mullenders, L.H. 2008. Transcription-coupled nucleotide excision repair in mammalian cells: molecular mechanisms and biological effects. Cell Res 18(1): 73-84.

Gong, F., Fahy, D., and Smerdon, M.J. 2006. Rad4-Rad23 interaction with SWI/SNF links ATP-dependent chromatin remodeling with nucleotide excision repair. Nat Struct Mol Biol 13(10): 902-907.

Gong, F., Kwon, Y., and Smerdon, M.J. 2005. Nucleotide excision repair in chromatin and the right of entry. DNA Repair (Amst) 4(8): 884-896.

Groisman, R., Polanowska, J., Kuraoka, I., Sawada, J., Saijo, M., Drapkin, R., Kisselev, A.F., Tanaka, K., and Nakatani, Y. 2003. The ubiquitin ligase activity in the DDB2 and CSA complexes is differentially regulated by the COP9 signalosome in response to DNA damage. Cell 113(3): 357-367.

Guzder, S.N., Sung, P., Prakash, L., and Prakash, S. 1999. Synergistic interaction between yeast nucleotide excision repair factors NEF2 and NEF4 in the binding of ultraviolet-damaged DNA. J Biol Chem 274(34): 24257-24262.

Hanawalt, P.C. 2002. Subpathways of nucleotide excision repair and their regulation. Oncogene 21(58): 8949-8956.

Hwang, B.J., Ford, J.M., Hanawalt, P.C., and Chu, G. 1999. Expression of the p48 xeroderma pigmentosum gene is p53-dependent and is involved in global genomic repair. Proc Natl Acad Sci U S A 96(2): 424-428.

Jackson, S. and Xiong, Y. 2009. CRL4s: the CUL4-RING E3 ubiquitin ligases. Trends Biochem Sci 34(11): 562-570.

Jansen, L.E., Verhage, R.A., and Brouwer, J. 1998. Preferential binding of yeast Rad4.Rad23 complex to damaged DNA. J Biol Chem 273(50): 33111-33114.

Jiang, Y., Wang, X., Bao, S., Guo, R., Johnson, D.G., Shen, X., and Li, L. 2010. INO80 chromatin remodeling complex promotes the removal of UV lesions by the nucleotide excision repair pathway. Proc Natl Acad Sci U S A 107(40): 17274-17279.

Jones, K.L., Zhang, L., Seldeen, K.L., and Gong, F. 2010. Detection of bulky DNA lesions: DDB2 at the interface of chromatin and DNA repair in eukaryotes. IUBMB Life 62(11): 803-811.

Kapetanaki, M.G., Guerrero-Santoro, J., Bisi, D.C., Hsieh, C.L., Rapic-Otrin, V., and Levine, A.S. 2006. The DDB1-CUL4ADDB2 ubiquitin ligase is deficient in xeroderma pigmentosum group E and targets histone H2A at UV-damaged DNA sites. Proc Natl Acad Sci U S A 103(8): 2588-2593.

Kirkin, V. and Dikic, I. 2010. Ubiquitin networks in cancer. Curr Opin Genet Dev 21(1): 21-28.

Kornberg, R.D. and Lorch, Y. 1999. Twenty-five years of the nucleosome, fundamental particle of the eukaryote chromosome. Cell 98(3): 285-294.

Kraemer, K.H., Patronas, N.J., Schiffmann, R., Brooks, B.P., Tamura, D., and DiGiovanna, J.J. 2007. Xeroderma pigmentosum, trichothiodystrophy and Cockayne syndrome: a complex genotype-phenotype relationship. Neuroscience 145(4): 1388-1396.

Lans, H., Marteijn, J.A., Schumacher, B., Hoeijmakers, J.H., Jansen, G., and Vermeulen, W. 2010. Involvement of global genome repair, transcription coupled repair, and chromatin remodeling in UV DNA damage response changes during development. PLoS Genet 6(5): e1000941.

Liu, L., Lee, S., Zhang, J., Peters, S.B., Hannah, J., Zhang, Y., Yin, Y., Koff, A., Ma, L., and Zhou, P. 2009. CUL4A abrogation augments DNA damage response and protection against skin carcinogenesis. Mol Cell 34(4): 451-460.

Luger, K., Mader, A.W., Richmond, R.K., Sargent, D.F., and Richmond, T.J. 1997. Crystal structure of the nucleosome core particle at 2.8 A resolution. Nature 389(6648): 251-260.

Luijsterburg, M.S., Goedhart, J., Moser, J., Kool, H., Geverts, B., Houtsmuller, A.B., Mullenders, L.H., Vermeulen, W., and van Driel, R. 2007. Dynamic in vivo interaction of DDB2 E3 ubiquitin ligase with UV-damaged DNA is independent of damage-recognition protein XPC. J Cell Sci 120(Pt 15): 2706-2716.

Matsuda, N., Azuma, K., Saijo, M., Iemura, S., Hioki, Y., Natsume, T., Chiba, T., and Tanaka, K. 2005. DDB2, the xeroderma pigmentosum group E gene product, is directly ubiquitylated by Cullin 4A-based ubiquitin ligase complex. DNA Repair (Amst) 4(5): 537-545.

Min, J.H. and Pavletich, N.P. 2007. Recognition of DNA damage by the Rad4 nucleotide excision repair protein. Nature 449(7162): 570-575.

Mueller, J.P. and Smerdon, M.J. 1995. Repair of plasmid and genomic DNA in a rad7 delta mutant of yeast. Nucleic Acids Res 23(17): 3457-3464.

Nag, A., Bondar, T., Shiv, S., and Raychaudhuri, P. 2001. The xeroderma pigmentosum group E gene product DDB2 is a specific target of cullin 4A in mammalian cells. Mol Cell Biol 21(20): 6738-6747.

Nag, R. and Smerdon, M.J. 2009. Altering the chromatin landscape for nucleotide excision repair. Mutat Res 682(1): 13-20.

Nichols, A.F., Itoh, T., Graham, J.A., Liu, W., Yamaizumi, M., and Linn, S. 2000. Human damage-specific DNA-binding protein p48. Characterization of XPE mutations and regulation following UV irradiation. J Biol Chem 275(28): 21422-21428.

Osley, M.A., Tsukuda, T., and Nickoloff, J.A. 2007. ATP-dependent chromatin remodeling factors and DNA damage repair. *Mutat Res* 618(1-2): 65-80.

Pines, A., Backendorf, C., Alekseev, S., Jansen, J.G., de Gruijl, F.R., Vrieling, H., and Mullenders, L.H. 2009. Differential activity of UV-DDB in mouse keratinocytes and fibroblasts: impact on DNA repair and UV-induced skin cancer. DNA Repair (Amst) 8(2): 153-161.

Ramsey, K.L., Smith, J.J., Dasgupta, A., Maqani, N., Grant, P., and Auble, D.T. 2004. The NEF4 complex regulates Rad4 levels and utilizes Snf2/Swi2-related ATPase activity for nucleotide excision repair. Mol Cell Biol 24(14): 6362-6378.

Reed, S.H., You, Z., and Friedberg, E.C. 1998. The yeast RAD7 and RAD16 genes are required for postincision events during nucleotide excision repair. In vitro and in vivo studies with rad7 and rad16 mutants and purification of a Rad7/Rad16-containing protein complex. J Biol Chem 273(45): 29481-29488.

Sadowski, M. and Sarcevic, B. 2010. Mechanisms of mono- and poly-ubiquitination: Ubiquitination specificity depends on compatibility between the E2 catalytic core and amino acid residues proximal to the lysine. Cell Div 5: 19.

Sancar, A. and Reardon, J.T. 2004. Nucleotide excision repair in E. coli and man. Adv Protein Chem 69: 43-71.

Sarkar, S., Kiely, R., and McHugh, P.J. 2010. The Ino80 chromatin-remodeling complex restores chromatin structure during UV DNA damage repair. *J Cell Biol* 191(6): 1061-1068.

Scrima, A., Konickova, R., Czyzewski, B.K., Kawasaki, Y., Jeffrey, P.D., Groisman, R., Nakatani, Y., Iwai, S., Pavletich, N.P., and Thoma, N.H. 2008. Structural basis of UV DNA-damage recognition by the DDB1-DDB2 complex. Cell 135(7): 1213-1223.

Setlow, R.B. 2001. Human cancer: etiologic agents/dose responses/DNA repair/cellular and animal models. Mutat Res 477(1-2): 1-6.

Shiyanov, P., Nag, A., and Raychaudhuri, P. 1999. Cullin 4A associates with the UV-damaged DNA-binding protein DDB. J Biol Chem 274(50): 35309-35312.

Smerdon, M.J. 1991. DNA repair and the role of chromatin structure. Curr Opin Cell Biol 3(3): 422-428.

Sugasawa, K. 2009. UV-DDB: A molecular machine linking DNA repair with ubiquitination. DNA Repair (Amst).

Sugasawa, K., Okuda, Y., Saijo, M., Nishi, R., Matsuda, N., Chu, G., Mori, T., Iwai, S., Tanaka, K., and Hanaoka, F. 2005. UV-induced ubiquitylation of XPC protein mediated by UV-DDB-ubiquitin ligase complex. Cell 121(3): 387-400.

Tang, J. and Chu, G. 2002. Xeroderma pigmentosum complementation group E and UV-damaged DNA-binding protein. DNA Repair (Amst) 1(8): 601-616.

Tang, J.Y., Hwang, B.J., Ford, J.M., Hanawalt, P.C., and Chu, G. 2000. Xeroderma pigmentosum p48 gene enhances global genomic repair and suppresses UV-induced mutagenesis. Mol Cell 5(4): 737-744.

Verhage, R., Zeeman, A.M., de Groot, N., Gleig, F., Bang, D.D., van de Putte, P., and Brouwer, J. 1994. The RAD7 and RAD16 genes, which are essential for pyrimidine dimer removal from the silent mating type loci, are also required for repair of the nontranscribed strand of an active gene in Saccharomyces cerevisiae. Mol Cell Biol 14(9): 6135-6142.

Wakasugi, M., Kawashima, A., Morioka, H., Linn, S., Sancar, A., Mori, T., Nikaido, O., and Matsunaga, T. 2002. DDB accumulates at DNA damage sites immediately after UV irradiation and directly stimulates nucleotide excision repair. J Biol Chem 277(3): 1637-1640.

Wang, H., Zhai, L., Xu, J., Joo, H.Y., Jackson, S., Erdjument-Bromage, H., Tempst, P., Xiong, Y., and Zhang, Y. 2006. Histone H3 and H4 ubiquitylation by the CUL4-DDB-ROC1 ubiquitin ligase facilitates cellular response to DNA damage. Mol Cell 22(3): 383-394.

Wang, Y. 2008. Bulky DNA lesions induced by reactive oxygen species. Chem Res Toxicol 21(2): 276-281.

Waters, R., Teng, Y., Yu, Y., Yu, S., and Reed, S.H. 2009. Tilting at windmills? The nucleotide excision repair of chromosomal DNA. DNA Repair (Amst) 8(2): 146-152.

Wood, R.D. 2010. Mammalian nucleotide excision repair proteins and interstrand crosslink repair. Environ Mol Mutagen 51(6): 520-526.

Zaidi, I.W., Rabut, G., Poveda, A., Scheel, H., Malmstrom, J., Ulrich, H., Hofmann, K., Pasero, P., Peter, M., and Luke, B. 2008. Rtt101 and Mms1 in budding yeast form a CUL4(DDB1)-like ubiquitin ligase that promotes replication through damaged DNA. EMBO Rep 9(10): 1034-1040.

Zamble, D.B., Mu, D., Reardon, J.T., Sancar, A., and Lippard, S.J. 1996. Repair of cisplatin--DNA adducts by the mammalian excision nuclease. Biochemistry 35(31): 10004-10013.

Zhang, L., Jones, K., and Gong, F. 2009a. The molecular basis of chromatin dynamics during nucleotide excision repair. Biochem Cell Biol 87(1): 265-272.

Zhang, L., Zhang, Q., Jones, K., Patel, M., and Gong, F. 2009b. The chromatin remodeling factor BRG1 stimulates nucleotide excision repair by facilitating recruitment of XPC to sites of DNA damage. Cell Cycle 8(23): 3953-3959.

Zhao, Q., Wang, Q.E., Ray, A., Wani, G., Han, C., Milum, K., and Wani, A.A. 2009. Modulation of nucleotide excision repair by mammalian SWI/SNF chromatin-remodeling complex. J Biol Chem 284(44): 30424-30432.

Zhu, Q., Wani, G., Arab, H.H., El-Mahdy, M.A., Ray, A., and Wani, A.A. 2009. Chromatin restoration following nucleotide excision repair involves the incorporation of ubiquitinated H2A at damaged genomic sites. DNA Repair (Amst) 8(2): 262-273.

15

TopBP1 in DNA Damage Response

Ewa Forma, Magdalena Brys and Wanda M. Krajewska
Department of Cytobiochemistry, University of Lodz
Poland

1. Introduction

DNA, the genetic material of cells, is constantly exposed to a range of endogenous and environmental damaging agents (Jungmichel & Stucki, 2010). DNA molecule is the target of endogenous cellular metabolites such as reactive oxygen species (ROS) (Ciccia & Elledge, 2010; Poehlmann & Roessner, 2010). ROS may cause different alterations in a genome, e.g. simple DNA mutations, DNA single and double strand breaks (SSBs and DSBs, respectively), or more complex changes, including deletions, translocations and fusions (Poehlmann & Roessner, 2010). Alterations may be generated spontaneously due to dNTP misincorporation during DNA replication, interconversion between DNA bases caused by deamination, loss of DNA bases following DNA depurination or depyrimidination and modification of DNA bases by alkylation. Hydrolytic deamination (loss of an amino group) can directly convert one base to another. For example, deamination of cytosine results in uracil and with much lower frequency converts adenine to hypoxanthine. In depurination or depyrimidination, purine or pyrimidine bases are completely removed, leaving deoxyribose sugar depurinated or depyrimidinated that may cause breakage in the DNA backbone (Ciccia & Elledge, 2010; Rastogi et al., 2010). Altogether, it has been estimated that every cell could experience up to 10^5 spontaneous DNA lesions per day (Ciccia & Elledge, 2010). Environmental DNA damage can be produced by physical or chemical sources, such as ionizing radiation (IR), ultraviolet (UV) light from sunlight and organic and inorganic chemical substances (Muniandy et al., 2010; Rastogi et al., 2010; Su et al., 2010). Exposure to ionizing radiation from, e.g. cosmic radiation and medical treatments employing X-rays or radiotherapy inflicts DNA single and double strand breaks, oxidation of DNA bases and DNA-protein crosslinks in the genomic DNA (Ciccia & Elledge, 2010; Su et al., 2010). Ionizing radiation provokes DNA damage directly by energy deposit on the DNA double helix and indirectly by reactive oxygen/nitrogen species (ROS/RNS) (Corre et al., 2010). Ultraviolet radiation (mainly UV-B) is a powerful agent that may lead to the formation of three major classes of DNA lesions, such as cyclobutane pyrimidine dimmers (CPDs), pyrimidine 6-4 pyrimidone photoproducts (6-4 PPs) and their Devar isomers (Rastogi et al., 2010). Cells may become transiently exposed to external sources of DNA damage, such as cigarette smoke or various toxic chemical compounds (Jungmichel & Stucki, 2010). Many antineoplastic drugs currently used in cancer treatment express their cytotoxic effects through their ability to directly or indirectly damage DNA and thus resulting in cell death. Major types of DNA damage induced by anticancer treatment include single and double strand breaks, interstrand, intrastrand and DNA-protein crosslinks, as well as interference with nucleotide metabolism and DNA synthesis (Pallis & Karamouzis, 2010). Alkylating agents, such as methyl methanesulphonate (MMS), tenozalamide, streptozotocin, procarbazine, dacarbazine, ethylnitrosourea, diethylnitrosamine and nitrosoureas attach alkyl groups to DNA bases, while crosslinking agents such as mitomycin (MMC), cisplatin, psoralen and nitrogen mustard induce covalent links between bases of the same DNA strand (intrastrand crosslinks) or of different DNA strands (interstrands crosslinks) (Ciccia & Elledge, 2010; Muniandy et al., 2010; Pallis & Karamouzis, 2010). Other chemical agents, such as topoisomerase inhibitors induce the formation of single or double strand breaks by trapping topoisomerase-DNA covalent complexes (Ciccia & Elledge, 2010). Camptothecin and novel noncamptothecins in clinical development target eukaryotic IB type topoisomerase (Topo I), whereas human IIA type topoisomerases (Topo IIα and Topo IIβ) are the targets of widely used anticancer agents, such as etoposide, anthracyclines (doxorubicin, daunorubicin) and mitoxantrone (Pommier et al., 2010).

The biochemical consequences of DNA lesions are diverse and range from obstruction of fundamental cellular pathways like transcription and replication to fixation of mutations. Cellular misfunctioning, cell death, aging and cancer are the phenotypical consequences of DNA damage accumulation in the genome. To counteract DNA damage, repair mechanisms specific for many types of lesions have evolved. Mispaired DNA bases are replaced with correct bases by mismatch repair (MMR) (Ciccia & Elledge, 2010). The bases excision repair (BER) exerts its biological role by removing bases that have been damaged by alkylation, oxidation, ring saturation, as well as a short strand that contains the damaged bases. BER also plays an important role in the repair of DNA single strand breaks generated spontaneously or induced by exogenous DNA-damaging factors such as cytotoxic anticancer agents (Pallis & Karamouzis, 2010). DNA single strand breaks may be also repaired by single strand break repair (SSBR) (Ciccia & Elledge, 2010). Nucleotide excision repair (NER) is a highly conserved pathway that repairs DNA damage caused by UV radiation, mutagenic chemicals or chemotherapeutic drugs that form bulky DNA adducts (Pallis & Karamouzis, 2010). The most toxic lesions in DNA are double strand breaks where the phosphate backbones of the two complementary DNA strands are broken simultaneously (Hiom, 2010). Double strand breaks are repaired by two major repair pathways depending on the context of DNA damage, i.e. homologous recombination (HR) and nonhomologous end-joining (NHEJ) (Hiom, 2010; Pallis & Karamouzis, 2010). While NHEJ promotes potential inaccurate relegation of double strand breaks, HR precisely restores genomic sequence of the broken DNA ends by using sister chromatids as template for repair (Ciccia & Elledge, 2010). Additionally, some specialized polymerases can temporarily take over lesion-arrested DNA polymerases during S phase, in a mutagenic mechanism called translesion synthesis (TLS). Such polymerases only work if a more reliable system, such as homologous recombination, cannot avoid stumbled DNA replication (Essers et al., 2006).

DNA repair is carried out by the plethora of enzymatic activities that chemically modify DNA to repair DNA damage, including nucleases, helicases, polymerases, topoisomerases, recombinases, ligases, glycosylases, demethylases, kinases and phosphatases. These repair tools must be precisely regulated, because each in its own right can wreak havoc on the integrity of DNA if misused or allowed to gain access to DNA at the inappropriate time or place (Ciccia & Elledge, 2010). The DNA repair mechanisms function in conjunction with an intricate machinery of damage sensors, responsible of a series of phosphorylations and chromatin modifications that signal to the rest of the cell the presence of lesions on DNA.

Together DNA repair mechanisms and DNA damage signaling system form a molecular shield against genomic instability.

2. DNA damage checkpoints

To maintain genomic integrity and faithful transmission of fully replicated and undamaged DNA during cell division, eukaryotic organisms evolved a complex DNA surveillance program (Reihardt & Yaffe, 2009). Apart from DNA repair mechanisms mentioned above, DNA damage response represents a complex network of multiple signaling pathways involving cell cycle checkpoints, transcriptional regulation, chromatin remodeling and apoptosis (Dai & Grant, 2010; Danielsen et al., 2009). In response to DNA damage, eukaryotic cells activate a complex protein kinase-based signaling network to arrest progression through the cell cycle. Activation of signaling cascade recruits repair machinery to the site of DNA damage, provides time for repair or if the genotoxic insult exceeds repair capacity, additional signaling pathways leading to cell death, presumably *via* apoptosis, are activated (Reinhardt et al., 2010; Reinhardt & Yaffe, 2009). When DNA damage occurs, distinct, albeit overlapping and cooperating checkpoint pathways are activated, which block S phase entry (the G1/S phase checkpoint), delay S phase progression (the S phase checkpoints) or prevent mitotic entry (the G2/M phase checkpoint). The primary G1/S cell cycle checkpoint controls the commitment of eukaryotic cells to transition through G1 phase and enter DNA synthesis phase. In G1 phase, cells have to make a decision between continuing proliferation or exiting the cell cycle to become quiescent differentiated, senescent or apoptotic (Dijkstra et al., 2009). The S phase checkpoints are activated when DNA damage occurs during DNA synthesis, or when DNA replication intermediates accumulate. Depending on the type and magnitude of damage, cells activate one of three distinct S phase checkpoint pathways: an intra-S phase checkpoint induced by double strand breaks, a replication checkpoint by the stalled replication fork and the S/M checkpoint blocking premature mitosis. The S/M checkpoint differs from the well-defined G2/M checkpoint. The S/M checkpoint is ATM-independent, it is measurable only several hours after DNA damage and is initiated in cells that were in S phase at the time of insult (Hurley & Bunz, 2009; Rodriguez-Bravo et al., 2007). When cells encounter DNA damage in G2, the G2/M checkpoint stops the cell cycle to prevent the cell from entering mitosis. Defects in cell cycle arrest at the respective checkpoint are associated with genome instability and oncogenesis (Houtgraaf et al., 2006).

3. Checkpoint signaling cascade

Proteins of checkpoint signaling pathways are classified as sensors, transducers and effectors (Fig. 1). Following DNA damage, sensor multiprotein complexes, e.g. MRN (MRE11-Rad50-NBS1) or 9-1-1 (Rad9-Rad1-Hus1) recognize damage and recruit proximal transducers, i.e. ATM (ataxia telangiectasia mutated) and ATR (ATM and Rad3-related) kinases to lesions where they are initially activated. ATM and ATR transduce signals to distal transducer, i.e. checkpoint kinases Chk1 and Chk2 (Dai & Grant, 2010; Niida & Nakanishi, 2006). Chk1 and Chk2 kinases, distal transducers, transfer the signal of DNA damage to effectors, such as Cdks (cyclin-dependent kinases), Cdc25 (cell division cycle 25) and p53 (Dai & Grant, 2010; Houtgraaf et al., 2006; Nakanishi, 2009; Nakanishi et al., 2009). The key difference between ATM and ATR is the signal that activates them. ATM is activated exclusively by DSBs, which can arise from endogenous (ROS, eroded telomeres, intermediates of immune and meiotic recombination) or exogenous (IR, genotoxic drugs) sources (Lopez-Contreras & Fernandez-Capetillo, 2010). In contrast, ATR responds to many types of DNA damage and replication stress including breaks, crosslinks and base adducts. ATR senses abnormally long stretches of single strand DNA that arise from the functional uncoupling of helicase and polymerase activities at replication forks or from the processing of DNA lesions such as the resection of DSBs (Mordes & Cortez, 2008). ATR but not ATM is essential for viability. The early embryonic death in ATR knockout mice shows that ATR is essential for cell growth and differentiation at an early stage of development (Smits et al., 2010). In addition, disruption of ATR in mouse or human cells results in cell cycle arrest or death, even without exogenous DNA damage (Cortez et al., 2001; Smits et al., 2010). Although complete inactivation of ATR is lethal, a hypomorphic mutation was found in humans suffering from the rare autosomal recessive disorder, Seckel syndrome, characterized by growth retardation and microcephaly. In homozygosity, that mutation affects ATR splicing which results in the reduction of ATR protein levels to almost undetectable, yet the remaining protein is sufficient for viability (Kerzendorfer & O'Driscoll, 2009; O'Driscoll et al., 2004; Smits et al., 2010).

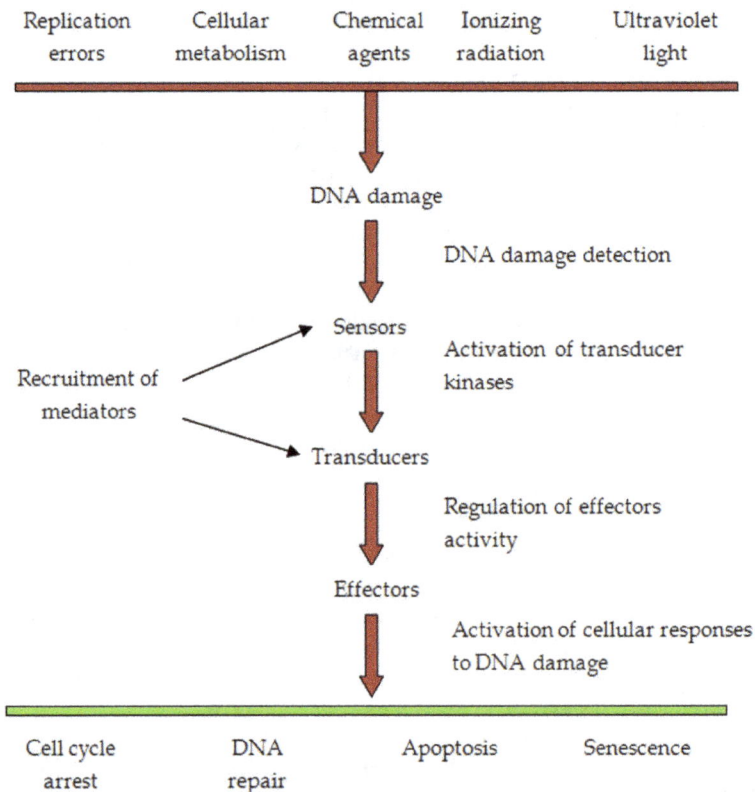

Fig. 1. Signal transduction of DNA damage response (DDR)

In addition to damage sensors and signal transducers, many other proteins called mediators are involved in DNA damage response. Mediators are mostly cell cycle specific proteins associated with damage sensors and signal transducers at particular phases of the cell cycle and, as a consequence, help provide signal transduction specificity. ATM and ATR phosphorylate most of these mediators. Well-known examples of mediators are 53BP1 (p53 binding protein 1), MDC1 (mediator of DNA damage checkpoint 1), BRCA1 (breast cancer 1), SMC1 (structural maintenance of chromosomes 1), FANCD2 (Fanconi anemia, complementation group D2), Claspin, Timeless, Tipin and histone H2AX (Dai & Grant, 2010; Houtgraaf et al., 2006; Yang et al., 2010). This group of regulators involves also TopBP1 protein (topoisomerase IIβ binding protein 1) (Cimprich & Cortez, 2008). Certain molecules may have multiple functions in this signal transduction pathway. For example ATM and ATR can simultaneously act as a sensor and a transducer. Consequently, signal transduction in DNA damage response is not one-dimensional but a complex network of interacting molecules (Poehlmann & Roessner, 2010).

4. Structure of TopBP1 and its similarity to BRCA1

Topoisomerase IIβ binding protein 1 (TopBP1) has been identified as a protein interacting with topoisomerase IIβ in a yeast two-hybrid screen (Morishima et al., 2007; Yamane et al., 1997). Interaction with topoisomerase IIβ is mediated by carboxyl-terminal region (aa 862-1522) of TopBP1 *in vitro* (Honda et al., 2002; Yamane et al., 1997). TopBP1 shares sequence and structural homologies with *Saccharomyces cerevisiae* Dpb11, *Schizosaccharomyces pombe* Cut5/Rad4, *Drosophila melanogaster* Mus101 and *Xenopus levis* Xmus101 (Araki et al., 1995; Garcia et al., 2005; Morishima et al., 2007; Ogiwara et al., 2006; Parrilla-Castellar & Karnitz, 2003; Taricani & Wand, 2006; van Hatten et al., 2002).

TopBP1 protein seems to be essential for maintenance of chromosomal integrity and cell proliferation. This protein appeared to be involved in DNA damage response, DNA replication checkpoint, chromosome replication and regulation of transcription (Bang et al., 2011; Garcia et al., 2005; Jeon et al., 2011). TopBP1 knockout mouse exhibits early embryonic lethality at the peri-implantation stage and TopBP1 deficiency induces cellular senescence in primary cells (Bang et al., 2011; Jeon et al., 2011).

TopBP1 gene comprising 28 exons is located on chromosome 3q22.1 and encodes a 1522 amino acid protein (180 kDa) (Karppinen et al., 2006; Xu & Leffak, 2010; Yan & Michael, 2009a,b). The structure of protein is characterized by the presence of interspersed throughout the whole molecule eight copies of the BRCT domain (C-terminal domain of BRCA1), originally identified as a tandemly repeated sequence motif in carboxyl-terminal region of BRCA1 (Fig. 2) (Glover, 2006; Lelung et al., 2010; Wright et al., 2006; Yamane et al., 1997; Yamane & Tsuruo, 1999). BRCT domains, about 90 amino acids in length, are hydrophobic and are involved in an interaction with other proteins and phosphorylated peptides, as well as in an interaction with single- and double-stranded DNA (Glover, 2006; Rodriquez et al., 2003; Wright et al., 2006). A sequence analysis has shown that BRCT repeats are present in a large family of proteins that are implicated in the cellular response to DNA damage. Next to BRCA1 and TopBP1, members of this family include several proteins that are directly linked to DNA repair and cell cycle checkpoints, such as XRCC1 (X-ray cross complementing protein 1), DNA ligase III and IV, MDC1, BARD1 (BRCA1 associated RING domain protein 1), Rad9, MCPH1 (microcephalin 1) (Glover, 2006; Glover et al., 2004; Hou et al., 2010; Yamane et al., 2002; Yamane & Tsuruo, 1999; Yang et al., 2008).

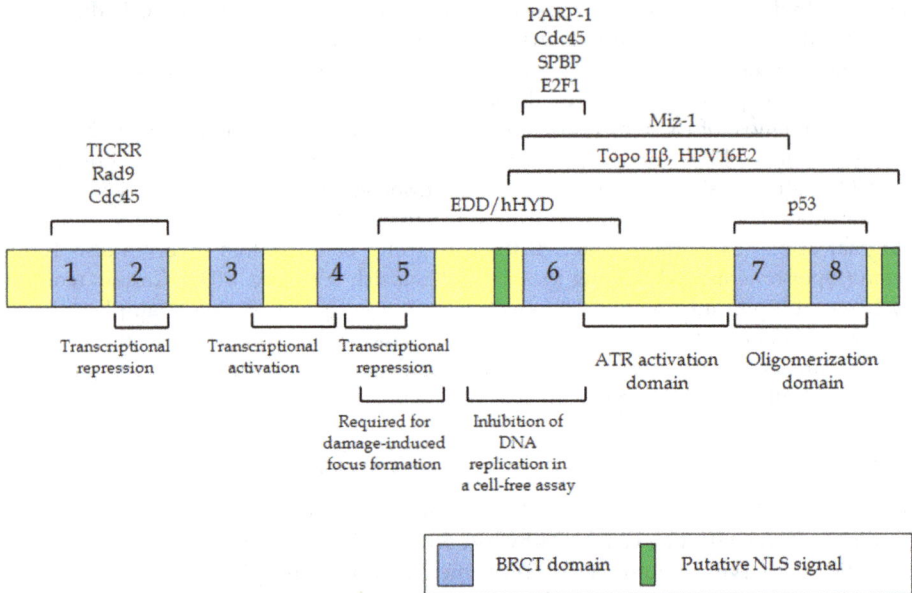

Fig. 2. TopBP1 functional domains and sites of interacting proteins

The carboxyl-terminal region of TopBP1 containing two BRCT domains shows considerable similarity to the corresponding part of BRCA1 (Going et al., 2007; Karppinen et al., 2006; Makiniemi et al., 2001; Morris et al., 2009; Yamane et al., 1997, 2003). Apart from structural similarity TopBP1 shares many other common features with BRCA1. The expression of both proteins is the highest in S phase cells. TopBP1 and BRCA1 are phosphorylated by ATM in response to DNA damage and DNA replication stress and they both colocalize with PCNA (proliferating cell nuclear antigen) at stalled replication forks (Makiniemi et al., 2001; Yamane et al., 2003). The localization patterns of TopBP1 and BRCA1 have similarities also during late mitosis, as well as in meiotic prophase I (Karppinen et al., 2006; Reini et al., 2004). Furthermore, the two proteins have been shown to possess overlapping functions in G2/M checkpoint regulation (Karppinen et al., 2006). Yamane et al. (2003) demonstrated that a BRCA1-mutant or a TopBP1-reduced background results in only partial abrogation at G2/M checkpoint, whereas the combined TopBP1-reduced and BRCA-mutant background result in the nearly complete abrogation. In response to ionizing radiation TopBP1 and BRAC1 colocalize with Rad50, ATM, Rad9, BLM (Bloom syndrome protein), PCNA, NBS1 (Nijmegen breakage syndrome 1) and γH2AX in IR-induced nuclear foci (Germann et al., 2010; Xu et al., 2003).

TopBP1 protein possesses transcriptional-activation domain and two surrounding repressor domains and can play a role in regulating transcription directly (Fig. 2). A transcriptional-activation domain is located between amino acids 460 - 591 and partly contains BRCT4 domain. This region essential for transactivation is rich in hydrophobic amino acids interspersed with acidic residues, typical of identified transcriptional domains. On amino-terminal side of the transcriptional activation domain, Wright et al., (2006) identified a repressor domain involving BRCT2 that is able to repress the TopBP1 transcriptional activation domain. Additionally, another repressor domain exists on the C terminus of the activation domain, which requires amino acids 586 – 675. TopBP1 protein exerts its function in the nucleus and the carboxyl-terminal region of TopBP1 contains two putative nuclear localization signals (Going et al., 2007; Liu et al., 2003; Sokka et al., 2010). Liu et al. (2003) showed that deletion of the BRCT7-8 and NLS region of TopBP1 induces cytoplasmic localization of the protein. Aberrant expression and intracellular localization of TopBP1 is observed immunohistochemically in breast cancer (Going et al., 2007).

5. TopBP1 as multifunctional protein

TopBP1 protein has been proposed as a transcriptional repressor of E2F1 and transcriptional co-activator with HPV16 E2 (Liu et al., 2004; Wright et al., 2006; Yoshida & Inoue, 2004). The E2F transcription factors E2F1 to E2F6 bind to E2F sites in promoters and regulate the expression of a large array of genes that encode proteins important for DNA replication and cell cycle progression. In response to growth signals, activated G1 cycline-dependent kinase phosphorylate retinoblastoma protein (Rb) and release E2F from Rb binding. This event is critical in controlling G1/S transition. Among the E2F family members, E2F1, E2F2 and E2F3 are transcriptional activators and are induced in response to growth stimulation, with peak accumulation at G1/S. Together, they are essential for cellular proliferation since a combined mutation of E2F1, E2F2 and E2F3 completely blocks cellular proliferation. In contrary, E2F4 and E2F5 act mainly as transcriptional repressors (Chen et al., 2009; Liu et al., 2003; Poznic, 2009). TopBP1 protein interacts with E2F1 through the sixth BRCT motif of TopBP1 and N terminus of E2F1 (Fig. 2) (Lelung et al., 2010; Liu et al., 2003). This interaction is induced by ATM-mediated phosphorylation of E2F1 at Ser31 during DNA damage. By this interaction, the transcriptional activity of E2F1 is repressed and E2F1 is recruited to DNA damage induced nuclear foci (Liu et al., 2003). Moreover, the interaction between TopBP1 protein and E2F1, as well as the repression of E2F1 activity, are specific for E2F1 but are not seen in E2F2, E2F3 and E2F4, suggesting that TopBP1 is E2F1 exclusive regulator (Liu et al., 2004). Liu et al. (2004) showed that E2F1 is also regulated by a novel Rb-independent mechanism, in which TopBP1 protein recruits Brg1/BRM (Brahma-related gene 1/Brahma protein), a central subunit of the SWI/SNF (SWItch/sucrose nonfermentable) chromatin modeling complex, to specifically inhibit E2F1 transcriptional activity. This regulation appeared to be critical for E2F1-dependent apoptosis control during S phase and DNA damage. On the other hand, TopBP1 is induced by E2F1 and interacts with E2F1 during G1/S transition. Thus, E2F1 and TopBP1 form a feedback regulation to prevent apoptosis during DNA replication (Liu et al., 2003).

Human papillomaviruses (HPVs) are causative agents in a number of human diseases the most common of which is cervical cancer. More than 95% of cervical carcinomas harbor HPV sequences and HPV16 is most frequently detected. The HPV16 E2 protein is a 43 kDa phosphoprotein that binds as a homodimer to 12 bp palindromic DNA sequences in the transcriptional control region of the viral genome. After binding, E2 can either upregulate or repress transcription from the adjacent promoter depending on cell type and protein levels and this regulation controls the expression of viral oncoproteins E6 and E7. The carboxyl-terminal portion of TopBP1 interacts with E2 and TopBP1 protein can enhance the ability of E2 to activate transcription and replication (Fig. 2) (Boner et al., 2002).

TopBP1 protein also interacts with SPBP (stromelysin-1 platelet-derived growth factor (PDGF) responsive element binding protein) and enhances the transcriptional activity of Ets1 on the *Myc* and *MMP-3* promoters *in vitro* and *in vivo* (Sjottem et al., 2007). This interaction is mediated by ePHD (extended plant homeodomain) domain of SPBP and the BRCT6 domain of TopBP1 (Sjottem et al., 2007). SPBP a 220 kDa ubiquitously expressed nuclear protein is shown to intensify or repress the transcriptional activity. Originally SPBP was identified as a protein involved in transcriptional activation of matrix metalloproteinase-3 (MMP3), stromelysin-1 promoter *via* the specific sequence element SPRE (stromelysin-1 PDGF responsive element) (Rekdal et al., 2000; Sonz et al., 1995). Later SPBP was found to act as a transcriptional coactivator since it enhanced the transcriptional activity of the positive cofactor and RING finger protein SNURF/RNF4 (small nuclear RING finger protein/RING finger protein 4) and of certain transcription factors, such as Sp1 (specificity protein 1), Ets (E-twenty-six specific), Pax6 (paired box gene 6) and Jun (Lyngso et al., 2000; Rekdal et al., 2000; Sjottem et al., 2007). On the other hand, SPBP appears to act as phosphoserine-specific repressor of estrogen receptor α (ERα) (Gburick et al., 2005; Sjottem et al., 2007).

In unstressed cells TopBP1 protein associates with Miz-1 (Myc interacting zinc finger protein 1). BRCT6 and BRCT7 of TopBP1 are required and largely sufficient to mediate the interaction with Miz-1 (Fig. 2) (Herold et al., 2002, 2008; Wenzel et al., 2003). This zinc finger protein that contains an amino-terminal POZ (poxvirus and zinc finger) was initially described as a protein that interacts with C terminus of Myc oncoprotein (Courapied et al., 2010; Herold et al., 2008). Miz-1 protein activates transcription of genes encoding the cell cycle inhibitors p15^{INK46} and p21^{Cip1}, leading to cell cycle arrest. Miz-1 can also repress transcription when it forms complexes with Myc and other transcription factors (Herold et al., 2002, 2008; Wenzel et al., 2003). In response to UV irradiation Miz-1 is released from an inhibitory complex formed with TopBP1 and binds to the start site of *p21^{Cip1}* promoter. Thus the dissociation of TopBP1 from Miz-1 may facilitate the induction of *p21^{Cip1}* (Herold et al., 2002, 2008; Wenzel et al., 2003). On the other hand, Miz-1 is required for the binding of TopBP1 to chromatin and to protect TopBP1 from proteasomal degradation. TopBP1 protein that is not bound to chromatin is ubiquitilated by HECTH9 (HUWE1) ligase. Expression of Myc leads to dissociation of TopBP1 from chromatin and reduces the amount of total TopBP1 (Herold et al., 2008). Furthermore, TopBP1 has been shown to be ubiquitilated by ubiquitin ligase EDD/hHYD (E3 identified by differential display/ human hyperplastic discs), another HECT (homologous to E6-AP C-terminus) domain E3 enzymes. The HECT E3 ubiquitin-protein ligases have been found from yeast to humans. They are characterized by the HECT domain. EDD/hHYD interacts with the minimal region of the amino acids 661 – 1080 including BRCT5 and BRCT6 of TopBP1 protein. TopBP1 was found to be usually ubiqitilated and degraded by the proteasome in intact cells. X-irradiation seems to abolish TopBP1 degradation and induce the stable complex formation of TopBP1 with other molecules in DNA double strand breaks (Honda et al., 2002; Scheffrer & Staub, 2007). Binding of the transcription factor Miz-1 and TopBP1 protein is also regulated by TopBP1 ADP-ribosylation (Table 1). ADP-ribosylation is one of the post-translational protein modifications. Polymers of ADP-ribose are formed from donor NAD$^+$ molecules and covalently attached to glutamic acid, aspartic acid or lysine residues of a target protein. The process is catalyzed by the poly(ADP-ribose) polymerase (PARP) family of proteins. The best known of these proteins is PARP1 which is implicated in transcription, chromatin remodeling, apoptosis and DNA repair (Sokka et al., 2010; Woodhouse & Dainov, 2008). TopBP1 and PARP-1 interact both *in vitro* and *in vivo*. The interaction depends on sixth BRCT domain of TopBP1 and on the fact that this domain is ADP-ribosylated by PARP-1. The post-translational ADP-ribosylation of TopBP1 by PARP1 may support the release of

Miz-1 from the complex with TopBP1 (Wollmann et al., 2007; Yamane et al., 1997; Yamane & Tsuruo, 1999).

Site(s)	Modification	Enzyme	Reference
Y in BRCT1-4 region	phosphorylation	c-Abl	Zeng et al., 2005
S214	phosphorylation	ATM/ATR	Matsuoka et al., 2007
S492	phosphorylation	ATM	Sokka et al., 2010
S405	phosphorylation	ATM/ATR	Matsuoka et al., 2007
S409	phosphorylation	ATM/ATR	Matsuoka et al., 2007
S554	phosphorylation	ATM	Sokka et al., 2010
K581	acetylation	N/D	Choudhary et al., 2010
S766	phosphorylation	ATM	Sokka et al., 2010
S805	phosphorylation	N/D	Beausoleil et al., 2006; Wang et al., 2008
T848	phosphorylation	N/D	Dephoure et al., 2008
S860	phosphorylation	N/D	Dephoure et al., 2008
S861	phosphorylation	N/D	Dephoure et al., 2008
S864	phosphorylation	N/D	Dephoure et al., 2008
S888	phosphorylation	N/D	Beausoleil et al., 2006; Dephoure et al., 2008; Wang et al., 2008
900-991 (BRCT6)	ADP-ribosylation	PARP-1	Wollmann et al., 2007; Yamane et al., 1997; Yamane & Tsuruo, 1999
T975	phosphorylation	ATM/ATR	Matsuoka et al., 2007
S1002	phosphorylation	N/D	Dephoure et al., 2008; Wang et al., 2008
S1051	phosphorylation	ATM/ATR	Matsuoka et al., 2007
T1062	phosphorylation	ATM	Sokka et al., 2010
T1086	phosphorylation	ATM/ATR	Matsuoka et al., 2007
S1138	phosphorylation	ATM	Yoo et al., 2007
S1159	phosphorylation	Akt	Liu et al., 2006

Table 1. Post-translation modifications of the human TopBP1 protein (N/D – not determined)

Apart from the mentioned above ADP-ribosylation, TopBP1 undergoes other post-translational modifications, such as acetylation and phosphorylation (Table 1). Lysine acetylation is a reversible post-translational modification, which neutralizes the positive charge of this amino acid changing protein function. Lysine acetylation preferentially targets large macromolecular complexes involved in diverse cellular processes, such as chromatin remodeling, cell cycle, splicing, nuclear transport and actin nucleation. Acetylation of TopBP1 protein occurs at position 581 but the exact role of this modification remains to be resolved (Choudhary et al., 2010).

TopBP1 is a phosphoprotein and is phosphorylated in response to DNA damage (Makiniemi et al., 2001; Yamane et al., 2003). After DNA damage, TopBP1 protein localizes at IR-induced nuclear foci and is phosphorylated by ATM kinase (Yamane et al.,

2003). Human TopBP1 is phosphorylated at several S/TQ sites, which are consensus sequences of PIKK (phosphatidylinositol 3-kinase-related kinase) targets (Hashimoto et al., 2006; Matsuoka et al., 2007). However, the phosphorylation of TopBP1 protein occurs mostly on serine and to a lesser extent on threonine (Makiniemi et al., 2001). TopBP1 protein is also phosphorylated by Akt *in vitro* and *in vivo* on Ser1159. Phosphorylation by Akt kinase induces oligomerization of TopBP1 through its seventh and eighth BRCT domains. The Akt-dependent oligomerization is crucial for TopBP1 to interact with E2F1 and repress its activity. TopBP1 phosphorylation by Akt is also required for interaction between TopBP1 and Miz-1 or HPV16 E2 and repression of Miz-1 transcriptional activity, suggesting a general role for TopBP1 oligomerization in the control of transcription factors (Liu et al., 2006a).

The other TopBP1 interacting proteins are PML (promyelocytic leukemia protein), TICRR (TopBP1-interacting, checkpoint and replication regulator) and p53. PML is a multifunctional protein that plays essential roles in cell growth regulation, apoptosis, transcriptional regulation and genome stability. PML tumor suppressor gene is consistently disrupted by t(15;17) in patients with acute promyelocytic leukemia. PML colocalizes and associates *in vivo* with TopBP1 in response to ionizing radiation and both proteins colocalize with Rad50, BRCA1, ATM, Rad9 and BLM. PML plays a role in regulation of TopBP1 functions by association and stabilization of the protein in response to IR-induced DNA damage (Xu et al., 2003). TICRR is required to prevent mitotic entry after treatment with ionizing radiation. TICRR deficiency is embryonic-lethal in the absence of exogenous DNA damage because it is essential for normal cell cycle progression. Specifically, the loss of TICRR impairs DNA replication and disrupts the S/M checkpoint, leading to premature mitotic entry and mitotic catastrophe. TICRR associates with TopBP1 *in vivo* and this interaction requires the two N-terminal BRCT domains. Sansam et al. (2010) showed that interaction between TICRR and TopBP1 is essential for replication preinitiation complex. TopBP1 is also involved in regulation of p53 activity. The regulation is mediated by an interaction between the seventh and eighth BRCT domains of TopBP1 and the DNA binding domain of p53, leading to inhibition of p53 promoter binding activity. Thus, TopBP1 may inhibit expression of several canonic p53 target genes including GADD45 (growth arrest and DNA damage protein 45), p21[Cip1], PUMA (p53 upregulated modulator of apoptosis), NOXA, BAX (Bcl-2 associated X protein), IGFBP3 (insulin-like growth factor binding protein 3). The repression of p53 proapoptotic genes such as NOXA, PUMA and BAX suggests that TopBP1 can inhibit p53-mediated apoptosis during DNA damage. Deregulation of this control may have pathological consequences (Liu et al., 2009).

TopBP1 also plays a role in DNA replication and S phase progression. Expression of TopBP1 mRNA and protein is induced concomitantly with S phase entry (Makiniemi et al., 2001). Neutralizing TopBP1 with a polyclonal antiserum raised against the sixth BRCT domain inhibits replicative DNA synthesis in HeLa cell nuclei *in vitro*. This may indicate that the sixth BRCT domain is critical for replication activity, possibly *via* interaction with crucial replication factors (Makiniemi et al., 2001; Schmidt et al., 2008). The physical interaction between TopBP1 and polymerase ε also implies an involvement of TopBP1 in replication (Makiniemi et al., 2001). The loading of Cdc45 (cell division cycle 45) onto chromatin is critical for loading various replication proteins, including DNA polymerase α, DNA polymerase ε, RPA (replication protein A) and PCNA. Human TopBP1 recruits Cdc45 to origins of DNA replication and is required for the formation of the initiation complex of replication in human cells. The first, second and sixth BRCT domains of TopBP1 interact with Cdc45 and this interaction inhibits transcriptional activity of TopBP1 (Schmidt et al., 2008; Sokka et al., 2010). Both proteins interact exclusively at the G1/S boundary of cell cycle. Only weak interaction could be found at the G2/M boundary (Schmidt et al., 2008).

6. TopBP1 and activation of ATR pathway

The major regulators of DNA damage response are the phosphoinositide 3-kinase (PI3K)-related proteins kinases (PIKKs), including ataxia telangiectasia mutated (ATM) and ATM and Rad3-related (ATR) (Cimprich & Cortez, 2008; Lopez-Contreras & Fernandez-Capetillo, 2010; Takeishi et al., 2010). Other members of this family comprise mTOR (mammalian target of rapamycin), which coordinates protein synthesis and cell growth, DNA-PKcs (DNA-dependent protein kinase catalytic subunit), which promotes DNA double strand break repair by nonhomologous end-joining and SMG1, which regulates nonsense-mediated mRNA decay (Cimprich & Cortez, 2008; Mordes et al., 2008). PIKKs are large proteins (2549 – 4128 amino acids) with common domain architecture. All of them contain a large region of repeated HEAT (Huntington, elongation factor 3, PR65/A, TOR) domains in the N terminus, highly conserved C-terminal kinase domain flanked by FAT (FRAP, ATM, TRAP /FKBP-rapamycin associated protein, ATM, *trp* RNA binding attenuation protein) and FATC (FAT C terminus) and PIKK regulatory domain (PRD) between the kinase and FATC domains (Cimprich & Cortez, 2008; Lopez-Contreras & Fernandez-Capetillo, 2010; Mordes et al., 2008). PRD, poorly conserved between family members but highly conserved within orthologous present in different organisms, is not essential for basal kinase activity but plays a regulatory role in at least ATM, ATR and mTOR (Cimprich & Cortez, 2008). PRD of ATM and mTOR is targeted for post-translational modifications that regulate their activity (Cimprich & Cortez, 2008; Mordes et al., 2008). The N-terminal regions of the kinases mediate interaction with the protein cofactors (Cimprich & Cortez, 2008). ATM and ATR are proteins of about 300 kDa, with a conserved C-terminal catalytic domain that preferably phosphorylates serine or threonine residues followed by a glutamine, i.e. SQ or TQ motif (Choi et al., 2009; Smits et al., 2010).

The initial step in ATR activation is recognition of DNA structures that are induced by the damaging agents (Smits et al., 2010). As mentioned, ATR responds to a wide variety of DNA damage that results in the generation of single-stranded DNA (ssDNA) (Takeishi et al., 2010). In eukaryotes, DNA damage-induced ssDNA is first detected by ssDNA binding protein complex RPA (Fig. 3) (Smits et al., 2010). RPA is a heterotrimeric protein complex composed of three subunits with a size of 70, 30 and 14 kDa, which are known as RPA70, RPA32 and RPA14 or alternatively RPA1, RPA2 and RPA3, respectively (Binz et al., 2004; Broderick et al., 2010; Fanning et al., 2006). RPA is identified to be a crucial component in DNA replication, DNA recombination and DNA repair (Ball et al., 2007; Broderick et al., 2010; Cimprich & Cortez, 2008). After binding to ssDNA either during DNA replication or in response to DNA damage, RPA is phosphorylated and this is thought to be an important event in DNA damage response (Binz et al., 2004; Broderick et al., 2010). Recent observations have shown the involvement of ATR in the RPA2 phosphorylation in response to stalled replication fork in S phase generated by genotoxic agents such as UV (Broderick et al., 2010; Olson et al., 2006).

RPA-coated ssDNA is necessary for ATR activation, but it is not sufficient, as at least several additional factors are also required. This kinase forms a stable complex with ATRIP (ATR-interacting protein) which regulates the localization of ATR to sites of replication stress and

DNA damage. Apart from ATRIP, activation of ATR requires the activator protein TopBP1 which plays dual role in the initiation of DNA replication and DNA damage response (Mordes & Cortez, 2008). ATRIP was identified as a 85 kDa an ATR binding partner that interacts directly with RPA to dock the ATR-ATRIP complex onto ssDNA (Ball et al., 2007; Choi et al., 2010; Kim et al., 2005; Warmerdam & Kanaar, 2010; Yan & Michael, 2009a,b). Independently, the Rad17-RFC complex is loaded onto these sites of damage in RPA-dependent manner (Burrows & Elledge, 2008; Lee & Dunphy, 2010). The Rad17-RFC complex consists of the Rad17 subunit and four additional subunits of replication factor C named from RFC2 to RFC5. During normal replication the RFC complex, containing RFC1 instead of Rad17, plays a role in the loading of PCNA onto DNA. PCNA is a processivity factor for DNA polymerases. Both the Rad17 and RFC complexes require RPA for their loading onto DNA (Majka et al., 2006; Medhurst et al., 2008; Warmerdam & Kanaar, 2010). However, Rad17-RFC requires 5′ dsDNA-ssDNA junctions, rather than the 3′ ended junctions preferred by PCNA. These types of structures are specifically created by the resection of DSBs, stalled replication forks and UV-induced ssDNA gaps. The Rad17-RFC protein complex facilitates the loading of the Rad9-Rad1-Hus1 (9-1-1) sliding clamp onto the DNA (Choi et al., 2010; Lopez-Contreras & Fernandez-Capetillo, 2010; Van et al., 2010; Warmerdam & Kanaar, 2010; Yan & Michael, 2009a). The necessity of the 9-1-1 complex in the ATR branch was explained by showing that Rad9 recruits the ATR-activator TopBP1 protein near sites of DNA damage, which was consistent with earlier reports showing interaction between Rad9 and TopBP1 protein (Greer et al., 2003; Makiniemi et al., 2001; Smits et al., 2010). The amino-terminal region of TopBP1 protein comprising BRCT1 and BRCT2 binds the C terminus of Rad9. More precisely, the interaction between Rad9 and TopBP1 depends on Ser373 phosphorylation in the C-terminal tail of Rad9 (Delacroix et al., 2007; Kumagai et al., 2006; Lee et al., 2007; Rappas et al., 2011; Smits et al., 2010; Takeishi et al., 2010). Then, TopBP1 protein binds ATR through its ATR activation domain (AAD), located between the sixth and seventh BRCT repeats, in an ATRIP-dependent manner and this interaction is required for ATR stimulation (Kumagai et al, 2006; Mordes et al., 2008; Smits et al., 2010; Takeishi et al., 2010). ATRIP contains a conserved TopBP1 interacting region, required for the association of TopBP1 and ATR and the subsequent TopBP1-mediated triggering of ATR activity (Mordes et al., 2008; Smits et al., 2010).

ATR-mediated activation of Chk1 in response to genotoxic stress requires another protein that binds independently of ATR or Rad17/9-1-1 named Claspin (Kumagai et al., 2004; Liu et al., 2006b; Scorah & McGowan, 2009; Smits et al., 2010). Claspin is proposed to function as adaptor molecule bringing ATR and Chk1 together (Kumagai & Dunphy, 2000; Smits et al., 2010). The Claspin-Chk1 interaction depends on ATR-mediated phosphorylation of Claspin and is required for Chk1 phosphorylation by ATR. Subsequent studies identified repeated phosphopeptide motifs in Claspin, which are required for association with phosphate binding sites in the N-terminal kinase domain of Chk1, resulting in full activation of Chk1 (Smits et al., 2010). In response to DNA damage or replication stress activated ATR and its effectors such as Chk1 ultimately slow origin firing and induce cell cycle arrest, as well as stabilize and restart stalled replication forks (Cimprich & Cortez, 2008).

The mechanism by which TopBP1 binding activates ATR is poorly defined. The primary binding site for the activation domain of TopBP1 on the ATR complex is within ATRIP and mutations in this region of ATRIP block activation (Cimprich & Cortez, 2008; Mordes et al., 2008). In addition, activation involves amino acids that are located between the ATR kinase domain and the FATC domain, of PIKK regulatory domain - PRD of ATR. Mutations in this region have no effect on the basal activity of ATR, although they prevent ATR activation by TopBP1 protein and cause checkpoint defects and mimic a complete deletion of ATR in human somatic cells (Cimprich & Cortez, 2008; Mordes et al., 2008). Thus, efficient activation of ATR by TopBP1 protein may be required to achieve sufficient signal amplification for the proper execution of cellular response to DNA damage (Sokka et al., 2010).

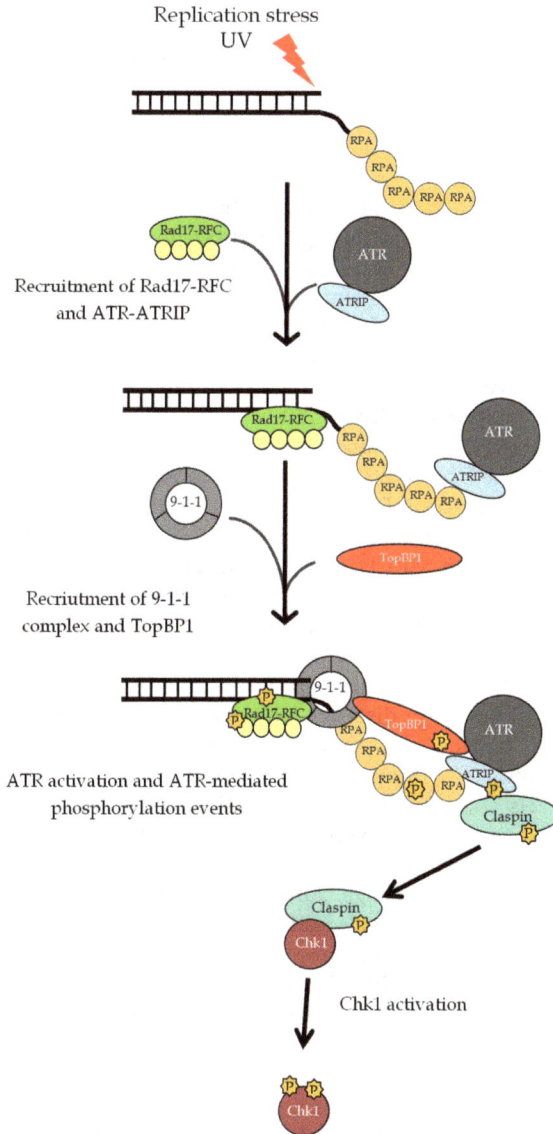

Fig. 3. Role of TopBP1 in activation of ATR pathway in response to replication stress and UV-induced DNA damage

7. Role of TopBP1 in DSB repair

TopBP1 protein also plays a direct and essential role in the pathway that connects ATM to ATR, specifically in response to the occurrence of DSBs in a genome (Yoo et al., 2007). DNA double strand breaks are among the most deleterious DNA lesions that threaten genomic integrity. DSBs are generated not only by exogenous DNA-damaging agents, but also by normal cellular processes, such as V(D)J recombination, meiosis and DNA replication. Furthermore, increased amounts of DSBs are induced by oncogenic stresses during the early stage at tumorigenesis (O'Driscoll & Jeggo, 2005; Shiotani & Zou, 2009; Williams et al., 2007). Two major forms of DSB repair are found within eukaryotic cells: nonhomologous end-joining (NHEJ) and homologous recombination (HR). NHEJ requires several complementary bases for repair and is the predominant form of DSB repair in G0/G1 cells. During NHEJ DNA ends are minimally processed to reveal short stretches of complementarity on either side of the break. NHEJ pathway is inherently mutagenic. In contrast, HR pathway predominates during S and G2 phases and repairs DNA with high fidelity by employing homologous chromosomal or sister chromatid DNA as a template to synthesize new error-free DNA (Williams et al., 2007). The main PIKK that responds to DSBs is ATM, the protein that is defective in the hereditary disorder ataxia telangiectasia (O'Driscoll & Jeggo, 2005). DSBs are recognized by the MRE11-RAD50-NBS1 complex, which promotes the activation of ATM and the preparation of DNA ends for homologous recombination (Fig. 4) (Ciccia & Elledge, 2010; O'Driscoll & Jeggo, 2005; Williams et al., 2007). RAD50 contains ATPase domains that interact with MRE11 (meiotic recombination 11) and associates with the DNA ends. MRE11 has endonuclease and exonuclease activities important for the initial step of DNA end resection that is essential for homologous recombination (Ciccia & Elledge, 2010; Williams et al., 2007). The third subunit of the MRN complex, NBS1, interacts with MRE11 and contains additional protein-protein interaction domains important for MRN function in DNA damage response. NBS1 associates with ATM *via* its C-terminal region, which promotes the recruitment of ATM to DSBs, where ATM is activated by the MRN complex (Ciccia & Elledge, 2010; Jazayeri et al., 2008; Kanaar & Wyman, 2008; Rupnik et al., 2010). Mutations in the human *NBS1* gene result in Nijmegen breakage syndrome (NBS), a rare disorder with abnormal responses to ionizing radiation that resemble those in patients with ataxia telangiectasia (Horton et al., 2011). DNA end resection is regulated by ATM through CtIP (C-terminal binding protein/CtBP interacting protein), which interacts with BRCA1 and MRN (Ciccia & Elledge, 2010). In addition, Exo1 (exonuclease 1), which is involved in the processive stage of DSB resection together with BLM following the initial resection carried out by CtIP, is also stimulated by ATM phosphorylation (Bolderson et al., 2010; Ciccia & Elledge, 2010; Shiotani et al., 2009; Smits et al., 2010). DSB resection and formation of 3' ssDNA ends leads to RPA accumulation. RPA-ssDNA complexes play a critical role in activation of ATR pathway, as described in detail above.

TopBP1 protein appeared to be involved in ATR-dependent DSB repair. In human cells, DSB induces formation of distinct TopBP1 foci that colocalize with BRCA1, PCNA, NBS1, 53BP1 and γH2AX (Germann et al., 2011). *In vitro* studies showed that in nuclear foci, TopBP1 protein physically associates with NBS1. Several of TopBP1 foci increased and colocalized with NBS1 after ionizing radiation, whereas these nuclear foci were not observed in Nijmegen breakage syndrome cells. The association between TopBP1 and NBS1 involves the first pair of BRCT repeats in TopBP1. In addition the two tandem BRCT repeats of NBS1 are required for this binding. Functional studies with mutated forms of TopBP1 and NBS1

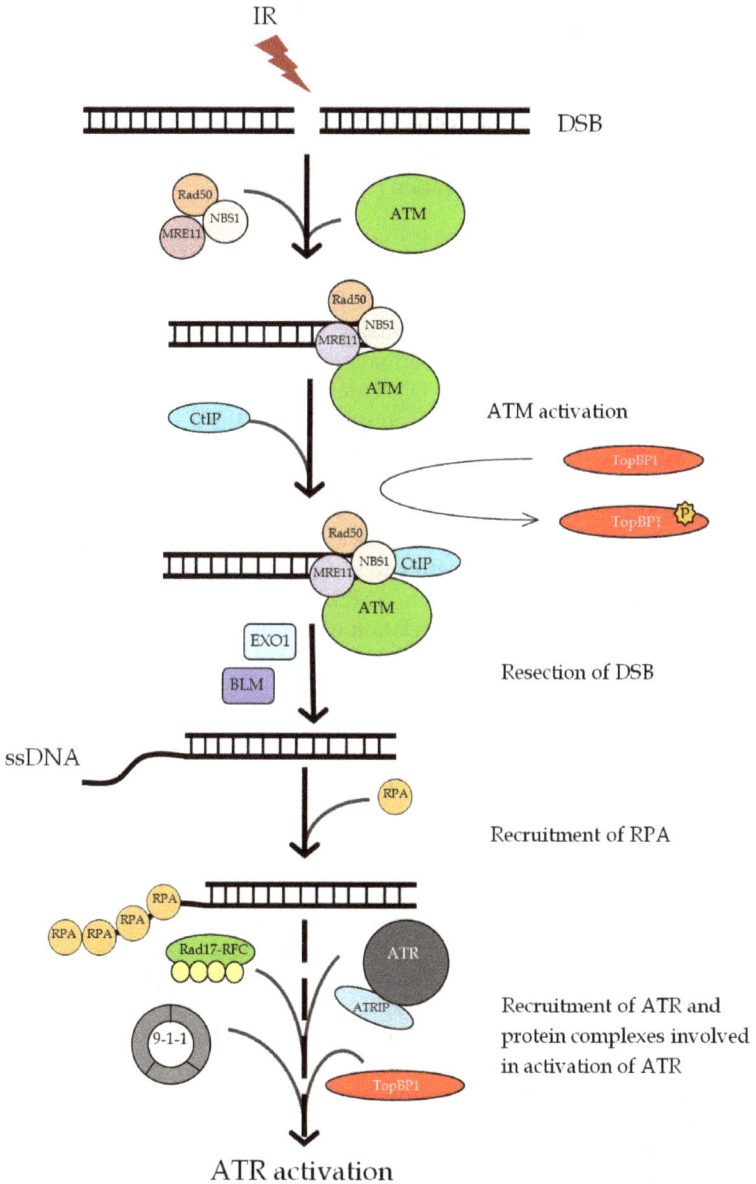

Fig. 4. Role of TopBP1 in ATR activation in response to DNA double strand breaks

suggest that the BRCT-dependent association of these proteins is critical for normal checkpoint response to DSB (Morishima et al., 2007; Yoo et al., 2009). The MRN complex is a crucial mediator in the process whereby ATM promotes the TopBP1-dependent activation of ATR-ATRIP in response to DSBs (Morishima et al., 2007; Yoo et al., 2009). In *Xenopus* egg extracts, ATM associates with TopBP1 protein and phosphorylates it on Ser1131. This phosphorylation enhances the capacity for TopBP1 protein to activate ATR-ATRIP (Yoo et al., 2009). Yoo et al. (2009) showed that ATM can no longer interact with TopBP1 protein in NBS1-depleted egg extracts. Thus, the MRN complex helps to bridge ATM and TopBP1 together. ATM contributes to the activation of ATR through two collaborating mechanisms. First, ATM helps to create appropriate DNA structures that trigger activation of ATR. Second, ATM strongly stimulates the function of TopBP1 protein *via* its phosphorylation that directly carries out the ATR activation (Yoo et al., 2007).

8. Conclusion

DNA is continuously exposed to a range of damaging agents, including reactive cellular metabolites, environmental chemicals, ionizing radiation and UV light. To prevent loss or incorrect transmission of genetic information and development of abnormalities and tumorigenesis all cells have evolved DNA damage response pathways to maintain their genome integrity. The DNA damage response involves the sensing of DNA damage signal to a network of cellular pathways, including cell cycle checkpoint, DNA repair and apoptosis. TopBP1 protein was first identified as an interacting partner for topoisomerase IIβ. This protein shares structural and functional similarities with BRCA1 and plays a critical role in the DNA damage response and checkpoint control. TopBP1 is essential for ATR activation in response to replication stress and UV-induced damage and also plays a direct role in the pathway that connects ATM to ATR in response to DSBs. The biological functions of TopBP1 protein, as well as its close relation with BRCA1 suggest a crucial role of TopBP1 in the maintenance of genome integrity and cell cycle regulation.

9. References

Araki H., Leem S.H., Phongdara A. & Sugino A. (1995). Dpb11, which interacts with DNA polymerase II (ε) in *Sacharomyces cerevisiae*, has a dual role in S-phase progression and at a cell cycle checkpoint. *Proceedings of the National Academy of Sciences of the United States of America*, Vol. 92, No. 25, (December 1995), pp. 11791-11795, ISSN: 0027-8424

Ball H.L., Ehrhardt M.R., Mordes D.A., Glick G.G., Chazin W.J. & Cortez D. (2007). Function of a conserved checkpoint recruitment domain in ATRIP proteins. *Molecular and Cellular Biology*, Vol. 27, No. 9, (May 2007), pp. 3367-3377, ISSN: 0270-7306

Bang S.W., Ko M.J., Kang S., Kim G.S., Kang D., Lee J.H. & Hwang D.S. (2011). Human TopBP1 localization to the mitotic centrosome mediates mitotic progression. *Experimental Cell Research*, Vol. 317, No. 7, (April 2011), pp. 994-1004, ISSN: 0014-4827

Beausoleil S.A., Villen J., Gerber S.A., Rush J. & Gygi S.P. (2006). A probability-based approach for high-throughput protein phosphorylation analysis and site localization. *Nature Biotechnology*, Vol. 24, No. 10, (October 2006), pp. 1285-1292, ISSN: 1087-0156

Binz S.K., Sheehan A.M. & Wold M.S. (2004). Replication protein A phosphorylation and the cellular response to DNA damage. *DNA Repair*, Vol. 3, No. 8-9, (August-September 2004), pp. 1015-1024, ISSN: 1568-7856

Bolderson E., Tomimatsu N., Richard D.J., Boucher D., Kumar R., Pandita T.K., Burma S. & Khanna K.K. (2009). Phosphorylation of Exo1 modulates homologous recombination repair of DNA double-strand breaks. *Nucleic Acids Research*, Vol. 38, No. 6, (December 2009), pp. 1821-1831, ISSN: 0305-1048

Boner W., Taylor E.R., Tsirimonaki E., Yamane K., Campo M.S. & Morgan I.M. (2002). A functional interaction between the human papillomavirus 16 transcription/replication factor E2 and the DNA damage response protein TopBP1. *The Journal of Biological Chemistry*, Vol. 277, No. 25, (June 2002), pp. 22297-22303, ISSN: 0021-9258

Broderic S., Rehmet K., Concannon C. & Nasheuer H.P. (2010), In: *Genome stability and human diseases*, Nasheuer H.P., pp. 143-164, Springer, ISBN: 978-90-481-3470-0, Dordrecht, Holland

Burrows A.E. & Elledge S.J. (2008). How ATR turns on: TopBP1 goes on ATRIP with ATR. *Genes and Development*, Vol. 22, No. 11, (June 2008), pp. 1416-1421, ISSN 0890-9369

Chen H.Z., Tsai S.Y. & Leone G. (2009). Emerging roles of E2Fs in cancer: an exit from cell cycle control. *Nature Reviews Cancer*, Vol. 9, No. 11, (November 2009), pp. 785-797, ISSN: 1474-175X

Choi J.H., Lindsey-Boltz L.A., Kemp M., Mason A.C., Wold M.S. & Sancar A. (2010). Reconsititution of RPA-covered single-stranded DNA activated ATR-Chk1 signaling. *Proceedings of the National Academy of Science of the United States of America*, Vol. 107, No. 31, (August 2010), pp. 13660-13665, ISSN: 0027-8424

Choi J.H., Sancar A., Lindsey-Boltz L.A. (2009). The human ATR-mediated DNA damage checkpoint in a reconstituted system. *Methods*, Vol. 48, No. 1, (May 2009), pp. 3-7, ISSN: 1046-2023

Choudhary C., Kumar C., Gnad F., Nielsen M.L., Rehman M., Walther T.C., Olsen J.V. & Mann M. (2009). Lysine acetylation targets protein complexes and co-regulates major cellular functions. *Science*, Vol. 325. No. 5942, (August 2009), pp. 834-840, ISSN: 0036-8075

Ciccia A. & Elledge S.J. (2010). The DNA damage response: making it safe to play with knives. *Molecular Cell*, Vol. 40, No. 2, (October 2010), pp. 179-204, ISSN: 1097-2765

Cimprich K.A. & Cortez D. (2008). ATR: an essential regulator of genome integrity. *Nature Reviews Molecular Cell Biology*, Vol. 9, No. 8, (August 2008), pp. 616-627, ISSN: 1471-0072

Corre I., Niaudet C. & Paris F. (2010). Plasma membrane signaling induced by ionizing radiation. *Mutation Research*, Vol. 704, No. 1-3, (April-June 2010), pp. 61-67, ISSN: 0027-5107

Cortez D., Gutunku S., Qin J. & Elledge S.J. (2001). ATR and ATRIP: partners in checkpoint signaling. *Science*, Vol. 294, No. 5547, (November 2001), pp. 1713-1716, ISSN: 0193-4511

Courapied S., Cherier J., Vigneron A., Troadec M.B., Giraud S., Valo I., Prigent C., Gamelin E., Coqueret O. & Barre B. (2010). Regulation of the Aurora-A gene following topoisomerase I inhibition: implication of the Myc transcription factor. *Molecular Cancer*, Vol. 9, No. 1, (August 2010), pp. 205, ISSN: 1476-4598

Dai Y. & Grant S. (2010). New insights into Checkpoint kinase 1 (Chk1) in the DNA damage response (DDR) signaling network: rationale for employing Chk1 inhibitors in cancer therapeutics. *Clinical Cancer Research*, Vol. 16, No. 2, (January 2010), pp. 376-383, ISSN: 1078-0432

Danielsen J.M.R., Larsen D.H., Schou K.B., Freire R., Falck J., Bartek J. & Lukas J. (2009). HCLK2 is required for activity of the DNA damage response kinase ATR. *The Journal of Biological Chemistry*, Vol. 284, No. 7, (February 2009), pp. 4140-4147, ISSN: 0021-9258

Delacroix S., Wagner J.M., Kobayashi M., Yamamoto K. & Karnitz L.M. (2007). The Rad9-Hus1-Rad1 (9-1-1) clamp activates checkpoint signaling *via* TopBP1. *Genes and Development*, Vol. 21, No. 12, (June 2007), pp. 1472-1477, ISSN: 0890-9369

Dephoure N., Zhou C., Villen J., Beausoleil S.A., Bakalarski C.E., Elledge S.J. & Gygi S.P.(2008). A quantitative atlas of mitotic phosphorylation. *Proceedings of the National Academy of Science of the United States of America*, Vol. 105, No. 31, (August 2008), pp. 10762-10767, ISSN: 0027-8424

Dijkstra K.K., Blanchetot C. & Boonstra J. (2009). Evasion of G1 checkpoint in cancer, In: *Checkpoint Controls and Targets in Cancer Therapy*, Siddik Z.H., pp. 3-26, Humana Press, ISBN: 978-1-60761-177-6, Totowa, New Jersey, USA

Essers J., Vermeulen W. & Houtsmeller A.B. (2006). DNA damage repair: anytime, anywhere? *Current Opinion in Cell Biology*, Vol. 18, No. 3, (June 2006), pp. 240-246, ISSN: 0955-0674

Fannig E., Klimovich V. & Nager A.R. (2006). A dynamic model for replication protein A (RPA) function in DNA processing pathways. *Nucleic Acids Research*, Vol. 34, No. 15, (August 2006), pp. 4126-4137, ISSN: 0305-1048

Garcia V., Furuya K. & Carr A.M. (2005). Identification and functional analysis of TopBP1 and its homologs. *DNA Repair*, Vol. 4, No. 11, (November 2005), pp. 1227-1239, ISSN: 1568-7856

Gburcik V., Bot N., Maggiolini M. & Picard D. (2005). SPBP is a phosphoserine-specific repressor of estrogen receptor α. *Molecular and Cellular Biology*, Vol. 25, No. 9, (May 2005), pp. 3421-3430, ISSN: 0270-7306

Germann S.M., Oestergaard V.H., Haas C., Salis P., Motegi A. & Lisby M. (2011). Dpb11/TopBP1 plays distinct roles in DNA replication, checkpoint response and homologous recombination. *DNA Repair*, Vol. 10, No. 2, (February 2011), pp. 210-224, ISSN: 1568-7856

Glover J.N.M. (2006). Insights into the molecular basis of human hereditary breast cancer from studies of the BRCA1 BRCT domain. *Familial Cancer*, Vol. 5, No. 1, (March 2006), pp. 89-93, ISSN: 1389-9600

Glover J.N.M., Williams R.S. & Lee M.S. (2004). Interactions between BRCT repeats and phosphoproteins: tangled up in two. *Trends in Biochemical Sciences*, Vol. 29, No. 11, (November 2004), pp. 579-585, ISSN: 0968-0004

Going J.J., Nixon C., Dornan E.S., Boner W., Donaldson M.M. & Morgan I.M. (2007). Aberrant expression of TopBP1 in breast cancer. *Histopathology*, Vol. 50, No. 4, (March 2007), pp. 418-424, ISSN: 0309-0167

Greer D.A., Besley B.D.A., Kennedy K.B. & Davey S. (2003). hRad9 rapidly binds DNA containing double-strand breaks and is required for damage-dependent topoisomerase IIβ binding protein 1 focus formation. *Cancer Research*, Vol. 63, No. 16, (August 2003), pp. 4829-4835, ISSN: 0008-5472

Hashimoto Y., Tsujimura T., Sugino A. & Takisawa H. (2006). The phosphorylated C-terminal domain of *Xenopus* Cut5 directly mediates ATR-dependent activation of Chk1. *Genes to Cells*, Vol. 11, No. 9, (September 2006), pp. 993-1007, ISSN: 1356-9597

Herold S., Hock A., Herkert B., Berns K., Mullenders J., Beijersbergen R., Bernards R. & Eilers M. (2008). Miz1 and HectH9 regulate the stability of the checkpoint protein, TopBP1. *EMBO Journal*, Vol. 27, No. 21, (November 2008), pp. 2851-2861, ISSN: 0261-4189

Herold S., Wenzel M., Beuger V., Frohme C., Beul D., Hillukkala T., Syvaoja J., Saluz H.P., Haenel F. & Eilers M. (2002). Negative regulation of the mammalian UV response by Myc through association with Miz-1. *Molecular Cell*, Vol. 10, No. 3, (September 2002), pp. 509-521, ISSN: 1097-2765

Hiom K. (2010). Coping with DNA double strand breaks. *DNA Repair*, Vol. 9, No. 12, (December 2010), pp. 1256-1263, ISSN: 1568-7856

Honda Y., Tojo M., Matsuzaki K., Anan T., Matsumoto M., Ando M., Saya H. & Nakao M. (2002). Cooperation of HECT-domain ubiquitin ligase hHYD and DNA topoisomerase II-binding protein for DNA damage response. *The Journal of Biological Chemistry*, Vol. 277, No. 5, (February 2002), pp. 3599-3605, ISSN: 0021-9258

Horton J.K., Stefanick D.F., Zeng J.Y., Carrozza M.J. & Wilson S.H. (2011). Requirement for NBS1 in the S phase checkpoint response to DNA methylation combined with PARP inhibition. *DNA Repair*, Vol. 10, No. 2, (February 2011), pp. 225-234, ISSN: 1568-7856

Houtgraaf J.H., Versmissen J. & van der Giessen W.J. (2006). A concise review of DNA damage checkpoints and repair in mammalian cells. *Cardiovascular Revascularization Medicine*, Vol. 7, No. 3, (July-September 2006), pp. 165-172, ISSN: 1553-8389

Huo Y., Bai L., Xu M. & Jiang T. (2010). Crystal structure of the N-terminal region of human Topoisomerase IIβ binding protein 1. *Biochemical and Biophysical Research Communications*, Vol. 401, No. 3, (October 2010), pp. 401-405, ISSN: 0006-291X

Hurley P.J. & Bunz F. (2009). Distinct pathways involved in S-phase checkpoint control, In: *Checkpoint Controls and Targets in Cancer Therapy*, Siddik Z.H., pp. 27-36, Humana Press, ISBN: 978-1-60761-177-6, Totowa, New Jersey, USA

Jazayeri A., Balestrini A., Garner E., Haber J.E. & Costanzo V. (2008). Mre11-Rad50-Nbs1-dependent processing of DNA breaks generates oligonucleotides that stimulate ATM activity. *EMBO Journal*, Vol. 27, No. 14, (July 2008), pp. 1953-1962, ISSN: 0261-4189

Jeon Y., Ko E., Lee K.Y., Ko M.J, Park S.Y., Kang J., Jeon C.H., Lee H. & Hwang D.S. (2011). TopBP1 deficiency causes an early embryonic lethality and induces cellular senescence in primary cells. *The Journal of Biological Chemistry*, Vol. 286, No. 7, (February 2011), pp. 5414-5422, ISSN: 0021-9258

Jungmichel S. & Stucki M. (2010). MDC1: The art of keeping things in focus. *Chromosoma*, Vol. 119, No. 4, (August 2010), pp. 337-349, ISSN: 0009-5915

Kanaar R. & Wyman C. (2008). DNA repair by the MRN complex: break it to make it. *Cell*, Vol. 135, No. 1, (October 2008), pp. 14-16, ISSN: 0092-8674

Karppinen S.M., Erkko H., Reini K., Pospiech H., Heikkinen K., Rapakko K., Syvaoja J.E. & Winqvist R. (2006). Identification of a common polymorphism in the TopBP1 gene associated with hereditary susceptibility to breast and ovarian cancer. *European Journal of Cancer*, Vol. 42, No. 15, (October 2006), pp. 2647-2652, ISSN: 0959-8049

Kerzendorfer C. & O'Driscoll M. (2009). Human DNA damage response and repair deficiency syndromes: Linking genomic instability and cell cycle checkpoint proficiency. *DNA Repair*, Vol. 8, No. 9, (September 2009), pp. 1139-1152, ISSN: 1568-7856

Kim J.E., McAvoy S.A., Smith D.I. & Chen J. (2005). Human TopBP1 ensures genome integrity during normal S phase. *Molecular and Cellular Biology*, Vol. 25, No. 24, (December 2005), pp. 10907-10915, ISSN: 0270-7306

Kumagai A., Kim S.M. & Dunphy W.G. (2004). Claspin and the activated from of ATR-ATRIP collaborate in the activation of Chk1. *The Journal of Biological Chemistry*, Vol. 279, No. 48, (November 2004), pp. 49599-49608, ISSN: 0021-9258

Kumagai A., Lee J., Yoo H.Y. & Dunphy W.G. (2006). TopBP1 activates the ATR-ATRIP complex. *Cell*, Vol. 124, No. 5, (March 2006), pp. 943-955, ISSN: 0092-8674

Lee J. & Dunphy W.G. (2010). Rad17 plays a central role in establishment of the interaction between TopBP1 and the Rad9-Hus1-Rad1 complex at stalled replication forks. *Molecular Biology of the Cell*, Vol. 21, No. 6, (March 2010), pp. 926-935, ISSN: 1059-1524

Lee J., Kumagai A. & Dunphy W.G. (2007). The Rad9-Hus1-Rad1 checkpoint clamp regulates interaction of TopBP1 with ATR. *The Journal of Biological Chemistry*, Vol. 282, No. 38, (September 2007), pp. 28036-28044, ISSN: 0021-9258

Lelung C.C.Y., Kellogg E., Kuhnert A., Hanel F., Baker D. & Glover J.N.M. (2010). Insights from the crystal structure of the sixth BRCT domain of topoisomerase IIβ binding protein 1. *Protein Science*, Vo. 19, No. 1, (January 2010), pp. 162-167, ISSN: 0961-8368

Liu K., Bellam N., Lin H.Y., Wang B., Stockard C.R., Grizzle W.E. & Lin W.C. (2009). Regulation of p53 by TopBP1: a potential mechanism for p53 inactivation in cancer, *Molecular and Cellular Biology*, Vol. 29, No. 10, (May 2009), pp. 2673-2693, ISSN: 0270-7306

Liu K., Paik J.C., Wang B., Lin F.T. & Lin W.C. (2006a). Regulation of TopBP1 oligomerization by Akt/PKB for cell survival. *EMBO Journal*, Vol. 25, No. 20, (October 2006), pp. 4795-4807, ISSN: 0261-4189

Liu S., Bekker-Jensen S., Mailand N., Lukas C., Bartek J. & Lukas J. (2006b). Claspin operates downstream of TopBP1 to direct ATR signaling towards Chk1 activation. *Molecular and Cellular Biology*, Vol. 26, No. 16, (August 2006), pp. 6056-6064, ISSN: 0270-7306

Liu K., Luo Y., Lin F.T. & Lin W.C. (2004). TopBP1 recruits Brg1/Brm to repress E2F1-induced apoptosis, a novel pRb-independent and E2F1-specific control for cell survival. *Genes and Development*, Vol. 18, No. 6, (March 2004), pp. 673-686, ISSN: 0890-9369

Liu K., Lin F.T., Ruppert J.M. & Lin W.C. (2003). Regulation of E2F1 by BRCT domain-containing protein TopBP1. *Molecular and Cellular Biology*, Vol. 23, No. 9, (May 2003), pp. 3287-3304, ISSN: 0270-7306

Lopez-Contreras A.J. & Fernandez-Capetillo O. (2010). The ATR barrier to replication-born DNA damage. *DNA Repair*, Vol. 9, No. 12, (December 2010), pp. 1249-1255, ISSN: 1568-7856

Lyngso C., Bouteiller G., Damgaard C.K., Ryom D., Sanchez-Munoz S., Norby P.L., Bonven B.J. & Jorgensen P. (2000). Interaction between the transcription factor SPBP and the positive cofactor RNF4. *The Journal of Biological Chemistry*, Vol. 275, No. 34, (August 2000), pp. 26144-26149, ISSN: 0021-9258

Majka J., Binz S.K., Wold M.S. & Burgers P.M. (2006). Replication protein A directs loading of the DNA damage checkpoint clamp to 5'-DNA junctions. *The Journal of Biological Chemistry*, Vol. 281, No. 38, (September 2006), pp. 27855-27861, ISSN: 0021-9258

Makiniemi M., Hillukkala T., Tuusa J., Reini K., Vaara M., Huang D., Pospiech H., Majuri I., Westerling T., Makela T.P. & Syvaoja J.E. (2001). BRCT domain-containing protein TopBP1 functions in DNA replication and damage response. *The Journal of Biological Chemistry*, Vol. 276, No. 32, (August 2001), pp. 30399-30406, ISSN: 0021-9258

Matsuoka S., Ballif B.A., Smogorzewska A., McDonald III E.R., Hurov K.E., Luo J., Bakalarski C.E., Zhao Z., Solimini N., Lerenthal Y., Shiloh Y., Gygi S.P. & Elledge

S.J. (2007). ATM and ATR substrate analysis reveals extensive protein networks responsive to DNA damage. *Science*, Vo. 316, No. 5828, (May 2007), pp. 1160-1166, ISSN: 0193-4511

Medhurst A.L., Warmerdam D.O., Akerman I., Verwayen E.H., Kanaar R., Smits V.A.J. & Lakin N.D. (2008). ATR and Rad17 collaborate in modulating Rad9 localisation at sites of DNA damage. *Journal of Cell Science*, Vol. 121, No. 23, (December 2008), pp. 3933-3940, ISSN: 0021-9533

Mordes D.A. & Cortez D. (2008). Activation of ATR and related PIKKs. *Cell Cycle*, Vol. 7, No. 18, (September 2008), pp. 2809-2812, ISSN: 1551-4005

Mordes D.A., Glick G.G., Zhao R. & Cortez D. (2008). TopBP1 activates ATR through ATRIP and a PIKK regulatory domain. *Genes and Development*, Vol. 22, No. 11, (June 2008), pp. 1478-1489, ISSN: 0890-9369

Morishima K., Sakamoto S., Kobayashi J., Izumi H., Suda T., Matsumoto Y., Tauchi H., Ide H., Komatsu K. & Matsuura S. (2007). TopBP1 associates with NBS1 and is involved in homologous recombination repair. *Biochemical and Biophysical Research Communications*, Vol. 362, No. 4, (November 2007), pp. 872-879, ISSN: 0006-291X

Morris J.S., Nixon C., King O.J.A., Morgan I.M. & Philbey A.W. (2009). Expression of TopBP1 in canine mammary neoplasia in relation to histological type, Ki67, ERα and p53. *The Veterinary Journal*, Vol. 179, No. 3, (March 2009), pp. 422-429, ISSN: 1090-0233

Muniandy P., Liu J., Majumdar A., Liu S. & Seidman M.M. (2010). DNA interstrand cross link repair in mammalian cells: step by step. *Critical Reviews in Biochemistry and Molecular Biology*, Vol. 45, No. 1, (February 2010), pp. 23-49, ISSN: 1040-9238

Nakanishi M. (2009). Chromatin modifications and orchestration of checkpoint response in cancer, In: *Checkpoint Controls and Targets in Cancer Therapy*, Siddik Z.H., pp. 83-94, Humana Press, ISBN: 978-1-60761-177-6, Totowa, New Jersey, USA

Nakanishi M., Niida H., Murakami H. & Shimada M. (2009). DNA damage responses in skin biology – Implication in tumor prevention and aging acceleration. *Journal of Dermatological Science*, Vol. 56, No. 2, (November 2009), pp. 76-81, ISSN: 0923-1811

Niida H. & Nakanishi M. (2006). DNA damage checkpoints in mammals. *Mutagenesis*, Vol. 21, No. 1, (January 2006), pp. 3-9, ISSN: 0267-8357

O'Driscoll M. & Jeggo P.A. (2006). The role of double-strand break repair – insights from human genetics. *Nature Reviews Genetics*, Vol. 7, No. 1, (January 2006), pp. 45-54, ISSN: 1471-0056

Ogiwara H., Ui A., Onoda F., Tada S., Enomoto T. & Seki M. (2006). Dpb11, the pudding yeast homolog of TopBP1, functions with the checkpoint clamp in recombination repair. *Nucleic Acids Research*, Vol. 34, No. 11, (July 2006), pp. 3389-3398, ISSN: 0305-1048

Olson E., Nievera C.J., Klimovivh V., Fanning E. & Wu X. (2006). RPA2 is a direct downstream target for ATR to regulate the S-phase checkpoint. *The Journal of Biological Chemistry*, Vol. 281, No. 51, (December 2006), pp. 39517-39533, ISSN: 0021-9258

Pallis A.G. & Karamouzis M.V. (2010). DNA repair pathways and their implication in cancer treatment. *Cancer and Metastasis Reviews*, Vol. 29, No. 4, (September 2010), pp. 677-685, ISSN: 0167-7659

Parrilla-Castellar E.R. & Karnitz L.M. (2003). Cut5 is required for the binding of Atr and DNA polymerase α to genotoxin-damaged chromatin. *The Journal of Biological Chemistry*, Vol. 278, No. 46, (November 2003), pp. 45507-45511, ISSN: 0021-9258

Poehlmann A. & Roessner A. (2010). Importance of DNA damage checkpoints in the pathogenesis of human cancers. *Pathology Research and Practice*, Vol. 206, No. 9, (September 2010), pp. 591-601, ISSN: 0344-0338

Pommier Y., Leo E., Zhang H.L. & Marchand C. (2010). DNA topoisomerases and their poisoning by anticancer and antibacterial drugs. *Chemistry and Biology*, Vol. 17, No. 5, (May 2010), pp. 421-433, ISSN: 1074-5521

Poznic M. (2009). Retinoblastoma protein: a central processing unit. *Journal of Biosciences*, Vol. 34, No. 2, (June 2009), pp. 305-312, ISSN: 0250-5991

Rappas M., Oliver A.W. & Pearl L.H. (2011). Structure and function of the Rad9-binding region of the DNA-damage checkpoint adaptor TopBP1. *Nucleic Acids Research*, Vol. 39, No. 1, (January 2011), pp. 313-324, ISSN: 0305-1048

Rastogi R.P., Richa, Kumar A., Tyagi M.B. & Sinha R.P. (2010). Molecular mechanisms of ultraviolet radiation-induced DNA damage and repair. *Journal of Nucleic Acids*, Vol. 2010:592980, (December 2010), ISSN: 2090-0201

Reinhardt H.C. & Yaffe M.B. (2009). Kinases that control the cell cycle in response to DNA damage: Chk1, Chk2, and MK2. *Current Opinion in Cell Biology*, Vol. 21, No. 2, (April 2009), pp. 245-255, ISSN: 0955-0674

Reinhardt H.C., Hasskamp P., Schmedding I., Morandell S., van Vugt M.A.T.M., Wang X.Z., Linding R., Ong S.E., Weaver D., Carr S.A. & Yaffe M.B. (2010). DNA damage activates a spatially distinct late cytoplasmic cell-cycle checkpoint network controlled by MK2-mediated RNA stabilization. *Molecular Cell*, Vol. 40, No. 1, (October 2010), pp. 34-49, ISSN: 1097-2765

Reini K., Uitto L., Perera D., Moens P.B., Freire R. & Syvaoja J.E. (2004). TopBP1 localises to centrosomes in mitosis and to chromosome cores in meiosis. *Chromosoma*, Vol. 112, No. 7, (May 2004), pp. 323-330, ISSN: 0009-5915

Rekdal C., Sjottem E. & Johansen T. (2000). The nuclear factor SPBP contains different functional domains and stimulates the activity of various transcriptional activators. *The Journal of Biological Chemistry*, Vol. 275, No. 51, (December 200), pp. 40288-40300, ISSN: 0021-9258

Rodriguez-Bravo V., Guaita-Esteruelas S., Salvador N., Bachs O. & Agell N. (2007). Different S/M checkpoint responses of tumor and non-tumor cell lines to DNA replication inhibition. *Cancer Research*, Vol. 67, No. 24, (December 2007), pp. 11648-11656, ISSN: 0008-5472

Rogriguez M., Yu X., Chen J. & Songyang Z. (2003). Phosphopeptide binding specificities of BRCA1 COOH-terminal (BRCT) domains. *The Journal of Biological Chemistry*, Vol. 278, No. 52, (December 2003), pp. 52914-52918, ISSN: 0021-9258

Rupnik A., Lownde N. & Grenon M. (2010). MRN and the race to the break. *Chromosoma*, Vol. 119, No. 2, (April 2010), pp. 115-135, ISSN: 0009-5915

Sansam C.L., Cruz N.M., Danielian P.S., Amsterdam A., Lau M.L., Hopkins N. & Lees J.A. (2010). A vertebrate gene, *ticrr*, is an essential checkpoint and replication regulator. *Genes and Development*, Vol. 24, No. 2, (January 2010), pp. 183-194, ISSN: 0890-9369

Sanz L., Moscat J. & Diaz-Meco M. (1995). Molecular characterization of a novel transcription factor that controls stromelysin expression. *Molecular and Cellular Biology*, Vol. 15, No. 6, (June 1995), pp. 3164-3170, ISSN: 0270-7306

Scheffner M. & Staub O. (2007). HECT E3s and human disease. *BioMed Central Biochemistry*, Vol. 8, Suppl. 1, (November 2007), pp. S6, ISSN: 1471-2091

Schmidt U., Wollmann Y., Franke C., Grosse F., Saluz H.P. & Hanel F. (2008). Characterization of the interaction between the human DNA topoisomerase IIβ-binding protein (TopBP1) and the cell division cycle 45 (Cdc45) protein. *The Biochemical Journal*, Vol. 409, No. 1, (January 2008), pp. 169-177, ISSN: 0264-6021

Scorah J. & McGowan C.H. (2009). Claspin and Chk1 regulate replication fork stability by different mechanisms. *Cell Cycle*, Vol. 8, No. 7, (April 2009), pp. 1036-1043, ISSN: 1551-4005

Shiotani B. & Zou L. (2009). Single-stranded DNA orchestrates an ATM-to-ATR switch at DNA breaks. *Molecular Cell*, Vol. 33, No. 5, (March 2009), pp. 547-558, ISSN: 1097-2765

Sjottem E., Rekdal C., Svineng G., Johansen S.S., Klenow H., Uglehus R.D. & Johansen T. (2007). The ePHD protein SPBP interacts with TopBP1 and together they co-operate to stimulate Ets1-mediated transcription. *Nucleic Acids Research*, Vol. 35, No. 19, (October 2007), pp. 6648-6662, ISSN: 0305-1048

Smits V.A.J., Warmerdam D.O., Martin Y. & Freire R. (2010). Mechanisms of ATR-mediated checkpoint signaling. *Frontiers in Bioscience*, Vol. 15, No. 3, (June 2010), pp. 840-853, ISSN: 1093-4715

Sokka M., Parkkinen S., Pospiech H. & Syvaoja J.E. (2010), Function of TopBP1 in genome stability, In: *Genome stability and human diseases*, Nasheuer H.P., pp. 119-141, Springer, ISBN: 978-90-481-3470-0, Dordrecht, Holland

Su Y., Meador J.A., Geard C.R. & Balajee A.S. (2010). Analysis of ionizing radiation-induced DNA damage and repair in three-dimensional human skin model system. *Experimental Dermatology*, Vol. 19, No. 8, (August 2010), pp. e16-22, ISSN: 0906-6705

Takeishi Y., Ohashi E., Ogawa K., Masai H., Obuse C. & Tsurimoto T. (2010). Casein kinase 2-dependent phosphorylation of human Rad9 mediates the interaction between human Rad9-Hus1-Rad1 complex and TopBP1. *Genes to Cells*, Vol. 15, No. 7, (June 2010), pp. 761-771, ISSN: 1356-9597

Taricani L. & Wang T.S.F. (2006). Rad4[TopBP1], a scaffold protein, plays separate roles in DNA damage and replication checkpoints and DNA replication. *Molecular Biology of the Cell*, Vol. 17, No. 8, (August 2006), pp. 3456-3468, ISSN: 1059-1524

Van C., Yan S., Michael W.M., Waga S. & Cimprich K.A. (2010). Continued primer synthesis at stalled replication forks contributes to checkpoint activation. *The Journal of Cell Biology*, Vol. 189, No. 2, (April 2010), pp. 233-246, ISSN: 0021-9525

Van Hatten R.A., Tutter A.V., Holway A.H., Khederian A.M., Walter J.C. & Michael W.M. (2002). The *Xenopus* Xmus101 protein is required for the recruitment of Cdc45 to origins of DNA replication. *The Journal of Cell Biology*, Vol. 159, No. 4, (November 2002), pp. 541-547, IISN: 0021-9525

Wang B., Malik R., Nigg E.A. & Korner R. (2008). Evaluation of the low-specificity protease elastase for large-scale phosphoproteome analysis. *Analytical Chemistry*, Vol. 80, No. 24, (December 2008), pp. 9536-9533, ISSN: 0003-2700

Wanzel M., Herold S. & Eilers M. (2003). Transcriptional repression by Myc. *Trends in Cell Biology*, Vol. 13, No. 3, (March 2003), pp. 146-150, ISSN: 0962-8924

Warmerdam D.O. & Kanaar R. (2010). Dealing with DNA damage: Relationships between checkpoint and repair pathways. *Mutation Research*, Vol. 704, No. 1-3, (April-June 2010), pp. 2-11, ISSN: 0027-5107

Williams R.S., Williams J.S. & Tainer J.A. (2007). Mre11-Rad50-Nbs1 is a keystone complex connecting DNA repair machinery, double-strand break signaling, and the chromatin template. *Biochemistry and Cell Biology*, Vol. 85, No. 4, (August 2007), pp. 509-520, ISSN: 0829-8211

Wollmann Y., Schmidt U., Wieland G.D., Zipfel P.F., Saluz H.P. & Hanel F. (2007). The DNA topoisomerase IIβ binding protein 1 (TopBP1) interacts with poly (ADP-ribose) polymerase (PARP-1). *Journal of Cellular Biochemistry*, Vol. 102, No. 1, (September 2007), pp. 171-182, ISSN: 0730-2312

Woodhouse B.C. & Dianov G.L. (2008). Poly ADP-ribose polymerase-1: An international molecule of mystery. *DNA Repair*, Vol. 7, No. 7, (July 2008), pp. 1077-1086, ISSN: 1568-7856

Wright R.H.G., Dornan E.S., Donaldson M.M. & Morgan I.M. (2006). TopBP1 contains a transcriptional activation domain suppressed by two adjacent BRCT domains. *The Biochemical Journal*, Vol. 400, No. 3, (December 2006), pp. 573-582, ISSN: 0264-6021

Xu Y. & Leffak M. (2010). ATRIP from TopBP1 to ATR – *in vitro* activation of a DNA damage checkpoint. *Proceedings of the National Academy of Science of the United States of America*, Vol. 107, No. 31, (August 2010), pp. 13561-13562

Xu Z.X., Timanova-Atanasova A., Zhao R.X. & Chang K.S. (2003). PML colocalizes with and stabilizes the DNA damage response protein TopBP1. *Molecular and Cellular Biology*, Vol. 23, No. 12, (June 2003), pp. 4247-4256, ISSN: 0270-7306

Yamane K., Chen J. & Kinsella T.J. (2003). Both DNA topoisomerase II-binding protein 1 and BRCA1 regulate the G2-M cell cycle checkpoint. *Cancer Research*, Vol. 63, No. 12, (June 2003), pp. 3049-3053, ISSN: 0008-5472

Yamane K., Wu X. & Chen J. (2002). A DNA damage-regulated BRCT-containing protein, TopBP1, is required for cell survival. *Molecular and Cellular Biology*, Vol. 22, No. 2, (January 2002), pp. 555-566, ISSN: 0270-7306

Yamane K. & Tsuruo T. (1999). Conserved BRCT regions of TopBP1 and of the tumor suppressor BRCA1 bind strand breaks and termini of DNA. *Oncogene*, Vol. 18, No. 37, (September 1999), pp. 5194-5203, ISSN: 0950-9232

Yamane K., Kawabata M. & Tsuruo T. (1997). A DNA-topoisomerase-II – binding protein with eight repeating regions similar to DNA-repair enzymes and to a cell-cycle regulator. *European Journal of Biochemistry*, Vol. 250, No. 3, (December 1997), pp. 794-799, ISSN: 0014-2956

Yan S. & Michael W.M. (2009). TopBP1 and DNA polymerase α-mediated recruitment of the 9-1-1 complex to stalled replication forks. *Cell Cycle*, Vol. 8, No. 18, (September 2009), pp. 2877-2884, ISSN: 1551-4005

Yan S. & Michael W.M. (2009). TopBP1 and DNA polymerase-α directly recruit the 9-1-1 complex to stalled DNA replication forks. *The Journal of Cell Biology*, Vol. 184, No. 6, (March 2009), pp. 793-804, ISSN: 0021-9525

Yang X., Wood P.A. & Hrushesky W.J.M. (2010). Mammalian TIMELESS is required for ATM-dependent CHK2 activation and G_2/M checkpoint control. *The Journal of Biological Chemistry*, Vol. 285, No. 5, (January 2010), pp. 3030-3034, ISSN: 0021-9258

Yang S.Z., Lin F.T. & Lin W.C. (2008). MCPH1/BRIT1 cooperates with E2F1 in the activation of checkpoint, DNA repair and apoptosis. *EMBO Journal*, Vol. 9. No. 9, (September 2008), pp. 907-915, ISSN: 1469-221X

Yoo H.Y., Kumagai A., Shevchenko A., Schevchenko A. & Dunphy W.G. (2009). The Mre11-Rad50-Nbs1 complex mediates activation of TopBP1 by ATM. *Molecular Biology of the Cell*, Vol. 20, No. 9, (May 2009), pp. 2351-2360, ISSN: 1059-1524

Yoo H.Y., Kumagai A., Schevchenko A., Schevchenko A. & Dunphy W.G. (2007). Ataxia-telangiectasia mutated (ATM)-dependent activation of ATR occurs through phosphorylation of TopBP1 by ATM. *The Journal of Biological Chemistry*, Vol. 282, No. 24, (June 2007), pp. 17501-17506, ISSN: 0021-9258

Yoshida K. & Inoue I. (2004). Expression of MCM10 and TopBP1 is regulated by cell proliferation and UV irradiation *via* E2F transcription factor. *Oncogene*, Vol. 23, No. 37, (August 2004), pp. 6250-6260, ISSN: 0950-9232

Zeng L., Hu Y. & Li B. (2005). Identification of TopBP1 as a c-Abl-interacting protein and a repressor for c-Abl expression. *The Journal of Biological Chemistry*, Vol. 280, No. 32, (August 2005), pp. 29374-29380, ISSN: 0021-9258

16

Post-Meiotic DNA Damage and Response in Male Germ Cells

Guylain Boissonneault et al.[*]
Department of Biochemistry, Faculty of Medicine and Health Sciences
Université de Sherbrooke
Canada

1. Introduction

Spermatids are haploid cells that differentiate into spermatozoa and may be considered as an interesting model of DNA damage response and repair. Key features, such a unique set of chromosomes, radioresistance to apoptosis, the presence of known end-joining DNA repair pathways and an underlying prerogative to limit the transmission of any mutation to the next generation, make them a unique cell type to provide new insights on similar pathways in somatic cells. Although DNA damage signaling and repair mechanisms have been extensively studied during meiosis, the contribution of post-meiotic germ cells to the genetic integrity of the male gamete have been overlooked. In this chapter we present clear evidences that the haploid phase of spermatogenesis, termed spermiogenesis, may represent an even greater challenge for the maintenance of the genetic integrity of the male gamete. Since transient DNA strand breaks are intrinsic to the differentiation program of spermatids (Leduc et al., 2008a; Marcon and Boissonneault, 2004), a better understanding of DNA repair pathways involved may shed some light on their potential contribution to male-driven *de novo* mutations and eventually to some unresolved cases of male infertility. This chapter will mainly focus on DNA breaks occurring in the post-meiotic phase of the spermatogenesis and how germ cells deal with it.

2. Spermatogenesis

In most mammals, testes are found in the scrotum and are maintained at lower temperature (2-8°C) than the core body (Harrison and Weiner, 1949; Setchell, 1998). In fact, spermatogenesis is known to work better at lower temperature and it was shown that fertility declines with scrotal hyperthermia. For example, higher scrotal temperature due to fever or lifestyle correlates with decreased semen quality in humans (reviewed in Jung and Schuppe, 2007).

To support germ cells in their development, Sertoli cells are located at the basal lamina, throughout the seminiferous tubules (Russell, 1990). They provide nutrients and essential

---

[*] Frédéric Leduc, Geneviève Acteau, Marie-Chantal Grégoire, Olivier Simard, Jessica Leroux, Audrey Carrier-Leclerc and Mélina Arguin
Department of biochemistry, Faculty of medicine and health sciences, Université de Sherbrooke, Canada

molecules to the differentiating germ cells and regulate the seminiferous tubular fluid (Griswold, 1998; Rato et al., 2010). The Sertoli cells are interconnected by different junctions, creating a unique barrier between surrounding blood vessels and differentiating germ cells that is known as the "blood-testis barrier" (Cheng and Mruk, 2002; Dym and Fawcett, 1970; Setchell, 1969; Vogl et al., 2008). This barrier restricts molecules to enter or exit the adluminal compartment, creating a microenvironment with diverse transporters and preventing immunological response against germ cells (reviewed in Mital et al., 2011).

Spermatogenesis is the cellular differentiation pathway leading to the production of male gametes. This process takes place in seminiferous tubules in the testis, which is a unique environment regulated by follicle-stimulating and luteinizing hormones, secreted by the pituitary gland (Russell, 1990). From birth to puberty, seminiferous tubules are composed of spermatogonia, the precursor stem cell of the germinal cells, and Sertoli cells. At the onset of puberty, spermatogonia undergo mitosis and commit to the differentiation pathway leading to male germ cells. As the germ cells differentiate, they migrate towards the lumen of the tubule, creating an organized stratified structure. Spermatogenesis can be divided in two phases, spermatocytogenesis and spermiogenesis.

Spermatocytogenesis is the process by which a spermatogonium differentiates into primary spermatocytes, which duplicate their DNA to undergo meiosis and become haploid spermatids. This meiotic division is important to create genetic variations by meiotic crossovers and random segregation of parental chromosomes. Spermiogenesis is characterized by the radical metamorphosis of the haploid spermatid into spermatozoa, requiring the reorganization of their organelles. The acrosome, a cap-like structure needed for enzymatic digestion of the oocyte outer membrane, is formed from the Golgi apparatus. At the opposite nuclear pole, the flagellum begins to grow from the centrioles and mitochondria groups at the mid-piece of the emerging flagellum to produce the required energy for its later motion. Finally, the spermatid is stripped of most of its cytoplasm and ultimately released in the lumen of the seminiferous tubule. Most interestingly, the nucleus of spermatids is also remodeled and condensed to protect the paternal genome as well as providing a more hydrodynamic shape. However, this nuclear reorganization is characterized by transient DNA strand breaks that may be necessary to relieve the torsional stress as outlined below.

2.1 Chromatin remodeling process

Through the chromatin remodeling process of spermatids, the paternal genome is condensed tenfold compared to somatic cells, forming a nucleus with an hydrodynamic-shape (Balhorn et al., 1984). To achieve such a high degree of compaction, chromatin must first rely on a set of abundant transition proteins (TPs) subsequently replaced by the protamines (PRMs) (Balhorn et al., 1984; Braun, 2001). The arginine-rich PRMs bind DNA and neutralize the phosphodiester backbone of the double helix (Balhorn, 1982), allowing for a tight compaction of the DNA into torroids (Ward, 1993). Although the onset of chromatin remodeling is poorly understood, incorporation of testis-specific histone variants (Churikov et al., 2004; Govin et al., 2007; Lu et al., 2009; Martianov et al., 2005; Yan et al., 2003) and regulated post-translational modifications of histones, such as acetylation (Christensen et al., 1984; Grimes and Henderson, 1984; Marcon and Boissonneault, 2004; Meistrich et al., 1992), ubiquitination (Baarends et al., 1999; Chen et al., 1998), methylation (Godmann et al., 2007; van der Heijden et al., 2006) and phosphorylation (Blanco-Rodríguez,

2009; Krishnamoorthy et al., 2006; Leduc et al., 2008a; Meyer-Ficca et al., 2005) are known to initiate and participate in the exchange from histones to the more basic proteins such as TPs and PRMs during the transition from the round to the elongated spermatids. These modifications are known to modulate the affinity of histones for DNA, but also the affinity of other proteins for histones, such as chromatin remodelers, DNA repair proteins or the transcription machinery. After spermiation, this unique protamine-based chromatin is further stabilized by the creation of intraprotamine disulfide bonds during the transit through the epididymis (Golan et al., 1996). Therefore, protamination provides both chemical and mechanical protection to the male haploid genome. Interestingly, protamination of the male genome is not complete and varies across species. In the mouse spermatozoon, about 1-2% of the genome remains organized by histones (Balhorn et al., 1977), whereas up to 15% of histones are still found in humans spermatozoa (Gatewood et al., 1990; Gusse et al., 1986; Tanphaichitr et al., 1978). This observation lead to hypothesize that these nucleosomes could serve as epigenetic markers for embryonic development (Arpanahi et al., 2009; Zalenskaya et al., 2000) (for a more detailed review on the sperm chromatin organization, see Johnson et al., 2011)

3. Nature of endogenous DNA damages during spermiogenesis

3.1 Single strand damage and repair

Depending on the type of damage, specific pathways achieve single strand damage repair (see Table 1). Mispaired DNA bases that primarily arise during replication are corrected by mismatch repair (MMR), while small chemical alterations of DNA bases such as alkylation, deamination and oxidative damage are repaired by base excision repair (BER) (Mukherjee et al., 2010; Robertson et al., 2009). More complex lesions such as those induced by UV (pyrimidine dimers and helix-distorting lesions) are corrected by nucleotide excision repair (NER), a multistep pathway that involves more than 30 proteins (Hoeijmakers, 2009; Nouspikel, 2009). DNA nicks are repaired by single-strand break repair (SSBR). These DNA repair pathways are known to be present and active during spermiogenesis (Olsen et al., 2001; Schultz et al., 2003). To our knowledge, single-strand damages do not present a major threat to spermatids. With the exception of exposures to toxicant that could challenge these pathways, in normal conditions, single-strand DNA damage during spermiogenesis is likely attributed to the massive transcription that is taking place at these steps and is efficiently resolved by spermatids (Olsen et al., 2001). DNA double-strand breaks were reported as part of the normal differentiation program of spermatids during spermiogenesis which may represent an important source of genetic instability and therefore we will focus on these pathways.

3.2 Double-strand breaks in spermatids
3.2.1 Possible origin of DNA breaks

Several hypotheses have been formulated to elucidate the origin and role of DNA strand breaks in spermatids. Sakkas and colleagues suggested that "abortive apoptosis" may be the cause since abnormal human spermatozoa presented some apoptotic-like features (Sakkas et al., 1999). Further investigation led to the demonstration that other biomarkers of apoptosis in sperm cells were present such as BCL-X, TP53, caspases, in addition to diverse structural defects (Baccetti et al., 1997; Donnelly et al., 2000; Gandini et al., 2000; Sakkas et al., 2002; Weng et al., 2002). Due to technical limitations at the time, DNA breaks were only observed

DNA repair pathways		DNA damages	Implicated proteins
Mismatch repair (MMR)		Mispaired DNA bases	MSH1-6, MLH1, MLH3, PMS1, PMS2, EXO1, RPA, PCNA, RFC
Base excision repair (BER)	Short-patch	Small DNA bases chemical alteration arising from alkylation, deamination and oxidative damage	UNG, APEX1, POL β, XRCC1, LIG3
	Long-patch		UNG, APEX1, POL β / δ, FEN1, PCNA, LIG1
Nucleotide excision repair (NER)	Transcription -coupled or not	Pyrimidine dimer	XPC complex, DDB complex, ERCC3 (TFIIH), XPA-RPA complex, ERCC5 (XPG), ERCC1-ERCC4 (XPF), LIG3, DNA polymerase δ
Single strand break repair (SSBR)	Short-patch or long-patch	Single strand break (SSB)	APE1, PNKP, APTX, TDP1, POL β/δ/ε, PCNA, XRCC1, LIG1/3, FENI, PARP

Table 1. Summary of the single strand DNA repair pathways in mammalian cells (Ciccia and Elledge, 2010; Hoeijmakers, 2009; Martin et al., 2010; Mukherjee et al., 2010; Nouspikel, 2009; Robertson et al., 2009).

in a subset of the whole population of elongating spermatids and therefore abortive apoptosis could represent a sound explanation. However, some studies demonstrated that round spermatids are radioresistant to apoptosis and may not have the proper machinery and checkpoints to trigger such process (Ahmed et al., 2010; Oakberg and Diminno, 1960). Furthermore, our group have demonstrated that transient DNA breaks were present in the whole population of elongating spermatids of fertile mice and humans during chromatin remodeling and were therefore part of the normal differentiation program of these cells (Marcon and Boissonneault, 2004). The persistence of these breaks beyond the chromatin remodeling steps in pathological conditions may explain the presence of DNA fragmentation found in spermatozoa of infertile men (Leduc et al., 2008b).

Generation of controlled DNA breaks either single- or double-stranded may be important to relieve the torsional stress induced by the withdrawal of histones (Boissonneault, 2002). The simple mechanical stress resulting from the accumulation of free supercoils could induce non-B DNA structures and possibly DNA breaks as the chromatin remodeling is extensive and takes place within many differentiation steps. However, enzymatic induction of DNA strand breaks is more likely, as their free ends can be end-labeled with polymerases that require a 3'OH as substrate, such as the terminal deoxynucleotidyl transferase (TdT) used in TUNEL labeling. Specific nucleases could be involved in this process, and it is not excluded that retrotransposon nucleases could play a role as they are expressed throughout spermatogenesis, including in the nucleus of spermatids (Branciforte and Martin, 1994; Ergün et al., 2004; Gasior et al., 2006). However, topoisomerases have long been considered likely candidates to support chromatin remodeling from bulky histone-bound chromatin to compact and transcriptionally inert protamine-bound DNA because of their ubiquitous role in chromosome dynamics during the somatic cell cycle (McPherson and Longo, 1993).

3.2.2 Topoisomerases as candidates to supercoiling removal

Change in DNA topology can be achieved by single-strand breaks (SSBs) generated by type I topoisomerase, which modifies the linking number in steps of one. Single-strand breaks would be considered a much smaller threat for the genome's integrity of spermatids than a DSB that could be generated by type II topoisomerases. However, chromatin remodeling in spermatids was clearly shown to be associated with an increase in type II topoisomerase (Chen and Longo, 1996; Laberge and Boissonneault, 2005; Leduc et al., 2008a; McPherson and Longo, 1992, 1993; Meyer-Ficca et al., 2011b; Roca and Mezquita, 1989). A possible link between type II topoisomerases and DNA breaks found in elongating spermatids was suggested by the elimination of DNA breaks in spermatids nuclei incubated with type II topoisomerase inhibitors such as suramin and etoposide (Laberge and Boissonneault, 2005). In mammal cells, the α and β isoforms of topoisomerase share more than 80% of homology and are differentially expressed. Topoisomerase IIα (TOP2A) is mostly found in replicating cells whereas topoisomerase IIβ (TOP2B) predominates in quiescent cells (Morse-Gaudio and Risley, 1994; Turley et al., 1997). Using immunofluorescence on mouse testis sections, we have observed TOP2B foci in nuclei of elongating spermatids whereas TOP2A remained undetected in these cells but highly present in spermatocytes (see Figure 1) (Leduc et al., 2008a). Detection of TOP2B in elongating spermatids is not surprising, as spermatids are non-replicative cells. Recent studies confirmed the involvement of TOP2B in elongating spermatids (Meyer-Ficca et al., 2011b) and also observed its presence further downstream of the male germ cells differentiation program as part of the nuclear matrix of sperm cells, supporting its earlier role in the chromatin remodeling of spermatids (Shaman et al., 2006).

3.2.3 Topoisomerases and DNA repair

Type II topoisomerase activity may be modulated by post-translational modifications, such as phosphorylation by kinases and poly (ADP-ribosyl)ation by poly (ADP-ribose) polymerases (PARPs), a well-known family of proteins involved in a multitude of nuclear events, such as DNA repair and chromatin remodeling. This complementary interaction between TOP2B and PARPs may be involved in numerous cellular processes. For example, TOP2B and PARP1 are known to modulate transcription in somatic cells (Ju et al., 2006). Furthermore, these proteins may be important constituents of the nuclear matrix; Zaalishvili and coworkers observed the stimulation of cleavage of nuclear matrix associated DNA loops of neuron and leukocyte nucleoids when incubated in buffer supporting topoisomerase and PARP activity (Zaalishvili et al., 2005). This stimulation was reversed by the addition of thymidine, a PARP inhibitor. The authors suggested that a PARP-modified topoisomerase II may cut efficiently but the (ADP-ribosyl)ation could inhibit the religation. Recently, Meyer-Ficca and colleagues demonstrated a possible modulation of TOP2B activity by PARP and PARG *in vitro* using recombinant proteins as well as *in vivo* during mouse spermiogenesis through the use of inhibitors and knockout mouse models (Meyer-Ficca et al., 2011b). According to their findings, there is a functional relationship between the DNA strand break activity of TOP2B and the DNA strand break-dependent activation of

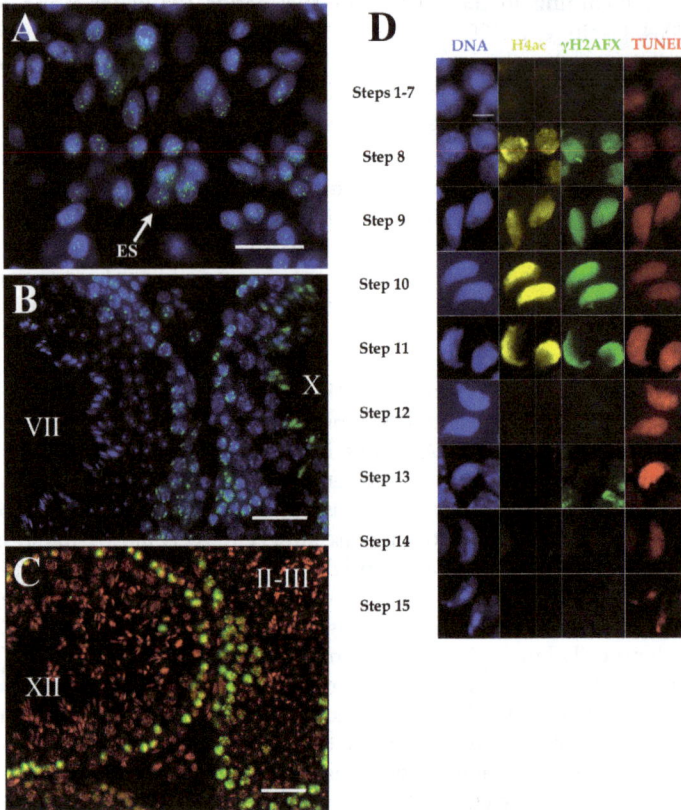

Fig. 1. Presence of type II topoisomerases, hyperacetylated histone H4, γH2AFX and DNA breaks during mouse spermiogenesis. (A) Overlay of TOP2B immunofluorescence (green) and DAPI nuclear staining (blue) of a stage IX tubule demonstrating the presence of TOP2B in nuclei of elongating spermatids (ES) at the onset of chromatin remodeling. (B) Overlay of TOP2B immunofluorescence (green) and DAPI (blue) nuclear staining of stages VII and X tubules. (C) Overlay of immunofluorescence of TOP2A (green) and TO-PRO3 (red) nuclear staining of stages XII and II-III demonstrating the nuclear presence of TOP2A in zygotene and pachytene spermatocytes but complete absence in spermatids. (D) Detection of hyperacetylated histone H4 and γH2AFX by immunofluorescence and DNA breaks by TUNEL during mouse spermiogenesis. DNA was counterstained by TO-PRO3. (A-C) Immunofluoresence on Bouin-fixed testis sections was done as previously described(Leduc et al., 2008a). (D) Squash preparation were done as previously described (Kotaja et al., 2004; Leduc et al., 2008a), fixed with ice-cold ethanol and processed for TUNEL and immunofluorescence. Bars = 10 μm (A and B), 20 μm (C) and 5μm (D).

PARP enzymes. Moreover, alteration in the PAR metabolism leads to a greater retention of histones in spermatozoa (Meyer-Ficca et al., 2011a). Whether PARP1 is involved directly in chromatin remodeling, DNA repair or combination of both in spermatids remains to be determined and will be discussed further in section 4.

4. DNA damage response and DNA repair of double-strand breaks

4.1 DNA damage signaling pathways

The first step following a DSB is the detection of the DNA damage by sensors (Lamarche et al., 2010). At least four independent sensors can detect DSBs: PARPs in all cases of SSBs and, to a lesser extent, DSBs, Ku70/80 in D-NHEJ, MRE11-RAD50-NBS1 (MRN) complex in all cases of DSBs and replication protein A1 (RPA) in HR (Ciccia and Elledge, 2010; Lamarche et al., 2010).

As previously stated in section 3.2.3, the presence and activity of PARP1 and PARP2 have been recently investigated during spermiogenesis of mouse and rat (Ahmed et al., 2010; Dantzer et al., 2006; Meyer-Ficca et al., 2005; Meyer-Ficca et al., 2011a; Meyer-Ficca et al., 2009; Meyer-Ficca et al., 2011b). Although the individual absence of these proteins leads only to subfertility in male, it is believed that they play a key role in the maintenance of genomic integrity of spermatids. As discussed previously, PARPs may be involved in DNA repair and signaling, in the drastic chromatin remodeling of spermatids and even in the repackaging of their genome with protamines (Quénet et al., 2009). However, the embryonic lethal phenotype of double knockout mouse prevent a better assessment of their critical role during spermiogenesis, as the absence of one can be compensated for by the other. The Ku heterodimer binds to DSB ends and is required for the non-homologous end-joining pathway (NHEJ). In addition to its role in DNA repair, Ku proteins are also required for the maintenance of telomeres and subtelomeric gene silencing (Celli et al., 2006; Lamarche et al., 2010). KU70 is present during the spermiogenesis of mouse (Goedecke et al., 1999; Hamer et al., 2003), human (Leduc *et al.*, unpublished observations), and grasshoppers (Cabrero et al., 2007), but seems to decrease as spermiogenesis proceeds, most notably after the expulsion of histones. Although initial analyses of the implication of MRN complex as sensor in non-homologous end-joining pathways produced conflicting results (Di Virgilio and Gautier, 2005; Huang and Dynan, 2002), recent studies showed that siRNA mediated knockdown of Mre11 results in reduced end-joining efficiency in both D-NHEJ and B-NHEJ pathways (Rass et al., 2009; Xie et al., 2009) and should be considered a good candidate for DNA breaks detection and signaling in spermatids. As for Ku proteins, Mre11 is also present during spermiogenesis (Goedecke et al., 1999). Contrary to these DNA break sensor proteins, RPA may not play such an important role during spermiogenesis as spermatids, being haploid, cannot rely on HR repair processes.

The detection of DNA damage by sensors activates proteins of the phosphatidylinositol 3-kinase-like protein kinase (PIKKs) family such as ATM, ATR, and DNA PKcs and members of the PARP family. These proteins post-translationnally modify key protein targets triggering a signal transduction cascades that forms the DNA damage response (DDR) (Lamarche et al., 2010). During mouse spermiogenesis, ATM and DNA PKcs are present and active (Ahmed et al., 2010; Scherthan et al., 2000). These kinases are responsible for the phosphorylation of the histone H2A variant, H2AFX, at serine 139 (γH2AFX, previously referred to as γH2AX), which quickly occurs after a DSB. This modification can spread up to a 2 Mbp region flanking all DSBs (Kinner et al., 2008) and it could help the recruitment of other proteins of the DDR (Celeste et al., 2003). Within minutes following DNA damage, γH2AFX appears at discrete nuclear foci that dissolve after the completion of DNA repair. It remains unclear whether γH2AFX is replaced completely with new H2AFX histones, or simply dephosphorylated, but strong evidences suggest that the latter mechanism is prominent (Chowdhury et al., 2005; Rogakou et al., 1999). Therefore, the implication of γH2AFX in all cases of DSBs makes it a novel biomarker for DSBs detection by immunoflorescence (Mah et al., 2010; Mah et al., 2011). Upon γH2AFX signaling, specific pathways are recruited according to cell type or the cell cycle phase (Shrivastav et al., 2008). The presence of γH2AFX during spermiogenesis has been first shown in rats (Meyer-Ficca et al., 2005) and we confirmed its presence at the corresponding steps during mouse spermiogenesis (Leduc et al., 2008a) (see Figure 1). As shown in Figure 1, the presence of γH2AFX and hyperacetylated histone H4, a biomarker of chromatin remodeling coincides with the presence of TOP2B. These results confirm the previously published strong TUNEL labeling of elongating spermatids during chromatin remodeling (Laberge and Boissonneault, 2005; Marcon and Boissonneault, 2004).

Therefore, spermatids undergo multiple transient DSBs, inducing a classic DDR signaling. In addition, as seen by immunofluorescence in Figure 1, γH2AFX is present in all spermatids throughout chromatin remodeling as part of the normal process of maturation of spermatids. The pattern of γH2AFX in spermatids as seen in Figures 1 and 2 is dependent on fixation and tissue processing; ethanol fixation provides a better context for TUNEL labeling but alters nuclear distribution of proteins. Furthermore, we have also found the presence of γH2AFX and DNA breaks during human spermiogenesis (see Figure 2), while other groups subsequently demonstrated similar DDR signaling in grasshoppers (Cabrero et al., 2007) and even in the algae *Charas vulgaris* (Wojtczak et al., 2008). Moreover, the presence of DNA breaks has also been found during spermiogenesis of drosophila (Rathke et al., 2007). Altogether, these results suggest that the DDR triggered by endogenous breaks in spermatids is evolutionary conserved and could represent a new source of male–driven genetic instability in species where gametogenesis requires condensation of the genetic material.

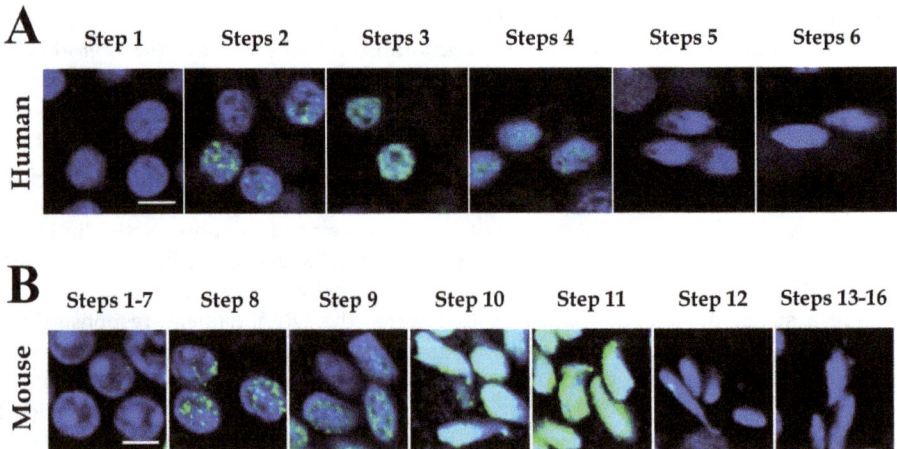

Fig. 2. Detection of γH2AFX during spermiogenesis of human (upper panel), and mouse (lower panel). DNA was counterstained by TO-PRO3. Bars = 5μm. Immunofluoresence on paraformaldehyde-fixed testis sections was done as previously described (Leduc et al., 2008a).

4.2 Evidences of an active DNA repair system during spermiogenesis

Although these DSBs are considered the most harmful genetic damage for a cell, we know from experimental data (Marcon and Boissonneault, 2004) that these breaks are repaired by the end of spermiogenesis in fertile animals. The disappearance of γH2AFX in mouse spermatids (step 13 to 16) shown in Figure 1 cannot be associated with completion of DNA repair or dephosphorylation as a majority of histones are expulsed from the nucleus to be replaced by PRMs. However, we obtained other evidences of an active DNA repair system at these steps by demonstrating incorporation of dNTPs *in situ* that is sustained through all the chromatin remodeling steps (see Figure 3) (Leduc et al., 2008a). Furthermore, as seen in Figure 1, the appearance and disappearance of TUNEL labeling is coincident with γH2AFX fluorescence. To confirm that the loss of TUNEL labeling was associated with DNA repair and not with the lack of penetrability of the TdT in the nuclei of condensed spermatids, we decondensed spermatids prior to TUNEL with similar results (Marcon and Boissonneault, 2004) (Acteau et al., unpublished observations). Therefore, DNA breaks are properly repaired by the end of the spermatids differentiation program. As previously stated, mammalian cells can rely on four DNA DSBs repair pathways, each of which having different degree of fidelity. As spermatids differentiate to spermatozoa with fertilizing potential, any errors due to faulty or incomplete DNA repair may be transmitted to the next generation. Severe alteration in the repair process may cause infertility or possibly be incompatible with embryonic development (Leduc et al., 2008b).

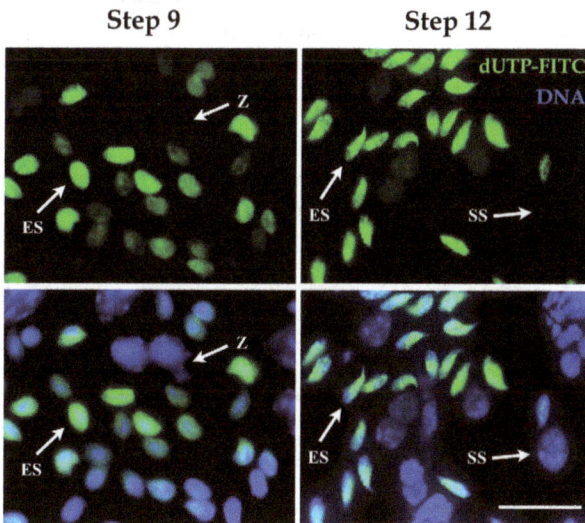

Fig. 3. In situ endogenous DNA polymerase activity assay (Leduc et al., 2008a) on squash preparation of stage IX and XII tubules. DNA was counterstained by DAPI. Bar = 10 μm

4.3 Towards identification of DNA repair pathways

Double-strand breaks are processed either by homologous recombination, single-strand annealing (SSA) or non-homologous end-joining (Caldecott, 2008). Two types of NHEJ are available to mammalian cells: the pathway that is dependent of DNA PKcs (referred to as D-

NHEJ) and the alternative (or "back-up") pathway (referred to as B-NHEJ), which is also known as microhomology-mediated end-joining (MMEJ) (Ciccia and Elledge, 2010; West, 2003). Therefore, we will discuss known somatic DNA repair pathways and their potential role in spermatids when supported by published data.

4.3.1 Homologous recombination and single-strand annealing

Given the haploid character spermatids, HR could not take place as sister chromatids or homologous chromosomes are not available for recombination. Considering that HR precisely restores the genomic sequence of the broken DNA ends by utilizing sister chromatids as template for DNA repair, HR usually occurs in late S2 and G phase of the cycle in mammals (Kass and Jasin, 2010), whereas spermatids are considered to be in a G1-like phase (Ahmed et al., 2010). Upon resection at DNA breaks by the MRN complex, two different pathways are usually possible: HR or SSA (Wold, 1997). The SSA pathway could use repetitive DNA sequences to promote the DNA repair of DSBs in spermatids (Hartlerode and Scully, 2009; Motycka et al., 2004). This pathway is known to introduce errors such as deletions, insertions and even be a substrate for chromosomal translocations (Griffin and Thacker, 2004). There is currently no evidence that spermatids use SSA rather than NHEJ to repair DSBs, but some key proteins of this pathway, although also part of the NER pathway (see Table 1 and Table 2), are present during spermiogenesis including ERCC1 (Hsia et al., 2003; Paul et al., 2007) as well as XPF (Shannon, 1999).

DNA repair pathways	Implicated proteins	Typical error
Homologous recombination (HR)	BRCA1/2, Rad51, XRCC2, LIG1, RPA	Error free
Single strand annealing (SSA)	ERCC1, ERCC4 (XPF), Rad52	Indels (++) Chromosomal translocation (++)
Non-homologous end-joining DNA PK dependant (D-NHEJ)	DNA PKcs, Ku70, ARTEMIS, XRCC4, LIG4, XLF (NHEJ1)	Indels (+) Chromosomal translocation (+)
Alternative non-homologous end joining (B-NHEJ)	PARP, XRCC1, LIG3	Indels (+) Chromosomal translocation (+)

Table 2. DNA double-strand break repair pathways and their typical error. (+) Occasional, (++) frequent (Ciccia and Elledge, 2010; Griffin and Thacker, 2004).

4.3.2 Non-homologous end joining

Besides SSA, B-NHEJ and D-NHEJ are potentially available for the repair of double-strand breaks during spermiogenesis (Leduc et al., 2008a; Leduc et al., 2008b). In somatic cells, NHEJ pathways promote the religation of DSBs, introducing small insertions and deletions. NHEJ pathways operates throughout the cell cycle but are most active during G1 phase because HR cannot proceed during that time (Daley et al., 2005). Spermatids provide a similar cellular context as G1 phase of somatic cells. However, dynamics of DNA repair by NHEJ pathways, as illustrated in irradiated round spermatids, are much slower (Ahmed et al., 2010). According to Ahmed and colleagues both pathways are present and active during mouse spermiogenesis: spermatids of SCID mice, lacking the D-NHEJ because of the absence of DNA PKcs, displayed slower repair than those from wild type mice (Ahmed et al., 2010). Further studies on the end-joining pathways in elongating spermatids will be required as these are known to be error-prone in somatic cells. This may also be the case in spermatids. Although an attenuation of the frequency of mutations may be found in the germ line (Walter et al., 1998), the chromatin remodeling in spermatids may still be the key differentiation steps where most of the new mutations repertoire is being produced for the transmission to the next generation.

5. Possible consequences on fertility and genetic integrity

5.1 Incomplete DNA repair

High level of sperm DNA fragmentation, sperm DNA damages and chromatin alterations decrease pregnancy rates in natural fertilization, intrauterine insemination and *in vitro* fertilization (Bungum et al., 2007; Duran, 2002; Evenson et al., 1999; Evenson and Wixon, 2006; Spano et al., 2000; Zini, 2011). Moreover, pregnancy loss following *in vitro* fertilization (IVF) or intracytoplasmic sperm injection (ICSI) treatments has also been attributed to poor sperm DNA integrity (Zini et al., 2008). Although sperm DNA fragmentation is more frequent in infertile men, sperm of fertile men display DNA fragmentation but to a lesser extent (Bellver et al., 2010; Brahem et al., 2011; Perrin et al., 2009; Rybar et al., 2009; Venkatesh et al., 2011; Watanabe et al., 2011). After fertilization, the oocyte can efficiently repair some paternal DNA damages (Brandriff and Pedersen, 1981; Marchetti et al., 2007), but in the case of highly damaged sperm DNA, this could exceed the DNA repair capacity of the oocyte leading to some genetic aberrations, developmental arrest or pregnancy loss.

5.2 Faulty repair

Structural aberrations

Chromosomal structural aberrations such as translocations, deletions and inversions, may originate from meiotic recombination involving non-allelic repeated DNA sequences (Heyer et al., 2010). However, since about 80% of chromosomal rearrangements are reported to be of paternal origin (Buwe et al., 2005; Thomas et al., 2006) and that male and female meiosis involves similar genetic mechanisms (Gu et al., 2008; Thomas et al., 2006), one can surmise that yet another process unique to male gametogenesis may be involved. We therefore hypothesize that the chromatin remodeling process in spermatids, generating transient double-strand breaks, may provide the proper context for faulty repair and induction of transgenerational polymorphism. In addition, it is tempting to speculate that, because chromatin condensation occurs, free DNA ends are brought in juxtaposition, increasing the chance of NHEJ repair involving two different chromosomes, which may lead to translocations. Interestingly, chromosomes possess their own territory within the nucleus of somatic cells and in sperm cells (Hazzouri et al., 2000; Manvelyan et al., 2008; Mudrak et al., 2005; Zalenskaya et al., 2000). Moreover, chromosomes known to have higher translocation rates have close chromosomal territories in somatic cells (Branco and Pombo, 2006; Brianna

Caddle et al., 2007; Manoj S Gandhi and Nikiforov, 2009). Thus, chromosomes with close chromosomal territories in spermatids could be more prone to interchromosomal translocation during chromatin remodeling. In addition, the potential for the spermatidal chromatin remodeling to produce non-B DNA structure may exacerbate the propensity for spermatids to produce translocation (Hidehito Inagaki and Kurahashi, 2009; Raghavan and Lieber, 2004). Further investigation is needed on the mechanism and potential involvement of chromatin remodeling in such events.

Insertions and deletions

As outlined above, NHEJ repair pathways are known to create insertions and deletions (indels) as they use microhomology to join the two DNA ends. This type of mutations may be particularly harmful in coding sequences as it may cause codon frameshifts or alteration of mRNA splicing. Moreover, Y chromosomes microdeletions, associated with increased male infertility, may exhibit the classical signature of micro-homology based DNA repair pathways such as SSA and B-NHEJ as the deletion occurs between repetitive, often palindromic sequences (Paulo Navarro-Costa and Plancha; Yen, 1998). Although SSA is available during most of spermatogenesis, the B-NHEJ signature on the highly repetitive Y chromosome may be indicative of a faulty DNA repair in spermatids as this pathway is inhibited during meiosis.

Dynamic mutations

Several diseases with dynamic mutation, characterized by the expansion over generation of a trinucleotidic repeat (TNRs), are associated with a paternal bias of expansion, such as Huntington disease (HD), spinocerebellar ataxia type 2 and 7 (Cancel et al., 1997; Stevanin et al., 1998; Zühlke et al., 1993). TNRs are microsatellites sequences in coding or non-coding region of the genome. Their stability, which is relative to the chance of adopting a secondary structure, is dependent of the nature of the sequence and the length of the TNR (Kovtun et al.; Tóth et al., 2000). The exact mechanism of TNR expansion or contraction is still not clear. However, studies show strong evidences that TNR expansion in the huntingtin gene occurs during spermiogenesis; in a transgenic mouse model carrying the mutated human gene, an increased length of the CAG repetition was observed in mature spermatozoa but not in early haploid spermatids and other tissues. Kovtun and McMurray also demonstrated the involvement of MSH2, a protein involved in the gap repair and mismatch repair pathways, as this expansion was absent in HD mice MSH2-/- (Albin and Tagle, 1995; Kovtun and McMurray, 2001). The remodeling chromatin of spermatids may promote secondary structure formation at TNRs, providing an ideal context for such mutations.

6. Conclusion

The chromatin remodeling in spermatids involves transient DNA-strand breaks. Our group has generated evidences that a significant number of double-strand breaks are generated. These DSBs seemingly trigger a damage response as H2AFX is phosphorylated and a DNA repair pathway yet to be identified. As a result, no such DSBs are found in the mature sperm unless a pathological condition prevails. The non-templated DNA repair of these transient DSBs are expected to introduce small mutations likely distributed randomly across the haploid genome although their distribution remains to be established. Meiosis is well known to produce new combination of alleles but is not a primary driver of sequence divergence (Noor, 2008). Potential new gene function must arise through point mutations or indels and the present review suggests chromatin remodeling of spermatids as an appropriate context for such induction of new polymorphism and possible translocations. Although the frequency of mutation may be lower in germ cells than in somatic cells (Walter et al., 1998), we hypothesize that most of the new mutations generated during spermatogenesis may be through the process of endogenous strand breaks and repair during spermiogenesis. Owing to the 1% chance for a random mutation to occur within genes due to exonic representation in the genome, most mutations are expected to be silent but, if within coding sequences, potential alteration of gene function may be transmitted. In summary, repair of the endogenous DSBs in spermatids may represent a new male-driven source of genetic variation.

7. References

Ahmed, E.A., de Boer, P., Philippens, M.E.P., Kal, H.B., and de Rooij, D.G. (2010). Parp1–XRCC1 and the repair of DNA double strand breaks in mouse round spermatids. Mutation Research *683*, 84-90.

Albin, R.L., and Tagle, D.A. (1995). Genetics and molecular biology of Huntington's disease. Trends in Neurosciences *18*, 11-14.

Arpanahi, A., Brinkworth, M., Iles, D., Krawetz, S., Paradowska, A., Platts, A., Saida, M., Steger, K., Tedder, P., and Miller, D. (2009). Endonuclease-sensitive regions of human spermatozoal chromatin are highly enriched in promoter and CTCF binding sequences. Genome Res.

Baarends, W., Hoogerbrugge, J., Roest, H., Ooms, M., Vreeburg, J., Hoeijmakers, J., and Grootegoed, J. (1999). Histone ubiquitination and chromatin remodeling in mouse spermatogenesis. Dev Biol *207*, 322-333.

Baccetti, B., Strehler, E., Capitani, S., Collodel, G., De Santo, M., Moretti, E., Piomboni, P., Wiedeman, R., and Sterzik, K. (1997). The effect of follicle stimulating hormone therapy on human sperm structure. Hum Reprod *12*, 1955-1968.

Balhorn, R. (1982). A model for the structure of chromatin in mammalian sperm. J Cell Biol *93*, 298-305.

Balhorn, R., Gledhill, B.L., and Wyrobek, A.J. (1977). Mouse sperm chromatin proteins: quantitative isolation and partial characterization. Biochemistry *16*, 4074-4080.

Balhorn, R., Weston, S., Thomas, C., and Wyrobek, A. (1984). DNA packaging in mouse spermatids. Synthesis of protamine variants and four transition proteins. Exp Cell Res *150*, 298-308.

Bellver, J., Meseguer, M., Muriel, L., García-Herrero, S., Barreto, M.A.M., Garda, A.L., Remohí, J., Pellicer, A., and Garrido, N. (2010). Y chromosome microdeletions, sperm DNA fragmentation and sperm oxidative stress as causes of recurrent spontaneous abortion of unknown etiology. Human Reproduction *25*, 1713-1721.

Blanco-Rodríguez, J. (2009). gammaH2AX marks the main events of the spermatogenic process. Microsc Res Tech.

Boissonneault, G. (2002). Chromatin remodeling during spermiogenesis: a possible role for the transition proteins in DNA strand break repair. FEBS Lett *514*, 111-114.

Brahem, S., Mehdi, M., Elghezal, H., and Saad, A. (2011). Analysis of Sperm Aneuploidies and DNA Fragmentation in Patients With Globozoospermia or With Abnormal Acrosomes. Urology.

Branciforte, D., and Martin, S.L. (1994). Developmental and cell type specificity of LINE-1 expression in mouse testis: implications for transposition. Mol Cell Biol *14*, 2584-2592.

Branco, M.R., and Pombo, A. (2006). Intermingling of Chromosome Territories in Interphase Suggests Role in Translocations and Transcription-Dependent Associations. PLoS biology *4*, e138.

Brandriff, B., and Pedersen, R. (1981). Repair of the ultraviolet-irradiated male genome in fertilized mouse eggs. Science *211*, 1431-1433.

Braun, R.E. (2001). Packaging paternal chromosomes with protamine. Nat Genet *28*, 10-12.

Brianna Caddle, L., Grant, J.L., Szatkiewicz, J., Hase, J., Shirley, B.-J., Bewersdorf, J., Cremer, C., Arneodo, A., Khalil, A., and Mills, K.D. (2007). Chromosome neighborhood composition determines translocation outcomes after exposure to high-dose radiation in primary cells. Chromosome research *15*, 1061-1073.

Bungum, M., Humaidan, P., Axmon, A., Spano, M., Bungum, L., Erenpreiss, J., and Giwercman, A. (2007). Sperm DNA integrity assessment in prediction of assisted reproduction technology outcome. Human reproduction *22*, 174-179.

Buwe, A., Guttenbach, M., and Schmid, M. (2005). Effect of paternal age on the frequency of cytogenetic abnormalities in human spermatozoa. Cytogenet Gen Res *111*, 213-228.

Cabrero, J., Palomino-Morales, R.J., and Camacho, J.P.M. (2007). The DNA-repair Ku70 protein is located in the nucleus and tail of elongating spermatids in grasshoppers. Chromosome Res *15*, 1093-1100.

Caldecott, K.W. (2008). Single-strand break repair and genetic disease. Nature Reviews Genetics *9*, 619-631.

Cancel, G., Dürr, A., Didierjean, O., Imbert, G., Bürk, K., Lezin, A., Belal, S., Benomar, A., Abada-Bendib, M., Vial, C., *et al.* (1997). Molecular and clinical correlations in spinocerebellar ataxia 2: a study of 32 families. Human Molecular Genetics *6*, 709-715.

Celeste, A., Fernandez-Capetillo, O., Kruhlak, M.J., Pilch, D.R., Staudt, D.W., Lee, A., Bonner, R.F., Bonner, W.M., and Nussenzweig, A. (2003). Histone H2AX phosphorylation is dispensable for the initial recognition of DNA breaks. Nature Cell Biology *5*, 675-679.

Celli, G.B., Denchi, E.L., and de Lange, T. (2006). Ku70 stimulates fusion of dysfunctional telomeres yet protects chromosome ends from homologous recombination. Nat Cell Biol *8*, 885-890.

Chen, H., Sun, J., Zhang, Y., Davie, J., and Meistrich, M. (1998). Ubiquitination of histone H3 in elongating spermatids of rat testes. J Biol Chem *273*, 13165-13169.

Chen, J., and Longo, F. (1996). Expression and localization of DNA topoisomerase II during rat spermatogenesis. Mol Reprod Dev *45*, 61-71.

Cheng, C.Y., and Mruk, D.D. (2002). Cell junction dynamics in the testis: Sertoli-germ cell interactions and male contraceptive development. Physiol rev *82*, 825-874.

Chowdhury, D., Keogh, M.-C., Ishii, H., Peterson, C.L., Buratowski, S., and Lieberman, J. (2005). gamma-H2AX dephosphorylation by protein phosphatase 2A facilitates DNA double-strand break repair. Molecular cell *20*, 801-809.

Christensen, M., Rattner, J., and Dixon, G. (1984). Hyperacetylation of histone H4 promotes chromatin decondensation prior to histone replacement by protamines during spermatogenesis in rainbow trout. Nucleic Acids Res *12*, 4575-4592.

Churikov, D., Siino, J., Svetlova, M., Zhang, K., Gineitis, A., Morton Bradbury, E., and Zalensky, A. (2004). Novel human testis-specific histone H2B encoded by the interrupted gene on the X chromosome. Genomics *84*, 745-756.

Ciccia, A., and Elledge, S.J. (2010). The DNA Damage Response: Making It Safe to Play with Knives. Molecular cell *40*, 179-204.

Daley, J.M., Laan, R.L.V., Suresh, A., and Wilson, T.E. (2005). DNA joint dependence of pol X family polymerase action in nonhomologous end joining. The Journal of biological chemistry *280*, 29030-29037.

Dantzer, F., Mark, M., Quenet, D., Scherthan, H., Huber, A., Liebe, B., Monaco, L., Chicheportiche, A., Sassone-Corsi, P., de Murcia, G., *et al.* (2006). Poly(ADP-ribose) polymerase-2 contributes to the fidelity of male meiosis I and spermiogenesis. Proc Natl Acad Sci USA *103*, 14854-14859.

Di Virgilio, M., and Gautier, J. (2005). Repair of double-strand breaks by nonhomologous end joining in the absence of Mre11. The Journal of cell biology *171*, 765-771.

Donnelly, E.T., O'Connell, M., McClure, N., and Lewis, S.E. (2000). Differences in nuclear DNA fragmentation and mitochondrial integrity of semen and prepared human spermatozoa. Hum Reprod *15*, 1552-1561.

Duran, E.H. (2002). Sperm DNA quality predicts intrauterine insemination outcome: a prospective cohort study. Human Reproduction *17*, 3122-3128.

Dym, M., and Fawcett, D.W. (1970). The blood-testis barrier in the rat and the physiological compartmentation of the seminiferous epithelium. Biol Reprod *3*, 308-326.

Ergün, S., Buschmann, C., Heukeshoven, J., Dammann, K., Schnieders, F., Lauke, H., Chalajour, F., Kilic, N., Strätling, W.H., and Schumann, G.G. (2004). Cell type-specific expression of LINE-1 open reading frames 1 and 2 in fetal and adult human tissues. J Biol Chem *279*, 27753-27763.

Evenson, D., Jost, L., Marshall, D., Zinaman, M., Clegg, E., Purvis, K., de Angelis, P., and Claussen, O. (1999). Utility of the sperm chromatin structure assay as a diagnostic and prognostic tool in the human fertility clinic. Hum Reprod *14*, 1039-1049.

Evenson, D., and Wixon, R. (2006). Clinical aspects of sperm DNA fragmentation detection and male infertility. Theriogenology *65*, 979-991.

Gandini, L., Lombardo, F., Paoli, D., Caponecchia, L., Familiari, G., Verlengia, C., Dondero, F., and Lenzi, A. (2000). Study of apoptotic DNA fragmentation in human spermatozoa. Hum Reprod *15*, 830-839.

Gasior, S.L., Wakeman, T.P., Xu, B., and Deininger, P.L. (2006). The human LINE-1 retrotransposon creates DNA double-strand breaks. J Mol Biol *357*, 1383-1393.

Gatewood, J.M., Cook, G.R., Balhorn, R., Schmid, C.W., and Bradbury, E.M. (1990). Isolation of four core histones from human sperm chromatin representing a minor subset of somatic histones. J Biol Chem *265*, 20662-20666.

Godmann, Auger, Ferraroni-Aguiar, Sauro, D., Sette, Behr, and Kimmins (2007). Dynamic Regulation of Histone H3 Methylation at Lysine 4 in Mammalian Spermatogenesis. Biol Reprod.

Goedecke, W., Eijpe, M., Offenberg, H., van Aalderen, M., and Heyting, C. (1999). Mre11 and Ku70 interact in somatic cells, but are differentially expressed in early meiosis. Nat Genet *23*, 194-198.

Golan, R., Cooper, T.G., Oschry, Y., Oberpenning, F., Schulze, H., Shochat, L., and Lewin, L.M. (1996). Changes in chromatin condensation of human spermatozoa during epididymal transit as determined by flow cytometry. Hum Reprod 11, 1457-1462.

Govin, J., Escoffier, E., Rousseaux, S., Kuhn, L., Ferro, M., Thévenon, J., Catena, R., Davidson, I., Garin, J., Khochbin, S., et al. (2007). Pericentric heterochromatin reprogramming by new histone variants during mouse spermiogenesis. J Cell Biol 176, 283-294.

Griffin, C.S., and Thacker, J. (2004). The role of homologous recombination repair in the formation of chromosome aberrations. Cytogenet Genome Res 104, 21-27.

Grimes, S., and Henderson, N. (1984). Hyperacetylation of histone H4 in rat testis spermatids. Exp Cell Res 152, 91-97.

Griswold, M.D. (1998). The central role of Sertoli cells in spermatogenesis. Seminars in cell & developmental biology 9, 411-416.

Gu, W., Zhang, F., and Lupski, J. (2008). Mechanisms for human genomic rearrangements. Patho Genetics 1, 4.

Gusse, M., Sautière, P., Bélaiche, D., Martinage, A., Roux, C., Dadoune, J.P., and Chevaillier, P. (1986). Purification and characterization of nuclear basic proteins of human sperm. Biochim Biophys Acta 884, 124-134.

Hamer, G., Roepers-Gajadien, H.L., van Duyn-Goedhart, A., Gademan, I.S., Kal, H.B., van Buul, P.P.W., Ashley, T., and de Rooij, D.G. (2003). Function of DNA-protein kinase catalytic subunit during the early meiotic prophase without Ku70 and Ku86. Biol Reprod 68, 717-721.

Harrison, R.G., and Weiner, J.S. (1949). Vascular patterns of the mammalian testis and their functional significance. The Journal of experimental biology 26, 304-316, 302 pl.

Hartlerode, A.J., and Scully, R. (2009). Mechanisms of double-strand break repair in somatic mammalian cells. The Biochemical journal 423, 157-168.

Hazzouri, M., Pivot-Pajot, C., Faure, A., Usson, Y., Pelletier, R., Sele, B., Khochbin, S., and Rousseaux, S. (2000). Regulated hyperacetylation of core histones during mouse spermatogenesis: involvement of histone deacetylases. Eur J Cell Biol 79, 950-960.

Heyer, W.-D., Ehmsen, K.T., and Liu, J. (2010). Regulation of homologous recombination in eukaryotes. Annual review of genetics 44, 113-139.

Inagaki H, Ohye T, Kogo H, Kato T, Bolor H, Taniguchi M, Shaikh TH, Emanuel BS and Kurahashi H (2009). Chromosomal instability mediated by non-B DNA: Cruciform conformation and not DNA sequence is responsible for recurrent translocation in humans. Genome research 19, 191.

Hoeijmakers, J.H.J. (2009). DNA damage, aging, and cancer. The New England journal of medicine 361, 1475-1485.

Hsia, K., Millar, M., King, S., Selfridge, J., Redhead, N., Melton, D., and Saunders, P. (2003). DNA repair gene Ercc1 is essential for normal spermatogenesis and oogenesis and for functional integrity of germ cell DNA in the mouse. Development 130, 369-378.

Huang, J., and Dynan, W.S. (2002). Reconstitution of the mammalian DNA double-strand break end-joining reaction reveals a requirement for an Mre11/Rad50/NBS1-containing fraction. Nucleic acids research 30, 667-674.

Johnson, G.D., Lalancette, C., Linnemann, A.K., Leduc, F., Boissonneault, G., and Krawetz, S.A. (2011). The sperm nucleus: chromatin, RNA, and the nuclear matrix. Reproduction 141, 21-36.

Ju, B., Lunyak, V., Perissi, V., Garcia-Bassets, I., Rose, D., Glass, C., and Rosenfeld, M. (2006). A topoisomerase IIbeta-mediated dsDNA break required for regulated transcription. Science *312*, 1798-1802.

Jung, A., and Schuppe, H.-C. (2007). Influence of genital heat stress on semen quality in humans. Andrologia *39*, 203-215.

Kass, E.M., and Jasin, M. (2010). Collaboration and competition between DNA double-strand break repair pathways. FEBS letters *584*, 3703-3708.

Kinner, A., Wu, W., Staudt, C., and Iliakis, G. (2008). Gamma-H2AX in recognition and signaling of DNA double-strand breaks in the context of chromatin. Nucleic acids research *36*, 5678-5694.

Kotaja, N., Kimmins, S., Brancorsini, S., Hentsch, D., Vonesch, J., Davidson, I., Parvinen, M., and Sassone-Corsi, P. (2004). Preparation, isolation and characterization of stage-specific spermatogenic cells for cellular and molecular analysis. Nat Methods *1*, 249-254.

Kovtun, I., and McMurray, C. (2001). Trinucleotide expansion in haploid germ cells by gap repair. Nat Genet *27*, 407-411.

Kovtun, I.V., Goellner, G., and McMurray, C.T. Structural features of trinucleotide repeats associated with DNA expansion. Biochem Cell Biol. *79*, 325-36.

Krishnamoorthy, T., Chen, X., Govin, J., Cheung, W., Dorsey, J., Schindler, K., Winter, E., Allis, C., Guacci, V., and Khochbin, S. (2006). Phosphorylation of histone H4 Ser1 regulates sporulation in yeast and is conserved in fly and mouse spermatogenesis. Genes & development *20*, 2580.

Laberge, R.-M., and Boissonneault, G. (2005). On the nature and origin of DNA strand breaks in elongating spermatids. Biol Reprod *73*, 289-296.

Lamarche, B.J., Orazio, N.I., and Weitzman, M.D. (2010). The MRN complex in double-strand break repair and telomere maintenance. FEBS letters *584*, 3682-3695.

Leduc, F., Maquennehan, V., Nkoma, G.B., and Boissonneault, G. (2008a). DNA damage response during chromatin remodeling in elongating spermatids of mice. Biol Reprod *78*, 324-332.

Leduc, F., Nkoma, G.B., and Boissonneault, G. (2008b). Spermiogenesis and DNA repair: a possible etiology of human infertility and genetic disorders. Syst Biol Reprod Med *54*, 3-10.

Loonie D. Russell, A.P.S.H., Robert Ettlin (1990). Histological and histopathological evaluation of the testis (Cache River Press).

Lu, S., Xie, Y.M., Li, X., Luo, J., Shi, X.Q., Hong, X., Pan, Y.H., and Ma, X. (2009). Mass spectrometry analysis of dynamic post-translational modifications of TH2B during spermatogenesis. Mol Hum Reprod *15*, 373-378.

Mah, L.-J., El-Osta, A., and Karagiannis, T.C. (2010). gammaH2AX: a sensitive molecular marker of DNA damage and repair. Leukemia : official journal of the Leukemia Society of America, Leukemia Research Fund, UK *24*, 679-686.

Mah, L.-J., Orlowski, C., Ververis, K., Vasireddy, R.S., El-Osta, A., and Karagiannis, T.C. (2011). Evaluation of the efficacy of radiation-modifying compounds using γH2AX as a molecular marker of DNA double-strand breaks. Genome integrity *2*, 3.

Gandhi MS, Stringer JR, Nikiforova MN, Medvedovic M and Nikiforov YE.. (2009). Gene position within chromosome territories correlates with their involvment in distinct rearrangement types in thyroid cancer cells. Genes Chromosomes Cancer *48*, 222.

Manvelyan, M., Hunstig, F., Bhatt, S., Mrasek, K., Pellestor, F., Weise, A., Simonyan, I., Aroutiounian, R., and Liehr, T. (2008). Chromosome distribution in human sperm - a 3D multicolor banding-study. Molecular cytogenetics 1, 25.

Marchetti, F., Essers, J., Kanaar, R., and Wyrobek, A.J. (2007). Disruption of maternal DNA repair increases sperm-derived chromosomal aberrations. Proc Natl Acad Sci USA 104, 17725-17729.

Marcon, L., and Boissonneault, G. (2004). Transient DNA strand breaks during mouse and human spermiogenesis new insights in stage specificity and link to chromatin remodeling. Biol Reprod 70, 910-918.

Martianov, I., Brancorsini, S., Catena, R., Gansmuller, A., Kotaja, N., Parvinen, M., Sassone-Corsi, P., and Davidson, I. (2005). Polar nuclear localization of H1T2, a histone H1 variant, required for spermatid elongation and DNA condensation during spermiogenesis. Proc Natl Acad Sci USA 102, 2808-2813.

Martin, L.M., Marples, B., Coffey, M., Lawler, M., Lynch, T.H., Hollywood, D., and Marignol, L. (2010). DNA mismatch repair and the DNA damage response to ionizing radiation: making sense of apparently conflicting data. Cancer treatment reviews 36, 518-527.

McPherson, S., and Longo, F. (1992). Localization of DNase I-hypersensitive regions during rat spermatogenesis: stage-dependent patterns and unique sensitivity of elongating spermatids. Mol Reprod Dev 31, 268-279.

McPherson, S., and Longo, F. (1993). Nicking of rat spermatid and spermatozoa DNA: possible involvement of DNA topoisomerase II. Dev Biol 158, 122-130.

Meistrich, M.L., Trostle-Weige, P.K., Lin, R., Bhatnagar, Y.M., and Allis, C.D. (1992). Highly acetylated H4 is associated with histone displacement in rat spermatids. Mol Reprod Dev 31, 170-181.

Meyer-Ficca, M., Scherthan, H., Burkle, A., and Meyer, R. (2005). Poly(ADP-ribosyl)ation during chromatin remodeling steps in rat spermiogenesis. Chromosoma 114, 67-74.

Meyer-Ficca, M.L., Ihara, M., Lonchar, J.D., Meistrich, M.L., Austin, C.A., Min, W., Wang, Z.-Q., and Meyer, R.G. (2011a). Poly(ADP-ribose) metabolism is essential for proper nucleoprotein exchange during mouse spermiogenesis. Biol Reprod 84, 218-228.

Meyer-Ficca, M.L., Lonchar, J., Credidio, C., Ihara, M., Li, Y., Wang, Z.-Q., and Meyer, R.G. (2009). Disruption of poly(ADP-ribose) homeostasis affects spermiogenesis and sperm chromatin integrity in mice. Biol Reprod 81, 46-55.

Meyer-Ficca, M.L., Lonchar, J.D., Ihara, M., Meistrich, M.L., Austin, C.A., and Meyer, R.G. (2011b). Poly(ADP-Ribose) Polymerases PARP1 and PARP2 Modulate Topoisomerase II Beta (TOP2B) Function During Chromatin Condensation in Mouse Spermiogenesis. Biology of reproduction.

Mital, P., Hinton, B.T., and Dufour, J.M. (2011). The Blood-Testis and Blood-Epididymis Barriers Are More Than Just Their Tight Junctions. Biology of reproduction.

Morse-Gaudio, M., and Risley, M.S. (1994). Topoisomerase II expression and VM-26 induction of DNA breaks during spermatogenesis in Xenopus laevis. J Cell Sci 107 (Pt 10), 2887-2898.

Motycka, T.A., Bessho, T., Post, S.M., Sung, P., and Tomkinson, A.E. (2004). Physical and functional interaction between the XPF/ERCC1 endonuclease and hRad52. The Journal of biological chemistry 279, 13634-13639.

Mudrak, O., Tomilin, N., and Zalensky, A. (2005). Chromosome architecture in the decondensing human sperm nucleus. Journal of Cell Science *118*, 4541-4550.

Mukherjee, S., Ridgeway, A.D., and Lamb, D.J. (2010). DNA mismatch repair and infertility. Current Opinion in Urology *20*, 525-532.

Noor, M.A.F. (2008). Mutagenesis from Meiotic Recombination Is Not a Primary Driver of Sequence Divergence between Saccharomyces Species. Molecular biology and evolution *25*, 2439-2444.

Nouspikel, T. (2009). DNA repair in mammalian cells : Nucleotide excision repair: variations on versatility. Cellular and molecular life sciences : CMLS *66*, 994-1009.

Oakberg, E.F., and Diminno, R.L. (1960). X-ray sensitivity of primary spermatocytes of the mouse.int. Int J Radiat Biol *2*, 196-209.

Olsen, A., Bjørtuft, H., Wiger, R., Holme, J., Seeberg, E., Bjørås, M., and Brunborg, G. (2001). Highly efficient base excision repair (BER) in human and rat male germ cells. Nucleic Acids Res *29*, 1781-1790.

Paul, C., Povey, J.E., Lawrence, N.J., Selfridge, J., Melton, D.W., and Saunders, P.T.K. (2007). Deletion of Genes Implicated in Protecting the Integrity of Male Germ Cells Has Differential Effects on the Incidence of DNA Breaks and Germ Cell Loss. PLoS ONE *2*, e989.

Paulo Navarro-Costa, J.G., and Plancha, C.E. The AZFc region of the Y chromosome: at the crossroads between genetic diversity and male infertility. Human Reproduction Update *16*, 525.

Perrin, A., Caer, E., Oliver-Bonet, M., Navarro, J., Benet, J., Amice, V., De Braekeleer, M., and Morel, F. (2009). DNA fragmentation and meiotic segregation in sperm of carriers of a chromosomal structural abnormality. Fertility and sterility *92*, 583-589.

Quénet, D., Mark, M., Govin, J., van Dorsselear, A., Schreiber, V., Khochbin, S., and Dantzer, F. (2009). Parp2 is required for the differentiation of post-meiotic germ cells: identification of a spermatid-specific complex containing Parp1, Parp2, TP2 and HSPA2. Exp Cell Res *315*, 2824-2834.

Raghavan, S.C., and Lieber, M.R. (2004). Chromosomal Translocations and Non-B DNA Structures in the Human Genome. Cell cycle *3*, 760-766.

Rass, E., Grabarz, A., Plo, I., Gautier, J., Bertrand, P., and Lopez, B.S. (2009). Role of Mre11 in chromosomal nonhomologous end joining in mammalian cells. Nature structural molecular biology *16*, 819-824.

Rathke, C., Baarends, W.M., Jayaramaiah-Raja, S., Bartkuhn, M., Renkawitz, R., and Renkawitz-Pohl, R. (2007). Transition from a nucleosome-based to a protamine-based chromatin configuration during spermiogenesis in Drosophila. J Cell Sci *120*, 1689-1700.

Rato, L., Socorro, S., Cavaco, J.E.B., and Oliveira, P.F. (2010). Tubular fluid secretion in the seminiferous epithelium: ion transporters and aquaporins in Sertoli cells. The Journal of membrane biology *236*, 215-224.

Robertson, A.B., Klungland, A., Rognes, T., and Leiros, I. (2009). DNA repair in mammalian cells: Base excision repair: the long and short of it. Cellular and molecular life sciences *66*, 981-993.

Roca, J., and Mezquita, C. (1989). DNA topoisomerase II activity in nonreplicating, transcriptionally inactive, chicken late spermatids. Embo J *8*, 1855-1860.

Rogakou, E.P., Boon, C., Redon, C., and Bonner, W.M. (1999). Megabase chromatin domains involved in DNA double-strand breaks in vivo. The Journal of cell biology *146*, 905-916.

Rybar, R., Markova, P., Veznik, Z., Faldikova, L., Kunetkova, M., Zajicova, A., Kopecka, V., and Rubes, J. (2009). Sperm chromatin integrity in young men with no experiences of infertility and men from idiopathic infertility couples. Andrologia *41*, 141-149.

Sakkas, D., Mariethoz, E., Manicardi, G., Bizzaro, D., Bianchi, P.G., and Bianchi, U. (1999). Origin of DNA damage in ejaculated human spermatozoa. Rev Reprod *4*, 31-37.

Sakkas, D., Moffatt, O., Manicardi, G., Mariethoz, E., Tarozzi, N., and Bizzaro, D. (2002). Nature of DNA damage in ejaculated human spermatozoa and the possible involvement of apoptosis. Biol Reprod *66*, 1061-1067.

Scherthan, H., Jerratsch, M., Dhar, S., Wang, Y., Goff, S., and Pandita, T. (2000). Meiotic telomere distribution and Sertoli cell nuclear architecture are altered in Atm- and Atm-p53-deficient mice. Mol Cell Biol *20*, 7773-7783.

Schultz, N., Hamra, F.K., and Garbers, D.L. (2003). A multitude of genes expressed solely in meiotic or postmeiotic spermatogenic cells offers a myriad of contraceptive targets. Proceedings of the National Academy of Sciences of the United States of America *100*, 12201-12206.

Setchell, B.P. (1969). Do Sertoli cells secrete fluid into the seminiferous tubules? J Reprod Fertil *19*, 391-392.

Setchell, B.P. (1998). The Parkes Lecture. Heat and the testis. J Reprod Fertil *114*, 179-194.

Shaman, J., Prisztoka, R., and Ward, W. (2006). Topoisomerase IIB and an extracellular nuclease interact to digest sperm DNA in an apoptotic-like manner. Biol Reprod *75*, 741-748.

Shannon, M. (1999). Characterization of the Mouse Xpf DNA Repair Gene and Differential Expression during Spermatogenesis. Genomics *62*, 427-435.

Shrivastav, M., de Haro, L.P., and Nickoloff, J.A. (2008). Regulation of DNA double-strand break repair pathway choice. Cell research *18*, 134-147.

Spano, M., Bonde, J.P., Hjøllund, H.I., Kolstad, H.A., Cordelli, E., and Leter, G. (2000). Sperm chromatin damage impairs human fertility. The Danish First Pregnancy Planner Study Team. Fertility and sterility *73*, 43-50.

Stevanin, G., Giunti, P., David, G., and Belal, S. (1998). De novo expansion of intermediate alleles in spinocerebellar ataxia 7. Human molecular ....

Tanphaichitr, N., Sobhon, P., Taluppeth, N., and Chalermisarachai, P. (1978). Basic nuclear proteins in testicular cells and ejaculated spermatozoa in man. Exp Cell Res *117*, 347-356.

Thomas, N.S., Durkie, M., Van Zyl, B., Sanford, R., Potts, G., Youings, S., Dennis, N., and Jacobs, P. (2006). Parental and chromosomal origin of unbalanced de novo structural chromosome abnormalities in man. Hum Genet *119*, 444-450.

Tóth, G., Gáspári, Z., and Jurka, J. (2000). Microsatellites in different eukaryotic genomes: survey and analysis. Genome research *10*, 967-981.

Turley, H., Comley, M., Houlbrook, S., Nozaki, N., Kikuchi, A., Hickson, I., Gatter, K., and Harris, A. (1997). The distribution and expression of the two isoforms of DNA topoisomerase II in normal and neoplastic human tissues. Br J Cancer *75*, 1340-1346.

van der Heijden, G., Derijck, A., Ramos, L., Giele, M., van der Vlag, J., and de Boer, P. (2006). Transmission of modified nucleosomes from the mouse male germline to the zygote and subsequent remodeling of paternal chromatin. Dev Biol *298*, 458-469.

Venkatesh, S., Kumar, R., Deka, D., Deecaraman, M., and Dada, R. (2011). Analysis of sperm nuclear protein gene polymorphisms and DNA integrity in infertile men. Systems Biology in Reproductive Medicine.

Vogl, A.W., Vaid, K.S., and Guttman, J.A. (2008). The Sertoli cell cytoskeleton. Adv Exp Med Biol *636*, 186-211.

Walter, C., Intano, G., McCarrey, J., McMahan, C., and Walter, R. (1998). Mutation frequency declines during spermatogenesis in young mice but increases in old mice. Proc Natl Acad Sci USA *95*, 10015-10019.

Ward, W.S. (1993). Deoxyribonucleic acid loop domain tertiary structure in mammalian spermatozoa. Biol Reprod *48*, 1193-1201.

Watanabe, S., Tanaka, A., Fujii, S., Mizunuma, H., Fukui, A., Fukuhara, R., Nakamura, R., Yamada, K., Tanaka, I., Awata, S., *et al.* (2011). An investigation of the potential effect of vacuoles in human sperm on DNA damage using a chromosome assay and the TUNEL assay. Human Reproduction.

Weng, S., Taylor, S., Morshedi, M., Schuffner, A., Duran, E., Beebe, S., and Oehninger, S. (2002). Caspase activity and apoptotic markers in ejaculated human sperm. Mol Hum Reprod *8*, 984-991.

West, S.C. (2003). Molecular views of recombination proteins and their control. Nature reviews Molecular cell biology *4*, 435-445.

Wojtczak, A., Popłońska, K., and Kwiatkowska, M. (2008). Phosphorylation of H2AX histone as indirect evidence for double-stranded DNA breaks related to the exchange of nuclear proteins and chromatin remodeling in Chara vulgaris spermiogenesis. Protoplasma *233*, 263-267.

Wold, M.S. (1997). Replication protein A: a heterotrimeric, single-stranded DNA-binding protein required for eukaryotic DNA metabolism. Annual review of biochemistry *66*, 61-92.

Xie, A., Kwok, A., and Scully, R. (2009). Role of mammalian Mre11 in classical and alternative nonhomologous end joining. Nature structural & molecular biology *16*, 814-818.

Yan, W., Ma, L., Burns, K., and Matzuk, M. (2003). HILS1 is a spermatid-specific linker histone H1-like protein implicated in chromatin remodeling during mammalian spermiogenesis. Proc Natl Acad Sci USA *100*, 10546-10551.

Yen, P. (1998). A Long-Range Restriction Map of Deletion Interval 6 of the Human Y Chromosome: A Region Frequently Deleted in Azoospermic Males. Genomics *54*, 5-12.

Zaalishvili, G.T., Tsetskhladze, Z.R., Margiani, D.O., Gabriadze, I.I., and Zaalishvili, T.M. (2005). ADP-ribosylation intensifies cleavage of DNA loops in the nuclear matrix. Mol Biol (Mosk) *39*, 317-320.

Zalenskaya, I.A., Bradbury, E.M., and Zalensky, A.O. (2000). Chromatin structure of telomere domain in human sperm. Biochem Biophys Res Commun *279*, 213-218.

Zini, A. (2011). Are sperm chromatin and DNA defects relevant in the clinic? Systems Biology in Reproductive Medicine *57*, 78-85.

Zini, A., Boman, J.M., Belzile, E., and Ciampi, A. (2008). Sperm DNA damage is associated with an increased risk of pregnancy loss after IVF and ICSI: systematic review and meta-analysis. Human Reproduction 23, 2663-2668.

Zühlke, C., Rless, O., Bockel, B., Lange, H., and Thies, U. (1993). Mitotic stability and meiotic variability of the (CAG) nrepeat in the Huntington disease gene. Human Molecular Genetics 2, 2063-2067.

17

Roles of MicroRNA in DNA Damage and Repair

Xinrong Chen and Tao Chen

National Center for Toxicological Research/US Food and Drug Administration

U.S.A.

1. Introduction

DNA damage mainly results from either endogenous metabolic activity, such as oxidative stress, or environmental exposure, such as ionizing irradiation. In human cells, endogenous and exogenous genotoxic agents produce as many as 1 million molecular lesions per cell per day. If the unrepaired lesions occur in certain critical genes, they can cause mutations that can lead to tumors (Lodish H, 2004).

There are several different types of DNA damage, including DNA hydrolysis, DNA adduction, DNA crosslinking, and DNA strand breakage. DNA hydrolysis is the breaking of DNA through the addition of water. Hydrolysis of DNA bases consists of deamination, depurination, and depyrimidination. A DNA adduct is a piece of DNA covalently bonded to a chemical. DNA crosslinks are links formed within a single (intrastrand) or between strands of DNA (interstrand). There are two types of DNA strand breaks, single strand breaks and double strand breaks. DNA double strand breaks are particularly hazardous to the cells because they can lead to genome rearrangements. (Rich et al., 2000).

Cells respond to DNA damage through a variety of different mechanisms, such as apoptosis, senescence, and DNA repair. Excessive DNA damage induces apoptosis, or programmed cell death, that eliminates cells with heavily damaged DNA, thus protecting the organism from the mutations potentially induced by the damage. Unrepaired DNA damage is a driving force for senescence. Senescence serves as a functional alternative to apoptosis in cases where the physical presence of cells is required for spatial reasons. If DNA replication occurs before DNA damage is repaired, mutations can be formed in the cells. To prevent mutation formation, cells have developed DNA repair mechanisms to correct DNA.

There are several different types of DNA repair. They are direct reversal, base excision repair (BER), nucleotide excision repair (NER), mismatch repair (MMR), non-homologous end-joining (NHEJ), and homologous recombination repair (HRR). Direct reversal can remove DNA damage by chemically reversing it. Since the correction only occurs in one of the four bases and not the phosphodiester backbone, this type of repair does not need any DNA template. For example, methylation of guanine bases can be directly reversed by methyl guanine methyl transferase (MGMT) that removes the methyl group. BER amends damage to single nucleotides produced by oxidation, alkylation, or hydrolysis. NER corrects ethylation products, bulky DNA adducts, helix-distorting changes, such as thymine dimers, and single-strand breaks. MMR repairs mismatched bases in double-stranded DNA (e.g., A:C or G:T). HRR is a mechanism for DNA double-strand repair that reconstitutes the original sequence using the sister chromatid as a template. NHEJ is a relatively simple way for DNA double-strand repair and it just rejoins two broken ends without correcting any deletions or rearrangements of DNA.

2. Biogenesis of miRNA

A microRNA gene can be located in an intron of another gene, in either the sense or antisense orientation. miRNA can be coordinately expressed with its host gene, or it can have its own promoter independent of its host gene (Ozsolak et al., 2008). The biogenesis of miRNA is a complex process as shown in Figure 1. miRNA is first transcribed as a long primary miRNA (pri-miRNA) by RNA polymerase II in the nucleus (Lee et al., 2004). Pri-miRNA is structurally similar to mRNA, but contains a stable stem-loop structure (Cai et al., 2004). Recognition of the hairpin and selection of a cleavage site are mediated by DGCR8. Nuclear RNase III (Drosha) then cleaves the pri-miRNA to release the hairpin-shaped precursor miRNA (pre-miRNAs). The pre-miRNA is exported from the nucleus to the cytoplasm by Exportin 5 (Exp5). In the cytoplasm, the pre-miRNA is subsequently cut by cytoplasmic RNase III (Dicer) in complex with Argonaute2 (Ago2) and TRBP, a double-stranded RNA-binding protein. This process cleaves the pre-miRNA hairpins to remove its hairpin loop, resulting in a miRNA duplex with the appropriate length (Gregory et al., 2005; Han et al., 2004; Lee et al., 2003). Normally, one strand of the duplex is then degraded. The mature miRNA are incorporated into an RNA-induced silencing complex (RISC) (Gregory, et al., 2005; Grishok et al., 2001; Hutvagner et al., 2001; Ketting et al., 2001; Maniataki and Mourelatos, 2005). RISC recognizes target mRNAs through full or partial base-pairing interactions between the miRNA and the to "3'-untranslated region (UTR) of the target mRNA. Depending on pairing interactions between miRNAs and their targets, miRNAs suppress their target gene expression by either mRNA cleavage or translational repression. If an mRNA target match perfectly or near-perfectly to the miRNA, the mRNA will be degraded; otherwise, the mRNA will be translationally suppressed (Meister and Tuschl, 2004).

3. Alteration of miRNA biogenesis in response to DNA damage and repair

Because miRNAs are actively involved in regulation of genes that are related to DNA damage and repair, it was not surprising to find that miRNA biogenesis changes in response to DNA damage and repair. Several studies demonstrated that both miRNA transcription and maturation process are altered in response to DNA damage and repair.

Recent studies show that transcription of miRNA can be directly affected by DNA damage. The P53 gene plays a critical role in this regulation. For example, miR-34a can be up-regulated by the P53 gene in response to DNA damage (Chang et al., 2007; Corney et al., 2007; He et al., 2007; Raver-Shapira et al., 2007; Welch et al., 2007). Up-regulation of miR-34a results in apoptosis, cell-cycle arrest, and DNA repair. miR-34a is a direct transcriptional target of P53 because the promoter region of miR-34a contains a canonical P53 binding site. When DNA damage activates the P53 gene, P53 protein binds to the promoter of miR-34a and up-regulates miRNA expression. In Caenorhabditis elegans, miR-34a expression was enhanced by irradiation in a P53 independent manner, and knocking down of the Cep1 gene (homolog of the P53 gene) had no effect on the miR-34a response to irradiation (Kato et al., 2009). Up-regulation of miR-34a in response to genotoxin exposure is also observed in different biological systems (Chen et al., 2011; Li et al., 2010;

Li et al., 2011; Zenz et al., 2009). miR-34c, another member of miR-34 family, is transcriptionally up-regulated by P53 following DNA damage (Cannell et al., 2010). In addition to miR-34a, P53 can also regulate the expression of miR-192, miR-194, and miR-215. These miRNAs are considered tumor suppressor miRNAs (Braun et al., 2008; Georges et al., 2008).

miRNA biogenesis is globally induced upon DNA damage in an ATM (ataxia telangiectasia mutated) dependent manner (Zhang et al., 2011). The ATM gene encodes a DNA damage-inducible kinase. ATM controls cell grow rate by interacting with other proteins, for example BRCA1, following DNA damage. In response to strand breaks or other type of DNA damage, the ATM protein coordinates DNA repair by activating other proteins. Because of its central role in cell division and DNA repair, the ATM protein is important in carcinogenesis. More than one-fourth of miRNAs were significantly upregulated after DNA damage, while loss of ATM activity abolished their induction. Their results show that DNA damage activates the ATM kinase that directly binds to and phosphorylates KH-type splicing regulatory protein (KSRP), leading to enhanced interaction between KSRP and pri-miRNAs and increased KSRP activity in miRNA processing. The increased activity, in turn, results in more pre-miRNAs from pri-miRNAs, so that more miRNA products are produced to respond to the DNA damage.

Other studies show a different mechanism by which DNA damage signaling is linked to the miRNA maturation processes. Several miRNAs with growth suppressive function, including miR-16-1, miR-143 and miR-145, were regulated at the post transcriptional level through a P53-mediated miRNA maturation process in response to DNA damage (Suzuki et al., 2009; Toledo and Bardot, 2009). The P53 tumor suppressor protein binds to Drosha to facilitate the processing of pri-miRNAs to pre-miRNAs. Mutation in the DNA-binding domain of P53 decreases processing of pri-miRNAs by Drosha, and reduces the expression of the related miRNAs. In silico analyses, all three component of the P53 tumor suppressor, P53, P63, and P73, can regulate the major components of miRNA processing, such as Drosha-DGCR8, Dicer-TRBP2, and Agronaute proteins. Thus, when DNA damage activates the P53 gene, the activated P53 gene can modulate miRNA expression by affecting the miRNA biogenesis processes.

miR-24 regulates the DNA damage response by down-regulation of H2AX, the initial sensor protein for the DNA damage response. miR-100, miR-101 and miR-421 suppress ATM, the chief transducer of the DNA damage response, by targeting the 3'-UTR of ATM. miR-16 can up-regulate ATM activity by suppressing levels of Wip1. DNA repair pathways are regulated by a number of miRNAs involved in different types of DNA damage correction. the NER protein RAD23B was down-regulated by miR-373. MMR protein MSH2 and MSH6 were down regulated by miR-21 and MLH1/MSH2 were suppressed by miR-155. The HRR protein BRCA1 was down-regulated by miR-182 and RAD52 was suppressed by miR-210 and miR-373. The NHEJ protein DNA-PKcs was suppressed by miR-101 (Yan, Ng. 2010).

4. miRNA regulation of signal transduction for DNA damage

miRNAs regulate multiple aspects of the DNA damage response pathway, including regulation of signal transduction of DNA damage, changing expression level of master regulatory proteins such as P53, modulating key protein expression in different types of DNA repair such as MMR, NER, NHEJ and HRR. Figure 2 and Table 1 summarize recently reported miRNAs associated with DNA damage and repair.

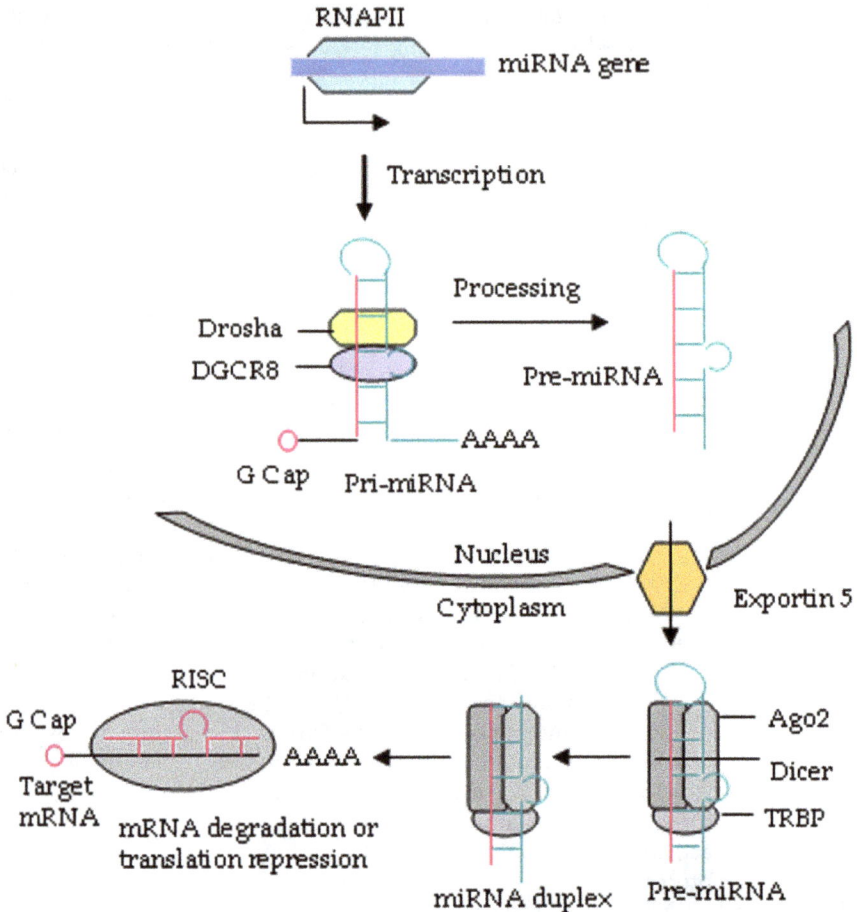

Fig. 1. MicroRNA biogenesis. A microRNA gene is transcribed by RNA polymerase II (RNAPII) to produce a pri-miRNA. The pri-miRNA is formed by RNase III family Drosha, cooperating in a complex with dsRNA-binding proteins DGCR8. The Drosha–DGCR8 complex processes the pri-miRNA into an ~70-nucleotide pre-miRNA, which is exported to the cytoplasm by expotin 5. The cytoplasm pre-miRNA is cleaved by Dicer, assisted by TRBP and AGO2, and yields an ~20-bp miRNA/miRNA* duplex. One strand of the miRNA/miRNA* duplex is preferentially incorporated into a miRNA-induced silencing complex (RISC), whereas the other strand is degraded (not shown). RISC recognizes target mRNAs and lets the miRNA binds to its target mRNA to suppress gene expression, either by mRNA cleavage or translational repression.

DNA damage activates the signal transduction process that leads to cell cycle arrest, which can lead to apoptosis or DNA repair. This DNA-damage response is mainly regulated at the transcriptional and posttranslational levels. Recent evidence suggests that miRNAs offer another degree of regulation at the posttranscriptional level in response to DNA damage. The DNA damage response to UV light was severely attenuated after the key components of

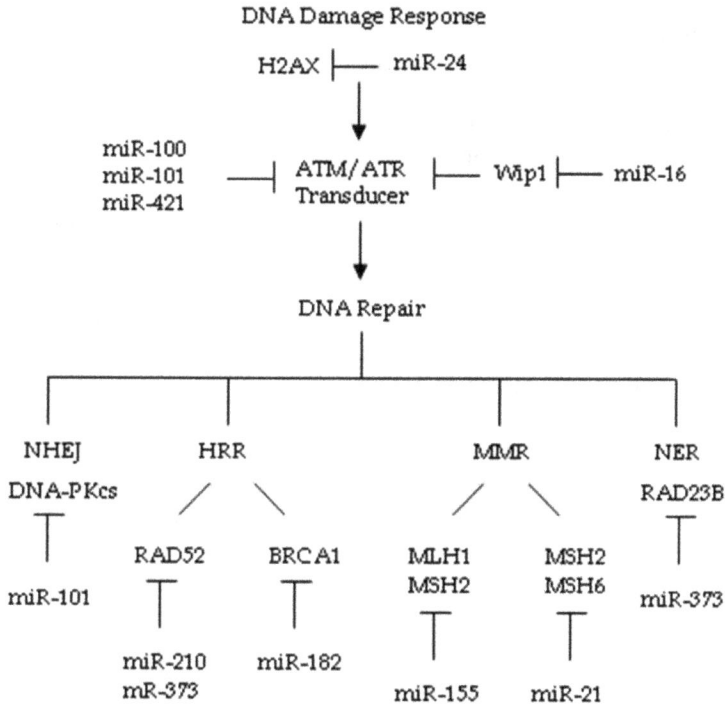

Fig. 2. miRNAs directly regulate DNA repair

the miRNA-processing pathway (Dicer and Ago2) were knocked down. miRNA mediated gene regulation operates earlier than most other transcriptional responses following genotoxic stress (Pothof et al., 2009).

H2AX, a histone variant, is an initial sensor protein for the DNA damage response. The function of H2AX is associated with DNA double strand break repair. miR-24 expression is up-regulated during hematopoietic cell differentiation into multiple lineages. miR-24 regulates H2AX expression through binding to its 3'-UTR. Both H2AX mRNA and protein levels are dramatically reduced by high levels of miR-24 in terminal differentiated human blood cells. miR-24 mediated suppression of H2AX in terminally differentiated blood cells renders them hypersensitive to gamma-irradiation, deficient in DSB repair, and susceptible to chromosomal instability (Lal et al., 2009).

Wild-type p53-induced phosphatase 1 (Wip1) is an oncogene with critical function in the ATM/ATR-p53 DNA damage signaling pathway. Wip1 reverses DNA damage–induced cell cycle checkpoints by dephosphorylating several key DNA damage responsive proteins. Recently, miRNAs are found to play an important role in suppressing Wip1 activity. Knockdown of miR-15a and miR-16 promotes survival, proliferation and invasiveness of untransformed prostate cells, and tumor formation in immunodeficient NOD-SCID mice. Conversely, reconstitution of miR-15a and miR-16 expression results in marked regression of prostate tumor xenografts. The function of miR-15a and miR-16 is considered through their regulation of Wip1 expression. miR-16 can down-regulate the expression level of Wip1 by targeting the 3′ UTR of Wip1. As a result, the Wip1 protein level is significantly deceased, which prevents a premature inactivation of ATM/ATR signaling and allows a functional completion of the early DNA damage response (Zhang et al., 2010).

miRNA	Pathway Involved	Target Protein	Net Effect	Reference
miR-24	DDR	H2AX	-	Lal, Pan. 2009
miR-16	DDR	Wip1	+	Zhang, Wan. 2010
miR-100	DDR	ATM	-	Ng, WL. 2010
miR-101	DDR	ATM	-	Yan, Ng. 2010
miR-421	DDR	ATM	-	Hu, Du. 2010)
miR-373	NER	RAD23B	-	Crosby, Kulshreshtha. 2009
miR-21	MMR	MSH2, MSH6	-	Valeri, Gasparini. 2010)
miR-155	MMR	MLH1, MSH2	-	Volinia, Calin. 2006
miR-182	HRR	BRCA1	-	Moskwa, Buffa. 2011
miR-210	HRR	RAD52	-	Crosby, Kulshreshtha. 2009
miR-373	HRR	RAD52	-	Crosby, Kulshreshtha. 2009
miR-101	NHEJ	DNA-PKcs	-	Yan, Ng. 2010
miR-29	P53	P85a, CDC42	+	Park, Lee. 2009
miR-34a	P53	SIRT1	+	Yamakuchi, Ferlito. 2008)
miR122	P53	Cyclin G1	+	Fornari, Gramantieri. 2009
miR-125b	P53	P53	-	Le, Teh. 2009
miR-504	P53	P53	-	Hu, Chan. 2010

Table 1. miRNAs involved in DNA repair (notes: - means inhibit and + means stimulate)

ATM is a serine/threonine kinase that transfers the DNA damage signals to down-steam events, such as cell cycle arrest, apoptosis and DNA repair (Lavin, 2008; Shiloh, 2003). ATM plays a critical role in the maintenance of genomic stability by activating cell cycle checkpoints and promoting DNA double-strand breaks repair. M059J is a human malignant glioma cell line with high sensitivity to ionizing radiation due to low-expression of ATM. The low-expression of ATM is related to miR-100 (Ng et al., 2010). Both computational analysis and luciferase reporter gene assay indicate that miR-100 can target the 3'-UTR of ATM. miR-100 was found to be highly-expressed in M059J cells by RNase protection assay and qRT-PCR. Up-regulation of miR-100 in M059K cells reduces ATM expression and renders them hypersensitive to ionizing radiation, while Knock-down of miR-100 promotes ATM expression in M059J cells. These results indicate that the low-expression of ATM in M059J cells is mainly due to the high expression of miR-100.

Another miRNA miR-421 is also involved in ATM regulation. miR-421 suppresses the expression of ATM by targeting the 3′ UTR of ATM. Ectopic expression of miR-421 lead to a deficient cell cycle checkpoint in S-phase and increased sensitivity to ionizing radiation (Hu et al., 2010a). Blocking the interaction between miR-421 and ATM with chemically synthesized oligonucleotides rescued the defective phenotype caused by miR-421 over expression, suggesting that ATM mediates the effect of miR-421 on cell-cycle checkpoints followed by radiation.

5. miRNA regulation of core components of DNA damage response

miRNAs are involved in DNA repair by regulating critical components of the DNA repair pathways, such as P53. As a transcription factor, the tumor suppressor P53 is a powerful regulator of diverse cellular processes including cell-cycle arrest, DNA repair, apoptosis and cellular senescence. P53 and its signaling pathway, play a pivotal role in maintaining genomic stability and tumor suppression (Levine et al., 2004; Levine et al., 2006). Recently, P53 activity was found to be widely regulated by a number of miRNAs. These miRNAs either directly target the 3' UTR of P53 or indirectly regulate P53 activity by modulating proteins associated with P53 (Figure 3). Among these miRNAs, miR-504 negatively regulate p53 expression through binding to two DNA *cis* element located in the P53 3' UTR. Ectopic expression of miR-504 reduces the protein level of P53 and impairs P53-mediated apoptosis and cell cycle arrest (Hu et al., 2010b). miR-125b is another negative regulator of P53 in both zebrafish and humans (Le et al., 2009). Knocking down of miR-125b increased the expression level of P53; and over-expression of miR-125b suppressed the expression of P53. Interestingly, miR-125b was down-regulated when the Zebrafish embryo was exposed to gamma irradiation, corresponding to the up-regulation of P53 protein induced by the irradiation exposure.

In addition to the direct binding to P53, several miRNA including miR-34a, miR-29 and miR-122 can indirectly modify P53 activity (Fornari et al., 2009; Park et al., 2009; Yamakuchi et al., 2008). miR-34a is a direct transcriptional target of P53 (Chang, et al., 2007; Corney, et al., 2007; Raver-Shapira, et al., 2007). P53 can up-regulate miR-34a expression by binding to a palindromic sequence located in miR-34a promoter region. miR-34a can positively regulate P53-dependent apoptosis through another intermediate protein, SIRT1 (Yamakuchi, et al., 2008). miR-34 inhibition of SIRT1 leads to an increase in acetylated P53. As a result, miR-34 suppression of SIRT1 ultimately leads to P53 mediated apoptosis in human colon cancer cells. miR-29 family members directly suppress P85a and CDC42, both of which negatively regulate P53. As a result, miR-29 positively up-regulates P53 level and induces apoptosis and DNA repair in a P53-dependent manner (Park, et al., 2009). miR-122 is a liver-specific miRNA accounting for 70% of the total miRNA population. miR-122 can down-regulate the expression of cyclin G1, which has the potential to inhibit P53 activity and promote cancer development. From a therapeutic perspective, miR-122 has potential to become a miRNA based therapy for hepatocellular carcinoma (HCC) patients (Fornari, et al., 2009).

6. Functions of miRNAs in mismatch repair (MMR)

MMR corrects erroneous deletion, insertion, or mis-incorporation of bases resulting from DNA replication, DNA recombination, or DNA damage. Human mutS homolog 2 (hMSH2) and mutL homolog 1 (hMLH1) function as core proteins in MMR. They form heterodimers with protein homologs hMSH3 or hMSH6 (Fishel, 2001). The over-expression of miR-21 is linked to progression of human colorectal cancer (Link et al., 2010; Ng et al., 2009). It was reported recently that miR-21 directly targeted the 3' UTRs of both the hMSH2 and hMSH6 mRNAs (Valeri et al., 2010a). Protein levels of hMSH2 and hMSH6 in the cells transfected with a locked nucleic acid (LNA) against miR-21 were significantly increased over the control cells. In addition, the over-expression of miR-21 was inversely correlated with the down regulation of hMSH2 in colorectal cancer tumors. Because the hMSH2-hMSH6 heterodimer is the key initiation component of MMR, the down regulation of hMSH2 is likely to suppress MMR, and ultimately enhance tumor progression.

miR-155 also plays a critical role in MMR. Over-expression of miR-155 reduced the levels of the human mismatch repair genes MLH1, MSH2 and MSH6 in a colorectal cancer cell line.

In addition, high expression of miR-155 was inversely correlated with the low expression of MLH1 and MSH2 protein in human colorectal cancer. More importantly, human tumors with unexplained MMR inactivation showed miR-155 over expression (Valeri et al., 2010b; Volinia et al., 2006). These results indicate that increased expression of miR-155 down-regulates MMR genes and results in an increase in genomic instability.

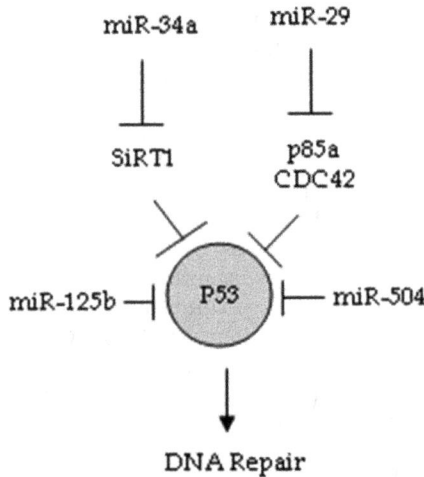

Fig. 3. miRNA indirectly regulates DNA repair through P53

miR-504 and miR-125b directly bind to the P53 3'-UTR and down-regulate P53 activity. miR-34a positively up-regulates P53 through SIRT1 inhibition, a negative regulator of P53. miR-29 down-regulates the P85a regulatory subunit of PI3K, which enhances P53 activity through the negative feedback loop between PI3K-AKT-MDM2 and P53.

7. Functions of miRNAs in nucleotide excision repair (NER)

NER recognizes bulky, helix distorting defects, such as cross-linking thymine dimmers. NER is particularly important for removing the vast majority of UV-induced DNA damage. Currently, only one miRNA is reported to be related with NER (Crosby et al., 2009). miR-373 suppresses the expression of a NER protein called RAD23B. RAD23B is a key component of the XPC/RAD23B complex that mediates damage recognition in the NER pathway (Batty et al., 2000). NER activity is functionally reduced in hypoxic cells (Yuan et al., 2000). A possible mechanism for the hpoxia-induced down-regulation of RAD23B is that hpoxia can up-regulate miR-373 expression, and the up-regulated miR-373 then suppresses RAD23B expression. This mechanism was supported by the fact that pre-treating cells with anti-miR-373 reversed the hypoxia-mediated down-regulation of RAD23B in hypoxic cells (Crosby, et al., 2009).

8. Functions of miRNAs in non-homologous end-joining (NHEJ)

NHEJ is a relatively simple but error prone DNA double strand break repair. It ligates broken ends, without the need for a homologous template. DNA protein kinase (DNA-PK) is a core component of mammalian NHEJ and involves a catalytic subunit (DNA-PKcs) that can act as a regulatory element. DNA-PKcs is a molecular sensor for DNA damage that enhances the signal via phosphorylation of many downstream targets. Therefore, DNA-PKcs is an essential factor for NHEJ. Yan et al. found that miR-101 could efficiently target DNA-PKcs and ATM via binding to their 3'- UTRs. Up-regulating miR-101 efficiently reduced the protein levels of DNA-PKcs and ATM in tumor cells, and most importantly, sensitized the tumor cells to radiation in vitro and in vivo (Yan et al., 2010). Radiotherapy kills tumor-cells by inducing DNA double strand breaks (DSBs). However, the efficient repair of double strand breaks in tumors frequently prevents successful treatment. Therefore, miR-101 could be used to target DNA DSB repair genes, in order to sensitize tumors to radiation and improve tumor radiotherapy.

9. Functions of miRNAs in homologous recombination repair (HRR)

HRR is the most widely used repair mechanism which can accurately repair DNA double strand breaks. HRR reconstitutes the genetic information using the sister chromatid as a template. Several proteins are involved in the HRR process. Rad 52 protein recognizes double-strand breaks and adheres to the free ends of the break while the Rad51 protein, together with tumor-suppressor protein BRCA1, searches the undamaged sister chromatid for homologous pairing (Haber, 2000; Orelli and Bishop, 2001).

Both miR-210 and miR-373 were up-regulated in hypoxic cells. Up-regulation of miR-210 significantly suppressed the expression level of RAD51, while up-regulation of miR-373 inhibited the expression of RAD52. The modulation of miR-210 to RAD51 and miR-373 to RAD52 were verified by microarray analysis and luciferase reporter gene assay. Both of the miRNAs can bind to the binding sites in the 3' UTRs of their respective target mRNAs (Crosby, et al., 2009). Thus, hypoxia-inducible miR-210 and miR-373 regulate HRR via targeting RAD51 and RAD52.

BRCA1 is a constituent of several different protein complexes and is a key protein for HRR. Expression of BRCA1 is commonly decreased in sporadic breast tumors, and this correlates with poor prognosis of breast cancer patients (Mueller and Roskelley, 2003). It was recently reported that miR-182 down-regulated BRCA1 expression. As a result, the HRR efficiency for DNA double strand break repair was impaired (Moskwa et al., 2011; Yao and Ventura, 2011). Antagonizing miR-182 enhanced BRCA1 protein level, which, in turn, protected cells from irradiation exposure. Over-expressing of miR-182 reduced BRCA1 protein level, which impaired HRR efficiency and rendered cells hypersensitive to irradiation. The impaired HRR phenotype due to miR-182 over-expression was able to be fully rescued by over-expressing of BRCA1. Thus, these data demonstrate miR-182-mediated down-regulation of BRCA1 suppresses HRR.

10. Conclusion

miRNAs appear to be involved in DNA damage and repair in many ways. miRNA biogenesis, including miRNA gene transcription and miRNA maturation processes, is readily altered in response to DNA damage. miRNAs regulate the ATM and P53 that are the regulators of the global induction of miRNA biogenesis upon DNA damage. miRNAs are also involved in signal transduction processes that leads to cell cycle arrest, apoptosis or DNA repair upon DNA damage. miR-100 and miR-421 can regulate expression of ATM, a critical protein in DNA damage signalling. miR-24 suppresses gene expression of H2AX, an initial sensor protein for DNA damage response. miR-16 down-regulates the expression level of Wip1, an inhibitor of ATM/ATR-p53 DNA damage signalling pathway. miRNAs can mediate the activity of P53, a core component of the DNA damage response. miR-504 and miR-125b negatively regulate p53 expression. miR-34a, miR-29 and miR-122 can indirectly modify P53 activity by regulating the P53-related factors. miRNAs play important roles in different types of DNA repair. miR-21 down-regulates MMR proteins, MSH2 and MSH6, while miR-155 reduced the expression of the MMS genes MLH1, MSH2 and MSH6. miR-373 suppresses expression of RAD23B, a key component of the NER. miR-101 down-regulates the protein level of DNA-PKcs, an essential factor for NHEJ. miR-210, miR-373 and miR-182 down-regulate the expression of RAD51, RAD52 and BRCA1, respectively. RAD51, RAD52 and BRCA1 are all key components of HRR. With increased studies of miRNAs' roles in DNA damage and repair, more miRNAs will be discovered to involve in the DNA damage and repair pathways.

11. Acknowledgements

The views presented in this article do not necessarily reflect those of the Food and Drug Administration. We would like to thank Dr. Barbara Parsons and Mr. Jian Yan for their review of this manuscript.

12. References

Batty, D., Rapic'-Otrin, V., Levine, A. S. and Wood, R. D. (2000). Stable binding of human XPC complex to irradiated DNA confers strong discrimination for damaged sites. J Mol Biol 300, 275-90.

Braun, C. J., Zhang, X., Savelyeva, I., Wolff, S., Moll, U. M., Schepeler, T., Orntoft, T. F., Andersen, C. L. and Dobbelstein, M. (2008). p53-Responsive micrornas 192 and 215 are capable of inducing cell cycle arrest. Cancer Res 68, 10094-104.

Cai, X., Hagedorn, C. H. and Cullen, B. R. (2004). Human microRNAs are processed from capped, polyadenylated transcripts that can also function as mRNAs. RNA 10, 1957-66.

Cannell, I. G., Kong, Y. W., Johnston, S. J., Chen, M. L., Collins, H. M., Dobbyn, H. C., Elia, A., Kress, T. R., Dickens, M., Clemens, M. J., Heery, D. M., Gaestel, M., Eilers, M., Willis, A. E. and Bushell, M. (2010). p38 MAPK/MK2-mediated induction of miR-34c following DNA damage prevents Myc-dependent DNA replication. Proc Natl Acad Sci U S A 107, 5375-80.

Chang, T. C., Wentzel, E. A., Kent, O. A., Ramachandran, K., Mullendore, M., Lee, K. H., Feldmann, G., Yamakuchi, M., Ferlito, M., Lowenstein, C. J., Arking, D. E., Beer, M. A., Maitra, A. and Mendell, J. T. (2007). Transactivation of miR-34a by p53 broadly influences gene expression and promotes apoptosis. Mol Cell 26, 745-52.

Chen, D. H., Li, Z. and Chen, T. (2011). Increased Expression of miR-34a in mouse spleen one day after exposure to N-ethyl-N-nitrosourea. Applied Journal of Toxicology.

Corney, D. C., Flesken-Nikitin, A., Godwin, A. K., Wang, W. and Nikitin, A. Y. (2007). MicroRNA-34b and MicroRNA-34c are targets of p53 and cooperate in control of cell proliferation and adhesion-independent growth. Cancer Res 67, 8433-8.

Crosby, M. E., Kulshreshtha, R., Ivan, M. and Glazer, P. M. (2009). MicroRNA regulation of DNA repair gene expression in hypoxic stress. Cancer Res 69, 1221-9.

Fishel, R. (2001). The selection for mismatch repair defects in hereditary nonpolyposis colorectal cancer: revising the mutator hypothesis. Cancer Res 61, 7369-74.

Fornari, F., Gramantieri, L., Giovannini, C., Veronese, A., Ferracin, M., Sabbioni, S., Calin, G. A., Grazi, G. L., Croce, C. M., Tavolari, S., Chieco, P., Negrini, M. and Bolondi, L. (2009). MiR-122/cyclin G1 interaction modulates p53 activity and affects doxorubicin sensitivity of human hepatocarcinoma cells. Cancer Res 69, 5761-7.

Georges, S. A., Biery, M. C., Kim, S. Y., Schelter, J. M., Guo, J., Chang, A. N., Jackson, A. L., Carleton, M. O., Linsley, P. S., Cleary, M. A. and Chau, B. N. (2008). Coordinated regulation of cell cycle transcripts by p53-Inducible microRNAs, miR-192 and miR-215. Cancer Res 68, 10105-12.

Gregory, R. I., Chendrimada, T. P., Cooch, N. and Shiekhattar, R. (2005). Human RISC couples microRNA biogenesis and posttranscriptional gene silencing. Cell 123, 631-40.

Grishok, A., Pasquinelli, A. E., Conte, D., Li, N., Parrish, S., Ha, I., Baillie, D. L., Fire, A., Ruvkun, G. and Mello, C. C. (2001). Genes and mechanisms related to RNA interference regulate expression of the small temporal RNAs that control C. elegans developmental timing. Cell 106, 23-34.

Haber, J. E. (2000). Partners and pathwaysrepairing a double-strand break. Trends Genet 16, 259-64.

Han, J., Lee, Y., Yeom, K. H., Kim, Y. K., Jin, H. and Kim, V. N. (2004). The Drosha-DGCR8 complex in primary microRNA processing. Genes Dev 18, 3016-27.

He, L., He, X., Lim, L. P., de Stanchina, E., Xuan, Z., Liang, Y., Xue, W., Zender, L., Magnus, J., Ridzon, D., Jackson, A. L., Linsley, P. S., Chen, C., Lowe, S. W., Cleary, M. A. and Hannon, G. J. (2007). A microRNA component of the p53 tumour suppressor network. Nature 447, 1130-4.

Hu, H., Du, L., Nagabayashi, G., Seeger, R. C. and Gatti, R. A. (2010a). ATM is down-regulated by N-Myc-regulated microRNA-421. Proc Natl Acad Sci U S A 107, 1506-11.

Hu, W., Chan, C. S., Wu, R., Zhang, C., Sun, Y., Song, J. S., Tang, L. H., Levine, A. J. and Feng, Z. (2010b). Negative regulation of tumor suppressor p53 by microRNA miR-504. Mol Cell 38, 689-99.

Hutvagner, G., McLachlan, J., Pasquinelli, A. E., Balint, E., Tuschl, T. and Zamore, P. D. (2001). A cellular function for the RNA-interference enzyme Dicer in the maturation of the let-7 small temporal RNA. Science 293, 834-8.

Kato, M., Paranjape, T., Muller, R. U., Nallur, S., Gillespie, E., Keane, K., Esquela-Kerscher, A., Weidhaas, J. B. and Slack, F. J. (2009). The mir-34 microRNA is required for the DNA damage response in vivo in C. elegans and in vitro in human breast cancer cells. Oncogene 28, 2419-24.

Ketting, R. F., Fischer, S. E., Bernstein, E., Sijen, T., Hannon, G. J. and Plasterk, R. H. (2001). Dicer functions in RNA interference and in synthesis of small RNA involved in developmental timing in C. elegans. Genes Dev 15, 2654-9.

Lal, A., Pan, Y., Navarro, F., Dykxhoorn, D. M., Moreau, L., Meire, E., Bentwich, Z., Lieberman, J. and Chowdhury, D. (2009). miR-24-mediated downregulation of

H2AX suppresses DNA repair in terminally differentiated blood cells. Nat Struct Mol Biol 16, 492-8.

Lavin, M. F. (2008). Ataxia-telangiectasia: from a rare disorder to a paradigm for cell signalling and cancer. Nat Rev Mol Cell Biol 9, 759-69.

Le, M. T., Teh, C., Shyh-Chang, N., Xie, H., Zhou, B., Korzh, V., Lodish, H. F. and Lim, B. (2009). MicroRNA-125b is a novel negative regulator of p53. Genes Dev 23, 862-76.

Lee, Y., Ahn, C., Han, J., Choi, H., Kim, J., Yim, J., Lee, J., Provost, P., Radmark, O., Kim, S. and Kim, V. N. (2003). The nuclear RNase III Drosha initiates microRNA processing. Nature 425, 415-9.

Lee, Y., Kim, M., Han, J., Yeom, K. H., Lee, S., Baek, S. H. and Kim, V. N. (2004). MicroRNA genes are transcribed by RNA polymerase II. EMBO J 23, 4051-60.

Levine, A. J., Finlay, C. A. and Hinds, P. W. (2004). P53 is a tumor suppressor gene. Cell 116, S67-9, 1 p following S69.

Levine, A. J., Hu, W. and Feng, Z. (2006). The P53 pathway: what questions remain to be explored? Cell Death Differ 13, 1027-36.

Li, Z., Branham, W. S., Dial, S. L., Wang, Y., Guo, L., Shi, L. and Chen, T. (2010). Genomic analysis of microRNA time-course expression in liver of mice treated with genotoxic carcinogen N-ethyl-N-nitrosourea. BMC Genomics 11, 609.

Li, Z., Fuscoe, J. C. and Chen, T. (2011). MicroRNAs and their predicted target messenger RNAs are deregulated by Exposure to a Carcinogenic Dose of Comfrey in Rat Liver. Environ Mol Mutagen.

Link, A., Balaguer, F., Shen, Y., Nagasaka, T., Lozano, J. J., Boland, C. R. and Goel, A. (2010). Fecal MicroRNAs as novel biomarkers for colon cancer screening. Cancer Epidemiol Biomarkers Prev 19, 1766-74.

Lodish H, B. A., Matsudaira P, Kaiser CA, Krieger M, Scott MP, Zipursky SL, Darnell J. (2004). Molecular Biology of the Cell, p963.

Maniataki, E. and Mourelatos, Z. (2005). A human, ATP-independent, RISC assembly machine fueled by pre-miRNA. Genes Dev 19, 2979-90.

Meister, G. and Tuschl, T. (2004). Mechanisms of gene silencing by double-stranded RNA. Nature 431, 343-9.

Moskwa, P., Buffa, F. M., Pan, Y., Panchakshari, R., Gottipati, P., Muschel, R. J., Beech, J., Kulshrestha, R., Abdelmohsen, K., Weinstock, D. M., Gorospe, M., Harris, A. L., Helleday, T. and Chowdhury, D. (2011). miR-182-mediated downregulation of BRCA1 impacts DNA repair and sensitivity to PARP inhibitors. Mol Cell 41, 210-20.

Mueller, C. R. and Roskelley, C. D. (2003). Regulation of BRCA1 expression and its relationship to sporadic breast cancer. Breast Cancer Res 5, 45-52.

Ng, E. K., Chong, W. W., Jin, H., Lam, E. K., Shin, V. Y., Yu, J., Poon, T. C., Ng, S. S. and Sung, J. J. (2009). Differential expression of microRNAs in plasma of patients with colorectal cancer: a potential marker for colorectal cancer screening. Gut 58, 1375-81.

Ng, W. L., Yan, D., Zhang, X., Mo, Y. Y. and Wang, Y. (2010). Over-expression of miR-100 is responsible for the low-expression of ATM in the human glioma cell line: M059J. DNA Repair (Amst) 9, 1170-5.

Orelli, B. J. and Bishop, D. K. (2001). BRCA2 and homologous recombination. Breast Cancer Res 3, 294-8.

Ozsolak, F., Poling, L. L., Wang, Z., Liu, H., Liu, X. S., Roeder, R. G., Zhang, X., Song, J. S. and Fisher, D. E. (2008). Chromatin structure analyses identify miRNA promoters. Genes Dev 22, 3172-83.

Park, S. Y., Lee, J. H., Ha, M., Nam, J. W. and Kim, V. N. (2009). miR-29 miRNAs activate p53 by targeting p85 alpha and CDC42. Nat Struct Mol Biol 16, 23-9.

Pothof, J., Verkaik, N. S., van, I. W., Wiemer, E. A., Ta, V. T., van der Horst, G. T., Jaspers, N. G., van Gent, D. C., Hoeijmakers, J. H. and Persengiev, S. P. (2009). MicroRNA-mediated gene silencing modulates the UV-induced DNA-damage response. EMBO J 28, 2090-9.

Raver-Shapira, N., Marciano, E., Meiri, E., Spector, Y., Rosenfeld, N., Moskovits, N., Bentwich, Z. and Oren, M. (2007). Transcriptional activation of miR-34a contributes to p53-mediated apoptosis. Mol Cell 26, 731-43.

Rich, T., Allen, R. L. and Wyllie, A. H. (2000). Defying death after DNA damage. Nature 407, 777-83.

Shiloh, Y. (2003). ATM and related protein kinases: safeguarding genome integrity. Nat Rev Cancer 3, 155-68.

Suzuki, H. I., Yamagata, K., Sugimoto, K., Iwamoto, T., Kato, S. and Miyazono, K. (2009). Modulation of microRNA processing by p53. Nature 460, 529-33.

Toledo, F. and Bardot, B. (2009). Cancer: Three birds with one stone. Nature 460, 466-7.

Valeri, N., Gasparini, P., Braconi, C., Paone, A., Lovat, F., Fabbri, M., Sumani, K. M., Alder, H., Amadori, D., Patel, T., Nuovo, G. J., Fishel, R. and Croce, C. M. (2010a). MicroRNA-21 induces resistance to 5-fluorouracil by down-regulating human DNA MutS homolog 2 (hMSH2). Proc Natl Acad Sci U S A 107, 21098-103.

Valeri, N., Gasparini, P., Fabbri, M., Braconi, C., Veronese, A., Lovat, F., Adair, B., Vannini, I., Fanini, F., Bottoni, A., Costinean, S., Sandhu, S. K., Nuovo, G. J., Alder, H., Gafa, R., Calore, F., Ferracin, M., Lanza, G., Volinia, S., Negrini, M., McIlhatton, M. A., Amadori, D., Fishel, R. and Croce, C. M. (2010b). Modulation of mismatch repair and genomic stability by miR-155. Proc Natl Acad Sci U S A 107, 6982-7.

Volinia, S., Calin, G. A., Liu, C. G., Ambs, S., Cimmino, A., Petrocca, F., Visone, R., Iorio, M., Roldo, C., Ferracin, M., Prueitt, R. L., Yanaihara, N., Lanza, G., Scarpa, A., Vecchione, A., Negrini, M., Harris, C. C. and Croce, C. M. (2006). A microRNA expression signature of human solid tumors defines cancer gene targets. Proc Natl Acad Sci U S A 103, 2257-61.

Welch, C., Chen, Y. and Stallings, R. L. (2007). MicroRNA-34a functions as a potential tumor suppressor by inducing apoptosis in neuroblastoma cells. Oncogene 26, 5017-22.

Yamakuchi, M., Ferlito, M. and Lowenstein, C. J. (2008). miR-34a repression of SIRT1 regulates apoptosis. Proc Natl Acad Sci U S A 105, 13421-6.

Yan, D., Ng, W. L., Zhang, X., Wang, P., Zhang, Z., Mo, Y. Y., Mao, H., Hao, C., Olson, J. J., Curran, W. J. and Wang, Y. (2010). Targeting DNA-PKcs and ATM with miR-101 sensitizes tumors to radiation. PLoS One 5, e11397.

Yao, E. and Ventura, A. (2011). A new role for miR-182 in DNA repair. Mol Cell 41, 135-7.

Yuan, J., Narayanan, L., Rockwell, S. and Glazer, P. M. (2000). Diminished DNA repair and elevated mutagenesis in mammalian cells exposed to hypoxia and low pH. Cancer Res 60, 4372-6.

Zenz, T., Mohr, J., Eldering, E., Kater, A. P., Buhler, A., Kienle, D., Winkler, D., Durig, J., van Oers, M. H., Mertens, D., Dohner, H. and Stilgenbauer, S. (2009). miR-34a as part of the resistance network in chronic lymphocytic leukemia. Blood 113, 3801-8.

Zhang, X., Wan, G., Mlotshwa, S., Vance, V., Berger, F. G., Chen, H. and Lu, X. (2010). Oncogenic Wip1 phosphatase is inhibited by miR-16 in the DNA damage signaling pathway. Cancer Res 70, 7176-86.

Zhang, X., Wan, G., Berger, F. G., He, X. and Lu, X. (2011). The ATM kinase induces microRNA biogenesis in the DNA damage response. Mol Cell 41, 371-83.

18

BRCA2 Mutations and Consequences for DNA Repair

Erika T. Brown
Medical University of South Carolina
United States

1. Introduction

The *BRCA2* gene was the second gene discovered to be associated with early-onset, familial breast cancer. The BRCA2 protein is expressed in breast, ovarian, prostate, and pancreatic tissues and is associated with cancer predisposition in all four, with breast cancer being the most predominant (Goggins, Schutte et al. 1996). BRCA2 is functionally defined as a tumor suppressor and is most critical in maintenance of genomic integrity and DNA repair fidelity. The importance of BRCA2 in maintaining genomic integrity is based on its function to specifically repair double-strand DNA breaks (DSBs) via the process of homologous recombination (HR). However, BRCA2 resolves genomic lesions in concert with a number of DNA repair proteins, the most significant being RAD51 (Sharan, Morimatsu et al. 1997),(Yuan, Lee et al. 1999). RAD51 is a recombinase that is highly conserved, having homologues in *E. coli* and yeast, as well as in mammals. BRCA2 modulates the activity of RAD51 during DNA repair, and they both are found in nuclear DNA damage-induced foci, which are complexes of DNA repair proteins bound to DNA during the process of repair (Roth, Porter et al. 1985; Roth and Wilson 1986; Derbyshire, Epstein et al. 1994; Jackson and Jeggo 1995; Takata, Sasaki et al. 1998; Johnson and Jasin 2000). The relationship between BRCA2 and RAD51 has been determined to be a fundamental interaction in the repair of DSBs.

The role of BRCA2 as a tumor suppressor has been established by its importance in maintaining genomic integrity. The inability of the cell to repair DSBs can potentially cause small-scale lesions in regions of the DNA that encode single genes and incite large-scale lesions, such as chromosomal anomalies. The consequence of such damage can disrupt the normal expression of gene products that are required to regulate cell growth and arrest and induce apoptosis, thereby establishing a cellular environment that can foster malignant transformation.

Cancer cells that express mutated BRCA2 have been shown to have elevated sensitivity to the anti-cancer therapeutics called PARP (Poly [ADP-ribose] polymerase) inhibitors. PARP inhibitors prevent the binding of PARP to sites of damaged DNA, which serves as a signal to initiate DNA repair (Schreiber, Dantzer et al. 2006); (Ratnam and Low 2007). The effectiveness of PARP inhibitors in BRCA2-mutated cells is based on the premise of synthetic lethality, which is when two pathway defects alone are innocuous, but combined become lethal (Ratnam and Low 2007). The unresolved DSBs of BRCA2-mutated cells combined with the inhibition of PARP activity are effective in promoting DNA damage-induced apoptosis. This finding has established mutated BRCA2 as a potential target in improving present anti-cancer therapeutic regimens.

The information that follows will provide a comprehensive understanding of BRCA2, starting from its functions at the molecular level in maintaining genomic integrity, to describing how deregulation can lead to disease predisposition and development, and concluding with the development of PARP inhibitors that use the DNA repair defects of BRCA2-mutations to improve the sensitivity of anti-cancer treatments towards BRCA2-tumors.

2. The role of BRCA2 in DNA repair

The BRCA2 protein specifically repairs double-strand DNA breaks (DSBs) via the process of homologous recombination (HR), thereby establishing its importance in maintaining genomic integrity. The BRCA2 gene is found on chromosome 13q12.3 and encodes a protein of 3,418 amino acids, resulting in a molecular weight of approximately 340 kDa. BRCA2 resolves genomic lesions in a complex with several additional DNA repair proteins, the most significant being RAD51 (Sharan, Morimatsu et al. 1997; Yuan, Lee et al. 1999). RAD51 is a highly conserved recombinase, having homologues in *E. coli* and yeast, as well as in mammals. BRCA2 modulates the activity of RAD51 during DNA repair and this relationship is determined to be a fundamental interaction in repair of DSBs.

2.1 The interaction between BRCA2 and the RAD51 recombinase

RAD51 catalyzes the strand exchange of DNA homologues to promote gene conversion and repair DSBs by HR (Ogawa, Yu et al. 1993) (Benson, Stasiak et al. 1994). HR is one of two pathways of repair of DSBs in mammals—the other being nonhomologous end-joining (NHEJ) (Derbyshire, Epstein et al. 1994), (Jackson and Jeggo 1995), (Roth, Porter et al. 1985), (Roth and Wilson 1986), (Takata, Sasaki et al. 1998) and (Johnson and Jasin 2000), (Figure 1). HR requires the damaged DNA molecule to use the undamaged homologue as a template in order to repair the DSB. NHEJ involves ligation of the DNA ends at the breakpoint junction regardless of whether the original genetic information is still present. As a result, HR confers greater accuracy in repair than NHEJ (Derbyshire, Epstein et al. 1994), (Jackson and Jeggo 1995), (Roth, Porter et al. 1985), (Roth and Wilson 1986), (Takata, Sasaki et al. 1998) and (Johnson and Jasin 2000). Studies performed in mice in which the Rad51 gene was either mutated or completely knocked out have shown its importance in genomic stability and cell viability (Taki, Ohnishi et al. 1996) and (Sonoda, Sasaki et al. 1998). Nonfunctional RAD51 does not repair chromosome breaks and other DNA lesions, thereby leading to an accumulation of DSBs and stalled replication forks (Taki, Ohnishi et al. 1996) and (Sonoda, Sasaki et al. 1998). In addition, inactivation of the RAD51 gene causes embryonic lethality (Tsuzuki, Fujii et al. 1996).

2.2 The structure of BRCA2

Yeast two-hybrid screening assays were used in the discovery of the interaction between RAD51 and BRCA2 (Mizuta, LaSalle et al. 1997), (Wong, Pero et al. 1997), (Chen, Chen et al. 1998) and (Marmorstein, Ouchi et al. 1998). And, studies examining the interaction between the two proteins have collectively shown that BRCA2 has two regions for RAD51 binding. The first region is in the mid-portion of BRCA2 and consists of eight highly conserved amino acid motifs called BRC repeats (Figure 2). The repeats have different binding affinities for RAD51:

repeats 1–4, 7, and 8 all interact with RAD51, with repeats 3 and 4 having the strongest affinity (Wong, Pero et al. 1997) (Bignell, Micklem et al. 1997). The second RAD51 binding site is located on the CTD (C-terminal domain) of BRCA2. This RAD51 binding site is described as playing a major role in the regulation of RAD51 recombination activity by displacing the single-strand DNA binding protein replication protein A (RPA) from the exonucleolytically processed 3'-single-strand overhangs of the DSBs, thus allowing RAD51 to bind and form nucleoprotein filaments (Yang, Jeffrey et al. 2002). The CTD portion of BRCA2 has been shown to be highly active in HR-mediated repair with RAD51 (Yang, Jeffrey et al. 2002). This region consists of five domains significant in DNA repair. The first is the α-helical domain, which interacts with the DMC1 protein—a meiosis specific paralog of RAD51 that forms nucleoprotein filaments and catalyzes strand exchange, and that BRCA2 requires for meiotic recombination (Thorslund, Esashi et al. 2007); (Jensen, Carreira et al.). The next three domains are the oligonucleotide–oligosaccharide binding domains (OB1, OB2, OB3) that have structural similarities with ssDNA binding proteins such as replication protein A (RPA). And, the fifth domain is the tower domain, which extends from OB2, and has structural similarities with the DNA binding domains of bacterial site-specific recombinases able to bind double-strand DNA (Yang, Jeffrey et al. 2002).

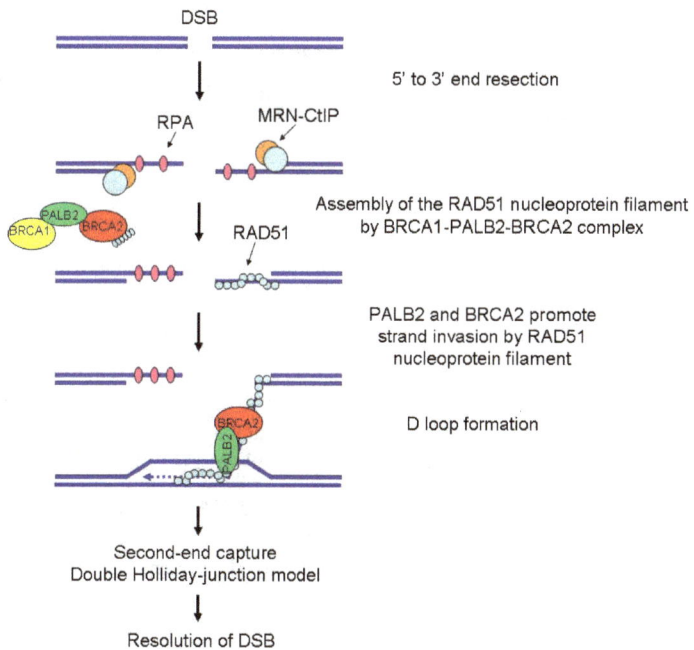

Fig. 1. Model of homologous recombination (HR)-mediated repair. After a DSB has occurred, the MRN-CtIP complex resects the 5'ends of the break. The 3'ssDNA overhangs are coated with replication protein A (RPA), which is displaced by RAD51. The BRCA1-PALB2-BRCA2 complex facilitates binding of RAD51 to form nucleoprotein filaments which invade the homologous strand, resulting in the D loop intermediate. This is followed by formation of the Holliday junction and resolution of the DSB.

Also located on the C-terminus of BRCA2 are its two nuclear localization signals (NLSs) (Spain, Larson et al. 1999) and (Yano, Morotomi et al. 2000). As a result, C-terminal mutations which disrupt, or truncations which remove, the NLSs are extremely detrimental to BRCA2 DNA repair functions, because they prevent nuclear localization. And, cell lines that have nonfunctional or absent BRCA2 NLSs primarily exhibit cytoplasmic localization of RAD51 after induction of DSBs by ionizing radiation (IR) (Spain, Larson et al. 1999). BRCA2 also interacts with RAD51 via a separate motif located at its C-terminus (Esashi, Christ et al. 2005). This interaction is regulated by cell cycle (CDK)-dependent phosphorylation of serine 3291 in exon 27 (and has been referred to, in some instances, as the "TR2" domain) and appears to function as a "switch" controlling recombinational repair activity during the transition from S/G2 to M phase in the cell cycle (Esashi, Christ et al. 2005). This phosphorylation site appears to be crucial in the checkpoint control mechanisms involved in the DNA repair pathway involving BRCA2.

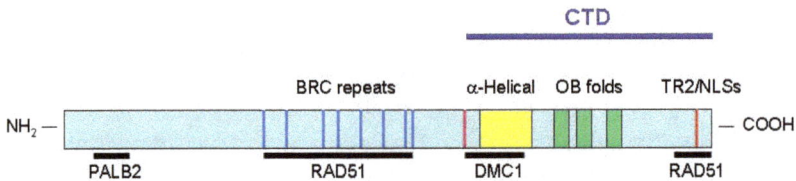

Fig. 2. Schematic of BRCA2. Starting at the N-terminus, the PALB2, the DMC1 and the two RAD51 binding sites on BRCA2 are indicated by black bars. The mid-portion contains eight highly-conserved BRC repeats. The CTD contains the α-helical domain, three OB-folds, the TR2 (location of S3291) domain, and putative nuclear localization signals (NLSs). The tower domain (not shown) extends from the second OB-fold (OB2).

The N-terminal region of BRCA2 does not bind RAD51; however, it does interact with a protein that is equally crucial to maintenance of genomic integrity, which is PALB2 (partner and localizer of BRCA2) (Xia, Sheng et al. 2006); (Rahman, Seal et al. 2007). PALB2 has been observed complexed with DNA damage-induced BRCA1/BRCA2 nuclear foci (Sy, Huen et al. 2009), (Zhang, Fan et al. 2009; Zhang, Ma et al. 2009). Subsequent studies have shown that PALB2 is crucial in the localization of BRCA2 to sites of DNA damage via associations with chromatin structures and in HR-mediated DNA repair ((Sy, Huen et al. 2009), (Zhang, Fan et al. 2009; Zhang, Ma et al. 2009). This indirectly influences the localization of RAD51 to sites of DNA damage, due to its reliance on BRCA2. During the process of HR-mediated repair, PALB2 appears to be crucial in "D-loop" formation, (Buisson, Dion-Cote et al.)2010) (figure 2). This is the step in which the 3' overhangs of the dsDNA break, resulting from resection of the 5' ends of the break, are coated with RAD51 protein to form nucleoprotein filaments which invade the homologous template and form a Holliday junction. In the absence of PALB2, cells exhibit genomic instability and treatment with drugs that cause inter-strand crosslinks leads to increased chromosome breakage.

3. Mutations of BRCA2, DNA repair fidelity and disease predisposition

BRCA2 and its predecessor, BRCA1, were the first genes to be discovered that were associated with early-onset, familial breast cancer. Furthermore, germline mutations of BRCA2 are also responsible for hereditary forms of ovarian, prostate and pancreatic cancer; however, the risk of acquiring breast cancer is most prevalent. Moreover, the risk for breast cancer is 50-80%, however, the degree of penetrance has been shown to vary (Tonin, Weber et al. 1996).

3.1 Mutated BRCA2 in familial cancers

Most mutations in BRCA2 are the result of small deletions and insertions. In fact, a BRCA2 mutation that has been of interest for almost two decades is the 6174delT mutation, in which the thymine at position 6174 is deleted. This mutation disrupts BRC repeats 5 and 6, and introduces a premature stop codon that abruptly truncates the protein (Neuhausen, Gilewski et al. 1996; Oddoux, Struewing et al. 1996; Roa, Boyd et al. 1996; Abeliovich, Kaduri et al. 1997; Levy-Lahad, Catane et al. 1997). The truncated form no longer possesses the CTD region, which comprises the domains required for DNA repair and recombination, the second RAD51 binding site, TR2/S3291, and the putative nuclear localization signals. As a consequence, cells with this mutation exhibit inefficient repair of DSBs, loss of genomic stability, and sensitivity to radiation and DNA crosslinking agents (Goggins, Schutte et al. 1996),(Ozcelik, Schmocker et al. 1997). The 6174delT is a founder mutation in the Ashkenazi Jewish population at a frequency of 1.36% ((Tonin, Weber et al. 1996)). And, it is the only BRCA2 mutation, along with three BRCA1 mutations, that is carried in 78-96% of Ashkenazi Jews with detectable mutations (Oddoux, Guillen-Navarro et al. 1999) (Mangold, Wang et al.)

Another BRCA2 mutation that was also discovered to have a founder's effect is the 999del5 mutation, which was discovered in an Icelandic population (Thorlacius, Olafsdottir et al. 1996). It is a five base-pair deletion that starts at nucleotide 999, codon 257 in exon 9. The mutation introduces a frame-shift that prematurely truncates the protein, and renders it nonfunctional, similar to the effect of the 6174delT founder mutation in the Ashkenazi Jewish population. Carriers of the mutation exhibit familial forms of male or female breast, prostate or pancreatic cancer. However, there are varying forms of penetrance, in which some carriers have never been diagnosed with cancer. In fact, there is either absolutely no phenotypic expression or diagnosis of varying forms of cancer (Thorlacius, Olafsdottir et al. 1996).

In a study of BRCA1/2 mutations performed in a Serbian population, one family was shown to carry a BRCA2 mutation that was an insertion of two nucleotides, c.4367_4368dupTT (Dobricic, Brankovic-Magic et al.). The mutation causes a frame-shift that alters codons 1381-1387 and introduces a premature stop codon at position 1388, resulting in a loss of > 2,000 amino acids at the C-terminus. The protein product lacks BRC repeats 3-8, as well as the crucial CTD and TR2 domains, rendering BRCA2 completely non-functional in regulating RAD51 activity, as well as in promoting HR-mediated repair of DSBs (Dobricic, Brankovic-Magic et al.)

3.2 Mutated BRCA2 in the development of Fanconi Anemia

Another inheritable condition resulting from mutated BRCA2 is the disorder Fanconi Anemia (FA). The disorder is rare and is characterized by aplastic anemia in childhood, susceptibility to leukemia and cancer, and hypersensitivity of FA cells to interstrand crosslinking agents, such as cisplatin (D'Andrea, 2010). The FA proteins are the products of 13 genes that comprise the following subtypes, FA-A, B, C, D1, D2, E, F, G, I, J, L, M, and N. And, eight of those gene products encoding proteins FANCA-C, FANCE-G, FANCL, and FANCM form a nuclear multi-protein core complex (the FA complex) that functions in the DNA repair pathway. Furthermore, it was discovered that genes underlying the FA-D1 (*FANCD1*) and FA-N (*FANCN*) subtypes were *BRCA2* and *PALB2*, respectively. Ultimately, the multi-protein core complex is responsible for monoubiquitylating FANCD2 on lysine 561 in order to activate the Fanconi Anemia pathway in response to S-phase progression or DNA damage (Zhi, Wilson et al. 2009), (Figure 3).

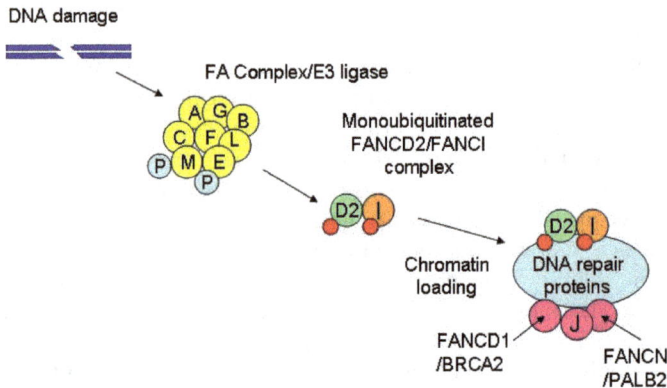

Fig. 3. Schematic of the Fanconi Anemia Pathway. After DNA damage, the ATR kinase phosphorylates and activates the FA core complex, comprised of FANCA, -B, -C, -E, -F, -G, -L, and –M. The complex functions as an E3 ligase and monoubiquitinates the FANCD2/FANCI complex, which then targets chromatin, where it assembles with other DNA repair proteins and FANCD1/BRCA2 and FANCN/PALB2 to repair damaged DNA.

The discovery of the FANCD1 protein being identified as BRCA2 was surprising, yet quite rational given the similarities between FANCD1 and BRCA2 mutated cells. They both exhibit chromosomal instabilities, sensitivity to ionizing radiation and crosslinking agents, and inefficient HR-mediated repair of DSBs. The role of FANCD1/BRCA2 in this protein complex is to act downstream in concert with the FA complex, additional FA members, and DNA repair proteins. However, Fanconi Anemia, subtype D1 is caused by biallelic inactivation of BRCA2; however, risk of breast, ovarian, prostate, and pancreatic cancers are associated with heterozygous BRCA2 mutations (Howlett, Taniguchi et al. 2002).

The FA proteins function in a DNA damage repair pathway, with the multi-protein core complex ultimately being responsible for monoubiquitylating the FANCD2 and FANC1 proteins, in response to DNA damage or entry into S phase of the cell cycle. Activation of the core complex is initiated by phosphorylation of FA proteins by the DNA damage sensing kinases ATM and ATR. After phosphorylation, the core complex assembles to form a nuclear ubiquitin E3 ligase complex. The complex proceeds to monoubiquitylate the FANCD2 and FANC1 proteins thereby causing them to move to chromatin structures and form nuclear foci at sites of DNA damage. The FA complex interacts with FA members, FANCD1/BRCA2, FANCN/PALB2 and FANCJ, along with other DNA repair proteins to promote HR-mediated resolution of DSBs. Given that this pathway is involved in HR-mediated repair, it was not surprising to discover the involvement of RAD51 and BRCA1 downstream in the FA DNA repair pathway. And, because of the involvement of BRCA2, along with BRCA1, this pathway is now referred to as the Fanconi Anemia/BRCA pathway or network (D'Andrea).

3.3 BRCA2 in cell cycle signaling and the DNA damage response

To activate BRCA2, a sequence of cell signaling events is initiated in response to DNA damage, called the DNA damage response (or DDR). When the cell has experienced DSBs, either by exogenous sources such as ionizing radiation or exposure to crosslinking agents such as cisplatin, or endogenous sources such as free radicals and stalled replication forks, the goal is to immediately arrest cell division and repair the damage. When efficient DNA repair does not occur, apoptosis is induced to prevent propagation of genetic mutations. The phosphoinositide 3-kinase related kinases (PIKKs), ataxia-telangiectasia mutated kinase (ATM), and ATM and Rad 3-related kinase (ATR), are crucial in the detection and subsequent resolution of DNA damage. Furthermore, they are also involved in the Fanconi Anemia pathway, as previously described. ATM and ATR "cross-talk" with each other, given that ATM activates ATR in response to ionizing radiation, and ATR activates ATM in response to ultraviolet light. With respect to the DDR pathway that involves BRCA2, resolution of DSBs is initiated by activation of ATR, after phosphorylation by ATM. ATR proceeds to phosphorylate and active Chk1, which then phosphorylates RAD51. Chk1 arrests the cells in S and G2 phases to ensure DNA is repaired before synthesis and cell division. At this point, RAD51 is now able to engage in HR-mediated repair of DSBs under the regulation of BRCA2 (McNeely, Conti et al.), (Connell, Shibata et al.).

BRCA2 appears to play a crucial role during S and G2 phases of the cell cycle. First, during S phase, replication forks can stall and collapse due to exogenous or endogenous sources of damage. A DNA strand break at a replication fork can mimic a DSB as a result of the nascent DNA chain that is being synthesized at the fork. At this point, activated RAD51 is required to repair the break and subsequently stalled fork. It has been proposed that deficient BRCA2, which functions to regulate RAD51 during HR-mediated repair, may be a major cause of diseases resulting from an accumulation of stalled replication forks and consequential DNA breaks that remain unrepaired (Lomonosov, Anand et al. 2003). And, with respect to G2 phase, in a study where the binding of BRCA2 with RAD51 was inhibited in cells expressing the BRC4 repeat, which competed against endogenous full-length BRCA2, there was a failure to initiate radiation-induced G2/M checkpoint arrest. These results implied that the interaction between BRCA2 and RAD51 was imperative for G2/M checkpoint control (Chen, Chen et al. 1999).

The majority of *BRCA2* mutations that are associated with cancer predisposition tend be truncations that remove substantial portions of the CTD, which is where the domains required for DNA repair are located. This region of the protein also appears to be significant in cell cycle changes due to the DNA damage response, via the TR2 domain. The TR2 domain contains a serine at 3291 that is CDK phosphorylated and appears to be one method in which binding between BRCA2 and RAD51 is regulated (Esashi, Christ et al. 2005). There is reduced phosphorylation at this site during S phase, which allows BRCA2 and RAD51 to interact, and engage in HR-mediated repair resulting from replication-induced DNA breaks. In addition, phosphorylation is reduced in response to ionizing radiation. However, phosphorylation of S3291 increases during G2/M to inactivate HR from occurring during mitosis. Further support for this region of the protein being a cancer-related mutation site is evidenced by the association of the P3292L mutation with breast cancer incidence (Esashi, Christ et al. 2005). The TR2 domain also only interacts with multimeric forms of RAD51, both in the presence and absence of DNA (Esashi, Galkin et al. 2007). And, RAD51 monomers bearing mutations that prevent self-association do not interact with the TR2 domain. The impact that this has on BRCA2 function is quite remarkable and has been elegantly summarized (Esashi, Galkin et al. 2007). In the absence of DNA damage, the TR2 domain is phosphorylated at S3291, preventing association of the C-terminus of BRCA2 with RAD51, as well as keeping BRCA2 inactive. However, concurrently, RAD51 is associated with BRCA2 via the BRC repeats in monomeric form. And, it has been noted that the BRC repeats may serve as a negative regulator of RAD51 by preventing it from forming nucleoprotein filaments with ssDNA until after damage has been detected and the DNA has been prepped for HR-mediated repair. After DNA damage has been detected, S3291 is dephosphorylated, now allowing BRCA2 to become activated. The C-terminus can now bind with RAD51 in multimeric form, and the OB folds which possess ssDNA binding activity, deliver RAD51 to sites of DNA damage. This change in BRCA2 function from negatively regulating RAD51 to mobilizing it to sites of damage may be driven by the self-assembly of RAD51 from a monomeric to a multimeric state in response to DNA damage (Esashi, Galkin et al. 2007). This detrimental function of the C-terminus of BRCA2 further substantiates how truncations of this region of the protein, which are commonly seen in BRCA2-cancers, incite genomic instability and subsequent malignant transformation.

The role of the BRC repeat region has been somewhat controversial. It has been described as the region of BRCA2 that is responsible for delivering RAD51 to ssDNA at sites of DNA damage (Carreira, Hilario et al. 2009), (Shivji, Davies et al. 2006) but, conversely, as a negative regulator of RAD51, which was described in the previous section (Nomme, Takizawa et al. 2008), (Davies and Pellegrini 2007). Results of a study investigating cancer-associated mutations of BRC repeats supported their role as a negative regulator that binds and inhibits RAD51 from engaging in HR. But, then releases RAD51 monomers upon detection of DNA damage, thus allowing RAD51 to multimerize and interact with the BRCA2 TR2 region for mobilization to sites of damage. At this point, RAD51 is ready to form nucleoprotein filaments on the 3′ ssDNA overhangs at the breakpoint junction, which will invade the DNA homologue to be used as the template for repair. Considering that the BRC repeats are important for modulating RAD51 activity, several cancer-associated mutations, primarily point mutations, have been identified in this highly conserved region of BRCA2. Cancer-associated mutations have been identified in BRC motifs 1(T1011R), 2(F1219L, S1221P), 4(G1529R), and 7(T1980I) (Esashi, Galkin et al. 2007). The effect of mutations in BRC motifs 2 and 4 on RAD51-mediated HR repair was assessed. The results determined that such mutations prevent binding of monomeric RAD51 to the BRC repeats, which prevents recruitment of RAD51 to DSBs, thereby inhibiting nucleoprotein filament formation and impairing HR-mediated repair (Tal, Arbel-Goren et al. 2009).

A great deal of attention has been focused on the role of the C-terminus and BRC repeat region of BRCA2 in HR-mediated repair. However, mutations of the N-terminus also have detrimental effects on protein function. The N-terminus of BRCA2 binds the protein PALB2 (partner and localizer of BRCA2) (Xia, Sheng et al. 2006); (Rahman, Seal et al. 2007), (Figure 2). PALB2 is also a member of the Fanconi Anemia pathway, denoted as FANCN, in the same manner in which BRCA2/FANCD1 is involved, as well (Figure 3). And, just as biallelic mutations of BRCA2/FANCD1 cause a subtype of Fanconi Anemia, and susceptibility to childhood cancers, biallelic mutations of PALB2/FANCN have a similar phenotype (D'Andrea). With respect to the interaction with BRCA2, PALB2 is responsible for localizing BRCA2 to the sites of DNA damage in order to promote repair (Xia, Sheng et al. 2006). Mutations of the PALB2 binding site on BRCA2 prevent this interaction, causing impaired formation of RAD51 damage-induced foci, and unresolved DSBs (Xia, Sheng et al. 2006). Furthermore, PALB2 is also able to bind DNA and enhance the recombination activity of RAD51 (Dray, Etchin et al.).

4. Therapeutic regimens designed to target BRCA2 defects

Cells that are defective in BRCA1 and BRCA2 retain unresolved DSBs. This attribute, which is detrimental in terms of genomic instability and risk for cancer, is actually a potent target for inhibitors of Poly(ADP-ribose) polymerase, or PARP, in the eradication of transformed cells.

4.1 Efficacy of PARP inhibitors in treating BRCA2-tumors

PARPs are a family of 17 enzymes, with PARP-1 and -2 having been shown to be involved in DNA repair. PARP-1 is a nuclear protein with a zinc-finger DNA binding domain (Amir, Seruga et al.). It is responsible for binding to the sites of single-strand breaks, signaling damage at the site, and the initiating repair. The zinc finger domain binds to ssDNA breaks, cleaves NAD+, and attaches multiple ADP-ribose units to the protein. This results in an extremely negatively charged target which causes unwinding of the damaged DNA, followed by repair by the Base-Excision Repair (BER) pathway (Schreiber, Dantzer et al. 2006); (Ratnam and Low 2007). However, PARP-1 has also been shown to serve as an anti-recombinogenic factor at sites of damage where it has bound, thereby having implications on inhibiting HR-mediated repair (Amir, Seruga et al.), (Sandhu, Yap et al.). BRCA1 and -2 mutant cells are defective in repair of DSBs, and as a consequence, are sensitive to agents that induce DSBs. PARP-1 inhibitors have been shown to be effective in selectively targeting BRCA1 and -2 defective cells by converting SSBs, which have been induced by the use of chemotherapeutic agents, ionizing radiation, or occurring in normal cellular processes, such as stalled replication forks, to DSBs. The SSBs would have normally been identified and resolved by PARP-1 binding and the BER pathway; however, PARP-1 inhibitors prevent such resolution, and during DNA synthesis, the SSBs are converted to DSBs. The DSBs are normally resolved by HR-mediated repair involving BRCA1, and most important BRCA2, with the recombinase RAD51. However, this is deficient in BRCA-mutant cells and the addition of PARP inhibition enhances DNA-damage induced cell cycle arrest and apoptosis. This process eradicates the tumor cells.

4.2 Development of PARP inhibitors

The first PARP-1 inhibitor created was 3-aminobenzamide (3-AB). It causes inhibition of PARP-1 by competing with NAD+ as a substrate. However, 3-AB showed poor specificity and inhibited de-novo purine synthesis (Purnell, Stone et al. 1980); (Drew and Plummer). Approximately, twenty years have passed since the synthesis of 3-AB, and the focus has been to create PARP-1 inhibitors with greater specificity for PARP-1 inhibition, only. In 2003, the PARP-1 inhibitor AG014699 was the first to enter clinical trials (Plummer and Calvert 2007), (Drew and Plummer). Xenograft studies showed significant delay of tumor growth when AG014699 was combined with irinotecan and irradiation and tumor regression when combined with temozolomide (Ratnam and Low 2007). There are presently at least eight PARP inhibitors in clinical trials (Drew and Plummer), (Amir, Seruga et al.), (Table 1). PARP inhibitors are effective at sensitizing tumor cells to other chemotherapeutic agents, and can be used as a combination therapy with platinums, temozolomide, topoisomerase I inhibitors, and γ-/X-radiation (Ratnam and Low 2007), (Curtin, Wang et al. 2004), (Miknyoczki, Jones-Bolin et al. 2003), (Nguewa, Fuertes et al. 2006), (Chalmers, Johnston et al. 2004), (Fernet, Ponette et al. 2000), (Veuger, Curtin et al. 2003). Due to PARP inhibitors effectively promoting cell cycle arrest and subsequent apoptosis, clinical trials are testing their efficacy as single-agents in the treatment of BRCA1- and BRCA2-tumors (Ratnam and Low 2007).

Agent	Single/combination therapy	Disease
Olaparib (AZD2281)	Single agent Combination trials	BRCA-related tumors Solid tumors
BSI-201	Single agent Combination trials (gemcitabine/carboplatin)	Triple negative breast cancer Advanced solid tumors
AG014699	Single agent Combination trials (temozolomide [TMZ]	Solid tumors Melanoma
ABT-888	Single agent	Solid tumors and lymphoid malignancies
INO-1001	Single agent Combination with TMZ	Melanoma Glioblastoma mutiforme
MK4827	Single agent	Solid tumors BRCA ovarian
GPI21016	Combination with TMZ	Solid tumors
CEP-9722	Combination with TMZ	Solid tumors

Table 1. PARP inhibitors presently in clinical trials

4.3 Clinical implications of PARP inhibitor use

In general, there is very high enthusiasm for the use of PARP inhibitors in the treatment of BRCA2-cancers. The requirement for specificity is met because the BRCA1/2-mutated cells are most sensitive to the inhibitors, due to their DNA repair defects, and the premise of "synthetic lethality", which is when two pathway defects alone are innocuous, but combined become lethal (Ratnam and Low 2007). The combination of impaired HR-mediated repair due to the BRCA1/2-mutation and the inhibition of PARP-1 to signal the DNA breaks provides the "synthetic lethality" that is necessary for the efficacy of PARP inhibitors in the treatment of BRCA-tumors. Furthermore, the therapeutic benefit of PARP inhibitors appears to greatly outweigh the undesirable side effects; however, there are areas of concern. First and foremost, PARP inhibitors are still in the early stages of clinical testing. Therefore, the optimal dosage and duration of treatment have not been definitively determined. And, although PARP inhibitors are effective against BRCA-tumors, there is the potential for possible toxicity in normal tissues. In the Olaparib phase I study, DSB accumulations were observed in normal tissues (eyebrow hair follicles), (Drew and Plummer). In addition to toxicity, the inhibitors may disrupt DNA repair pathways in normal tissue from DNA damage acquired through sun exposure or other environmental agents (Ratnam and Low 2007). And, the potential for secondary cancers to occur through genomic instability from inhibition of PARP-1 is possible. In an in vivo study of PARP-1 deficiency, female mice developed mammary carcinoma (Tong, Yang et al. 2007); (Drew and Plummer). Furthermore, secondary mutations after PARP inhibitor treatment may lead to drug resistance. Previous reports have observed intragenic secondary mutations/deletions of BRCA2 occurring after treatment with PARP-1, and the anti-cancer agent cisplatin, which restored the open-reading frame and led to the expression of new BRCA2 isoforms. This resulted in reversal of the original BRCA2 mutation and resistance to PARP inhibitors (Edwards, Brough et al. 2008), (Sakai, Swisher et al. 2008), (Drew and Plummer).

5. Conclusion

Overall, the use of PARP inhibitors appears to be very promising in the treatment of BRCA-tumors as a single agent, and as a chemotherapeutic/radiation sensitizer when used in combination with anticancer therapeutics or γ-radiation. The on-going clinical trials will provide more information about the aspects of PARP inhibitor usage that are presently vague, such as proper dosage and duration of treatment, possible effects on DNA repair mechanisms in normal cells, possible induction of secondary mutations, and acquired resistance of tumors over the course of treatment.

6. References

Abeliovich, D., L. Kaduri, et al. (1997). "The founder mutations 185delAG and 5382insC in BRCA1 and 6174delT in BRCA2 appear in 60% of ovarian cancer and 30% of early-onset breast cancer patients among Ashkenazi women." *Am J Hum Genet* 60(3): 505-14.

Amir, E., B. Seruga, et al. "Targeting DNA repair in breast cancer: a clinical and translational update." *Cancer Treat Rev* 36(7): 557-65.

Benson, F. E., A. Stasiak, et al. (1994). "Purification and characterization of the human Rad51 protein, an analogue of E. coli RecA." *Embo J* 13(23): 5764-71.

Bignell, G., G. Micklem, et al. (1997). "The BRC repeats are conserved in mammalian BRCA2 proteins." *Hum Mol Genet* 6(1): 53-8.

Buisson, R., A. M. Dion-Cote, et al. "Cooperation of breast cancer proteins PALB2 and piccolo BRCA2 in stimulating homologous recombination." *Nat Struct Mol Biol* 17(10): 1247-54.

Carreira, A., J. Hilario, et al. (2009). "The BRC repeats of BRCA2 modulate the DNA-binding selectivity of RAD51." *Cell* 136(6): 1032-43.

Chalmers, A., P. Johnston, et al. (2004). "PARP-1, PARP-2, and the cellular response to low doses of ionizing radiation." *Int J Radiat Oncol Biol Phys* 58(2): 410-9.

Chen, C. F., P. L. Chen, et al. (1999). "Expression of BRC repeats in breast cancer cells disrupts the BRCA2-Rad51 complex and leads to radiation hypersensitivity and loss of G(2)/M checkpoint control." *J Biol Chem* 274(46): 32931-5.

Chen, P. L., C. F. Chen, et al. (1998). "The BRC repeats in BRCA2 are critical for RAD51 binding and resistance to methyl methanesulfonate treatment." *Proc Natl Acad Sci U S A* 95(9): 5287-92.

Connell, C. M., A. Shibata, et al. "Genomic DNA damage and ATR-Chk1 signaling determine oncolytic adenoviral efficacy in human ovarian cancer cells." *J Clin Invest* 121(4): 1283-97.

Curtin, N. J., L. Z. Wang, et al. (2004). "Novel poly(ADP-ribose) polymerase-1 inhibitor, AG14361, restores sensitivity to temozolomide in mismatch repair-deficient cells." *Clin Cancer Res* 10(3): 881-9.

D'Andrea, A. D. "Susceptibility pathways in Fanconi's anemia and breast cancer." *N Engl J Med* 362(20): 1909-19.

Davies, O. R. and L. Pellegrini (2007). "Interaction with the BRCA2 C terminus protects RAD51-DNA filaments from disassembly by BRC repeats." *Nat Struct Mol Biol* 14(6): 475-83.

Derbyshire, M. K., L. H. Epstein, et al. (1994). "Nonhomologous recombination in human cells." *Mol Cell Biol* 14(1): 156-69.

Dobricic, J., M. Brankovic-Magic, et al. "Novel BRCA1/2 mutations in Serbian breast and breast-ovarian cancer patients with hereditary predisposition." *Cancer Genet Cytogenet* 202(1): 27-32.

Dray, E., J. Etchin, et al. "Enhancement of RAD51 recombinase activity by the tumor suppressor PALB2." *Nat Struct Mol Biol* 17(10): 1255-9.

Drew, Y. and R. Plummer "The emerging potential of poly(ADP-ribose) polymerase inhibitors in the treatment of breast cancer." *Curr Opin Obstet Gynecol* 22(1): 67-71.

Edwards, S. L., R. Brough, et al. (2008). "Resistance to therapy caused by intragenic deletion in BRCA2." *Nature* 451(7182): 1111-5.

Esashi, F., N. Christ, et al. (2005). "CDK-dependent phosphorylation of BRCA2 as a regulatory mechanism for recombinational repair." *Nature* 434(7033): 598-604.

Esashi, F., V. E. Galkin, et al. (2007). "Stabilization of RAD51 nucleoprotein filaments by the C-terminal region of BRCA2." *Nat Struct Mol Biol* 14(6): 468-74.

Fernet, M., V. Ponette, et al. (2000). "Poly(ADP-ribose) polymerase, a major determinant of early cell response to ionizing radiation." *Int J Radiat Biol* 76(12): 1621-9.

Goggins, M., M. Schutte, et al. (1996). "Germline BRCA2 gene mutations in patients with apparently sporadic pancreatic carcinomas." *Cancer Res* 56(23): 5360-4.

Howlett, N. G., T. Taniguchi, et al. (2002). "Biallelic inactivation of BRCA2 in Fanconi anemia." *Science* 297(5581): 606-9.

Jackson, S. P. and P. A. Jeggo (1995). "DNA double-strand break repair and V(D)J recombination: involvement of DNA-PK." *Trends Biochem Sci* 20(10): 412-5.

Jensen, R. B., A. Carreira, et al. "Purified human BRCA2 stimulates RAD51-mediated recombination." *Nature* 467(7316): 678-83.

Johnson, R. D. and M. Jasin (2000). "Sister chromatid gene conversion is a prominent double-strand break repair pathway in mammalian cells." *Embo J* 19(13): 3398-407.

Levy-Lahad, E., R. Catane, et al. (1997). "Founder BRCA1 and BRCA2 mutations in Ashkenazi Jews in Israel: frequency and differential penetrance in ovarian cancer and in breast-ovarian cancer families." *Am J Hum Genet* 60(5): 1059-67.

Lomonosov, M., S. Anand, et al. (2003). "Stabilization of stalled DNA replication forks by the BRCA2 breast cancer susceptibility protein." *Genes Dev* 17(24): 3017-22.

Mangold, K. A., V. Wang, et al. "Detection of BRCA1 and BRCA2 Ashkenazi Jewish founder mutations in formalin-fixed paraffin-embedded tissues using conventional PCR and heteroduplex/amplicon size differences." *J Mol Diagn* 12(1): 20-6.

Marmorstein, L. Y., T. Ouchi, et al. (1998). "The BRCA2 gene product functionally interacts with p53 and RAD51." *Proc Natl Acad Sci U S A* 95(23): 13869-74.

McNeely, S., C. Conti, et al. "Chk1 inhibition after replicative stress activates a double strand break response mediated by ATM and DNA-dependent protein kinase." *Cell Cycle* 9(5): 995-1004.

Miknyoczki, S. J., S. Jones-Bolin, et al. (2003). "Chemopotentiation of temozolomide, irinotecan, and cisplatin activity by CEP-6800, a poly(ADP-ribose) polymerase inhibitor." *Mol Cancer Ther* 2(4): 371-82.

Mizuta, R., J. M. LaSalle, et al. (1997). "RAB22 and RAB163/mouse BRCA2: proteins that specifically interact with the RAD51 protein." *Proc Natl Acad Sci U S A* 94(13): 6927-32.

Neuhausen, S., T. Gilewski, et al. (1996). "Recurrent BRCA2 6174delT mutations in Ashkenazi Jewish women affected by breast cancer." *Nat Genet* 13(1): 126-8.

Nguewa, P. A., M. A. Fuertes, et al. (2006). "Poly(ADP-ribose) polymerase-1 inhibitor 3-aminobenzamide enhances apoptosis induction by platinum complexes in cisplatin-resistant tumor cells." *Med Chem* 2(1): 47-53.

Nomme, J., Y. Takizawa, et al. (2008). "Inhibition of filament formation of human Rad51 protein by a small peptide derived from the BRC-motif of the BRCA2 protein." *Genes Cells* 13(5): 471-81.

Oddoux, C., E. Guillen-Navarro, et al. (1999). "Mendelian diseases among Roman Jews: implications for the origins of disease alleles." *J Clin Endocrinol Metab* 84(12): 4405-9.

Oddoux, C., J. P. Struewing, et al. (1996). "The carrier frequency of the BRCA2 6174delT mutation among Ashkenazi Jewish individuals is approximately 1%." *Nat Genet* 14(2): 188-90.

Ogawa, T., X. Yu, et al. (1993). "Similarity of the yeast RAD51 filament to the bacterial RecA filament." *Science* 259(5103): 1896-9.

Ozcelik, H., B. Schmocker, et al. (1997). "Germline BRCA2 6174delT mutations in Ashkenazi Jewish pancreatic cancer patients." *Nat Genet* 16(1): 17-8.

Plummer, E. R. and H. Calvert (2007). "Targeting poly(ADP-ribose) polymerase: a two-armed strategy for cancer therapy." *Clin Cancer Res* 13(21): 6252-6.

Purnell, M. R., P. R. Stone, et al. (1980). "ADP-ribosylation of nuclear proteins." *Biochem Soc Trans* 8(2): 215-27.

Rahman, N., S. Seal, et al. (2007). "PALB2, which encodes a BRCA2-interacting protein, is a breast cancer susceptibility gene." *Nat Genet* 39(2): 165-7.

Ratnam, K. and J. A. Low (2007). "Current development of clinical inhibitors of poly(ADP-ribose) polymerase in oncology." *Clin Cancer Res* 13(5): 1383-8.

Roa, B. B., A. A. Boyd, et al. (1996). "Ashkenazi Jewish population frequencies for common mutations in BRCA1 and BRCA2." *Nat Genet* 14(2): 185-7.

Roth, D. B., T. N. Porter, et al. (1985). "Mechanisms of nonhomologous recombination in mammalian cells." *Mol Cell Biol* 5(10): 2599-607.

Roth, D. B. and J. H. Wilson (1986). "Nonhomologous recombination in mammalian cells: role for short sequence homologies in the joining reaction." *Mol Cell Biol* 6(12): 4295-304.

Sakai, W., E. M. Swisher, et al. (2008). "Secondary mutations as a mechanism of cisplatin resistance in BRCA2-mutated cancers." *Nature* 451(7182): 1116-20.

Sandhu, S. K., T. A. Yap, et al. "Poly(ADP-ribose) polymerase inhibitors in cancer treatment: a clinical perspective." *Eur J Cancer* 46(1): 9-20.

Schreiber, V., F. Dantzer, et al. (2006). "Poly(ADP-ribose): novel functions for an old molecule." *Nat Rev Mol Cell Biol* 7(7): 517-28.

Sharan, S. K., M. Morimatsu, et al. (1997). "Embryonic lethality and radiation hypersensitivity mediated by Rad51 in mice lacking Brca2." *Nature* 386(6627): 804-10.

Shivji, M. K., O. R. Davies, et al. (2006). "A region of human BRCA2 containing multiple BRC repeats promotes RAD51-mediated strand exchange." *Nucleic Acids Res* 34(14): 4000-11.

Sonoda, E., M. S. Sasaki, et al. (1998). "Rad51-deficient vertebrate cells accumulate chromosomal breaks prior to cell death." *Embo J* 17(2): 598-608.

Spain, B. H., C. J. Larson, et al. (1999). "Truncated BRCA2 is cytoplasmic: implications for cancer-linked mutations." *Proc Natl Acad Sci U S A* 96(24): 13920-5.

Sy, S. M., M. S. Huen, et al. (2009). "PALB2 is an integral component of the BRCA complex required for homologous recombination repair." *Proc Natl Acad Sci U S A* 106(17): 7155-60.

Takata, M., M. S. Sasaki, et al. (1998). "Homologous recombination and non-homologous end-joining pathways of DNA double-strand break repair have overlapping roles in the maintenance of chromosomal integrity in vertebrate cells." *Embo J* 17(18): 5497-508.

Taki, T., T. Ohnishi, et al. (1996). "Antisense inhibition of the RAD51 enhances radiosensitivity." *Biochem Biophys Res Commun* 223(2): 434-8.

Tal, A., R. Arbel-Goren, et al. (2009). "Cancer-associated mutations in BRC domains of BRCA2 affect homologous recombination induced by Rad51." *J Mol Biol* 393(5): 1007-12.

Thorlacius, S., G. Olafsdottir, et al. (1996). "A single BRCA2 mutation in male and female breast cancer families from Iceland with varied cancer phenotypes." *Nat Genet* 13(1): 117-9.

Thorslund, T., F. Esashi, et al. (2007). "Interactions between human BRCA2 protein and the meiosis-specific recombinase DMC1." *Embo J* 26(12): 2915-22.

Tong, W. M., Y. G. Yang, et al. (2007). "Poly(ADP-ribose) polymerase-1 plays a role in suppressing mammary tumourigenesis in mice." *Oncogene* 26(26): 3857-67.

Tonin, P., B. Weber, et al. (1996). "Frequency of recurrent BRCA1 and BRCA2 mutations in Ashkenazi Jewish breast cancer families." *Nat Med* 2(11): 1179-83.

Tsuzuki, T., Y. Fujii, et al. (1996). "Targeted disruption of the Rad51 gene leads to lethality in embryonic mice." *Proc Natl Acad Sci U S A* 93(13): 6236-40.

Veuger, S. J., N. J. Curtin, et al. (2003). "Radiosensitization and DNA repair inhibition by the combined use of novel inhibitors of DNA-dependent protein kinase and poly(ADP-ribose) polymerase-1." *Cancer Res* 63(18): 6008-15.

Wong, A. K., R. Pero, et al. (1997). "RAD51 interacts with the evolutionarily conserved BRC motifs in the human breast cancer susceptibility gene brca2." *J Biol Chem* 272(51): 31941-4.

Xia, B., Q. Sheng, et al. (2006). "Control of BRCA2 cellular and clinical functions by a nuclear partner, PALB2." *Mol Cell* 22(6): 719-29.

Yang, H., P. D. Jeffrey, et al. (2002). "BRCA2 function in DNA binding and recombination from a BRCA2-DSS1-ssDNA structure." *Science* 297(5588): 1837-48.

Yano, K., K. Morotomi, et al. (2000). "Nuclear localization signals of the BRCA2 protein." *Biochem Biophys Res Commun* 270(1): 171-5.

Yuan, S. S., S. Y. Lee, et al. (1999). "BRCA2 is required for ionizing radiation-induced assembly of Rad51 complex in vivo." *Cancer Res* 59(15): 3547-51.

Zhang, F., Q. Fan, et al. (2009). "PALB2 functionally connects the breast cancer susceptibility proteins BRCA1 and BRCA2." *Mol Cancer Res* 7(7): 1110-8.

Zhang, F., J. Ma, et al. (2009). "PALB2 links BRCA1 and BRCA2 in the DNA-damage response." *Curr Biol* 19(6): 524-9.

Zhi, G., J. B. Wilson, et al. (2009). "Fanconi anemia complementation group FANCD2 protein serine 331 phosphorylation is important for fanconi anemia pathway function and BRCA2 interaction." *Cancer Res* 69(22): 8775-83.

Part 2

Evolution of DNA Repair

19

Meiosis as an Evolutionary Adaptation for DNA Repair

Harris Bernstein[1], Carol Bernstein[1] and Richard E. Michod[2]
[1]*Department of Cellular and Molecular Medicine, University of Arizona*
[2]*Department of Ecology and Evolutionary Biology, University of Arizona*
USA

1. Introduction

The adaptive function of sex remains, today, one of the major unsolved problems in biology. Fundamental to achieving a resolution of this problem is gaining an understanding of the function of meiosis. The sexual cycle in eukaryotes has two key stages, meiosis and syngamy. In meiosis, typically a diploid cell gives rise to haploid cells. In syngamy (fertilization), typically two haploid gametes from different individuals fuse to generate a new diploid individual. A unique feature of meiosis, compared to mitosis, is recombination between non-sister homologous chromosomes. Usually these homologous chromosomes are derived from different individuals. In mitosis, recombination can occur, but it is ordinarily between sister homologs, the two products of a round of chromosome replication. Birdsell & Wills (2003) have reviewed the various hypotheses for the origin and maintenance of sex and meiotic recombination, including the hypothesis that sex is an adaptation for the repair of DNA damage and the masking of deleterious recessive alleles. Recently, we presented evidence that among microbial pathogens, sexual processes promote repair of DNA damage, especially when challenged by the oxidative defenses of their biologic hosts (Michod et al., 2008). Here, we present evidence that meiosis is primarily an evolutionary adaptation for DNA repair. Since our previous review of this topic (Bernstein et al., 1988), there has been a considerable increase in relevant information at the molecular level on the DNA repair functions of meiotic recombination, and this new information is emphasized in the present chapter.

2. Meiosis in protists and simple multicellular eukaryotes is induced in response to stressful conditions that likely cause DNA damage

Eukaryotes appeared in evolution more than 1.5 billion years ago (Javaux et al., 2001). Among extant eukaryotes, meiosis and sexual reproduction are ubiquitous and appear to have been present early in eukaryote evolution. Malik et al. (2008) found that 27 of 29 tested meiotic genes were present in *Trichomonas vaginalis*, and 21 of these 29 genes were also present in *Giardia intestinalis*, indicating that most meiotic genes were present in a common ancestor of these species. Since these lineages are highly divergent among eukaryotes, these authors concluded that each of these meiotic genes were likely present in the common ancestor of all eukaryotes. Dacks and Roger (1999) also proposed that sex has a single evolutionary origin and was present in the last common ancestor of eukaryotes. Recently, this view received further support from a study of amoebae. Although amoebae generally have been assumed to be asexual, Lahr et al. (2011) showed that the majority of amoeboid lineages were likely anciently sexual, and that most asexual groups have probably arisen recently and independently.

Eukaryotes arose in evolution from prokaryotes, and eukaryotic meiosis may have arisen from bacterial transformation, a naturally occurring sexual process in prokaryotes. The fundamental similarities between transformation and meiosis have been explored (H. Bernstein & C. Bernstein, 2010). Bacterial transformation, like meiosis, involves alignment and recombination between non-sister homologous chromosomes (or parts of chromosomes) originating from different parents. Both during transformation and meiosis, homologs of the bacterial *recA* gene play a central role in the strand transfer reactions of recombination, indicating a mechanistic similarity. Also, bacterial transformation is induced by environmental stresses that are similar to those that induce meiosis in protists and simple multicellular eukaryotes, suggesting that there was continuity in the evolutionary transition from prokaryotic sex to eukaryotic sex. Evidence indicates that bacterial transformation is an adaptation for repairing DNA (Michod et al., 1988; Hoelzer & Michod, 1991; Michod & Wojciechowski, 1994; reviewed by Michod et al., 2008). Thus meiosis may have emerged from transformation as an adaptation for repairing DNA.

Among extant protists and simple multicellular eukaryotes sexual reproduction is ordinarily facultative. Meiosis and sex in these organisms is usually induced by stressful conditions. The paramecium tetrahymena can be induced to undergo conjugation leading to meiosis by washing, which causes rapid starvation (Elliott & Hayes, 1953). Depletion of the nitrogen source in the growth medium of the unicellular green alga *Chlamydomonas reinhardi* leads to differentiation of vegetative cells into gametes (Sager & Granick 1954). These gametes can then mate, form zygotes and undergo meiosis. Upon nitrogen starvation or desiccation, the human fungal pathogen *Cryptococcus neoformans* undergoes mating or fruiting, both processes involving meiosis (Lin et al., 2005).

In addition to starvation, oxidative stress is another condition that induces meiosis and sex. The haploid fission yeast *Schizosaccharomyces pombe* is induced to undergo sexual development and mating when the supply of nutrients becomes limiting (Davey et al., 1998). Moreover, treatment of late-exponential-phase *S. pombe* vegetative cells with hydrogen peroxide, which causes oxidative stress, increases the frequency of mating and production of meiotic spores by 4- to 18-fold (C. Bernstein & Johns, 1989). The oomycete *Phytophthora cinnamomi* is induced to undergo sexual reproduction by exposure to the oxidizing agent hydrogen peroxide or mechanical damage to hyphae (Reeves & Jackson, 1974). In the simple multicellular green algae *Volvox carteri*, sex is induced by heat shock (Kirk & Kirk, 1986). This effect can be inhibited by antioxidants, indicating that the induction of sex by heat shock is mediated by oxidative stress (Nedelcu & Michod, 2003). Furthermore, induction of oxidative stress by an inhibitor of the mitochondrial electron transport chain also induced sex in *V. carteri* (Nedelcu et al., 2004). The budding yeast *Saccharomyces cerevisiae* reproduces as mitotically dividing diploid cells when nutrients are plentiful, but undergoes meiosis to form haploid spores when starved (Herskowitz, 1988). When *S. cerevisiae* are starved, oxidative stress is increased and DNA double-strand breaks (DSBs) and apurinic/apyrimidinic sites accumulate (Steinboeck et al., 2010). Perhaps, in *S. cerevisiae*, the induction of sex by starvation is mediated by oxidative stress, analogous to the way induction of sex by heat is mediated by oxidative stress in *V. carteri*.

These observations suggest that meiosis is an adaptation for dealing with stress, particularly oxidative stress. It is well established that oxidative stress induces a variety of DNA damages including DNA DSBs, single-strand breaks and modified bases (Slupphaug et al., 2003). Thus we hypothesize that, in facultative sexual protists and simple multicellular eukaryotes, sex, with the central feature of meiosis, is an adaptive response to DNA damage, particularly oxidative DNA damage.

3. DNA damages induced by exogenous agents cause increased meiotic recombination

If recombination during meiosis is an adaptation for repairing DNA damages, then it would be expected that exposure to DNA damaging treatments would increase the frequency of recombination, as measured by crossovers between allelic markers. Stimulation of allelic recombination was reported in the fruitfly *Drosophila melanogaster* in response to exposure to the DNA damaging agents UV light (Prudhommeau & Proust, 1973), X-rays (Suzuki & Parry, 1964), and mitomycin C (Schewe et al., 1971). X-rays induce recombination in meiotic cells not only of *D. melanogaster* females, but also of males, which normally display no recombination during meiosis (Hannah-Alava, 1964).

Increased meiotic recombination in response to X-irradiation has also been reported in *Caenorhabditis elegans* (Kim & Rose, 1987), and in *S. cerevisiae* (Kelly et al., 1983).

4. During mitosis and meiosis, DNA damages caused by diverse exogenous agents can be repaired by homologous recombination

Molecular recombination (that is homologous physical exchange or informational exchange) during mitosis and meiosis functions as a DNA repair process designated homologous recombinational repair (HRR). Many of the gene products employed in mitotic HRR are also employed in recombination during meiosis. It is this consistent function of recombination across meiosis and mitosis in eukaryotes and transformation in prokaryotes that we seek to understand through the repair hypothesis. Mutants defective in HRR genes in *D. melanogaster* and yeast have reduced ability to repair DNA damages arising from a variety of exogenous sources. These mutants are also defective in recombination during meiosis. In general, loss of HRR capability causes increased sensitivity to killing by agents that harm cells primarily through induction of DNA damage. These agents are listed in Table 1. There have been no reports, that we know of, that HRR defective cells are sensitive to agents that harm cells by mechanisms other than primarily causing DNA damage.

In *D. melanogaster*, mutants defective in genes *mei-41, mei-9, hdm, spnA* and *brca2* have reduced spontaneous allelic recombination (crossing over) during meiosis and increased sensitivity to killing by exposure to numerous DNA damaging agents (Table 1). The Mei-41 protein is a structural and functional homolog of the human Atm (ataxia telangiectasia) protein (Hari et al., 1995), which plays a central role in HRR. The Mei-9 and Hdm proteins are components of a multiprotein complex that resolves meiotic recombination intermediates (Joyce et al., 2009). The SpnA protein is a homolog of yeast Rad51 (Staeva-Vieira et al., 2003), and Rad51 plays a central role in strand-exchange during HRR. The *D. melanogaster* Brca2 protein, a homolog of the human Brca2 protein that protects against breast cancer, regulates the activity of Rad51 protein in HRR. The Brca2 protein is required for HRR of DSBs during meiosis (Klovstad et al., 2008).

In *S. cerevisiae,* numerous mutant genes have been identified that confer sensitivity to radiation and/or genotoxic chemicals (Haynes & Kunz, 1981). Several of these mutant genes are also defective in meiotic recombination. For instance, the *rad52* gene is required for meiotic recombination (Game et al., 1980) as well as for mitotic recombination (Malone & Esposito, 1980). Mutants defective in the *rad52* gene are sensitive to killing by several DNA damaging agents (Table 1). Diploid cells of *S. cerevisiae* are able to repair DNA DSBs introduced by ionizing radiation, and this ability is lost in mutant strains defective in the *rad52* gene (Resnick & Martin, 1976). The Rad52 protein promotes the DNA strand exchange reaction of recombination during meiosis and mitosis (Mortensen et al., 2009).

Taken as a whole, these findings indicate that the products of genes *mei-41, mei-9, hdm, spnA, and brca2* in *D. melanogaster* and the *rad52* gene of yeast are required in meiosis for recombination and in somatic cells for HRR of potentially lethal DNA damages. Since the gene products that function in mitotic HRR are able to repair DNA damages from different sources, it can be reasonably assumed that these genes serve a similar DNA repair function during recombination in meiosis.

In the nematode *C. elegans* gonad, oocyte nuclei in the pachytene stage of meiosis, the stage in which HRR occurs, are hyper-resistant to X-ray irradiation compared to oocytes in the subsequent diakinesis stage of meiosis (Takanami et al., 2000). This hyper-resistance depends on expression of gene *ce-rdh-51*, a homolog of yeast *rad51* and *dmc1* that play a central role in meiotic HRR. Meiotic pachytene nuclei are also more resistant to heavy ion particle irradiation than the subsequent meiotic diplotene or diakinesis stages (Takanami et al., 2003). This resistance also depends on the *ce-rdh-51* gene, as well as on gene *ce-atl-1*. *ce-atl-1* is related to *atm* (ataxia –telangiectasia mutated), a gene necessary for repair of DSBs by HRR.

Coogan & Rosenblum (1988) measured repair of DSBs following γ-irradiation of rat spermatogenic cells during successive stages of germ cell formation. The stages were spermatagonia and preleptotene spermatocytes, pachytene spermatocytes and spermatid spermatocytes. The greatest repair capability was observed in pachytene, the stage of meiosis when HRR occurs. These findings indicate that HRR of γ-ray-induced DSBs occurs during meiosis. Several mammalian germ cell stages, including pachytene spermatocytes, produce levels of reactive oxygen species (ROS) sufficient to cause oxidative stress (Fisher & Aitkin, 1997). This observation suggests that HRR during meiosis may also remove DNA damages caused by natural endogenously produced ROS.

The results reviewed in this section indicate that, in both meiosis and mitosis, DNA damages caused by different exogenous agents are repaired by HRR, suggesting that DNA damages from natural endogenous sources (e.g. ROS) are similarly repaired. In general, DNA damage appears to be a fundamental problem for life. As noted by Haynes (1988), DNA is composed of rather ordinary molecular subunits, which are not endowed with any peculiar kind of quantum mechanical stability. He observed that its very "chemical vulgarity" makes DNA subject to all the "chemical horrors" that might befall any such molecule in a warm aqueous medium. The average amount of oxidative DNA damage occurring per cell per day is estimated to be about 10,000 in humans, and in rat, with a higher metabolic rate, about 100,000 (Ames et al., 1993). Most of these damages affect only one strand of the DNA, but a fraction, about 1-2%, are double-strand damages such as DSBs (Massie et al., 1972). These damages can be repaired accurately by HRR.

Organism	Mutant gene	Meiotic recomb-ination	DNA damaging agent(s)	Sensitivity to killing by agent(s)	Reference
D. melano-gaster	mei-41	Reduced	X-rays, UV, methyl methanesulfonate, nitrogen mustard, benzo(s)pyrene, 2-acetyl-aminofluorene	Increased	Baker et al., 1976; Boyd, 1978; Rasmuson, 1984
	mei-9	Reduced	X-rays, UV, methyl methanesulfonate, nitrogen mustard, benzo(s)pyrene, 2-acetyl-aminofluorene	Increased	Baker et al., 1976; Boyd, 1978; Rasmuson, 1984
	hdm	Reduced	methyl methanesulfonate	Increased	Joyce et al., 2009
	spnA	Reduced	X-rays	Increased	Staeva-Vieira et al., 2003
	brca2	HRR of DSBs is reduced	X-rays, methyl methanesulfonate	Increased	Klovstad et al., 2008
S. cerevisiae	rad52	Reduced	X-rays, methyl methanesulfonate, crosslinking agent 8-methoxypsoralen plus UV light	Increased	Haynes & Kunz, 1981; Henriques & Moustacchi, 1980; Game et al., 1980

Table 1. Mutants with reduced meiotic recombination and sensitive to killing by specific DNA damaging agents.

5. In humans and rodents, defects in HRR enzymes lead to infertility, as would be expected if removal of DNA damages is an essential function of meiosis

About 15% of all couples in the US are infertile, and an important cause of male infertility appears to be oxidative stress during gametogenesis (Makker et al., 2009). During spermatogenesis in the mouse, DNA repair capability declines after meiosis is complete, allowing accumulation of DNA damage (Marchetti & Wyrobek, 2008). Lewis & Aitken (2005) reviewed evidence that DNA damages in the germ line of men are associated with poor semen quality, low fertilization rates, impaired pre-implantation development, increased abortion, and elevated incidence of disease in the offspring including childhood cancer. They noted that the natural causes of this DNA damage are uncertain, but the major candidate is oxidative stress. On the hypothesis that meiosis is an adaptation for DNA repair, it is expected that loss of ability to repair DNA damages during meiosis would have adverse effects, including infertility. Although the finding of such adverse effects is expected on the hypothesis that meiosis is an adaptation for repairing naturally caused DNA damages, this finding does not prove the hypothesis. Another possibility is that during meiosis damages are introduced in a programmed fashion, leading to HRR. Such HRR may be necessary for proper pairing and segregation of chromosomes, and this process may be required for fertility (see section 8 below).

Inherited mutations in genes that specify proteins necessary for HRR cause infertility (Table 2) indicating that production of functional gametes depends on HRR. Genes *brca1*, *atm*, and *mlh1* are expressed in mitosis, but at a higher level in meiosis, and gene *dmc1* is expressed exclusively in meiosis (Table 2).

Gene	Species	Fold-increased expression in testes vs. somatic cells	Infertility in mutant females/males	References
brca1	Mouse	3×	male mice are infertile	Galetzka et al., 2007; Cressman et al., 1999
atm	human, mouse	4×	females and males in both humans and mice are infertile	Galetzka et al., 2007; Barlow et al., 1998
mlh1	Mouse	1.7×	female and male mice are infertile	Galetzka et al., 2007; Wei et al., 2002
dmc1	Mouse	specific for meiotic cells	female and male mice are infertile	Pittman et al., 1998

Table 2. Mutant genes defective in HRR that cause infertility in human and/or mouse

Brca1 functions during both meiotic and mitotic recombination. The inheritance of a mutant *brca1* allele substantially increases a woman's lifetime risk for developing breast or ovarian cancer due to a deficiency in HRR of DNA DSBs in somatic cells. Male *brca1* defective mice are infertile due to meiotic failure during spermatogenesis (Table 2), indicating that HRR is necessary during meiosis.

The Atm protein acts during both meiotic and mitotic recombination in detection and signaling of DSBs, and is necessary for fertility of females and males in both humans and mice (Table 2). Gametogenesis is severely disrupted in Atm-deficient mice as early as the leptonema stage of prophase I, resulting in apoptotic degeneration (Barlow et al., 1998).

Mismatch repair protein Mlh1 (homolog of *E. coli* MutL) is necessary for meiotic recombination (Wei et al., 2002). Mutation in the *mlh1* gene causes blockage at the pachytene stage of meiosis and female and male infertility (Table 2).

Dmc1 is a meiosis specific gene. Dmc1 protein (a homolog of *E. coli* RecA protein) functions during meiotic recombination to promote recognition of homologous DNA and to catalyze strand exchange. *Dmc1* deficient female and male mice are infertile due to arrest of gametes in meiotic prophase (Table 2).

The evidence reviewed in this section indicates that defective HRR of DNA damages during meiosis causes infertility.

6. Non-crossover (NCO) recombination during meiosis is likely an adaptation for DNA repair

Meiotic recombination appears to be a near universal feature of meiosis [although it may be absent in some situations, such as in *Drosophila* males (Chovnick et al., 1970)]. There are two major classes of meiotic recombination. If, during recombination, the chromosome arms on opposite sides of a DSB exchange partners, the recombination event is referred to as a crossover (CO). If the original configuration of chromosome arms is maintained, the recombination event is referred to as a non-crossover (NCO) (see Figure 1). The relative occurrence of NCO or CO recombination events is relevant to evolutionary theories of meiosis which assume producing genetic variation is the function of meiosis. NCO events have little effect on linkage disequilibrium (the statistical association of genes at different loci) and so produce very little genetic variation in terms of new combinations of genes. However, CO and NCO events are equivalent from the point of view of HRR.

Data based on tetrad analysis from several species of fungi indicates that the majority (about 2/3) of recombination events during meiosis are NCOs [see Whitehouse (1982), Tables 19 and 38, for summaries of data from *S. cerevisiae, Podospora anserine, Sordaria fimicola* and *Sordaria brevicollis*]. More recent work also supports a bias towards NCOs during meiosis. In mouse meiosis there are $\geq$ 10-fold more DSBs than CO recombinants (Moens et al., 2002), suggesting that most DSBs are repaired by NCO recombination. In *D. melanogaster* there is at least a 3:1 ratio of NCOs to COs (Mehrotra & McKim, 2006). These observations indicate that the majority of recombination events are NCOs. These NCOs involve informational exchange between two homologs but not physical exchange, and little genetic variation is created. Thus explanations for the adaptive function of meiosis that focus exclusively on crossing over are inadequate to explain the majority of recombination events.

Andersen & Sekelsky (2010) have argued that a common mechanism called "synthesis dependent strand annealing" (see section 7, below) is employed in both meiotic HRR of the NCO type and mitotic HRR (which is largely of the NCO type), and thus meiotic and mitotic NCOs probably have a similar function. Substantial evidence indicates that HRR during mitosis is an adaptation to repair DNA damages that originate from diverse endogenous and exogenous sources (e.g. endogenous ROS from oxidative metabolism and exogenous X-rays, UV, chemical carcinogens) (see examples in Table 1; also Lisby & Rothstein, 2009). Thus NCO recombination during meiosis, as in mitosis, likely functions to repair of DNA damages from diverse sources.

7. NCO recombination likely occurs by synthesis-dependent strand annealing

Molecular models of meiotic recombination have evolved over the years as relevant evidence accumulated. The model that has been most influential in recent decades has been the Double-Strand Break Repair model (Szostak et al. 1983). By this model, during each recombination event two Holliday Junctions (HJs) are formed and resolved (see Figure 1). Thus the Double-Strand Break Repair model can also be referred to as the Double Holliday Junction (DHJ) model. The DHJ model was considered to provide an explanation for both CO and NCO types of recombination events. However, Allers & Lichten (2001) showed that, although CO recombinants are likely formed by a pathway involving resolution of Holliday junctions, NCO recombinants arise by a different pathway that acts earlier in meiosis. Allers & Lichten (2001), McMahill et al. (2007) and Andersen & Sekelsky (2010) have presented evidence that NCO recombinants are generated during meiosis by an HRR repair process referred to as "Synthesis-Dependent Strand Annealing" or "SDSA" (see Figure 1). During SDSA the invading strand from a chromosome with a DSB is displaced from the D-loop structure of an intact chromosome and its newly synthesized sequence anneals to the other side of the break on the chromosome with the original DSB. This process can accurately repair DNA DSBs by copying the information lost in the damaged homolog from the other intact homolog without the need for physical exchange of DNA. This process contributes little to genetic variation since the arms of the chromosomes flanking the recombination event remain in the parental position.

Youds et al. (2010) presented evidence that the RTEL-1 protein of *C. elegans* physically dissociates strand invasion events, thereby promoting NCO repair by SDSA (Figure 1). HRR events initiated by DSBs consequently divide into two subsets, a larger subset which undergoes SDSA forming NCO recombinants, and a smaller subset which undergo DHJ repair and form CO recombinants. Perhaps SDSA is the preferred mode of HRR for unprogrammed double-strand damages, and DHJ repair is used primarily for programmed DSBs to promote proper chromosome segregation.

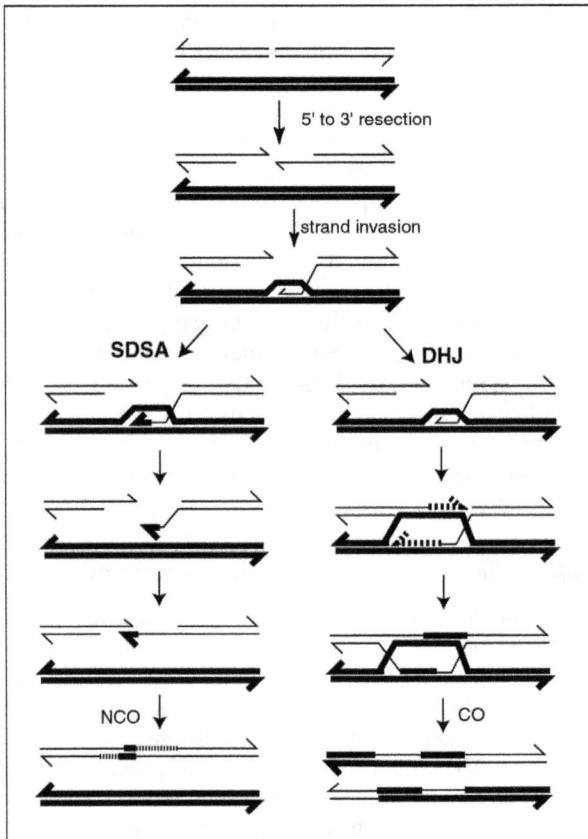

Fig. 1. Current models of meiotic recombination are initiated by a double-strand break or gap, followed by pairing with an homologous chromosome and strand invasion to initiate the recombinational repair process. Repair of the gap can lead to crossover (CO) or non-crossover (NCO) of the flanking regions. CO recombination is thought to occur by the Double Holliday Junction (DHJ) model, illustrated on the right, above. NCO recombinants are thought to occur primarily by the Synthesis Dependent Strand Annealing (SDSA) model, illustrated on the left, above. Most recombination events appear to be the SDSA type.

Although the SDSA model starts with a DSB, it would also be applicable to other types of double-strand damages such as interstrand-crosslinks, or a single-strand damage (e.g. an altered base) opposite a break in the other strand. In principle, both of these types of double-strand damages could be converted by nucleases to a DSB that would then be subject to SDSA.

8. The role of Spo11 in promoting accurate DNA repair can also facilitate proper chromosome segregation

In the budding yeast *S. cerevisiae*, synapsis (pairing of homologous chromosomes) and synaptonemal complex formation depend on Spo11, a nuclease related to type II topoisomerases. Spo11 induces DSBs leading to HRR events of the CO type that form the physical association between homologs (chiasmata) needed for synaptonemal complex formation and proper disjunction of non-sister homologs at the first meiotic division. On the basis of these properties of Spo11, it is sometimes assumed that the primary function of meiotic recombination is to promote synapsis. However, as reviewed by Barzel & Kupiec (2008), this theme cannot be generalized, as synapsis occurs independently of Spo11 induced recombination in the nematode worm *C. elegans* and the fruitfly *D. melanogaster*. In *C. elegans*, synapsis between homologs occurs normally in a *spo-11* mutant (Dernburg et al., 1998). The *D. melanogaster* gene *mei-W68* encodes a *spo11* homolog (McKim & Hayashi-Hagihara, 1998). In *D. melanogaster* females, meiotic chromosome synapsis occurs in the absence of *mei-W68* mediated CO recombination (McKim et al., 1998). Electron microscopy of oocytes from females homozygous for *mei-W68* mutations that eliminated meiotic recombination revealed normal synaptonemal complex formation. In *D. melanogaster* females, meiotic recombination does not appear to be necessary for synapsis. Since the role of Spo11 is of substantial interest in current discussions of the adaptive significance of meiotic recombination, we offer a speculation on its possible role consistent with the DNA repair hypothesis. As shown in Figure 1, both the DHJ and SDSA models for HRR start with a DSB. During meiosis in *S. cerevisiae*, DSBs are formed by a process that usually depends on Spo11. In *S. pombe*, Spo11 homolog Rec12 generates meiotic recombinants and meiosis specific DSBs. In *C. elegans*, a Spo11 homolog seems to have a similar role. We propose that DNA damages of various types are converted to DSBs, a "common currency," in order to initiate their recombinational repair (see also H. Bernstein et al., 1988). Spo11 appears to be employed in this process. Our reasoning is based on the precedents of the well-established pathways of nucleotide excision repair and base excision repair. In nucleotide excision repair, the initial steps of the pathway involve recognition of a wide variety of bulky damages followed by their removal to generate a single-strand gap, the "common currency" which is then repaired by a gap filling process. In base excision repair, a variety of altered bases are recognized by a corresponding variety of DNA glycosylases that generate an intermediate apurinic/apyrimidinic site, the "common currency" for further repair. On this reasoning, formation of DSBs by a Spo11-dependent process is part of an overall DNA repair sequence. In those species where the resolution of meiotic HRR by CO recombination is beneficial in promoting proper chromosome segregation at the first meiotic division, we think this benefit arose secondarily to the primary benefit of accurate DNA repair.

The function of recombination as a repair process may have arisen very early in the evolution of life [perhaps in the RNA world (H. Bernstein et al., 1984)], and the function of promoting synapsis during meiosis probably arose later in evolution in some eukaryotic lineages. If, in mammals, a major function of meiotic CO recombination, as distinct from NCO recombination, is to promote synapsis and proper chromosome segregation, then one might expect CO events to be localized to specific hot-spot sequences. Hot-spot determinants may also include specific proteins that bind to hot-spot sequences and facilitate CO recombination such as Prdm9 (Hochwagen and Marals, 2010). It is estimated that, in humans, the average number of endogenous DNA DSBs per somatic cell occurring at each cell generation is about 50 (Vilenchik & Knudson, 2003). This rate of DSB formation likely reflects unprogrammed damages, such as may be caused by ROS, and can be taken as an indication of the level of unprogrammed DSBs present in cells undergoing meiosis as well. In the human genome 25,000 hotspots for meiotic recombination have been identified (Myers et al., 2006). The average number of CO recombination events per hotspot is one CO event per 1,300 meioses. The large number of recombination hotspots is consistent with a wide distribution of sites vulnerable to unprogrammed DNA damage as well as specific sites where recombination would need to be induced to promote synapsis. A challenge for future research is the identification of the types of natural damages and programmed damages, and their frequencies, that are removed by CO recombinational repair during meiosis.

9. During meiosis, CO recombination can repair DNA damages independently of Spo11

In a *spo11* mutant of *S. cerevisiae*, the meiotic defects in recombination and synapsis are alleviated by X-irradiation, indicating that X-ray induced DNA damages can initiate CO recombination leading to synapsis independently of Spo11 (Thorne & Byers, 1993). Also, in *C. elegans*, Spo11 is required for meiotic recombination, but radiation induced-breaks alleviate this dependence (Dernberg et al., 1998). These findings indicate that unprogrammed DNA damages induced by X-rays can be repaired by HRR during meiosis independently of Spo11. In both *S. pombe* and *C. elegans*, mutants deficient for Spo11 undergo meiotic CO recombination when single base lesions of the type dU:dG are produced in their DNA (Pauklin et al., 2009). This recombination does not involve production of large numbers of DSBs, but does require uracil DNA-glycolylase, an enzyme that removes uracil from the DNA backbone and initiates base excision repair. These authors proposed that base excision repair of a uracil base, an abasic site, or a single-strand nick are sufficient to initiate meiotic CO recombination in *S pombe* and *C. elegans*.

In a Rec12 (Spo11 homolog) mutant strain of *S. pombe*, meiotic recombination can be restored to near normal levels by a deletion in *rad2* that encodes an endonuclease involved in Okazaki fragment processing (Farah et al., 2005). Both CO and NCO recombination were increased, but DSBs were undetectable. On the basis of the biochemical properties of Rad 2, these authors proposed that meiotic recombination can be initiated by non-DSB lesions, such as nicks and gaps, which accumulate during premeiotic DNA replication when Okasaki fragment processing is deficient.

In general the findings reviewed in this section indicate that DNA damages arising from a variety of sources can be repaired by meiotic HRR of the CO type, and that this repair may occur independently of Spo11.

10. DNA repair likely provides the strong short-term advantage that maintains meiosis, while genetic variation may provide a long-term advantage

Evolutionary explanations for sex have often assumed that the adaptive advantage of meiosis arises from the genetic variation produced. A variety of models and reviews have been presented in this active area of research (e.g. Barton & Charlesworth, 1998; Otto & Gerstein, 2006; Agrawal, 2006). However, Otto & Gerstein (2006) have also pointed out that in a fairly stable environment, individuals surviving to reproductive age have genomes that function well in their current environment. They raise the question of why such individuals should risk shuffling their genes with those of another individual, as happens during meiotic recombination. This consideration, and others, have led many investigators to question whether production of genetic diversity is the principal adaptive advantage of sex. Heng (2007) and Gorelick & Heng (2010) reviewed evidence that sex actually decreases most genetic variation. Their view is that sex acts like a coarse filter, weeding out major changes, such as chromosomal rearrangements, but allowing minor variation, such as changes at the nucleotide or gene level (that are often neutral), to flow through the sexual sieve. Thus, they consider that sex acts as a constraint on genomic variation, thereby limiting adaptive evolution.

We consider that the major adaptive advantage of meiosis is enhanced recombinational repair. In contrast to the variation hypothesis, DNA repair provides an appropriate explanation for the adaptive advantage of sex (and meiosis) in the short-term, since its benefits are large enough (removal of DNA damages that would be deleterious/lethal to gametes or progeny) to plausibly balance the large costs of sex. The large costs of sex include the "cost of males" (Maynard Smith, 1978; Williams, 1975), "recombinational load" that arises from the randomization of genetic information during sex and loss of coadapted gene complexes (Shields, 1982), the cost of mating (Bernstein et al., 1985b), and cost of sexually transmitted disease (Michod et al., 2008).

The hypothesis that meiosis is an adaptation for DNA repair can be consistently applied to all organisms that have sex, including the facultative sexual organisms discussed above, as well as species that undergo meiosis but experience little or no outcrossing, as described below. If, in the long-term, the genetic variation produced by sex increases the rate of adaptation, as proposed by a number of authors (Goddard et al., 2005; Colegrave et al., 2002; Kaltz & Bell, 2002; Cooper et al., 2005; de Visser & Elena, 2007; Peters & Otto, 2003), this would be an added benefit. However, in the short-term, we consider it unlikely that the benefit of variation is large enough to maintain sex.

In nature, many organisms that undergo meiosis outcross only rarely or not at all. In these cases, meiosis generates little or no genetic variation. In the budding yeast *S. cerevisiae*, outcrossing sex, in contrast to inbreeding sex, appears to be very infrequent in nature. Ruderfer et al. (2006) estimated that the ancestors of three *S. cerevisiae* strains outcrossed in nature only about once every 50,000 generations. On the other hand, mating between closely related yeast cells is likely to have been much more common in nature. Mating can occur when haploid cells of opposite mating types, MATa and MATα, come into contact. As pointed out by Zeyl & Otto (2007), mating between closely related cells is common for two reasons; (1) the close physical proximity of cells of opposite mating type from the same ascus (the sac that contains the products from a single meiosis), and (2) homothallism, the ability of haploid cells of one mating type to produce daughter cells of the opposite mating type. Thus, in nature, the meiotic events that produce little or no recombinational variation are much more frequent than meiotic events that do produce recombinational variation. This disparity is consistent with the idea that the primary adaptive function of meiosis in *S. cerevisiae* is HRR of DNA damages, since this benefit is realized in meiosis resulting from either inbreeding or outcrossing. If the primary adaptive function of meiosis were to generate genetic variation, it is difficult to understand how the complex process of meiosis could be selectively maintained in *S. cerevisiae* during the many generations in which there is no outcrossing.

Various levels of inbreeding due to consanguineous mating are known in many species. One extreme, but well studied, example among vertebrate species is the Mangrove Killifish, *Kryptolebias marmoratus*, which inhabits brackish water mangrove habitats from Brazil to Florida. These fish produce sperm and eggs by meiosis and reproduce routinely by self-fertilization. Each hermaphroditic individual normally fertilizes itself when a sperm and egg that it has produced by an internal organ unite inside the fish's body (Sakakura et al., 2006; for review see Avise, 2008). In this highly inbred hermaphroditic species meiotic recombination does not produce significant allelic variation, suggesting that meiosis is retained for some other adaptive benefit.

In higher plants, outcrossing sexual reproduction is the most common mode of reproduction, but about 15% of plants undergo meiosis and are principally self-fertilizing (C. Bernstein & H. Bernstein, 1991). We infer from these examples that the generation of genetic variation is not likely to be the adaptive benefit maintaining meiosis in these organisms. However, meiosis may be maintained by the adaptive benefit of HRR of DNA damage, since this benefit does not depend on outcrossing, nor that the participating chromosomes carry different alleles.

The meiotic function of repairing DNA damages primarily acts to preserve the existing genome. The generation of new genomic variants, a consequence of recombinational repair processes, appears to be a secondary effect that may provide a benefit in the long-term.

As discussed above, most HRR events during meiosis are of the NCO type, which generate minimal genetic variation compared to the CO type. This is consistent with the DNA repair hypothesis, since both the CO and NCO types of recombination can repair DNA. On the assumption that the generation of variation is the primary benefit of meiosis, the majority of HRR events, those of the NCO type, provide no significant benefit and hence are wasteful.

Even though, during meiosis, the frequency of CO recombination is ordinarily substantially less than the frequency of NCO recombination, during mitosis the frequency of CO compared to NCO recombination is even lower (e.g. Virgin et al., 2001; Prado et al., 2003). The higher frequency of CO recombinants during meiosis compared to mitosis may reflect the role of CO recombinants in promoting synapsis during meiosis (see section 8, above), a process distinct to meiosis.

11. During meiosis, HRR may remove a class of damages that cannot be accurately repaired during mitosis

HRR during meiosis offers unique advantages compared to HRR during mitosis, based on the opportunity for non-sister homologs to pair and recombine during meiosis, which does not happen during mitosis. In mitosis, HRR involves interaction between the sister-chromosomes formed upon DNA replication. Thus, in mitosis, HRR is limited to the phases of the cell cycle during DNA replication (S phase) and after DNA replication (G2/M). Prior to DNA replication (G1 phase) in mitosis, double-strand DNA damages, such as DSBs, are repaired by an inaccurate process, non-homologous end-joining (NHEJ), which generates mutation. Double strand damages arising after DNA replication, may be repaired during mitosis by HRR between sisters (Tichy et al., 2010). However, meiotic recombination can cope in a non-mutagenic way with double strand damages which arise at any point in the cell cycle.

Meiotic G1 phase cells appear to be more resistant to the lethal effects of X-irradiation than mitotic G1 phase cells (Kelly et al., 1983). This finding suggests that repair of DSBs is more efficient during meiotic than mitotic G1 phase, as DSBs are a common consequence of X-irradiation. We speculate that during meiosis, in contrast to mitosis, double-strand damages occurring prior to DNA replication may be accurately repaired by HRR because pairing occurs between non-sister chromosomes. If this is so, meiotic cells have the advantage, compared to mitotic cells, of being able to accurately and efficiently repair double-strand damages that occur both before and after replication. As a result, germ cells would tend to be protected against the mutagenic effect of inaccurate NHEJ that typically occurs prior to replication in mitotic cells.

Mao et al. (2008) presented evidence that one type of somatic cell, human fibroblasts, utilizes error-prone NHEJ as the major DSB repair pathway at all cell cycle stages. In these cells, HRR is nearly absent prior to replication (G1 phase) and is used, when it occurs, primarily in the S phase. Even after the S phase when two sister-chromosomes are present (the G2/M phase), NHEJ is elevated and HRR is in decline.

The situation is somewhat different in mammalian embryonic stem (ES) cells compared to differentiated somatic cells (Tichy et al., 2010). ES cells give rise to all of the cell types of an organism. Because mutations at this early embryonic stage are passed on to all clonal descendents, they can be seriously detrimental to the organism as a whole. Therefore robust mechanisms are needed in ES cells for reducing DNA damages (or eliminating damaged cells) in order to reduce mutations. Mouse ES cells were found to predominantly use high fidelity HRR to repair DSBs, compared to somatic cells that predominantly used NHEJ (Tichy et al., 2010). Furthermore mouse ES cells lack a G1 checkpoint and do not undergo cell-cycle arrest upon receiving DNA damage prior to DNA replication. Rather, they undergo p53-independent apoptosis in response to DNA damage (Aladjem et al., 1998). Consistent with these findings, mouse ES stem cells have a mutation frequency about 100-fold lower than that of isogenic mouse somatic cells (Cervantes et al., 2002), but, as discussed next, at a likely cost resulting from somatic selection against cells with unrepairable DSBs which arise before DNA replication.

These results imply that a low mutation rate is achievable in mitotic cells by using apoptosis to remove cells with DNA damages that are present prior to replication, and using HRR, rather than NHEJ, to remove double-strand damages present subsequent to DNA replication. The non-sister chromosomes present in every diploid somatic cell during mitosis, in principal, might pair and undergo accurate HRR (as in meiosis), but this does not ordinarily occur, presumably because, in somatic cells, the benefit is outweighed by costs [e.g. loss of heterozygosity and expression of deleterious recessive alleles including those leading to cancer]. Meiosis is therefore unique, in that DNA damages occurring both prior to and after DNA replication can be subject to high fidelity HRR between non-sister homologs. This would avoid the high costs of both deleterious mutation and loss of potential gametes due to apoptosis.

In humans at each cell division, 30,000-50,000 DNA replication origins are activated (Mechali et al., 2010). Thus the chromosome is ordinarily replicated in segments. We postulate that any segment containing a DSB will fail to complete its replication until the DSB is repaired. This limited and temporary blockage of replication may result directly from the break itself, or occur as a response to regulatory events set off by proteins that specifically bind to the broken ends. In any case, HRR can be carried out during the subsequent prophase I stage of meiosis, when the segment containing a DSB pairs with a non-sister homologue. This repair would then allow chromosome replication to be completed.

12. DNA damage during the mitotic divisions of the germ line in multicellular organisms

In multicellular eukaryotes there are typically many mitoses during germ line development, and only a single final meiosis leading to gamete formation. During the mitotic cell divisions in the germ line, DSBs and other double-strand damages occurring after DNA replication are likely repaired by HRR or eliminated from the cell lineage by death and/or apoptosis of the damaged cell. We have argued above (section 11) that because of the lack of pairing of non-sister homologs during mitosis, HRR is unable to accurately repair double-strand damages occurring before replication. Thus when double-strand damages occur prior to replication during the mitotic divisions in the germ line the consequence will be either increased mutation or increased apoptosis. By analogy with the strategy used by somatic stem cells (section 11, above), we think that the preferred strategy during these mitotic divisions is likely to be apoptosis, since this avoids mutations in the germ line that could be passed on to progeny. However, double-strand damages occurring prior to replication during meiosis need not lead to apoptosis (which would likely decrease fecundity), since these can be accurately repaired by HRR between non-sister chromosomes. The consequence will be enhanced gamete viability and fecundity, that is, enhanced fitness. In the mitotic divisions of the germ-line prior to meiosis, loss of cells due to DNA damage-induced apoptosis need not be very costly to organism fitness, since such losses could be made up by extra cell divisions of undamaged cells. However, the loss of sperm or egg cells due to unrepaired DNA damage would likely have substantial costs to fitness due to loss of fertility and progeny, as discussed above in section 5.

13. Why is meiosis frequently associated with outcrossing?

While the focus of this article is on the adaptive benefit of meiosis itself, we briefly consider why meiosis is frequently associated with outcrossing, where the chromosomes involved in recombination come from different unrelated parents in a prior generation. Previously, we discussed examples of meiosis occurring in association with inbreeding and self-fertilization. Meiosis with inbreeding will be favored when the costs of mating are high (e.g. the cost of finding a mate at low population density). These examples of inbred meiosis were presented to illustrate our argument that meiosis provides an adaptive advantage (accurate DNA repair) independent of whether significant recombinational variation is also produced. However, meiosis is often associated with outcrossing, and we now consider why.

A disadvantage of inbreeding, especially of self-fertilization, is expression of deleterious recessive mutations, resulting in inbreeding depression. Analysis of the effects of masking deleterious recessive mutations (genetic complementation) using heuristic modes and arguments indicated that complementation provides benefits sufficient to maintain outcrossing (H. Bernstein et al., 1985a, 1987; Michod, 1995). However, more explicit population genetic models have raised some issues that are in need of further clarification. In population genetics terms, the basic effect of outcrossing is to bring populations to Hardy-Weinberg (HW) equilibrium. Thus, outcrossing can be beneficial if there is another force that pushes the population away from HW equilibrium (generating either an excess or a deficit of heterozygotes) and if it's advantageous to go closer to HW equilibrium. One possible force that generates departure from HW equilibrium is dominance: for example if deleterious alleles tend to be recessive, after selection there will be an excess of heterozygotes (and a deficit of homozygotes). However in this case outcrossing is costly in the short term (because it tends to expose deleterious alleles), but beneficial in the long term (because purging them becomes more efficient). Otto (2003) showed that under this scenario high rates of outcrossing are favored only if deleterious alleles are weakly recessive (dominance close to 0.5). Another potential force pushing away from HW equilibrium considered by Roze and Michod (2010) is gene conversion which creates homozygosity. Gene conversion could result from mitotic HRR between sister chromosomes as discussed above. In this case (and if deleterious alleles tend to be partially recessive) outcrossing is beneficial in the short term (because it masks deleterious alleles) but disadvantageous in the long term (because purging is less efficient). The magnitude of this force may be estimated from rates of loss of heterozygosity during development [discussed in Roze and Michod (2010)]. The few estimates which exist indicate that the loss of heterozygosity is low, and thus this selective force for outcrossing may be weak. Clearly, we need more estimates of this critical parameter to know how large this force for outcrossing may be.

Another consequence of outcrossing is the generation of new genetic variants which may provide an additional long-term advantage.

14. The special case of asexual bdelloid rotifers

Bdelloid rotifers are common invertebrate animals. They are apparently obligate asexuals that reproduce by parthenogenesis. These organisms are extraordinarily resistant to ionizing radiation (Gladyshev and Meselson, 2008). This resistance appears to be a consequence of an evolutionary adaptation to survive desiccation in ephemerally aquatic habitats. Such desiccation causes extensive DNA breakage, which they are able to repair. Bdelloid primary oocytes are in the G1 phase of the cell cycle and thus lack sister chromatids. Welch et al. (2008) proposed a mechanism of repair involving interaction of non-sister co-linear chromosome pairs, which are maintained as templates for repair of DNA DSBs caused by the frequent desiccation and rehydration. Thus although these organisms apparently lack sex and meiosis, an essential feature of meiosis, HRR between non-sister homologs appears to be retained.

15. Conservation among eukaryotes of RecA-like proteins as key components of the HRR machinery acting during meiosis

Sex appears to be universally based on RecA-like proteins. RecA-like proteins play a key role in HRR, and the HRR machinery and its mechanism of action appear to be highly conserved among eukaryotes. The rad51 and dmc1 genes in the eukaryotic yeasts S. cerevisiae and S. pombe are orthologs of the bacterial recA gene. The dmc1 gene is found in many different eukaryote species, and has been reported, for instance, in the protists *Giardia, Trypanosoma, Leishmania, Entamoeba* and *Plasmodium* (Ramesh et al., 2005). Rad51 and Dmc1 proteins are recombinases that interact with single-stranded DNA to form filamentous intermediates called presynaptic filaments, and these filaments initiate HRR (Sauvageau et al., 2005; San Filippo et al., 2008). Dmc1 recombinase functions only during meiosis, whereas Rad51 recombinase acts in both somatic HRR and in meiosis. When it functions in meiosis, Rad51 mainly uses a sister chromosome for HRR. In contrast, Dmc1 mainly uses the non-sister homologous chromosome. The yeast Rad51 recombinase catalyzes ATP-dependent homologous DNA pairing and strand exchange, as does the bacterial RecA recombinase (Sung, 1994). The tertiary structure of the Dmc1 recombinase has an overall similarity to the bacterial RecA recombinase (Story et al., 1993). These observations suggest that the bacterial RecA that functions in the bacterial sexual process of transformation, and the yeast Rad51 and Dmc1 recombinases that act in meiosis have similar functions, consistent with the idea that meiotic recombination evolved from simpler sexual processes in bacteria

We next consider evidence that RecA orthologs play a key role in meiosis, not only in protists, but also in multicellular eukaryotes. RecA orthologs act in meiosis in a range of animals (e.g. nematodes, chickens, humans and mice) and plants (e.g. *Arabidopsis*, rice and lilies). The *rad51* gene is expressed at a high level in mouse testis and ovary, suggesting that Rad51 protein is involved in meiotic recombination (Shinohara et al., 1993). In mice, mutations in the *dmc1* gene cause sterility, failure to undergo intimate pairing of homologous chromosomes and an inability to complete meiosis (Pittman et al., 1998; Yoshida et al., 1998; see also Table 2). In the nematode *C. elegans*, resistance to DNA damage caused by X-irradiation in the meiotic pachytene nuclei depends on a RecA-like gene (Takanami et al., 2000). *RecA* gene orthologs are also expressed in chicken testis and ovary and in human testis. In humans, Dmc1, the meiosis-specific recombinase, forms nucleoprotein complexes on single-stranded DNA that promote a search for homology and carry out strand exchange, the two necessary steps of genetic recombination (Sehorn et al, 2004; Bugreev et al., 2005).

In lily plants, genes *lim15* and *rad51* are orthologs, respectively, of the *dmc1* and *rad51* genes of yeast. The lily proteins Lim15 and Rad51 colocalize on chromosomes in various stages of meiotic prophase I, and form discrete foci (Terasawa et al., 1995). The proteins of these foci are considered to participate in the search for, and pairing of, homologous sequences of DNA. In another plant, *Arabidopsis thaliana*, meiotic recombination requires Dmc1 (Couteau et al., 1999) and Rad51 (Li et al., 2004). In the rice plant, an ortholog of dmc1 is necessary for meiosis and has a key function in the pairing of homologous chromosomes (Deng and Wang, 2007).

In general, both animals and plants have RecA-like proteins that appear to have a central function in meiotic HRR. Furthermore, bacterial RecA and its animal and plant orthologs have very similar roles in the HRR events during the sexual processes of bacterial transformation and eukaryotic meiosis. In all cases, the RecA protein or RecA-like protein assembles on single-stranded DNA to form a pre-synaptic filament. This filament then attaches to a duplex DNA molecule and searches for homology in its target. When the presynaptic molecule locates an homologous sequence in the duplex molecule, it is able to form a DNA joint [Figure 2]. These joints are then processed further to complete the HRR event.

Transformation in Bacteria   Meiosis in a Eukaryote

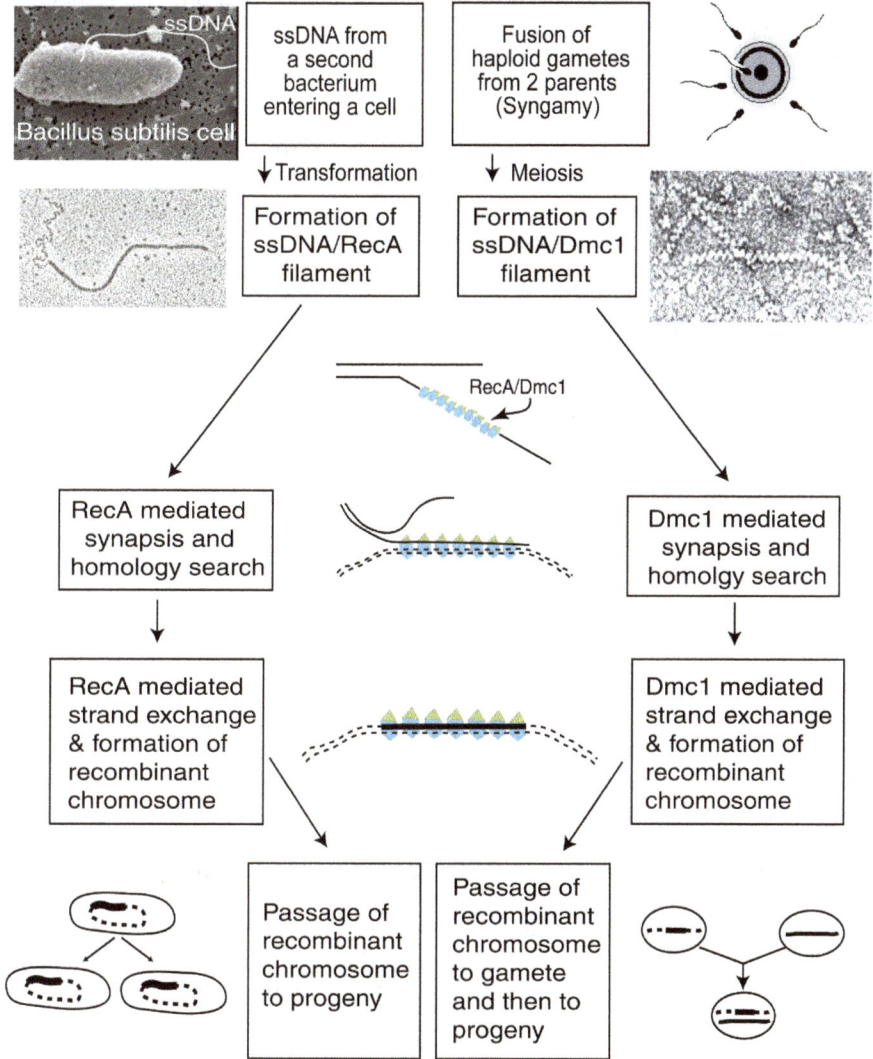

Fig. 2. Conservation of the key components of the HRR machinery during the sexual process of transformation in bacteria and during meiosis in eukaryotes. The bacterial RecA protein or the eukaryotic RecA-like protein, Dmc1, assembles on single-stranded DNA to form a pre-synaptic filament. This filament then attaches to a duplex DNA molecule and searches for homology in its target. When the pre-synaptic molecule locates an homologous sequence in the duplex molecule, it is able to form a DNA joint. These joints are then processed further to complete the HRR event.

16. Summary

Currently there is no general agreement among biologists on the adaptive function of sex. Meiosis, a key stage of the sexual cycle, involves close pairing and physical recombination and information exchange between homologous chromosomes ordinarily derived from two different parents. Fundamental to solving the problem of why sex exists is achieving an understanding of the function of meiosis.

A primitive form of meiosis was likely present early in the evolution of eukaryotes, perhaps in the single-celled ancestor of all eukaryotes that arose from ancestral bacteria over 1.5 billion years ago. Meiosis may be derived from bacterial transformation, a prokaryotic sexual process that promotes homologous recombinational repair of DNA as shown in Figure 2. Among extant single-cell eukaryotes, meiosis and facultative sex are ubiquitous. Entry into the sexual cycle ordinarily occurs in response to stressful conditions, such as oxidative stress, that tend to be associated with DNA damage. Thus meiosis may be an adaptation for dealing with such stresses and the resulting DNA damages. Consistent with this idea, exposure of eukaryotes to various DNA damaging agents increases meiotic recombination. Both in mitosis and meiosis, DNA damages caused by different exogenous agents are repaired by HRR, suggesting that DNA damages from natural sources (e.g. ROS) are also repaired by HRR. The consistent function of recombination in DNA repair across meiosis and mitosis in eukaryotes, and transformation in prokaryotes, is what we seek to understand through the repair hypothesis.

Defective HRR during meiosis causes infertility in humans and rodents, suggesting that removal of DNA damages is an essential function of meiosis. The majority of HRR events during both mitosis and meiosis are of the NCO type. NCO recombination is able to repair DNA damages from diverse sources. Furthermore NCO recombination likely occurs by synthesis-dependent strand annealing, a mechanism that involves a small exchange of information between two chromosomes but not physical exchange of DNA. Explanations of the adaptive function of meiosis that focus exclusively on crossing over, the minority of recombination events, are inadequate to explain the majority, the NCO type.

The Spo11 protein, a nuclease, produces DSBs that can initiate recombination and promote proper chromosome segregation. We speculate that Spo11 is part of a process that converts a variety of types of DNA damages to a "common currency," the DSB, which is then subject to HRR. During meiosis, DNA damages arising from a variety of sources can be repaired by HRR of the CO type, and this repair may occur independently of Spo11.

Genetic variation produced by meiotic recombination may provide a long-term benefit at the population level by reducing linkage disequilibrium and providing gene combinations on which selection can more effectively act, but the short-term adaptive benefit that maintains the machinery of meiosis is likely DNA repair. In contrast to mitosis, meiosis may allow greater accuracy in the repair of DNA damages, since double-strand damages occurring prior to DNA replication can, in principle, be accurately removed by HRR between non-sister homologous chromosomes, a process that is largely unavailable during mitosis.

Among different species, meiosis is frequently associated with outcrossing. This probably reflects the benefit of masking deleterious recessive alleles. However, numerous species that undergo meiosis are largely inbreeding or self-fertilizing. This implies that meiosis provides a benefit (accurate DNA repair) independently of the benefit of outcrossing and masking deleterious recessive alleles.

Animals and plants have RecA-like proteins that have key functions in meiotic recombination involving homology recognition and strand exchange. The function of these eukaryotic proteins is similar to the bacterial RecA protein that acts during the bacterial sexual process of transformation, further suggesting that eukaryotic meiosis may have evolved from simpler sexual processes in bacteria.

17. Conclusion

DNA damages appear to be a ubiquitous and serious problem for all of life. We consider that the heightened ability of meiosis to repair such damages in the DNA to be passed on to the next generation is a capability sufficient to explain its widespread occurrence.

18. Acknowledgment

We thank Deborah Shelton, Denis Roze and Mike Berman for their thoughtful and very helpful comments on drafts of the manuscript. This work was supported in part by Arizona Biomedical Research Commission Grant #0803 and the Department of Veterans Affairs (VA), Veterans Health Administration, Office of Research and Development, VA Merit Review Grant 0142 of the Southern Arizona Veterans Affairs Health Care System. This work was also supported by National Science Foundation grant DEB-0742383 and the College of Science at the University of Arizona.

19. References

Agrawal, A.P. (2006). Evolution of sex: why do organisms shuffle their genotypes? *Current Biology*, Vol.16, pp. R696-R704.

Aladjem, M.I., Spike, B.T., Rodewald, L.W., Hope, T.J., Klemm, M., Jaenishc, R. & Wahl, G.M. (1998). ES cells do not activate p53-dependent stress responses and undergo p53-independent apoptosis in response to DNA damage. *Current Biology*, Vol. 8, pp. 145-155.

Allers, T. & Lichten, M. (2001). Differential timing and control of noncrossover and crossover recombination during meiosis. *Cell*, Vol. 106, pp. 47-57.

Ames, B.N., Shigenaga, M.K. & Hagen, T.M. (1993). Oxidants, antioxidants, and the degenerative diseases of aging. *Proceedings of the National Academy of Sciences USA*, Vol. 90, pp. 7915-7922.

Andersen, S.L. & Sekelsky. J. (2010). Meiotic versus mitotic recombination: two different routes for double-strand break repair. *Bioessays*, Vol. 32, pp. 1058-1066.

Avise, J.C. (2008). Clonality: *The genetics, ecology, and evolution of sexual abstinence in vertebrate animals*. Oxford University Press, New York, NY

Baker, B.S., Boyd, J.B., Carpenter, A.T.C., Green, M.M., Nguyen, T.D., Ripoll, P. & Smith, P.D. (1976). Genetic controls of meiotic recombination and somatic DNA metabolism in *Drosophila melanogaster*. *Proceedings of the National Academy of Sciences USA*, Vol. 73, pp. 4140-4144.

Barlow, C., Liyanage, M., Moens, P.B., Tarsounas, M., Nagashima, K., Brown, K., Rottinghaus, S., Jackson, S.P., Tagle, D., Ried, T. & Wynshaw-Boris, A. (1998). Atm deficiency results in severe meiotic disruption as early as leptonema of prophase I. *Development*, Vol. 125, pp. 4007-4017.

Barton, N.H. & Charlesworth, B. (1998). Why sex and recombination? *Science*, Vol. 281, pp. 1986-1989.

Barzel, A. & Kupiec, M. (2008). Finding a match: how do homologous sequences get together for recombination? *Nature Reviews/Genetics*, Vol. 9, pp. 27-37.

Bernstein, C & Johns, V. (1989). Sexual reproduction as a response to H_2O_2 damage in *Schizosaccharomyces pombe*. *Journal of Bacteriology*, Vol. 171, pp.1893-1897.

Bernstein, C. & Bernstein, H. (1991). *Aging, Sex and DNA Repair*. Academic Press, San Diego

Bernstein, H., Byerly, H.C., Hopf, F.A. & Michod, R.E. (1984). Origin of sex. *Journal of Theoretical Biology*, Vol. 110, pp. 323-351.

Bernstein, H., Byerly, H.C., Hopf, F.A. & Michod, R.E. (1985a). Genetic damage, mutation and the evolution of sex. *Science*, Vol. 229, pp.1277-1281.

Bernstein, H., Byerly, H.C., Hopf, F.A. & Michod, R.E. (1985b). Sex and the emergence of species. *Journal of Theoretical Biology*, Vol. 117, pp. 665-690.

Bernstein, H., Hopf, F.A. & Michod, R.E. (1987). The molecular basis of the evolution of sex. *Advances in Genetics*, Vol. 24, pp. 323-370.

Bernstein, H., Hopf, F.A. & Michod, R.E. (1988). Is meiotic recombination an adaptation for repairing DNA, producing genetic variation, or both? In *The Evolution of Sex: An Examination of Current Ideas* R.E. Michod. & B.R. Levin (Eds.). pp. 139-160 Sinauer, Sunderland, Mass.

Bernstein, H. & Bernstein, C. (2010). Evolutionary origin of recombination during meiosis. *BioScience*, Vol. 60, No. 7, pp. 498-505.

Birdsell, J.A. & Wills, C. (2003) The evolutionary origin and maintenance of sexual recombination: A review of contemporary models. In: *Evolutionary Biology*. R.G. MacIntyre, M.T. Clegg, (Eds.), pp. 27-138. Kluwer Academic/Plenum Publishers.

Boyd J.B. (1978). DNA repair in *Drosophila*. In *DNA Repair Mechanisms* P.C. Hanawalt, E.C. Friedberg & C.F. Fox. (Eds.) pp. 449-452. Academic Press, New York.

Bugreev, D.V., Golub, E.I., Stasiak, A.Z., Stasiak, A. & Mazin A.V. (2005). Activation of human meiosis-specific recombinase Dmc1 by Ca2+. *Journal of Biological Chemistry*, Vol. 280, pp. 26886-26895.

Cervantes, R.B., Stringer, J.R., Shao, C., Tischfield, J.A. & Stambrook, P.J. (2002) Embryonic stem cells and somatic cells differ in mutation frequency and type. *Proceedings of the National Academy of Sciences USA*, Vol. 99, pp.3586-3590.

Chovnick, A., Ballantyne, G.H., Baillie, D.L., Holm, D.G. (1970). Gene conversion in higher organisms: half-tetrad analysis of recombination within the rosy cistron of *Drosophila melanogaster*. *Genetics*, Vol. 66, pp 315-329.

Colegrave, N., Kaltz, O. & Bell, G. (2002). The ecology and genetics of fitness in *Chlamydomonas*. VIII. The dynamics of adaptation to novel environments after a single episode of sex. *Evolution*, Vol. 56, pp. 14-21.

Cooper, T.F., Lenski, R.E. & Elena, S.P. (2005). Parasites and mutational load: an experimental test of a pluralistic theory for the evolution of sex. *Proceedings of the Royal Society B*, Vol. 272, pp. 311-317.

Coogan, T.P. & Rosenblum, I.Y. (1988). DNA double-strand damage and repair following gamma-irradiation in isolated spermatogenic cells. *Mutation Research*, Vol. 194, pp. 183-191.

Couteau, F., Belzile, F., Horlow, C., Granjean, O., Vezon, D. & Doutriaux, M.P. (1999). Random chromosome segregation without meiotic arrest in both male and female meiocytes of a Dmc1 mutant of Arabidopsis. *Plant Cell*, Vol. 11, 1623-1634.

Cressman, V.L., Backlund, D.C., Avrutskaya, A.V., Leadon, S.A., Godfrey, V. & Koller, B.H. (1999). Growth retardation, DNA repair defects, and lack of spermatogenesis in BRCA1-deficient mice. *Molecular and Cellular Biology*, Vol. 19, No. 10, pp. 7061-7075.

Dacks, J. & Roger, A.J. (1999). The first sexual lineage and the relevance of facultative sex. *Journal of Molecular Evolution*, Vol. 48, pp. 779-783.

Davey, J. (1998). Fusion of a fission yeast. *Yeast*, Vol. 14, pp. 1529-1566.

Deng, Z.Y. & Wang, T. (2007). OsDMC1 is required for homologous pairing in Oryza sativa. Plant Molecular Biology, Vol. 65, pp. 31-42.

Dernberg, A.F., McDonald, K., Moulder, G., Barstead, R., Dresser, M. & Villeneuve, A.M. (1998). Meiotic recombination in *C. elegans* initiates by a conserved mechanism and is dispensable for homologous chromosome synapsis. *Cell*, Vol. 94, pp. 387-398.

DeVisser, J.A. & Elena, S.F. (2007). The evolution of sex: empirical insights into the roles of epistasis and drift. *Nature Reviews/Genetics*, Vol. 8, pp.139-149.

Elliott, A.M. & Hayes, R.E. (1953). Mating types in tetrahymena. *Biological Bulletin*, Vol., 105, pp. 269-284.

Farah, J.A., Cromie, G., Davis, L., Steiner, W.W. & Smith, G.R. (2005). Activation of an alternative, Rec12 (Spo11)-independent pathway of fission yeast meiotic recombination in the absence of a DNA flap endonuclease. *Genetics*, Vol. 171, pp. 1499-1511.

Fisher, H.M. & Aitken, R.J. (1997) Comparative analysis of the ability of precursor germ cells and epididymal spermatozoa to generate reactive oxygen metabolites. *The Journal of Experimental Zoology*, Vol. 277, pp. 390-400.

Galetzka, D., Weis, E., Kohlschmidt, N., Bitz, O., Stein, R. & Haaf, T. (2007). Expression of somatic DNA repair genes in human testes. *Journal of Cellular Biochemistry*, Vol. 100, pp. 1232-1239.

Game, J.C., Zamb, T.J., Braun, R.J., Resnick, M. & Roth, R.M. (1980). The role of radiation (rad) genes in meiotic recombination in yeast. *Genetics*, Vol. 94, pp. 51-68.

Gladyshev, E. & Meselson, M. (2008). Extreme resistance of bdelloid rotifers to ionizing radiation. *Proceedings of the National Academy of Sciences USA*, Vol. 105, pp. 5139-5144.

Goddard, M.R., Godfray, H.C. & Burt, A. (2005). Sex increases the efficacy of natural selection in experimental yeast populations. *Nature*, Vol. 434, pp. 636-640.

Gorelick, R. & Heng, H.H.Q. (2011). Sex reduces genetic variation: A multidisciplinary review. *Evolution*, Vol. 65, pp. 1088-1098.

Hannah-Alava, A (1964). The brood pattern of X-ray-induced mutational damage in the germ cells of *Drosophila melanogaster* males. *Mutation Research*, Vol. 1, pp. 414-436.

Hari, K.L., Santerre, A., Sekelsky, J.J., McKim, K.S., Boyd, J.B. & Hawley, R.S. (1995). The *mei-41* gene of *D. melanogaster* is a structural and functional homolog of the human ataxia telangiectasia gene. *Cell*, Vol. 82, pp. 815-821.

Haynes, R.H. (1988). Biological context of DNA repair. In: *Mechanisms and Consequences of DNA Damage Processing*, E.C. Friedberg & P.C. Hanawalt (Eds), pp. 577-584. Alan R. Liss, New York

Haynes, R.H. & Kunz, B.A. (1981)/ DNA Repair and mutagenesis in yeast. In: *The Molecular Biology of the Yeast Saccharomyces. Life Cycle and Inheritance*, J. Strathern, E. Jones & J. Broach (Eds.) Cold Spring Harbor Laboratory, Cold Spring Harbor, N.Y.

Heng, H.H.Q. (2007) Elimination of altered karyotypes by sexual reproduction preserves species identity. *Genome*, Vol. 50, pp. 517-524.

Henriques, J.A.P. & Moustacchi, E. (1980). Sensitivity to photoaddition of mono- and bifunctional furocoumarins of X-ray sensitive mutants of *Saccharomyces cerevisiae*. *Photochemistry and Photobiology*, Vol. 31, pp. 557-561.

Herskowitz, I. (1988). Life cycle of the budding yeast *Saccharomyces cerevisiae*. *Microbiological Reviews*, Vol. 52, pp. 536-553.

Hochwagen A & Marais G.A.B. (2010). Meiosis: APRDM9 guide to the hotspots of recombination. *Current Biology*, Vol. 20, pp. R271-R274.

Hoelzer, M.A. & Michod, R.E. (1991). DNA repair and the evolution of transformation in *Bacillus subtilis*. III Sex with damaged DNA. *Genetics*, Vol.128, pp. 215-223.

Javaux, E.J., Knoll, A.H. & Walter, M.R. (2001). Morphological and ecological complexity in early eukaryote ecosystems. *Nature*, Vol. 412, pp. 66-69.

Joyce, E.F., Tanneti, S.N. & McKim, K.S. (2009). Drosophila Hold'em is required for a subset of meiotic crossovers and interacts with the DNA repair endonuclease complex subunits MEI-9 and ERCC1. *Genetics*, Vol. 181, pp. 335-340.

Kaltz, O. & Bell, G. (2002). The ecology and genetic of fitness in Chlamydomonas. XII. Repeated sexual episodes increase rates of adaptation to novel environments. *Evolution*, Vol. 56, pp. 1743-1753.

Kelly, S.L., Merrill, C. & Parry, J.M. (1983). Cyclic variations in sensitivity to X-irradiation during meiosis in *Saccharomyces cerevisiae*. *Molecular and General Genetics*, Vol. 191, pp. 314-318.

Kim, J.S. & Rose, A.M. (1987). The effect of gamma radiation on recombination frequency in *Caenorhabditis elegans*. *Genome*, Vol. 29, pp. 457-462.

Kirk, D.L. & Kirk, M. (1986). Heat shock elicits production of sexual inducer in *Volvox*. *Science*, Vol. 231, pp. 51-54.

Klovstad, M., Abdu, U. & Schupbach, T. (2008). Drosophila *brca2* is required for mitotic and meiotic DNA repair and efficient activation of the meiotic recombination checkpoint. *PloS Genetics*, Vol. 4, No. 2: e31.doi:10.1371/journal.pgen.0040031

Lahr, D.J.G., Parfrey, L.W., Mitchell, E.A.D., Katz, L.A. & Lara, E. (2011). The chastity of amoebae: re-evaluating evidence for sex in amoeboid organisms. *Proceedings of the Royal Society B: Biological Sciences*. Published online 23 March 2011. doi: 10.1098/rspb.so11.0289

Lewis, S.E.M. & Aitken, R.J. (2005). DNA damage to spermatozoa has impacts on fertilization and pregnancy. *Cell Tissue Research*, Vol. 322, pp. 33-41.

Li, W., Chen, C., Markmann-Mulisch, U., Timofejeva, I., Schmeizer, E., Ma, H. & Reiss, B. (2004). The Arabidopsis AtRAD51 gene is dispensable for vegetative development but required for meiosis. *Proceedings of the National Academy of Sciences*, Vol. 101, 10596-10601.

Lin, X., Hull, C.M. & Heitman, J. (2005). Sexual reproduction between partners of the same mating type in *Cryptococcus neoformans*. *Nature*, Vol. 434, pp. 1017-1021.

Lisby, M. & Rothstein, R. (2009). Choreography of recombination proteins during DNA damage response. *DNA Repair*, Vol. 8, pp. 1068-1076.

Makker, K., Agarwal, A. & Sharma, R. (2009). Oxidative stress & male infertility. *Indian Journal of Medical Research*, Vol. 129, pp. 357-367.

Malik, S-B., Pightling, A.W., Stefaniak, L.M., Schurko, A.M. & Logsdon, J.M. Jr. (2008). An expanded inventory of conserved meiotic genes provides evidence for sex in *Trichomonas vaginalis*. *PloS ONE*, Vol. 3, No. 8: e2879. doi:10.1371/journal.pone.0002879

Malone, R.E. & Esposito, R.E. (1980). The RAD52 gene is required for homothallic interconversion of mating types and spontaneous mitotic recombination in yeast. *Proceedings of the National Academy of Sciences USA*, Vol. 77, No. 1, pp. 503-507.

Mao, Z., Bozzella, M., Seluanov, A. & Gorbunova, V. (2008). DNA repair by nonhomologous end joining and homologous recombination during cell cycle in human cells. *Cell Cycle*, Vol. 7, pp. 2902-2906.

Marchetti, F. & Wyrobek, A.J. (2008). DNA repair decline during mouse spermiogenesis results in the accumulation of heritable DNA damage. *DNA Repair*, Vol. 7, pp. 572-581.

Massie, H.R., Samis, H.V. & Baird, M.B. (1972) The kinetics of degradation of DNA and RNA by H_2O_2. *Biochimica et Biophysica Acta*, Vol. 272, pp. 539-548.

Maynard Smith, J. (1978). *The Evolution of Sex*, Cambridge University Press, London

McKim, K.S., Green-Marroquin, B.L., Sekelsky. J.J., Chin, G., Steinberg, C., Khodosh, R. & Hawley, R.S. (1998). Meiotic synapsis in the absence of recombination. *Science*, Vol. 279, pp. 876-878.

McKim, K.S. & Hayashi-Hagihara, A. (1998). *mei-W68* in *Drosophila melanogaster* encodes a SPO11 homolog: evidence that the mechanism for initiating meiotic conversations is conserved. *Genes & Development*, Vol. 12, No. 18, pp. 2932-2942.

McMahill, M.S., Sham, C.W. and Bishop, D.K. (2007). Synthesis-dependent strand annealing in meiosis. *PloS Biology*, Vol. 5, No. 11: e299. doi:10.1371/journal.pbio.0050299

Mechali, M. (2010). Eukaryotic DNA replication origins: many choices for appropriate answers. *Nature Reviews: Molecular Cell Biology*, Vol. 11, pp. 728-738.

Mehrotra, S. & McKim, K.S. (2006). Temporal analysis of meiotic DNA double-strand break formation and repair in Drosophila females. *PloS Genetics*, Vol. 2, No.11: e200.doi:10.1371/journal.pgen.0020200

Michod, R.E., Wojciechowski, M.F. & Hoelzer, M.A. (1988). DNA repair and the evolution of transformation in the bacterium *Bacillus subtilis*. *Genetics*, Vol. 118, pp. 31-39.

Michod, R.E. & Wojciechowski, M.F. (1994). DNA repair and evolution of transformation IV. DNA damage increases transformation. *Journal of Evolutionary Biology*, Vol. 7, pp. 147-175.

Michod, R.E. (1995). *Eros and Evolution: A Natural Philosophy of Sex*. Addison-Wesley, Reading, MA.

Michod, R.E., Bernstein, H. & Nedelcu, A.M. (2008). Adaptive value of sex in microbial pathogens. *Infection, Genetics and Evolution*, Vol. 8, pp. 267-285.

Moens, P.B., Kolas, N.K., Tarsounas, M., Marcon, E., Cohen, P.E. & Spyropoulos, B. (2002). The time course and chromosomal localization of recombination-related proteins at meiosis in the mouse are compatible with models that can resolve the early DNA-DNA interactions without reciprocal recombination. *Journal of Cell Science*, Vol. 115, pp.1611-1622.

Mortensen, U.H., Lisby, M. & Rothstein, R. (2009). Rad52. *Current Biology*, Vol. 19, No.16, pp. R676-677.

Myers, S., Spencer, C.C.A., Auton, A., Bottolo, L., Freeman, C., Donnelly, P. & McVean, G. (2006). The distribution and causes of meiotic recombination in the human geneome. *Biochemical Society Transactions*, Vol. 34, No. 4, pp. 526-530.

Nedelcu, A.M. & Michod, R.E. (2003). Sex as a response to oxidative stress: The effect of antioxidants on sexual induction in a facultatively sexual lineage. *Proceedings of the Royal Society B (suppl.)*, Vol. 270, pp. S136-S139.

Nedelcu, A.M., Marcu, O. & Michod, R.E. (2004). Sex as a response to oxidative stress: a two-fold increase in cellular reactive oxygen species activates sex genes. *Proceedings of the Royal Society B*, Vol. 271, 1591-1596.

Otto, S.P. (2003). The advantages of segregation and the evolutionof sex. Genetics, Vol. 164, pp. 1099-1118.

Otto, S.P. & Gerstein, A.C. (2006). Why have sex? The population genetics of sex and recombination. *Biochemical Society Transactions*, Vol. 34, pp. 519-522.

Pauklin, S., Burkert, J.S., Martin, J., Osman, F., Weller, S., Boulton, S.J., Whitby, M.C. & Petersen-Mahrt, S.K. (2009). Alternative induction of meiotic recombination from single-base lesions of DNA deaminases. *Genetics*, Vol. 182, pp. 41-54.

Peters, A.D. & Otto, S.P. (2003). Liberating genetic variance through sex. *BioEssays*, 25: 533-537.

Pittman, D.L., Cobb, J., Schimenti, K.J., Wilson, L.A., Cooper, D.M., Brignull, E., Handel, M.A. & Schimenti, J.C. (1998). Meiotic prophase arrest with failure of chromosome synapsis in mice deficient for *Dmc1*, a germline-specific RecA homolog. *Molecular Cell*, Vol.1, pp. 697-705.

Prado, F., Cortes-Ledesma, F., Huertas, P. & Aguilera, A. (2003). Mitotic recombination in *Saccharomyces cerevisiae. Current Genetics*, Vol. 42, pp. 185-198.

Prudhommeau, C. & Proust, J. (1973). UV irradiation of polar cells of *Drosophila melanogaster* embryos. V. A study of the meiotic recombination in females with chromosomes of different structure. *Mutation Research*, Vol. 23, pp. 63-66.

Ramesh, M.A., Malik, S.B. & Logsdon, J.M. Jr. (2005). A phylogenomic inventory of meiotic genes: Evidence for sex in *Giardia intestinalis. Science*, Vol. 319, pp. 185-191.

Rasmuson, A. (1984). Effects of DNA-repair-deficient mutants on somatic and germ line mutagenesis in the UZ system of *Drosophila melanogaster. Mutation Research*, Vol. 141, pp. 29-33.

Reeves, R.J. & Jackson, R.M. (1974). Stimulation of sexual reproduction in *Phytophthora* by damage. *Journal of General Microbiology*, Vol. 84, pp. 303-310.

Resnick, M.A. & Martin, P. (1976). The repair of double-strand breaks in the nuclear DNA of *Saccharomyces cerevisiae* and its genetic control. *Molecular and General Genetics*, Vol. 143, pp. 119-129.

Roze, D. & Michod, R.E. (2010). Deleterious mutations and selection for sex in finite diploid populations. *Genetics*, Vol. 184, pp. 1095-1112.

Ruderfer, D.M., Pratt, S.C., Seidel, H.S. & Kruglyak, L. (2006). Population genomic analysis of outcrossing and recombination in yeast. *Nature Genetics*, Vol. 38, pp.1077-1081.

Sager, R. & Granick, S. (1954). Nutritional control of sexuality in *Chlamydomonas reinhardi*. *Journal of General Physiology*, Vol. 37, pp. 729-742.

Sakakura, Y., Soyano, K., Noakes, D.L.G. & Hagiwara, A. (2006). Gonadal morphology in the self-fertilizing mangrove killifish, *Kryptolebias marmoratus*. *Ichthyological Research*, Vol. 53, pp. 427-430.

San Filippo, J., Sung, P., & Klein H. (2008). Mechanism of eukaryotic homologous recombination. *Annual Review of Biochemistry*, Vol. 77: pp. 229-257.

Sauvageau, S., Stasiak, A.Z., Banville, I, Ploquin, M., Stasiak, A. & Masso, J.Y. (2005). Fission yeast RAD51 and Dmc1, two efficient DNA recombinases forming helical nucleoprotein filaments. *Molecular and Cellular Biology*, Vol. 25, pp. 4377-4387.

Schewe, M.J., Suzuki, D.T. & Erasmus, U. (1971). The genetic effects of mitomycin C in *Drosophila melanogaster*. II. Induced meiotic recombination. *Mutation Research*, Vol. 12, pp. 269-279.

Sehorn, M.G., Sigurdsson, S., Bussen, W., Unger, V.M. & Sung P. (2004). Human meiotic recombinase Dmc1 promotes ATP-dependent homologous DNA strand exchange. *Nature*, Vol. 429, 433-437.

Shields, W.M. (1982). *Philopatry, Inbreeding, and the Evolution of Sex*, State University of NY Press, Albany, NY.

Shinohara, A., Ogawa, H., Matsuda, Y., Ushio,N., Ikeo, K. & Ogawa, T. (1993). Cloning of human, mouse and fission yeast recombination genes homologous to RAD51 and RecA. *Nature Genetics*, Vol. 4, pp.239-243.

Slupphaug, G., Kavli, B. & Krokan, H.E. (2003). The interacting pathways for prevention and repair of oxidative DNA damage. *Mutation Research*, Vol. 531, pp. 231-251.

Staeva-Vieira, E., Yoo, S. & Lehmann, R. (2003). An essential role of DmRad51/SpnA in DNA repair and meiotic check point control. *The EMBO Journal*, Vol. 22, No. 21, pp. 5863-5874.

Steinboeck, F., Hubmann, M., Bogusch, A., Dorninger, P., Lengheimer, T. & Heidenreich, E. (2010). The relevance of oxidative stress and cytotoxic DNA lesions for spontaneous mutagenesis in non-replicating yeast cells. *Mutation Research*, Vol. 688, pp. 47-52.

Story, R.M., Bishop, D.K., Kleckner, N. & Steitz, T.A. (1993). Structural relationship of bacterial RecA proteins to recombination proteins from bacteriophage T4 and yeast. *Science*, Vol. 259, pps. 1892-1896.

Sung, P. (1994). Catalysis of ATP-dependent homologous DNA pairing and strand exchange by yeast RAD51 protein. *Science*, Vol. 265, pp.1241-1243.

Suzuki, D.T. & Parry, D.M. (1964). Crossing over near the centromere of chromosome 3 in *Drosophila melanogaster* females. *Genetics*, Vol. 50, pp.1427-1432.

Szostak, J.W., Orr-Weaver, T.L., Rothstein, R.J. & Stahl, F.W. (1983). The double-strand-break repair model for recombination. *Cell*, Vol. 33, pp. 25-35.

Takanami, T., Mori, A., Takahashi, H. & Higashitani, A. (2000). Hyper-resistance of meiotic cells to radiation due to a strong expression of a single recA-like gene in *Caenorhabditis elegans*. *Nucleic Acids Research*, Vol. 28, No. 21, pp. 4232-4236.

Takanami, T., Zhang, Y., Aoki, H., Abe, T., Yoshida, S., Takahashi, H., Horiuchi, S. & Higashitani, A. (2003). Efficient repair of DNA damage induced by heavy ion particles in meiotic prophase I nuclei of *Caenorhabditis elegans*. *Journal of Radiation Research*, Vol. 44, pp. 271-276.

Terasawa, M., Shinohara, A., Hotta, Y., Ogawa, H. & Ogawa, T. (1995). Localization of RecA-like recombination protein on chromosomes of the lily at various meiotic stages. *Genes and Development*, Vol. 9, pp. 925-934.

Thorne, L.W. & Byers, B. (1993). Stage-specific effects of X-irradiation on yeast meiosis. *Genetics*, Vol. 134, pp. 29-42.

Tichy, E.D., Pillai, R., Deng, L., Liang, L., Tischfield, J., Schwemberger, S.J., Babcock, G.F. & Stambrook, P.J. (2010). Mouse embryonic stem cells, but not somatic cells, predominantly use homologous recombination to repair double-strand breaks. *Stem Cells and Development*, Vol. 19, No. 11, pp. 1699-1711.

Vilenchik, M.M. & Knudson, A.G. (2003). Endogenous DNA double-stand breaks: Production, fidelity of repair and induction of cancer. *Proceedings of the National Academy of Sciences USA*, Vol. 100, pp. 12871-12876.

Virgin, J.B., Bailey, J.P., Hasteh, F., Neville, J., Cole, A. & Tromp, G. (2001). Crossing over is rarely associated with mitotic intragenic recombination in *Schizosaccharomyces pombe*. *Genetics*, Vol. 157, pp. 63-77.

Wei, K., Kucherlapati, R. & Edelmann, W. (2002). Mouse models for human DNA mismatch-repair gene defects. *TRENDS in Molecular Medicine*, Vol. 8, No. 7, pp. 346-353.

Welch, D.B.M., Welch, J.L.M. & Meselson, M. (2008). Evidence for degenerate tetraploidy in bdelloid rotifers. *Proceedings of the National Academy of Sciences USA* , Vol. 105, pp. 5145-5149.

Whitehouse, H.L.K. (1982). *Genetic Recombination*, Wiley, New York.

Williams, G.C. (1975). *Sex and Evolution*, Princeton University Press, Princeton, NJ

Yoshida, K., Kondoh, G., Matsuda, Y., Habu, T., Nishimune, Y. & Morita, T. (1998). The mouse RecA-like gene Dmc1 is required for homologous chromosome synapsis during meiosis. *Molecular Cell*, Vol. 1, pp. 707-718.

Youds, J.L., Mets, D.G., McIlwraith, M.J., Martin, J.S., Ward, J.D., Oneil, N.J., Rose, A.M., West, S.C., Meyer, B.J. & Boulton, S.J. (2010). RTEL-1 enforces meiotic crossover interference and homeostasis. *Science*, Vol. 327, pp.1254-1258.

Zeyl, C.W. & Otto, S.P. (2007). A short history of recombination in yeast. *Trends in Ecology and Evolution*, Vol. 22, No. 5, pp. 223-225.

20

From Seed to Tree: The Functioning and Evolution of DNA Repair in Plants

Jaana Vuosku[1,2], Marko Suokas[1], Johanna Kestilä[1],
Tytti Sarjala[2] and Hely Häggman[1]
[1]Department of Biology, University of Oulu, Oulu
[2]Finnish Forest Research Institute, Parkano Research Unit, Parkano
Finland

1. Introduction

In order to alleviate harmful effects of DNA damage and maintain genome integrity, all living organisms have developed a complex network of DNA repair mechanisms. However, the biochemical and genetic studies of DNA repair pathways have hitherto focused mostly on bacterial, yeast and mammalian systems (Sancar et al., 2004; Pan et al., 2006; Goosen & Moolenaar, 2008; Jackson & Bartek, 2009), whereas plants have been somewhat neglected in this respect. In plant cells, DNA damages can be generated "spontaneously" by reactive metabolites and by mistakes that occur during DNA replication and recombination processes or they can arise from exposure to environmental DNA damaging agents (Tuteja et al., 2001 & 2009). Plants are sessile organisms, which are continuously exposed to a wide variety of biotic and abiotic stresses, which can cause DNA damages directly or indirectly via the generation of reactive oxygen species (ROS) (Roldán-Arjona & Ariza, 2009). In plants, mutations, which initially arise in somatic cells, may also be present in gametes because plants lack a reserved germline and produce meiotic cells late in development (Walbot and Evans, 2003). However, the mutation rate in long-lived coniferous forest trees, such as pines, is not unexpectedly high, which indicates that the activities responsible for maintaining genome integrity must be efficient in somatic cells (Willyard et al., 2007).

This chapter gives an overview of the special requirement of DNA repair in plants particularly from the point of view of longevity and the lifestyle of plants. We introduce the sequences of the Scots pine (*Pinus sylvestris* L.) putative *RAD51* and *KU80* genes which are involved in the repair of double-strand breaks (DSBs) by homologous recombination (HR) and non-homologous end-joining (NHEJ), respectively. The novel sequence data is used in the reconstruction of the evolutionary history of the *RAD51* and *KU80* genes in eukaryotes. In addition, the use of the HR and NHEJ pathways is demonstrated during the Scots pine seed development. From its early stages of development in the mother plant onwards, a pine seed is exposed to developmentally programmed as well as environmental stresses which are potentially damaging to the genome. Furthermore, the pine seed represents an interesting inheritance of seed tissues as well as anatomically well-described sequences of embryogenesis. Thus, we consider the pine seed to be a model system for studying the DNA repairing mechanisms, yet not solely within plants, but in wider use – for eukaryotes in general.

2. Searching for a fountain of youth in pines

Organismal ageing is generally connected to deterioration. With the passage of time, organisms accumulate stochastic damage to DNA, proteins and other macromolecules (Rattan, 2008). If damages are left unrepaired, they impair important biological functions and, furthermore, result in age-related physiological changes, an increased susceptibility to diseases and environmental stress, reduced fertility, and finally, to increased mortality (Watson & Riha, 2011). The rate of damage accumulation should be approximately equal in all organisms. However, both the rate of senescence and the length of lifespan vary largely among organisms, which suggests that they are genetically determined (Finch, 2001).

Plants have adopted many survival strategies that are totally different from those of animals, and in relation to plants, even the terms individual, aging and lifespan may sometimes be difficult to define (Thomas, 2002; Munné-Bosch, 2007). Furthermore, vegetative propagation is common in plants, and even entire forests can consist of one tree clone. In quaking aspen (*Populus tremuloides* Michx.), clones are formed by sprouting of stems from the root systems of aspens that originally are derived from a seed (Lanner, 2002). The development of plants differs completely from the development of animals, which must be taken into account in inquiries into age-related changes in plants. In plants, only a fundamental body plan is established during embryogenesis, and practically all structures and organs are formed by the proliferation of meristematic cells throughout adult life (Watson & Riha, 2011). In plants, new organs develop asynchronously during a plant's life and these have shorter lifespans than the plant as a whole (Aphalo, 2010). Concerning plant ageing, it is essential to underline that senescence can also be a highly regulated physiological process, such as a development-related physiological cell death, which is significant when compared to the death of the whole organism. In annual plants, leaf senescence is connected with the death of the whole plant, whereas in perennials, leaf senescence is a regulated physiological process that contributes to nutrient recycling and allows the rest of the plant to benefit from the nutrients which have accumulated in leafs (Lim, 2007). In trees, the biomass may mostly consist of dead cells that form a supporting structure for a thin layer of newly emerged organs (Watson & Riha, 2011).

A walk through a park is enough to show that plants age as well and that the rate of senescence and the length of lifespan are species-specific. Plants can live from a few weeks to as long as millennia (Thomas, 2002; Lanner 2002). Monocarpic plants flower, set seed and die. The monocarpic habit is well exemplified by the model plant *Arabidopsis thaliana* (L.), which may go through its entire life cycle in 8 to 10 weeks, but may nevertheless produce thousands of offspring during that time (Hensel et al., 1993). In association with massive reproductive effort, the leaves, stems and fruits of the adult Arabidopsis plant undergo progressive senescence that ultimately results in the death of the plant (Hensel et al., 1993). Despite the fact that Arabidopsis is considered to be a mere weed, due to its small size, small genome, quick generation time, ease of genetic transformation, and the availability of mutant plants, it has been found to be useful both as a model for plants in general and for the study of a variety of fundamental biological processes (Meyerowitz, 1989; Swarbreck et al., 2008). In contrast to Arabidopsis, trees are examples of long-living organisms. Trees usually remain reproductive into great old age, and hence, the characteristics that prolong life are thought to be naturally selected because they increase fitness by multiplying reproductive opportunities (Lanner, 2002). In fact, the oldest living individual organism known on earth is a tree – a Great Basin bristlecone pine (*Pinus longaeva*), which has attained at least 4862 years (Lanner, 2002). While several Great Basin bristlecone pines have exceeded 4000 years of age, they do not show evident signs of senescence (Lanner & Connor, 2001). The grafting experiments with Scots pine indicated that age-related regulation in the growth is mainly caused by physical factors and not by the age itself (Vanderklein et al., 2007).Thus, the lifetime of trees seems to be mostly limited by external factors such as the activities of pests, the frequency and intensity of fires, and ultimately, by how long it takes for the soil to erode away from their roots (Lanner, 2002).

The two major groups of seed plants, angiosperms and gymnosperms, shared a common ancestor approximately 285 million years ago (Bowe et al. 2000). For several decades, Arabidopsis has provided the leading model for angiosperms (Meyerowitz, 1989), whereas pines, *Pinus* species, have been suggested as a model for gymnosperms and woody plants (Lev-Yadun & Sederoff 2000). The genus *Pinus* has a rich history of phylogenetic analysis, and the relationships between the approximately 120 extant species are well documented (Gernandt et al. 2005), as are the development, reproduction, ecology and genetics of many pine species (Lev-Yadun & Sederoff 2000). Although pines and other gymnosperms are generally considered to be difficult subjects for genetic studies e.g. due to their long generation times, large genome size and outbred mating system, they have one remarkable advantage: the haploid megagametophyte tissue represents a single meiotic product and makes the direct analysis of inheritance of genetic loci possible without the use of controlled crosses (Devey et al. 1995). Five pines were ranked to be the most interesting on the basis of their biological, geographical or economical importance. The economically dominant pines are loblolly pine (*P. taeda*), Monterey pine (*P. radiata*) and Scots pine (Lev-Yadun & Sederoff 2000). Scots pine is the most widely distributed Eurasian conifer and one of the keystone species in the Eurasian boreal forest zone, growing in a range of environments from Spain and Turkey to the subarctic forests of northern Scandinavia and Siberia (Mirov, 1967). Additionally, two bristlecone pines, *P. aristata* and *P. longaeva*, were selected to the top five due to their greatest longevity (Lev-Yadun & Sederoff 2000). Several reports have suggested that the activities responsible for the maintenance of genome integrity must be efficient in pines. Despite the long lifetime, the observed mutation rates in the somatic cells of pines were not unexpectedly high (Willyard et al., 2007). Furthermore, no age-dependent decline was detected in the telomeres of extremely long-lived bristlecone pines, although a positive correlation was found between telomere length and life expectancy in a study in which six tree species were compared (Flanary and Kletetschka, 2005). The results suggested that answers to many intriguing questions about the maintenance of genomic integrity during organismal ageing may be found in pine trees.

3. A future vision: From weed to seed

The seed represents the main vector of plant propagation and thus, in a plant's life, it is a critical stage with many special characteristics (Rajjou & Debeaujon, 2008). According to the practical instructions for plant seed storage (Bonner, 2008), plant seeds can be classified into five types: true orthodox, sub-orthodox, intermediates class between orthodox and recalcitrant (Ellis et al. 1990), temperate recalcitrant, and tropical-recalcitrant. The seeds of most tree species with high economic value (e.g. *Abies, Betula, Pinus, Picea*) at the Northern Temperate Zone as well as many tree species (e.g. *Cauarina, Eucalyptus, Tectona*) at tropics and subtropics are true orthodox. The water content of a seed is determined by seed composition and, in addition, it is in equilibrium with the prevailing relative humidity.

Orthodox seeds are able to withstand the reduction of moisture content to around 5% (Berjak & Pammenter, 2002) and they can be stored for long periods (10 to 50+ years) at subfreezing temperatures (Bonner, 2008). Embryo development, reserve accumulation and maturation / drying are the three typical stages of orthodox seed development, leading from a zygotic embryo to a mature, quiescent seed. The maturation drying causes severe stress, and a wide range of mechanisms such as protection, detoxification and repair are needed for the surviving of a seed during the dry state and to preserve the high germination ability (Buitink & Leprince, 2008; Rajjou & Debeaujon, 2008). The longevity of seeds during storage has a major ecological, agronomical as well as economical importance (Rajjou & Debeaujon, 2008), and seed conservation is one of the useful strategies to conserve plant genetic diversity (Cochrane, 2007). Furthermore, the seeds of particular plant species such as canna (*Canna compacta*), sacred lotus (*Nelumbo nucifera* Gaertn.) and date palm (*Phoenix dactylifera* L.) represent the most impressive examples of organismal longevity (Lerman & Cigliano, 1971; Shen-Miller, 2002; Sallon et al., 2008).

Seeds are subjected to DNA damage during maturation drying, but also during seed storage. Due to the fairly easy detection of chromosome breakage or translocations, DNA lesions during seed ageing has been demonstrated for a long time. As early as in 1969, it was shown that, in the seeds of crop species such as barley (*Hordeum vulgare* L.), broad bean (*Vicia faba* L.) and pea (*Pisum sativum* L.), chromosomal damages appeared as a result of the cumulative effects of temperature, moisture and oxygen during the ageing of seeds (Abdallah & Roberts, 1969). Later, the accumulation of chromosomal aberrations appeared to be a significant factor by its contribution to the loss of seed viability during storage (Cheah & Osborne, 1978). In maize (*Zea mays* L.) seed, the maturation drying / rehydration cycle creates thousands of single strand breaks (SSBs) in the genome of each cell (Dandoy et al., 1987). During germination, a seed recovers physically from maturation drying, resumes a sustained intensity of metabolism, completes essential cellular events to allow the embryo to emerge, and induces subsequent seedling growth (Nonogaki et al., 2010). Quantitative trait loci (QTL) mapping in Arabidopsis (Clerkx et al., 2004) and rice (*Oryza sativa* L.) (Miura et al., 2002) revealed that seed longevity during storage and germination is controlled by several genetic factors. In particular, the maintenance of genetic information during the seed dehydration and rehydration cycle has been found to be essential for plant survival (Osborne et al., 2002). It has been suggested that the capability to restore genetic integrity during rehydration in an embryo whose DNA is damaged is a major factor in the determining of the seed desiccation tolerance (Boubriak et al., 1997).

In seeds, DNA repair mechanisms improve emergence and germination, particularly under stress conditions. Artificially, DNA repair can be facilitated by seed priming, that is, by controlled hydration of seeds (Rajjou & Debeaujon, 2008). Due to incomplete hydration, seeds remain desiccation-tolerant and can be re-dried after treatment (Heydecker et al., 1973). For example, in *Artemisia sphaerocephala* and *Artemisia ordosia*, DNA repair during seed priming improves seed viability under harsh desert conditions (Huang et al., 2008). Although DNA repair has been demonstrated to occur during seed priming, the molecular mechanisms involved in DNA repair in seeds are still poorly known. In Arabidopsis seed, the activities of poly (ADP-ribose) polymerases (PARP enzymes) that are implicated in DNA base-excision repair are important for germination (Hunt et al., 2007). Also, DNA ligase VI (Waterworth et al., 2010) and one of the three *RAD21* gene homologues, *AtRAD21.1* (da Costa-Nunes et al., 2006), play critical roles in the recovery from DNA damage during Arabidopsis seed imbibition, prior to germination.

4. The lifestyle of plants - living hard, repairing smart

Although ageing may involve damage to various cellular constituents, the imperfect maintenance of genetic information has been suggested to be a critical contributor to ageing (Lombard et al., 2005). Thus, the necessity of appropriate and effective responses to potential mutagenic events is emphasized by several features in the plant's lifestyle which expose them to both external and internal sources of DNA damage. As sessile organisms, plants are continuously exposed to a wide variety of abiotic stresses such as infection by various pathogens, the ultraviolet (UV) component of sunlight, ozone, dehydration and wounding which may cause DNA damages directly or indirectly via the generation of reactive oxygen species (ROS) (Roldán-Arjona & Ariza, 2009). Plants and algae are the only photosynthetic eukaryotes able to capture energy from sun light. Thus, ROS are continuously produced within plant cells also as a result of normal oxidative cellular processes such as photosynthesis and mitochondrial respiration, and they may treat the integrity and viability of cells if they are not removed (Mittler et al., 2004). Oxidative stress, a situation in which ROS exceed cellular antioxidant defenses, can cause lipid peroxidation, protein damage as well as several types of DNA lesions (Lombard et al., 2005). Although ROS are toxic molecules, they also control many different processes in plants. Therefore, the level of ROS in plant cells is tightly regulated, and the intensity, duration and localization of different ROS signals are determined by interplay between the ROS production and ROS scavenging pathways (Mittler et al., 2004). Plant cells respond to persistent DNA stress by losing their competence to divide, which may lead to meristem arrest, but normally, meristems proliferate for the entire plant's lifetime which can be even millennia in some long-lived trees. That is, meristematic cells may divide thousands of times, which inevitably results in a replication-dependent loss of telomeres if their maintance is impaired (Watson & Riha, 2011).

Exogenous and endogenous genotoxic agents may produce various kinds of DNA lesions such as altered base, missing base, mismatch base, deletion, insertion, linked pyrimidines, single (SSB) and double strand breaks (DSB) as well as intra- and inter-strands cross-links (Tuteja et al., 2001). Therefore, organisms have developed a complex network of DNA repair mechanisms both to alleviate harmful effects of DNA damage and to maintain genome integrity (Hakem, 2008). In many cases, the same type of DNA lesions can be processed by several repairing mechanisms (Boyko et al., 2006). Depending on the severity and type of the DNA damage, cellular response can either be the activation of DNA repair pathways, but also a cell cycle arrest or a programmed cell death (PCD) (Barzilai et al., 2004), which indicates that DNA repair systems are tightly connected with other fundamental cellular processes. Particularly, DSBs can be extremely deleterious lesions. Even a single unprocessed DSB can cause a cell death (Rich et al., 2000) by inactivating key genes or by leading serious chromosomal aberrations (van Gent et al., 2001). On the other hand, cellular processes such as DNA replication and the repair of other kinds of DNA lesions give rise to DSBs, and thus, the consequences of DSBs are not always solely harmful to the cell (Bleuyard et al., 2006). Diploid cells can use homology-directed repair (HDR) in DSB repair. The most common form of HDR is homologous recombination (HR), which involves extensive sequence homology between the interacting DNA molecules (Lieber, 2010). In non-dividing haploid cells or in diploid cells that are not in S-phase, a homology donor is not nearby, but they can get over DSBs by non-homologous recombination (NHEJ), which acts independently of significant homology and simply rejoins the two ends of the break

(Bleuyard et al., 2006, Lieber, 2010). These two pathways have different repair fidelity: HR has been considered to be a more accurate pathway that ensures the repair of DSB without any loss of genetic information (Bleuyard et al., 2006), whereas NHEJ results in various mutations varying from single nucleotide substitutions to deletions or insertions of several nucleotides (Pelczar et al., 2003, Kovalchuk et al., 2004). However, HR has frequently found to lead to large segmental duplication, gene duplication, gene loss, or gene inactivation (Boyko et al., 2006). Thus both HR and NHEJ may have roles in genome evolution due to genome rearrangements. Especially in plants, genetic change in somatic cells is relevant for evolutionary considerations because mutations in meristematic cells can be transferred to the offspring (Walbot, 1996). Kirik et al. (2000) analyzed the formation of deletions during DSB repair in two dicotyledonous plant species, Arabidopsis and tobacco (*Nicotiana tabacum* L.), which differ over 20-fold in genome size. They found a putative inverse correlation between genome size and the average length of deletions, which suggested that species-differences in DSB repair may influence genome evolution in plants (Kirik et al., 2000). Pelczar et al. (2003) studied genome maintenance strategies of organisms belonging to different kingdoms (animals versus plants) but of similar genome size. They found that in human HeLa cells, 50–55% DSBs were repaired precisely – a high percentage when compared to as little as 15–30% in tobacco cells – and, moreover, the DSB repair in plants resulted in 30–40% longer deletions and significantly shorter insertions. The findings suggested that the strategies for DSB repair and genome maintenance may be different in plants and animals (Pelczar et al., 2003).

The molecular components of HR and NHEJ pathways are highly conserved amongst eukaryotes and both of the pathways are required for the repairing of DSB also in plants (Bray and West, 2005; Bleuyard et al., 2006). One of the central proteins in HR is RAD51, which ensures high fidelity DNA repair by facilitating strand exchange between damaged and undamaged homologous DNA segments (Baumann & West, 1998). In addition, several RAD51-like proteins such as XRCC2 appear to help with this process (Tambini et al., 2010). In the mediation of NHEJ, a DNA dependent protein kinase (DNA-PK) complex which comprises a KU70-KU80 heterodimer and a catalytic subunit (PKcs) plays a central role (Tamura et al., 2002). The key regulatory mechanisms that direct which pathway is used for DSB repair are still poorly known if they exist at all (Boyko et al., 2006). The suggestion that HR and NHEJ compete for available DNA ends at break sites is based at the molecular level on the equilibrium between RAD52 (HR) and KU70-KU80 dimer (NHEJ) in animals (Ray and Langer, 2002). However, Arabidopsis genome contains no *RAD52* homolog (Bleuyard et al., 2006), whereas *RAD51* homolog has been identified (Doutriaux et al., 1998). Thus, the availability of the key proteins, such as RAD51 and KU proteins, at the time of DSB repair may also be one of the regulatory mechanisms. In Arabidopsis, the rate of HR decreased with plant age, whereas the frequency of strand breaks and point mutations increased. These events were parallel by a decrease in the abundance of *RAD51* transcripts as well as increase in the abundance of *KU70* transcripts and KU70 protein (Boyko et al., 2006). These results of Boyko et al. (2006) suggest that the involvement of HR and NHEJ in DSB repair may be developmentally controlled in plants.

5. DNA fragmentation and repair during Scots pine seed development

As an orthodox seed, a developing pine seed goes through maturation drying during which metabolic activity is gradually reduced and the seed enters into a quiescent state. In addition to this, the development of a viable pine seed includes the strictly co-ordinated action of several cell death programs. A characteristic feature of the Scots pine seed development is the presence of more than one embryo in the developing seed (Fig. 1A). In the beginning of the seed development, the fertilization of many egg nuclei results in several embryos of the same ovule (Buchholz 1926). Later, polyzygotic embryos undergo cleavage polyembryony (Sarvas 1962). However, only the dominant embryo survives and completes its development (Fig. 1B), while subordinate embryos, as well as suspensor tissue, are deleted by programmed cell death (PCD) during the progress of seed development (Filonova et al. 2002). Megagametophyte cells in the embryo surrounding region (ESR) die through necrotic-like cell death (Vuosku et al., 2009), and in addition, the maternal cells of the nucellar layers face destruction during early embryogenesis (Hiratsuka et al., 2002; Vuosku et al., 2009).

In a gymnosperm seed, the megagametophyte tissue develops from a haploid megaspore before the actual fertilization of the eggs (Singh 1978). The megagametophyte houses the majority of the storage reserves of a seed (King & Gifford, 1997) and provides nutrition for the developing embryo during seed development as well as for the young seedling during early germination (Fig.1C). We have shown that, in Scots pine seed, the megagametophyte tissue stays alive from the early phases of embryo development until the imbibition phase of early germination of mature seed, except for the cells in the ESR (Vuosku et al., 2009). Positive signals in TUNEL (terminal deoxynucleotidyl transferase-mediated dUTP nick end labelling) assay indicate DNA fragmentation in the nuclei of the megagametophyte cells at the late embryogeny (Fig.1D). However, the megagametophyte cells do not show other morphological signs of cell death, but appear to be viable with the active gene expression. The decreasing expression of the PCD-related metacaspase (*MCA*) and Tat-D nuclease (*TAT-D*) genes during Scots pine seed development confirms that no large-scale PCD or nucleic acid fragmentation occur in the megagametophyte tissue. Instead, the DNA fragmentation may be a consequence of DNA strand breaks caused by maturation drying or by the DNA breaks with free 3'-OH ends that appear during DNA repair. During the seed development, the expression of *RAD51* gene decrease, whereas the expression of the *KU80* and DNA ligase (*LIG*) genes remain constant, which suggests that the proportion of mitotic cells decrease and the DNA breaks are mainly repaired by NHEJ pathway (Vuosku et al., 2009).

Nuclear DNA fragmentation is currently one of the most frequently used sign of PCD. However, in the Scots pine seed, the megagametophyte cells remain metabolically active until the imbibition phase of germination despite DNA fragmentation in the nuclei already during late seed development (Vuosku et al., 2009). In plants, both the tolerance of DNA fragmentation and effective DNA repair mechanisms may be adaptations to the special energy metabolism as well as to a sessile life style which exposes cells to various endogenous and exogenous stresses. Thus, in plants, DNA fragmentation can also be a temporary process and does not always proceed to cell death.

6. Evolution of DNA repair related *recA/RAD51* gene family and *KU80* gene in eukaryotes

Previously, the homologs of both *recA* and *RAD51* genes have been identified from several prokaryotes and eukaryotes (Eisen, 1995; Bishop et al., 1992; Shinohara et al., 1992). In Arabidopsis, nuclear genome codes four recA-like proteins, RECA1, RECA2, REC3 and DRT100 that have been located in mitochondria and chloroplasts (Cao et al., 1997; Pang et al., 1992; Shedge et al., 2007). In addition to RAD51, Arabidopsis genome encodes seven

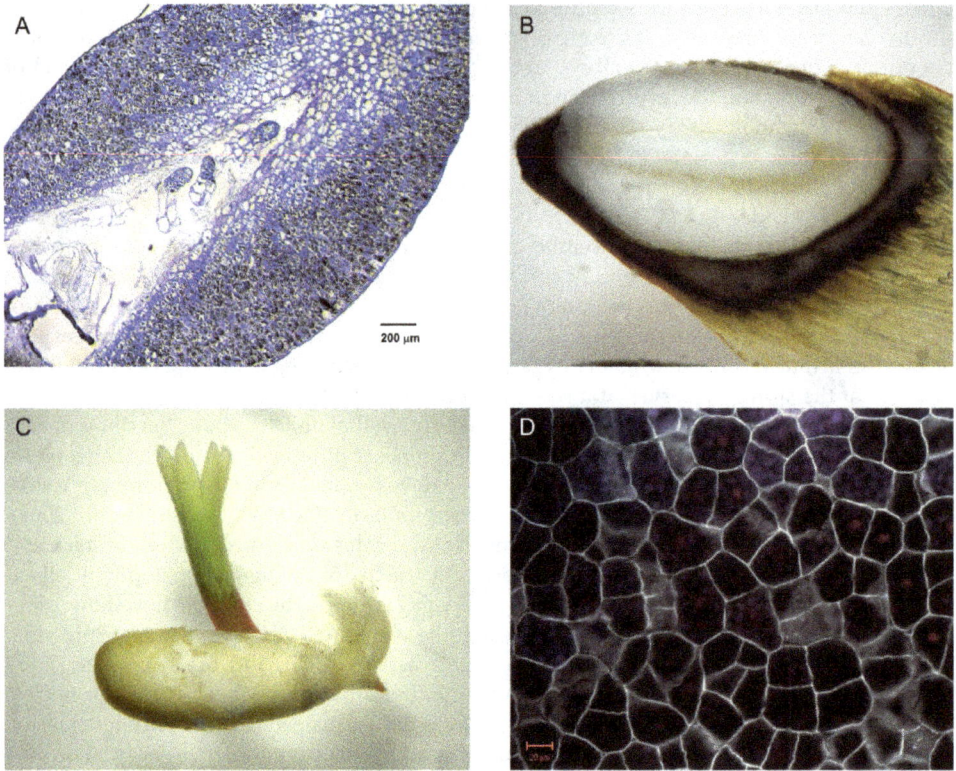

Fig. 1. Scots pine seed development. (A) The dominant embryo and subordinate embryos in the corrosion cavity surrounded by the megagametophyte. (B) A mature Scots pine seed. (C) A young Scots pine seedling. (D) TUNEL positive nuclei in the megagametophyte cells during seed development.

RAD51-like proteins, DMC1, RAD51B, RAD51C, RAD51D, DMC1, XRCC2 and XRCC3 which indicates that Arabidopsis contains the same family of RAD51-like proteins as vertebrates (Klimyuk & Jones, 1997; Doutriaux et al., 1998; Osakabe et al., 2002; Bleuyard et al., 2005). Also, the functions of RAD51 paralogs as well as the different requirements for the RAD51 paralogs in meiosis and DNA repair have been found to be conserved between plants and vertebrates (Bleuyard, et al., 2005). The presence of duplicated intron-free *RAD51* genes in the model moss *Physcomitrella patens* is unique among eukaryotes and may indicate the presence of unusual recombination apparatus in this organism (Markmann-Mulish, 2002). However, NHEJ, rather than HR, has been suggested to be the major pathway for repair DSBs in organisms with complex genomes, including vertebrates and plants (Gorbunova & Levy, 1999). The NHEJ pathway is mediated by KU70-KU80 heterodimer that shows evolutionary conserved functions (Critchlow & Jackson, 1998; Tamura et al., 2002). The KU70 and KU80 proteins of Arabidopsis share about 29% and 23% amino acid sequence identity with human KU70 and KU80 proteins, respectively (Tamura et al., 2002).

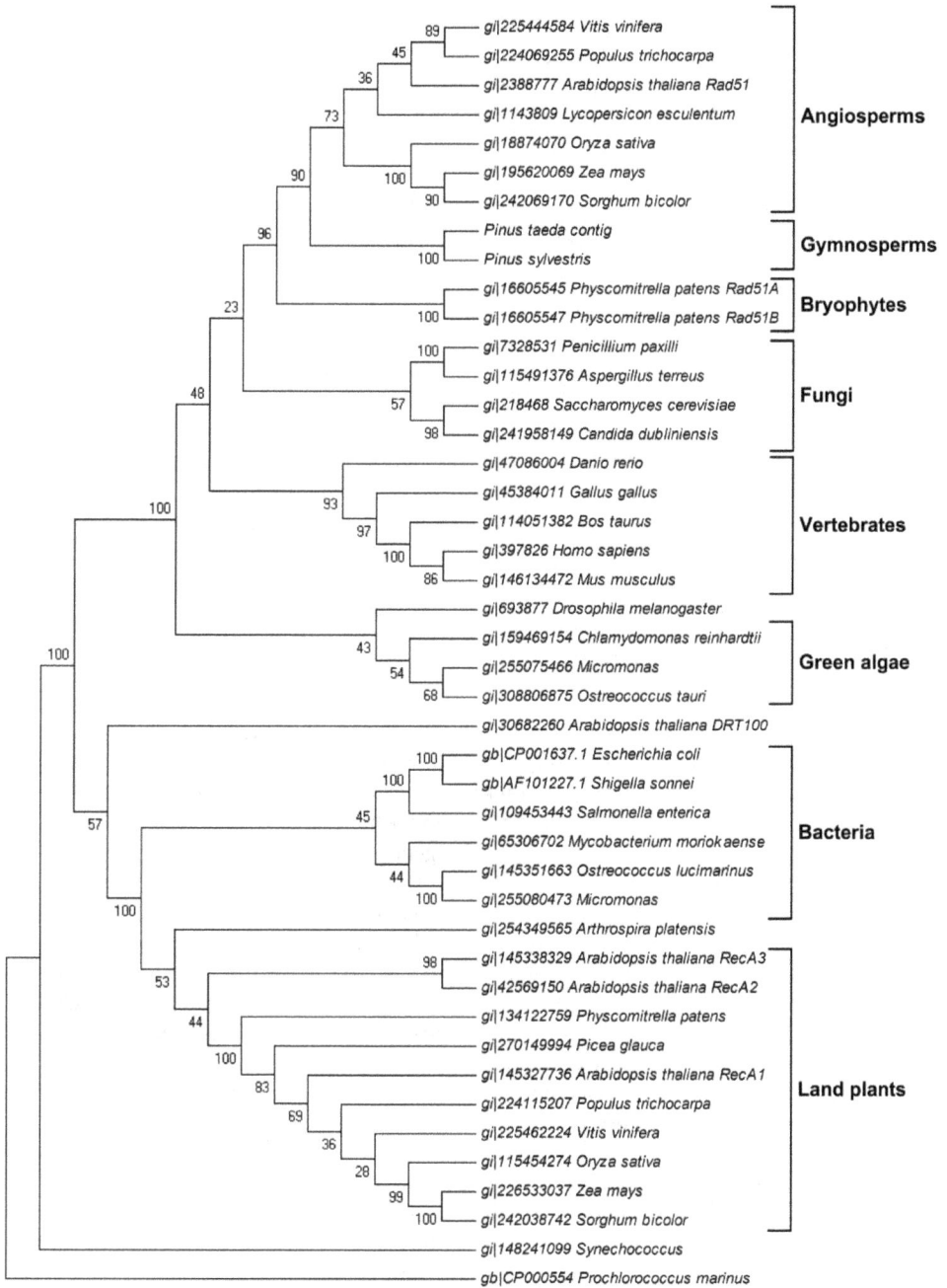

Fig. 2. Phylogenetic analysis of *recA* and *RAD51* sequences.

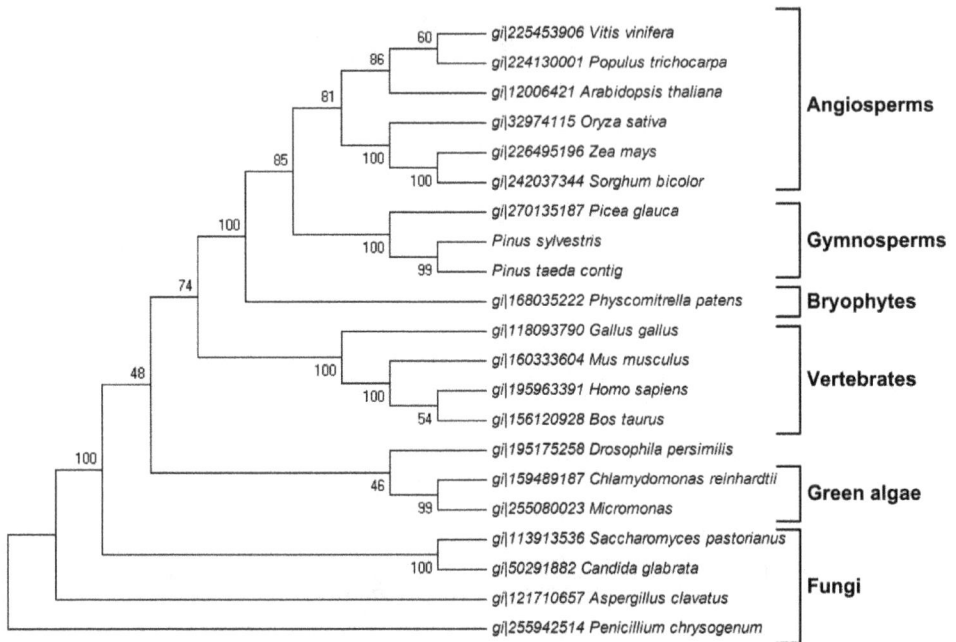

Fig. 3. Phylogenetic analysis of *KU80* sequences.

In the present study, we sequenced the coding regions of the Scots pine putative *RAD51* (GeneBank accession number: JN566226) and *KU80* (GeneBank accession number: JN566225) genes. The predicted amino acid sequences of the Scots pine RAD51 and KU80 proteins showed 77% and 41% identity to the Arabidopsis RAD51 and KU80 proteins, respectively. Blast searches in NCBI databases (http://www.ncbi.nlm.nih.gov) were performed for *recA/RAD51*-like genes as well as for KU80-like genes from various organisms, particularly from the species whose genomes have been completely sequenced. The nucleotide sequences were used for the reconstruction of the evolutionary history of the *recA/RAD51* gene family and *KU80* genes. In the case of other conifers, for which no unigene sequences were available, expressed sequence tag (EST) information was employed to reconstruct a contig containing the complete coding sequence. The nucleotide sequence alignments were performed with ClustalX (Thompson et al. 1997). Phylogenetic analyses were conducted in MEGA4 (Tamura et al. 2007) using the maximum parsimony (MP) method with close-neighbor-interchange algorithm (Nei and Kumar 2000). The bootstrap method (Felsenstein 1985) with 500 replicates was used to evaluate the confidence of the reconstructed trees.

In the phylogenetic tree, *recA* and *RAD51* sequences formed separate branches that were supported by 100% of the bootstraps (Fig. 2). Thus, the result supported the view that eukaryotic *recA* and *RAD51* genes have different evolutionary histories. The phylogenetic analysis suggested a common eukaryotic ancestor for *RAD51* genes, whereas eukaryotes seem to have acquired *recA* genes through horizontal gene transfer from bacteria. Endosymbiotic transfer of *recA* genes may have occurred from mitochondria and chloroplasts to nuclear genomes of ancestral eukaryotes (Lin et al., 2006). Both *RAD51* and *KU80* sequence-based phylogenies (Fig. 2 and 3) were in accordance with the current view of the evolution of green plants (Qiu and Palmer 1999). That is, morphologically simple plants such as *Physcomitrella* are followed by more complex flowering forms with highly developed breeding mechanisms at the top of the plant phylogeny tree. The novel gymnosperm sequences between bryophytes and angiosperms form the link that has been missing until now in the DNA repair genes based phylogenies.

7. Conclusions

Plants are sessile organisms, which are continuously exposed to a wide variety of biotic, abiotic or developmental stresses, which can cause DNA damages directly or indirectly via generation of ROS. In pines, the mechanisms maintaining genomic integrity must be efficient because the observed mutation rates in somatic cells are not high despite the long lifetime of the organisms. In pines, seed development includes developmentally programmed stresses as well as the strictly co-ordinated action of several cell death programs. Furthermore, pine seed represents an interesting inheritance of seed tissues and anatomically well-described sequences of embryogenesis. Thus, the pine seed provides a favorable model for the study of the effects of a variety of endogenous DNA damaging agents as well as developmentally regulated and environmental stresses on genome integrity. Due to the high evolutionary conservation of the DNA repair mechanisms, the pine seed, as a model system, may also shed light on the mechanisms that contribute to longevity and ageing in eukaryotes in general – things of great interest also with regard to the health of human beings.

8. Acknowledgements

We thank the personnel of the Finnish Forest Research Institute at the Punkaharju Research Unit and at the Parkano Research Unit for conducting the collections of the research material. The Research was funded by the Academy of Finland (Project 121994 to TS)

9. References

Abdalla, F.H. & Roberts H (1969) Effects of Temperature, Moisture, and Oxygen on the Induction of Chromosome Damage in Seeds of Barley, Broad Beans, and Peas during Storage. *Annals of Botany*, Vol.32, No.1, pp. 119-136

Aphalo, P.J. (2010) On How to Disentangle the Contribution of Different Organs And Processes to the Growth of Whole Plants. *Journal of Experimental Botany*, Vol.61, No.3, pp. 626-628

Barzilai, A. & Yamamoto, K.I. (2004) DNA Damage Responses to Oxidative Stress. *DNA repair*, Vol.3, No.8-9, pp. 1109-1115

Baumann, P. & West, S. (1998) Role of the Human RAD51 Protein in Homologous Recombination and Double-Stranded-Break Repair. *Trends in Biochemical Sciences*, Vol.23, No.7, pp. 247-251

Berjak, P. & Pammenter, N.M. (2002) Orthodox and Recalcitrant Seeds. In: *Tropical Tree Seed Manual*, J.A. Vozzo, (Ed.), 137-147, USDA Forest Service, Washington DC, USA

Bishop, D.K.; Park, D.; Xu L. & Kleckner, N. (1992) DMC1: A Meiosis-Specific Yeast Homolog of *E. coli recA* Required for Recombination, Synaptonemal Complex Formation, and Cell Cycle Progression. *Cell*, Vol.69, No.3, pp. 439-456

Bleuyard, J-Y.; Gallego, M.E., Savigny, F. & White, S.I (2005) Different Requirements for the Arabidopsis Rad51 Paralogs in Meiosis and DNA Repair. *The Plant Journal*, Vol.41, pp. 533-545

Bleuyard, J-Y.; Gallego, M.E. & White, C.I. (2006) Recent Advances in Understanding of the DNA Double-Strand Break Repair Machinery of Plants. *DNA repair*, Vol.5, pp. 1-12

Bonner, F.T. (2008) Storage of Seeds. In: *The Woody Plant Seed Manual*, F.T. Bonner, R.P. Karrfalt (Ed.), 85-95, USDA FS Agriculture Handbook 727

Bowe, L.M.; Coat, G. & dePamphilis, C.W. (2000) Phylogeny of Seed Plants based on All Three Genomic Compartments: Extant Gymnosperms Are Monophyletic and Gnetales' Closest Relatives Are Conifers. *Proceedings of the National Academy of Sciences of the United States of America*, Vol.97, No.8, pp 4092-4097

Bray, C.M. & West, C.E. (2005) DNA Repair Mechanisms in Plants: Crucial Sensors and Effectors for the Maintenance of Genome Integrity. *New Phytologist*, Vol.168, pp. 511-528

Boubriak, I.; Kargiolaki, H.; Lyne, L. & Osborne, D.J. (1997) The Requirement for DNA Repair in Desiccation Tolerance of Germinating Embryos. *Seed Science Research*, Vol.7, No.2, pp. 97-105

Boyko, A.; Zemp, F.; Filkowski, J. & Kovalchuk, I. (2006) Double-Strand Break Repair in Plants Is Developmentally Regulated. *Plant Physiology*, Vol.141, pp. 488-497

Buchholz, J.T. (1926) Origin of Cleavage Polyembryony in Conifers. *Botanical Gazette*, Vol.81, No.1, pp. 55-71

Buitink, J. & Leprince, O. (2008) Intracellular Glasses and Seed Survival in the Dry State. *Comptes Rendus Biologies*, Vol.331, No.10, pp. 788-795

Cao, J.; Combs, C. & Jagendorf, A.T. (1997) The Chloroplast-Located Homolog of Bacterial DNA Recombinase. *Plant and Cell Physiology*, Vol.38, No.12, pp.1319-1325

Cheah, K.S.E. & Osborne, D.J. (1978) DNA Lesions Occur With Loss of Viability in Embryos of Ageing Rye Seed. *Nature*, Vol.272, pp.593-599

Clerkx, E.J.; El-Lithy, M.E.; Vierling, E.; Ruys, G.J.; Blankestijn-De Vries, H.; Groot, S.P.C.; Vreugdenhil, D & Koornneef, M. (2004) Analysis of Natural Allelic Variation of Arabidopsis Seed Germination and Seed Longevity Traits Between the Accessions Landsberg *erecta* and Shakdara, Using a New Recombinant Inbred Line Population. *Plant Physiology*, Vol.135, pp. 432-443

Cochrane, J.A.; Crawford, A.D. & Monks, L.T. (2007) The Significance of Ex Situ Seed Conservation to Reintroduction of Threatened Plants. *Australian Journal of Botany*, Vol.55, No.3, pp. 356-361

Critchlow, S.E. & Jackson, S.P. (1998) DNA End-Joining: from Yeast to Man. *Trends in Biochemical Sciences.* Vol.23, No.10, pp. 394-398

da Costa-Nunes, J.A.; Bhatt, A.M.; O'Shea, S.; West, C.E.; Bray, C.M.; Grossniklaus, U & Dickinson, H.G. (2006) Characterization of the three *Arabidopsis thaliana* RAD21 cohesins reveals differential responses to ionizing radiation. *Journal of Experimental Botany*, Vol.57, No.4, pp. 971-983

Dandoy, E.; Schyns, R.; Deltour, L. & Verly, W.G. (1987) Appearance and Repair of Apurinic/Apyrimidinic Sites in DNA during Early Germination. *Mutation Research*, Vol.181, No.1, pp. 57-60

Devey, M.E.; Delfinomix, A.; Kinloch, B.B. & Neale, D.B. (1995) Random Amplified Polymorphic DNA Markers Tightly Linked to a Gene for Resistance to White-Pine Blister Rust in Sugar Pine. *The Proceeding of the National Academy of Sciences of the United States of America*, Vol.92, No.6, pp. 2066-2070

Doutriaux, M.; Couteau, F.; Bergounioux, C. & White, C. (1998) Isolation and Characterization of the Rad51 and DMC1 Homologs from *Arabidopsis Thaliana*. *Molecular and General Genetics*, Vol.257, No.3, pp. 283-291

Eisen, J.A. (1995) The RecA Protein as a Model Molecule for Molecular Systematic Studies of Bacteria: Comparison of Trees of RecA and 16S rRNAs from the Same Species. *Journal of Molecular Evolution*, Vol.41, No.6, pp. 1105-1123

Ellis, R.H.; Hong, T.D. & Roberts, E.H. (1990) An Intermediate Category of Seed Storage Behavior? I. Coffee. *Journal of Experimental Botany*, Vol.41, No.9, pp. 1167–1174

Felsenstein, J. (1985) Confidence Limits on Phylogenies: an Approach Using the Bootstrap. *Evolution*, Vol.39, No.4, pp. 783-791

Filonova, L.H; von Arnold, S.; Daniel, G & Bozhkov, P.V. (2002) Programmed Cell Death Eliminates All but One Embryo in a Polyembryonic Plant Seed. *Cell Death and Differentiation*, Vol.9, No.10, pp. 1057-1062

Finch, C.E. & Austad, S.N. (2001) History and Prospects: Symposium on Organisms with Slow Aging. *Experimental Gerontology*, Vol.36, No.4-6, pp. 593-597

Flanary, B.E. & Kletetschka, G. (2005) Analysis of Telomere Length and Telomerase Activity in Tree Species of Various Life-Spans, and with Age in the Bristlecone Pine *Pinus longaeva*. *Biogerontology*, Vol.6, No.2, pp.101-111

Gernandt, D.S.; Lopez, G.G.; Garcia, S.O. & Liston, A. (2005) Phylogeny and Classification of Pinus. *Taxon*, Vol.54, No.1, pp. 29-42

Goosen, N. & Moolenaar G.F. (2008) Repair of UV Damage in Bacteria. *DNA Repair*, Vol.7, pp. 353-379

Gorbunova, V. & Levy, A.A. (1999) How Plants Make Ends Meet: DNA Double Strand Break Repair. *Trends in Plants Sciences*, Vol.4, No.7, pp. 263-269

Hakem, R. (2008) DNA-Damage Repair; the Good, the Bad, and the Ugly. *The EMBO Journal*, Vol.27, pp. 589-605

Hensel, L.L.; Grbić, V.; Baumgarten, D.A. & Bleecker, A.B. (1993) Developmental and Age-Related Processes That Influence the Longevity and Senescence of Photosynthetic Tissues in *Arabidopsis*. *The Plant Cell*, Vol.5, pp.553-564

Heydecker, W.; Higgins, J. & Gulliver, R.L. (1973) Acclerated Germination by Osmotic Seed Treatment. *Nature*, Vol.246, No.2, pp.42-44

Hiratsuka, R.; Yamada, Y. & Terasaka, O. (2002) Programmed Cell Death of *Pinus* Nucellus in Response to Pollen Tube Penetration. *Journal of Plant Research*, Vol.115, No.2, pp. 141-8

Huang, Z.; Boubriak, I.; Osborne, D.J.; Dong, M. & Gutterman, Y. (2008) Possible Role of Pectin-Containing Mucilage and Dew in Repairing Embryo DNA of Seeds Adapted to Desert Conditions. *Annals of Botany*, Vol.101, pp. 277-283

Hunt, L.; Holdsworth, M.J. & Gray, J.E. (2007) Nicotinamidase Activity is Important for Germination. *The Plant Journal*, Vol.51, pp. 341-351

Jackson, S.P. & Bartek, J. (2009) The DNA-Damage Response in Human Biology and Disease. *Nature*, Vol.461, pp.1071-1078

King, J.E. & Gifford, D.J. (1997) Amino Acid Utilization in Seeds of Loblolly Pine during Germination and Early Seedling Growth (I. Arginine and Arginase Activity), *Plant Physiology*, Vol.113, No.4, pp. 1125-1135

Kirik, A.; Salomon S. & Puchta H. (2000) Species-Specific Double-strand Break Repair and Genome Evolution in Plants. *The EMBO Journal*, Vol.19, No.20, pp. 5562-5566

Klimyuk, V.I. & Jones, J.D. (1997) AtDMC1, the Arabidopsis Homologue of the Yeast DMC1 Gene: Characterization, Transposon-Induced Allelic Variation and Meiosis-Associated Expression. *Plant Journal*, Vol.11, No.1, pp. 1-14

Kovalchuk, I.; Pelczar, P. & Kovalchuk, O. (2004) High Frequency of Nucleotide Misincorporations upon the Processing of the Double-Strand Breaks. *DNA Repair*, Vol.3, No.3, pp. 217-223

Lanner, R.M. (2002) Why do Trees Live so Long? *Ageing Research Reviews*, Vol.1, No.4, pp. 653-671

Lanner, R.M. & Connor, K.F. (2001) Does Bristlecone Pine Senesce? *Experimental Gerontology*, Vol.36, No.4-6, pp. 675-685

Lerman, J.C. & Cigliano, E.M. (1971) New Carbon-14 Evidence for Six Hundred Years Old *Canna compacta* Seed, *Nature*, Vol.232, pp. 568-570

Lev-Yadun, S. & Sederoff, R. (2000) Pines as Model Gymnosperms to Study Evolution, Wood Formation, and Perennial Growth. *Journal of Plant Growth Regulation*, Vol.19, pp. 290-305

Lieber, M.R. (2010) The Mechanism of Double-Strand DNA Break Repair by the Nonhomologous DNA End Joining Pathway. *Annual Review of Biochemistry*, Vol.79, pp. 181-211

Lim, P.O.; Kim, H.J. & Nam, H.G. (2007) Leaf Senescence. *Annual Review of Plant Biology*, Vol.58, pp. 115-136

Lin, Z.; Kong, H.; Nei, M. & Ma, H. (2006) Origins and Evolution of the *recA/RAD51* Gene Family: Evidence for Ancient Gene Duplication and Endosymbiotic Gene Transfer. *The Proceeding of the National Academy of Sciences of the United States of America*, Vol.103, No.27, pp. 10328-10333

Lombard, D.B.; Chua, K.F.; Mostoslavsky, R.; Franco, S.; Gostissa, M. & Alt, F.W. (2005) DNA Repair, Genome Stability, and Aging, *Cell*, Vol.120, pp. 497-512

Markmann-Mulisch, U.; Hadi, M.Z.; Koepchen, K.; Alonso, J.C.; Russo, V.E.A.; Schell, J. & Reiss, B. (2002) The Organization of *Physcomitrella patens Rad51* Genes is Unique among Eukaryotic Organisms. *The Proceeding of the National Academy of Sciences of the United States of America*, Vol.99, No.5, pp. 2959-2964

Meyerowitz, E.M. (1989) Arabidopsis, a Useful Weed. *Cell*, Vol.56, pp. 263-269

Mittler, R.; Vanderauwera, S.; Gollery, M. & Van Breusegem F. (2004) Reactive Oxygen Gene Network of Plants. *Trends in Plant Sciences*, Vol.9, No.10, pp. 1360-1385

Mirov NT (1967) *The Genus Pinus*, The Ronald Press Company, New York

Miura, K.; Lin, Y.; Yano, M. & Nagamine, T. (2002) Mapping Quantitative Trait Loci Controlling Seed Longevity in Rice (*Oryza sativa* L.). *Theoretical and Applied Genetics*, Vol.104, No.6-7, pp.981-986

Munné-Bosch, S. (2007) Aging in Perennials. *Critical Review in Plant Sciences*, Vol.26, 123-138

Nei, M. & Kumar, S (2000) *Molecular Evolution and Phylogenetics*, Oxford University Press, New York

Nonogaki, H.; Bassel, G.W. & Bewley, J.D. (2010) Germination - Still a Mystery. *Plant Science*, Vol.179, No.6, pp. 574-581

Osakabe, K.; Yoshioka, T.; Ichikawa, H. & Toki, S. (2002) Molecular Cloning and Characterization of RAD51-like Genes from *Arabidopsis thaliana*. *Plant Molecular Biology*, Vol.50, No.1, pp. 71-81

Osborne, D.J.; Boubriak, I.; Leprince, O. (2002) Rehydration of Dried Systems: Membranes and the Nuclear Genome. In: *Desiccation and survival in plants: drying without dying*, M. Black & H.W. Pritchard, (Ed), 343-364, CABI Publishing. Wallingford, Oxon, UK

Pan, X.; Ye, P.; Yuan, D.S.; Wang, X.; Bader, J.S.; Boeke, J.D. (2006) A DNA Integrity Network in the Yeast *Saccharomyces cerevisiae*. *Cell*, Vol.124, No.5, pp.1069-1081

Pang, Q.; Hays, J.B. & Rajagopal, I. (1992) A Plant cDNA that Partially Complements *Escherichia coli recA* Mutations Predicts a Polypeptide not Strongly Homologous to RecA Proteins. *Proceeding of the National Academy of Sciences USA*, Vol.89, No.17, pp. 8073-8077

Pelczar, P.; Kalck, V. & Kovalchuk, I. (2003) Different Genome Maintenance Strategies in Human and Tobacco Cells. *Journal of Molecular Biology*, Vol.331, No.4, pp. 771-779

Qiu, Y-L & Palmer, J.D. (1999) Phylogeny of Early Land Plants: Insights from Genes and Genomes. *Trends in Plant Sciences*, Vol.4, No.1, pp. 26-29

Rajjou, L. & Debeaujon I (2008) Seed Longevity: Survival and Maintenance of High Germination Ability of Dry Seeds. *Comptes Rendus Biologies*, Vol.331, No.10, pp. 796-805

Rattan, S.I.S. (2008) Increased Molecular Damage and Heterogeneity as the Basis of Aging. *Biological Chemistry*, Vol.389, pp. 267-272

Ray, A. & Langer, M. (2002) Homologous Recombination: Ends as the Means. *Trends in Plant Science*, Vol.7, No.10, pp. 435-440

Rich, T; Allen, R.L. & Wyllie, A.H. (2000) Defying Death After DNA Damage. *Nature*, Vol.407, pp. 777-783

Roldán-Arjona, T. & Ariza, R.R. (2009) Repair and Tolerance of Oxidative DNA Damage in Plants. *Mutation Research*, Vol.681, No.2-3, pp. 169-179

Sancar, A; Lindsey-Boltz, L.A.; Ünsal-Kaçmaz, K. & Linn, S. (2004) Molecular Mechanisms of Mammalian DNA Repair and the DNA Damage Checkpoints. *Annual Review of Biochemistry*, Vol.73, pp.39-85

Sallon, S.; Solowey, E.; Cohen, Y.; Korchinsky, R.; Egli, M.; Woodhatch, I.; Simchoni, O. & Kislev, M (2008) Germination, Genetics, And Growth of an Ancient Date Seed. *Science*, Vol.320, No.5882, pp. 1464

Sarvas, R. (1962) Investigations on the Flowering and Seed Crop of *Pinus sylvestris*. *Communicationes Instituti Forestale Fennica*, Vol.53, pp. 1-198

Shedge, V.; Arrieta-Montiel, M., Christensen, A.C. & Mackenzie, S.A. (2007) Plant Mitochondrial Recombination Surveillance Requires Unusual *RecA* and *MutS* Homologs. *The Plant Cell*, Vol.19, No.4, pp. 1251-1264

Shen-Miller, J. (2002) Sacred Lotus, the Long-Living Fruits of China Antique. *Seed Science Research*, Vol.12, No.3, pp. 131-143

Shinohara, A.; Ogawa, H. & Ogawa, T. (1992) Rad51 Protein Involved in Repair and Recombination in *S. cerevisiae* is a RecA-like Protein. *Cell*, Vol.69, No.3, pp. 457-470

Singh H (1978) *Embryology of Gymnosperms*, Borntrager, Berlin

Swarbreck, D.; Wilks, C.; Lamesch, P.; Berardini, T.Z; Garcia-Hernandez, M.; Foerster, H.; Li, D.; Meyer, T.; Muller, R.; Ploetz, L.; Radenbaugh, A.; Singh, S.; Swing, V.; Tissier, C.; Zhang, P. & Huala, E. (2008) The Arabidopsis Information Resource (TAIR): Gene Structure and Function Annotation. *Nucleic Acids Research*, Vol.36, pp. D1009-D1014

Tambini, C.E.; Spink, K.G.; Ross, C.J.; Hill, M.A. & Thacker, J. (2010) The Importance of XRCC2 in RAD51-related DNA Damage Repair. *DNA Repair*, Vol.9, No.5, pp. 517-525

Tamura, K.; Adachi, Y.; Chiba, K.; Oguchi, K. & Takahashi, H. (2002) Identification of *Ku70* and *Ku80* Homologues in *Arabidopsis thaliana*: Evidence for a Role in the Repair of DNA Double-Strand Breaks. *The Plant Journal*, Vol.29, No.6, pp. 771-781

Tamura, K.; Dudley, J.; Nei, M. & Kumar, S. (2007) MEGA4: Molecular Evolutionary Genetics Analysis (MEGA) Software Version 4.0. *Molecular Biology and Evolution*, Vol.24, No.8, pp.1596-1599

Thomas, H. (2002) Ageing in Plants. *Mechanisms of Ageing and Development*. Vol.123, No.7, pp. 747-753

Thompson, J.D.; Gibson, T.J.; Plewniak, F.; Jeanmougin, F. & Higgins, D.G. (1997) The CLUSTAL_X Windows Interface: Flexible Strategies for Multiple Sequence Alignment Aided by Quality Analysis Tools. *Nucleic Acid Research*, Vol.25, No.24, pp. 4876-4882

Tuteja, N.; Ahmad, P.; Panda, B.B. & Tuteja, R. (2009) Genotoxic Stress in Plants: Shedding Light on DNA Damage, Repair and DNA Repair Helicases. *Mutation Research*, Vol.681, No.2-3, pp. 134-149

Tuteja, N.; Singh,M.B.; Misra, M.K.; Bhalla, P.L. & Tuteja, R. (2001) Molecular Mechanisms of DNA Damage and Prepair: Progress in Plants. *Critical Reviews in Biochemistry and Molecular Biology*, Vol.36, No.4, pp. 337-397

van Gent, DC.; Hoeijmakers, J.H.J.& Kanaar, R. (2001) Chromosomal Stability and the DNA Double-Stranded Break Connection. *Nature Review Genetics*, Vol.2, pp. 196-206

Vanderklein, D.; Martínez-Vilalta, J.; Lee, S. & Mencuccini, M. (2007) Plant Size, not Age, Regulates Growth and Gas Exchange in Grafted Scots Pine Trees. *Tree Physiology*, Vol.24, pp. 71-79

Vuosku, J.; Sarjala, T.; Jokela, A.; Sutela, S.; Sääskilahti, M.; Suorsa, M.; Läärä, E. & Häggman, H. (2009) One Tissue, Two Fates: Different Roles of Megagametophyte Cells during Scots Pine Embryogenesis, *Journal of Experimental Botany*, Vol.60, No.4, pp. 1375-86

Walbot, V. (1996) Sources and Consequences of Phenotypic and Genotypic Plasticity in Flowering Plants. *Trends in Plant Science*, Vol.1, No.1, pp. 27-33

Walbot, V. & Evans, M.M. (2003) Unique Features of the Plant Life Cycle and Their Consequences. *Nature Reviews Genetics*, Vol.4, pp. 369-379

Waterworth, W.M.; Masnavi, G.; Bhardwaj, R.M.; Jiang, Q.; Bray, C.M. & West, C.E. (2010) A Plant DNA Ligase is an Important Determinant of Seed Longevity. *The Plant Journal*, Vol.63, No.5, pp. 848-860

Watson, J.M. & Riha, K. (2011) Telomeres, Aging, and Plants: From Weeds to Methuselah – A Mini-Review. *Gerontology*, Vol.57, pp. 129-136

Willyard, A.; Syring, J.; Gernandt, D.S.; Liston, A. & Cronn, R. (2007) Fossil Calibration of Molecular Divergence Infers a Moderate Mutation Rate and Recent Radiations for *Pinus*. *Molecular Biology and Evolution*, Vol.24, No.1, pp. 90-101

Permissions

The contributors of this book come from diverse backgrounds, making this book a truly international effort. This book will bring forth new frontiers with its revolutionizing research information and detailed analysis of the nascent developments around the world.

We would like to thank Inna Kruman, for lending her expertise to make the book truly unique. She has played a crucial role in the development of this book. Without her invaluable contribution this book wouldn't have been possible. She has made vital efforts to compile up to date information on the varied aspects of this subject to make this book a valuable addition to the collection of many professionals and students.

This book was conceptualized with the vision of imparting up-to-date information and advanced data in this field. To ensure the same, a matchless editorial board was set up. Every individual on the board went through rigorous rounds of assessment to prove their worth. After which they invested a large part of their time researching and compiling the most relevant data for our readers. Conferences and sessions were held from time to time between the editorial board and the contributing authors to present the data in the most comprehensible form. The editorial team has worked tirelessly to provide valuable and valid information to help people across the globe.

Every chapter published in this book has been scrutinized by our experts. Their significance has been extensively debated. The topics covered herein carry significant findings which will fuel the growth of the discipline. They may even be implemented as practical applications or may be referred to as a beginning point for another development. Chapters in this book were first published by InTech; hereby published with permission under the Creative Commons Attribution License or equivalent.

The editorial board has been involved in producing this book since its inception. They have spent rigorous hours researching and exploring the diverse topics which have resulted in the successful publishing of this book. They have passed on their knowledge of decades through this book. To expedite this challenging task, the publisher supported the team at every step. A small team of assistant editors was also appointed to further simplify the editing procedure and attain best results for the readers.

Our editorial team has been hand-picked from every corner of the world. Their multi-ethnicity adds dynamic inputs to the discussions which result in innovative outcomes. These outcomes are then further discussed with the researchers and contributors who give their valuable feedback and opinion regarding the same. The feedback is then collaborated with the researches and they are edited in a comprehensive manner to aid the understanding of the subject.

Apart from the editorial board, the designing team has also invested a significant amount of their time in understanding the subject and creating the most relevant covers. They scrutinized every image to scout for the most suitable representation of the subject and create an appropriate cover for the book.

The publishing team has been involved in this book since its early stages. They were actively engaged in every process, be it collecting the data, connecting with the contributors or procuring relevant information. The team has been an ardent support to the editorial, designing and production team. Their endless efforts to recruit the best for this project, has resulted in the accomplishment of this book. They are a veteran in the field of academics and their pool of knowledge is as vast as their experience in printing. Their expertise and guidance has proved useful at every step. Their uncompromising quality standards have made this book an exceptional effort. Their encouragement from time to time has been an inspiration for everyone.

The publisher and the editorial board hope that this book will prove to be a valuable piece of knowledge for researchers, students, practitioners and scholars across the globe.

Effrossyni Boutou
Prenatal Diagnosis Lab, Laiko Hospital, Athens, Greece

Vassiliki Pappa
2nd Propaedeutic Pathology Clinic, Medical School, Athens University, Athens, Greece

Horst-Werner Stuerzbecher
Molecular Cancer Biology Group, Institute of Pathology, UK-SH, Luebeck, Germany

Yulin Zhang and Dexi Chen
Department of infectious diseases, Beijing You'an Hospital, Beijing liver disease research institute, Capital Medical University, Beijing, China

Lina Dagnino, Randeep Kaur Singh and David Judah
University of Western Ontario, Canada

Milena Popova, Sébastien Henry and Fabrice Fleury
Unité U3B, UMR 6204 CNRS 2, rue de la Houssinière, University of Nantes, France

Dorota Rybaczek
Department of Cytophysiology, University of Łódź, Poland

Adisorn Ratanaphan
Prince of Songkla University, Thailand

Inna I. Kruman
Texas Tech University Health Sciences Center, USA

Kristi L. Jones, Ling Zhang and Feng Gong
Department of Biochemistry and Molecular Biology, University of Miami, USA

Ewa Forma, Magdalena Brys and Wanda M. Krajewska
Department of Cytobiochemistry, University of Lodz, Poland

Guylain Boissonneault
Department of Biochemistry, Faculty of Medicine and Health Sciences, Université de Sherbrooke, Canada

Frédéric Leduc, Geneviève Acteau, Marie-Chantal Grégoire, Olivier Simard, Jessica Leroux, Audrey Carrier-Leclerc and Mélina Arguin
Department of biochemistry, Faculty of medicine and health sciences, Université de Sherbrooke, Canada

Xinrong Chen and Tao Chen
National Center for Toxicological Research/US Food and Drug Administration, U.S.A.

Erika T. Brown
Medical University of South Carolina, United States

List of Contributors

Ge Wang and Robert J. Maier
Department of Microbiology, University of Georgia, Athens, Georgia

Chuck C.-K. Chao
Department of Biochemistry and Molecular Biology, Graduate Institute of Biomedical Sciences, Chang Gung University, Taiwan, Republic of China

Gabriel Kaufmann
Tel Aviv University, Israel

Elena Davidov, Emmanuelle Steinfels-Kohn, Ekaterina Krutkina, Daniel Klaiman, Tamar Margalit, Michal Chai-Danino and Alexander Kotlyar
Tel Aviv University, Israel

Mikio Shimada, Akihiro Kato and Junya Kobayashi
Radiation Biology Center, Kyoto University, Japan

Mikio Shimada, Akihiro Kato and Junya Kobayashi
Radiation Biology Center, Kyoto University, Japan

Gang Liu
Clinical Medicine Research Center, Affiliated Hospital of Guangdong Medical College, PR China

David W. Kamp
Department of Medicine, Northwestern University Feinberg School of Medicine and Jesse Brown VA Medical Center, USA

Lonnie R. Welch
School of Electrical Engineering and Computer Science, Ohio University, Athens, Ohio, USA

Laura M. Koehly
Social and Behavioral Research Branch, National Human Genome Research Institute, NIH, Bethesda, Maryland, USA

Laura Elnitski
Genome Technology Branch, National Human Genome Research Institute, NIH, Bethesda, Maryland, USA

Effrossyni Boutou and Constantinos E. Vorgias
Department of Biochemistry & Molecular Biology, Faculty of Biology, School of Sciences Athens University, Athens, Greece

Harris Bernstein and Carol Bernstein
Department of Cellular and Molecular Medicine, University of Arizona, USA

Richard E. Michod
Department of Ecology and Evolutionary Biology, University of Arizona, USA

Jaana Vuosku, Marko Suokas, Johanna Kestilä and Hely Häggman
Department of Biology, University of Oulu, Oulu, Finland

Jaana Vuosku and Tytti Sarjala
Finnish Forest Research Institute, Parkano Research Unit, Parkano, Finland